ASTRONOMICAL DATA ANALYSIS SOFTWARE AND SYSTEMS VIII

A SERIES OF BOOKS ON RECENT DEVELOPMENTS IN ASTRONOMY AND ASTROPHYSICS

Managing Editor, D. Harold McNamara
Production Manager, Enid L. Livingston

Printed by BookCrafters, Inc.

First published 1999

ISBN 1-886733-94-5 ISSN 1080-7926

Please contact proper address for information on:

PUBLISHING:
Managing Editor
PO Box 24463
211 KMB
Brigham Young University
Provo, UT 84602-4463 USA

Phone: 801-378-2298
Fax: 801-378-4049
E-mail: pasp@astro.byu.edu

ORDERING BOOKS:
Astronomical Society of the Pacific
CONFERENCE SERIES
390 Ashton Avenue
San Francisco, CA 94112 - 1722 USA

Phone: 415-337-2624
Fax: 415-337-5205
E-mail: catalog@aspsky.org
Web Site: http://www.aspsky.org

A SERIES OF BOOKS ON RECENT DEVELOPMENTS IN ASTRONOMY AND ASTROPHYSICS

Vol. 1-Progress and Opportunities in Southern Hemisphere Optical Astronomy: CTIO 25th Anniversary Symposium
ed. V. M. Blanco and M. M. Phillips ISBN 0-937707-18-X

Vol. 2-Proceedings of a Workshop on Optical Surveys for Quasars
ed. P. S. Osmer, A. C. Porter, R. F. Green, and C. B. Foltz ISBN 0-937707-19-8

Vol. 3-Fiber Optics in Astronomy
ed. S. C. Barden ISBN 0-937707-20-1

Vol. 4-The Extragalactic Distance Scale: Proceedings of the ASP 100th Anniversary Symposium
ed. S. van den Bergh and C. J. Pritchet ISBN 0-937707-21-X

Vol. 5-The Minnesota Lectures on Clusters of Galaxies and Large-Scale Structure
ed. J. M. Dickey ISBN 0-937707-22-8

Vol. 6-Synthesis Imaging in Radio Astronomy: A Collection of Lectures from the Third NRAO Synthesis Imaging Summer School
ed. R. A. Perley, F. R. Schwab, and A. H. Bridle ISBN 0-937707-23-6

Vol. 7-Properties of Hot Luminous Stars: Boulder-Munich Workshop
ed. C. D. Garmany ISBN 0-937707-24-4

Vol. 8-CCDs in Astronomy
ed. G. H. Jacoby ISBN 0-937707-25-2

Vol. 9-Cool Stars, Stellar Systems, and the Sun. Sixth Cambridge Workshop
ed. G. Wallerstein ISBN 0-937707-27-9

Vol. 10-Evolution of the Universe of Galaxies: Edwin Hubble Centennial Symposium
ed. R. G. Kron ISBN 0-937707-28-7

Vol. 11-Confrontation Between Stellar Pulsation and Evolution
ed. C. Cacciari and G. Clementini ISBN 0-937707-30-9

Vol. 12-The Evolution of the Interstellar Medium
ed. L. Blitz ISBN 0-937707-31-7

Vol. 13-The Formation and Evolution of Star Clusters
ed. K. Janes ISBN 0-937707-32-5

Vol. 14-Astrophysics with Infrared Arrays
ed. R. Elston ISBN 0-937707-33-3

Vol. 15-Large-Scale Structures and Peculiar Motions in the Universe
ed. D. W. Latham and L. A. N. da Costa ISBN 0-937707-34-1

Vol. 16-Proceedings of the 3rd Haystack Observatory Conference on Atoms, Ions and Molecules: New Results in Spectral Line Astrophysics
ed. A. D. Haschick and P. T. P. Ho ISBN 0-937707-35-X

Vol. 17-Light Pollution, Radio Interference, and Space Debris
ed. D. L. Crawford ISBN 0-937707-36-8

Vol. 18-The Interpretation of Modern Synthesis Observations of Spiral Galaxies
ed. N. Duric and P. C. Crane ISBN 0-937707-37-6

Vol. 19-Radio Interferometry: Theory, Techniques, and Applications, IAU Colloquium 131
ed. T. J. Cornwell and R. A. Perley ISBN 0-937707-38-4

Vol. 20-Frontiers of Stellar Evolution: 50th Anniversary McDonald Observatory (1939-1989)
ed. D. L. Lambert ISBN 0-937707-39-2

Vol. 43-Sky Surveys: Protostars to Protogalaxies
ed. B. T. Soifer ISBN 0-937707-62-7

Vol. 44-Peculiar Versus Normal Phenomena in A-Type and Related Stars, IAU Colloquium 138
ed. M. M. Dworetsky, F. Castelli, and R. Faraggiana ISBN 0-937707-63-5

Vol. 45-Luminous High-Latitude Stars
ed. D. D. Sasselov ISBN 0-937707-64-3

Vol. 46-The Magnetic and Velocity Fields of Solar Active Regions, IAU Colloquium 141
ed. H. Zirin, G. Ai, and H. Wang ISBN 0-937707-65-1

Vol. 47-Third Decennial US-USSR Conference on SETI
ed. G. Seth Shostak ISBN 0-937707-66-X

Vol. 48-The Globular Cluster-Galaxy Connection
ed. Graeme H. Smith and Jean P. Brodie ISBN 0-937707-67-8

Vol. 49-Galaxy Evolution: The Milky Way Perspective
ed. Steven R. Majewski ISBN 0-937707-68-6

Vol. 50-Structure and Dynamics of Globular Clusters
ed. S. G. Djorgovski and G. Meylan ISBN 0-937707-69-4

Vol. 51-Observational Cosmology
ed. G. Chincarini, A. Iovino, T. Maccacaro, and D. Maccagni ISBN 0-937707-70-8

Vol. 52-Astronomical Data Analysis Software and Systems II
ed. R. J. Hanisch, R. J. V. Brissenden, and Jeannette Barnes ISBN 0-937707-71-6

Vol. 53-Blue Stragglers
ed. Rex A. Saffer ISBN 0-937707-72-4

Vol. 54-The First Stromlo Symposium: The Physics of Active Galaxies
ed. Geoffrey V. Bicknell, Michael A. Dopita, and Peter J. Quinn ISBN 0-937707-73-2

Vol. 55-Optical Astronomy from the Earth and Moon
ed. Diane M. Pyper and Ronald J. Angione ISBN 0-937707-74-0

Vol. 56-Interacting Binary Stars
ed. Allen W. Shafter ISBN 0-937707-75-9

Vol. 57-Stellar and Circumstellar Astrophysics
ed. George Wallerstein and Alberto Noriega-Crespo ISBN 0-937707-76-7

Vol. 58-The First Symposium on the Infrared Cirrus and Diffuse Interstellar Clouds
ed. Roc M. Cutri and William B. Latter ISBN 0-937707-77-5

Vol. 59-Astronomy with Millimeter and Submillimeter Wave Interferometry, IAU Colloquium 140
ed. M. Ishiguro and Wm. J. Welch ISBN 0-937707-78-3

Vol. 60-The MK Process at 50 Years: A Powerful Tool for Astrophysical Insight: A Workshop of the Vatican Observatory
ed. C. J. Corbally, R. O. Gray, and R. F. Garrison ISBN 0-937707-79-1

Vol. 61-Astronomical Data Analysis Software and Systems III
ed. Dennis R. Crabtree, R. J. Hanisch, and Jeannette Barnes ISBN 0-937707-80-5

Vol. 62-The Nature and Evolutionary Status of Herbig Ae / Be Stars
ed. P. S. Thé, M. R. Pérez, and E. P. J. van den Heuvel ISBN 0-937707-81-3

Vol. 63-Seventy-Five Years of Hirayama Asteroid Families: The role of Collisions in the Solar System History
ed. R. P. Binzel, Y. Kozai, and T. Hirayama ISBN 0-937707-82-1

Vol. 64-Cool Stars, Stellar Systems, and the Sun, Eighth Cambridge Workshop
ed. Jean-Pierre Caillault ISBN 0-937707-83-X

Vol. 65-Clouds, Cores, and Low Mass Stars
ed. Dan P. Clemens and Richard Barvainis ISBN 0-937707-84-8

Vol. 66- Physics of the Gaseous and Stellar Disks of the Galaxy
ed. Ivan R. King ISBN 0-937707-85-6

Vol. 67-Unveiling Large-Scale Structures Behind the Milky Way
ed. C. Balkowski and R. C. Kraan-Korteweg ISBN 0-937707-86-4

Vol. 68-Solar Active Region Evolution: Comparing Models with Observations
ed. K. S. Balasubramaniam and George W. Simon ISBN 0-937707-87-2

Vol. 69-Reverberation Mapping of the Broad-Line Region in Active Galactic Nuclei
ed. P. M. Gondhalekar, K. Horne, and B. M. Peterson ISBN 0-937707-88-0

Vol. 70-Groups of Galaxies
ed. Otto G. Richter and Kirk Borne ISBN 0-937707-89-9

Vol. 71-Tridimensional Optical Spectroscopic Methods in Astrophysics, IAU Colloquium 149
ed. G. Comte and M. Marcelin ISBN 0-937707-90-2

Vol. 72-Millisecond Pulsars: A Decade of Surprise.
ed. A. A. Fruchter, M. Tavani, and D. C. Backer ISBN 0-937707-91-0

Vol. 73-Airborne Astronomy Symposium on the Galactic Ecosystem: From Gas to Stars to Dust
ed. M. R. Haas, J. A. Davidson, and E. F. Erickson ISBN 0-937707-92-9

Vol. 74-Progress in the Search for Extraterrestrial Life: 1993 Bioastronomy Symposium
ed. G. Seth Shostak ISBN 0-937707-93-7

Vol. 75-Multi-Feed Systems for Radio Telescopes
ed. D. T. Emerson and J. M. Payne ISBN 0-937707-94-5

Vol. 76-GONG '94: Helio- and Astero-Seismology from the Earth and Space
ed. Roger K. Ulrich, Edward J. Rhodes, Jr., and Werner Däppen ISBN 0-937707-95-3

Vol. 77-Astronomical Data Analysis Software and System IV
ed. R. A. Shaw, H. E. Payne, and J. J. E. Hayes ISBN 0-937707-96-1

Vol. 78-Astrophysical Applications of Powerful New Databases: Joint Discussion No. 16 of the 22nd General Assembly of the IAU
ed. S. J. Adelman and W. L. Wiese ISBN 0-937707-97-X

Vol. 79-Robotic Telescopes: Current Capabilities, Present Developments, and Future Prospects for Automated Astronomy
ed. Gregory W. Henry and Joel A. Eaton ISBN 0-937707-98-8

Vol. 80-The Physics of the Interstellar Medium and Intergalactic Medium
ed. A. Ferrara, C. F. McKee, C. Heiles, and P. R. Shapiro ISBN 0-937707-99-6

Vol. 81-Laboratory and Astronomical High Resolution Spectra
ed. A. J. Sauval, R. Blomme, and N. Grevesse ISBN 1-886733-01-5

Vol. 82-Very Long Baseline Interferometry and the VLBA
ed. J. A. Zensus, P. J. Diamond, and P. J. Napier ISBN 1-886733-02-3

Vol. 83-Astrophysical Applications of Stellar Pulsation. IAU Colloquium 155
ed. R. S. Stobie and P. A. Whitelock ISBN 1-886733-03-1

Vol. 84-The Future Utilisation of Schmidt Telescopes, IAU Colloquium 148
ed. Jessica Chapman, Russell Cannon, Sandra Harrison, and Bambang Hidayat ISBN 1-886733-05-8

Vol. 85-Cape Workshop on Magnetic Cataclysmic Variables
ed. D. A. H. Buckley and B. Warner ISBN 1-886733-06-6

Vol. 86-Fresh Views of Elliptical Galaxies
ed. Alberto Buzzoni, Alvio Renzini, and Alfonso Serrano ISBN 1-886733-07-4

Vol. 87-New Observing Modes for the Next Century
ed. Todd Boroson, John Davies, and Ian Robson ISBN 1-886733-08-2

Vol. 88- Clusters, Lensing, and the Future of the Universe
ed. Virginia Trimble and Andreas Reisenegger ISBN 1-886733-09-0

Vol. 89-Astronomy Education: Current Developments, Future Coordination
ed.John R. Percy ISBN 1-886733-10-4

Vol. 90-The Origins, Evolution, and Destinies of Binary Stars in Clusters
ed. E. F. Milone and J. C. Mermilliod ISBN 1-886733-11-2

Vol. 91-Barred Galaxies, IAU Colloquium 157
ed. R. Buta, D. A. Crocker, and B. G. Elmegreen ISBN 1-886733-12-0

Vol. 92-Formation of the Galactic Halo–Inside and Out
ed. H. L. Morrison and A. Sarajedini ISBN 1-886733-13-9

Vol. 93-Radio Emission from the Stars and the Sun
ed. A. R. Taylor and J. M. Paredes ISBN 1-886733-14-7

Vol. 94-Mapping, Measuring, and Modelling the Universe
ed. Peter Coles, Vincent Martinez, and Maria-Jesus Pons-Borderia ISBN 1-886733-15-5

Vol. 95-Solar Drivers of Interplanetary and Terrestrial Disturbances: Proceedings of 16th International Workshop, National Solar Observatory/Sacramento Peak
ed. K. S. Balasubramaniam, S. L. Keil, and R. N. Smartt ISBN 1-886733-16-3

Vol. 96- Hydrogen-Deficient Stars
ed. C. S. Jeffery and U. Heber ISBN 1-886733-17-1

Vol. 97-Polarimetry of the Interstellar Medium
ed.W. G. Roberge and D. C. B. Whittet ISBN 1-886733-18-X

Vol. 98-From Stars to Galaxies: The Impact of Stellar Physics on Galaxy Evolution
ed. Claus Leitherer, Uta Fritze-von Alvensleben, and John Huchra ISBN 1-886733-19-8

Vol. 99-Cosmic Abundances: Proceedings of the 6th Annual October Astrophysics Conference
ed. Stephen S. Holt and Geroge Sonneborn ISBN 1-886733-20-1

Vol. 100-Energy Transport in Radio Galaxies and Quasars
ed. P. E. Hardee, A. H. Bridle, and J. A. Zensus ISBN 1-886733-21-X

Vol. 101-Astronomical Data Analysis Software and Systems V
ed. George H. Jacoby and Jeannette Barnes ISSN 1080-7926

Vol. 102-The Galactic Center, 4th ESO/CTIO Workshop
ed. Roland Gredel ISBN 1-886733-22-8

Vol. 103-The Physics of Liners in View of Recent Observations
ed. M. Eracleous, A. Koratkar, C. Leitherer, and L. Ho ISBN 1-886733-23-6

Vol. 104-Physics, Chemistry, and Dynamics of Interplanetary Dust, IAU Colloquium 150
ed. Bo A. S. Gustafson and Martha S. Hanner ISBN 1-886733-24-4

Vol. 105-Pulsars: Problems and Progress, IAU Colloquium 160
ed. M. Bailes, S. Johnston, and M. A. Walker ISBN 1-886733-25-2

Vol. 106-Minnesota Lectures on Extragalactic Neutral Hydrogen
ed. Evan D. Skillman ISBN 1-886733-26-0

Vol. 107-Completing the Inventory of the Solar System: A Symposium held in conjuunction with the 106th Annual Meeting of the ASP
ed. Terrence W. Rettig and Joseph M. Hahn ISBN 1-886733-27-9

Vol. 108-M. A. S. S. Model Atmospheres and Spectrum Synthesis: 5th Vienna Workshop
ed. S. J. Adelman, F. Kupka, and W. W. Weiss ISBN 1-886733-28-7

Vol. 109-Cool Stars, Stellar Systems, and the Sun, Ninth Cambridge Workshop
ed. Roberto Pallavicini and Andrea K. Dupree ISBN 1-886733-29-5

Vol. 110-Blazar Continuum Variability
ed.H. R. Miller, J. R. Webb, and J. C. Noble ISBN 1-886733-30-9

Vol. 111-Magnetic Reconnection in the Solar Atmosphere: Proceedings of a Yohkoh Conference
ed. R. D. Bentley and J. T. Mariska ISBN 1-886733-31-7

Vol. 112-The History of the Milky Way and Its Satellite System
ed. A. Burkert, D. H. Hartmann, and S. R. Majewski ISBN 1-886733-32-5

Vol. 113-Emission Lines in Active Galaxies: New Methods and Techniques, IAU Colloquium 159
ed. B. M. Peterson, F. Z. Cheng, and A. S. Wilson ISBN 1-886733-33-3

Vol. 114-Young Galaxies and QSO Absorption-Line Systems
ed. Sueli M. Viegas, Ruth Gruenwald, and Reinaldo R. de Carvalho ISBN 1-886733-34-1

Vol. 115-Galactic and Cluster Cooling Flows
ed. Noam Soker ISBN 1-886733-35-X

Vol. 116-The Second Stromlo Symposium: The Nature of Elliptical Galaxies
ed. M. Arnaboldi, G. S. Da Costa, and P. Saha ISBN 1-886733-36-8

Vol. 117- Dark and Visible Matter in Galaxies
ed. Massimo Persic and Paolo Salucci ISBN 1-886733-37-6

Vol. 118-First Advances in Solar Physics Euroconference: Advances in the Physics of Sunspots
ed. B. Schmieder, J. C. del Toro Iniesta, and M. Vázquez ISBN 1-886733-38-4

Vol. 119-Planets Beyond the Solar System and the Next Generation of Space Missions
ed. David R. Soderblom ISBN 1-886733-39-2

Vol. 120-Luminous Blue Variables: Massive Stars in Transition
ed. Antonella Nota and Henny J. G. L. M. Lamers ISBN 1-886733-40-6

Vol. 121-Accretion Phenomena and Related Outflows, IAU Colloquium 163
ed. D. T. Wickramasinghe, G. V. Bicknell and L. Ferrario ISBN 1-886733-41-4

Vol. 122-From Stardust to Planetesimals: Symposium held as part of the 108th Annual Meeting of the ASP
ed. Yvonne J. Pendleton and A. G. G. M. Tielens ISBN 1-886733-42-2

Vol. 123-The 12th 'Kingston Meeting': Computational Astrophysics
ed. David A. Clarke and Michael J. West ISBN 1-886733-43-0

Vol. 124-Diffuse Infrared Radiation and the IRTS
ed. Haruyuki Okuda, Toshio Matsumoto, and Thomas L. Roellig ISBN 1-886733-44-9

Vol. 125- Astronomical Data Analysis Software and Systems VI
ed. Gareth Hunt and H. E. Payne ISBN 1-886733-45-7

Vol. 126-From Quantum Fluctuations to Cosmological Structures
ed. D. Vallis-Gabaud, M. A. Hendry, P. Molaro, and K. Chamcham ISBN 1-886733-46-5

Vol. 127-Proper Motions and Galactic Astronomy
ed. Roberta M. Humphreys ISBN 1-886733-47-3

Vol. 128- Mass Ejection from AGN (Active Galactic Nuclei)
ed. N. Arav, I. Shlosman, and R. J. Weymann ISBN 1-886733-48-1

Vol. 129-The George Gamow Symposium
ed. E. Harper, W. C. Parke, and G. D. Anderson ISBN 1-886733-49-X

Vol. 130-The Third Pacfic Rim Conference on Recent Development on Binary Star Research
ed. Kam-Ching Leung ISBN 1-886733-50-3

Vol. 131-Boulder-Munich II: Properties of Hot, Luminous Stars
ed. Ian D. Howarth ISBN 1-886733-51-1

Vol. 132-Star Formation with the Infrared Space Observatory (ISO)
ed. João L. Yun and René Liseau ISBN 1-886733-53-X

Vol. 133-Science with the NGST
ed. Eric P. Smith and Anuradha Koratkar ISBN 1-886733-53-8

Vol. 134-Brown Dwarfs and Extrasolar Planets
ed. Rafael Rebolo, Eduardo L. Martin,
and Maria Rosa Zapatero Osorio ISBN 1-886733-54-6

Vol. 135-A Half Century of Stellar Pulsation Interpretations: A Tribute to Arthur N. Cox
ed. P. A Bradley and J. A. Guzik ISBN 1-886733-55-4

Vol. 136- Galactic Halos: A UC Santa Cruz Workshop
ed. Dennis Zaritsky ISBN 1-886733-56-2

Vol. 137-Wild Stars in the Old West: Proceedings of the 13th North American Workshop
on Cataclysmic Variables and Related Objects
ed. S. Howell, E.Kuulkers, and C. Woodward ISBN 1-886733-57-0

Vol. 138-1997 Pacific Rim Conference on Stellar Astrophysics
ed. Kwing L. Chan, K. S. Cheng, and Harinder P. Singh ISBN 1-886733-58-9

Vol. 139-Preserving the Astronomical Windows, Proceedings of Joint Discussion
No. 5 of the 23rd General Assembly of the IAU
ed. Syuzo Isobe and Tomohiro Hirayama ISBN 1-886733-59-7

Vol. 140-Synoptic Solar Physics – 18th NSO/Sacramento Peak Summer Workshop
ed. K. S. Balasubramaniam, J. W. Harvey, and D. M. Rabin ISBN 1-886733-60-0

Vol. 141-Astrophysics from Antarctica
ed. Giles Novak and Randall H. Landsberg ISBN 1-886733-61-9

Vol. 142-The Stellar Initial Mass Function, 38th Herstmonceux Conference
ed. Gerry Gilmore and Debbie Howell ISBN 1-886733-62-7

Vol. 143-The Scientific Impact of the Goddard High Resolution Spectrograph
ed. John C. Brandt, Thomas B. Ake III, and Carolyn Collins Petersen ISBN 1-886733-63-5

Vol. 144- Radio Emission from Galactic and Extragalactic Compact Sources,
IAU Colloquium 164
ed. J. Anton Zensus, G. B. Taylor, and J. M. Wrobel ISBN 1-886733-64-3

Vol. 145-Astronomical Data Analysis Software and Systems VII
ed. Rudolf Albrecht, Richard N. Hook, and Howard A. Bushouse ISBN 1-886733-65-1

Vol. 146-The Young Universe: Galaxy Formation and Evolution at
Intermediate and High Redshift
ed. S. D'Odorico, A. Fontana, and E. Giallongo ISBN 1-886733-66-X

Vol. 147-Abundance Profiles: Diagnostic Tools for Galaxy History
ed. Daniel Friedli, Mike Edmunds, Carmelle Robert,
and Laurent Drissen ISBN 1-886733-67-8

Vol. 148-Origins
ed. Charles E. Woodward, J. Michael Shull,
and Harley A. Thronson, Jr. ISBN 1-886733-68-6

Vol. 149-Solar System Formation and Evolution
ed. D. Lazzaro, R. Vieira Martins, S. Ferraz-Mello,
J. Fernández, and C. Beaugé ISBN 1-886733-69-4

Vol. 150-New Perspectives on Solar Prominences, IAU Colloquium 167
ed. David Webb, David Rust, and Brigitte Schmieder ISBN 1-886733-70-8

Vol. 151-Cosmic Microwave Background and Large Scale Structure of the Universe
ed. Yong-Ik Byun and Kin-Wang Ng ISBN 1-886733-71-6

Vol. 152-Fiber Optics in Astronomy III
ed. S. Arribas, E. Mediavilla, and F. Watson ISBN 1-886733-72-4

Vol. 153-Library and Information Services in Astronomy III, (LISA III)
ed. Uta Grothkopf, Heinz Andernach, Sarah Stevens-Rayburn,
and Monique Gomez ISBN 1-886733-73-2

Vol. 154-Cool Stars, Stellar Systems, and the Sun, Tenth Cambridge Workshop
ed. Robert A. Donahue and Jay A. Bookbinder ISBN 1-886733-74-0

Vol. 155-Second Advances in Solar Physics Euroconference:
Three-Dimensional Structure of Solar Active Regions
ed. Costas E. Alissandrakis and Brigitte Schmieder ISBN 1-886733-75-9

Vol. 156-Highly Redshifted Radio Lines
ed. C. L. Carilli, S. J. E. Radford, K. M. Menten and
G. I. Langston ISBN 1-886733-76-7

Vol. 157-Annapolis Workshop on Magnetic Cataclysmic Variables
ed. Coel Hellier and Koji Mukai ISBN 1-886733-77-5

Vol. 158-Solar and Stellar Activity: Similarities and Differences
ed. C. J. Butler and J. G. Doyle ISBN 1-886733-78-3

Vol. 159-BL Lac Phenomenon
ed. Leo O. Takalo and Aimo Sillanpää ISBN 1-886733-79-1

Vol. 160-Astrophysical Discs, An EC Summer School
ed. J. A. Sellwood and Jeremy Goodman ISBN 1-886733-80-5

Vol. 161-High Energy Processes in Accreting Black Holes
ed. Juri Poutanen and Roland Svensson ISBN 1-886733-81-3

Vol. 162-Quasars and Cosmology
ed. Gary Ferland and Jack Baldwin ISBN 1-886733-83-X

Vol. 163-Star Formation in Early Type Galaxies
ed. Jordi Cepa and Patricia Carral ISBN 1-886733-84-8

Vol. 164-Ultraviolet–Optical Space Astronomy Beyond HST
ed. Jon A. Morse, J. Michael Shull and Anne L. Kinney ISBN 1-886733-85-6

Vol. 165-The Third Stromlo Symposium: The Galactic Halo
ed. Brad K. Gibson, Tim S. Axelrod, and Mary E. Putman ISBN 1-886733-86-4

Vol. 166-Stromlo Workshop on High-Velocity Clouds
ed. Brad K. Gibson and Mary E. Putman ISBN 1-886733-87-2

Vol. 167-Harmonizing Cosmic Distance Scales in a Post-HIPPARCOS Era
ed. Daniel Egret and André Heck ISBN 1-886733-88-0

Vol. 168-New Perspectives on the Interstellar Medium
ed. A. R. Taylor, T. L. Landecker, and G. Joncas ISBN 1-886733-89-9

Vol. 169-11th European Workshop on White Dwarfs
ed. J.-E. Solheim and E. G. Meištas ISBN 1-886733-91-0

Vol. 170-The Low Surface Brightness Universe, IAU Colloquium 171
ed. J. I. Davies, C. Impey, and S. Phillipps ISBN 1-886733-92-9

Vol. 171-LiBeB, Cosmic Rays, and Related X-and Gamma-Rays
ed. Reuven Ramaty, Elisabeth Vangioni-Flam,
Michel Cassé, and Keith Olive ISBN 1-886733-93-7

Vol. 172-Astronomical Data Analysis Software and Systems VIII
ed. David M. Mehringer, Raymond L. Plante, and
Douglas A. Roberts ISBN 1-886733-94-5

Book orders or inquiries concerning these volumes should be directed to the:

Astronomical Society of the Pacific Conference Series
390 Ashton Avenue
San Francisco, CA 94112-1722 USA

Phone: 415-337-2126
Fax: 415-337-5205

E-mail: catalog@aspsky.org
Web Site: http://www.aspsky.org

ASTRONOMICAL SOCIETY OF THE PACIFIC
CONFERENCE SERIES

Volume 172

ASTRONOMICAL DATA ANALYSIS SOFTWARE AND SYSTEMS VIII

**Proceedings of a meeting held at
Urbana, Illinois, USA
1-4 November, 1998**

Edited by
**David M. Mehringer, Raymond L. Plante,
and Douglas A. Roberts**

Contents

Part 3. Observatory Planning and Scheduling

Part 4. Telescope and Instrument Control Systems

Part 5. Data Compression

Part 6. Data Pipelines

Part 7. Archiving and Information Services

Part 8. Data Analysis and Processing Techniques

Part 9. Astrostatistics and Databases

Part 10. User Interfaces and Visualization

Part 11. Software Systems and Tools

Preface

This volume contains papers presented at the eighth annual conference for Astronomical Data Analysis Software and Systems, also known as ADASS 98. It was held November 1–4, 1998 on the campus of the University of Illinois at Urbana–Champaign (UIUC) and hosted jointly by the UIUC Astronomy Department and the National Center for Supercomputing Applications (NCSA). There were 219 registered participants, including 72 people representing 15 countries outside the United States and Canada.

1. Conference Overview

Sunday afternoon featured three concurrent tutorials. NCSA's Beth Richardson presented a workshop on parallel computing which provided participants with strategies for cache tuning, data distribution, parallel debugging, and profiling. NCSA staff also presented a two-part tutorial entitled "Topics in Java Application Development." In the first part, Ed Grossman and Terry McLaren described how to create collaborative applications using a Java package called NCSA Habanero; in the second part, Steve Pietrowitz gave an introduction to 3D visualization via the Java 3D API. A third tutorial was presented by Jean-Luc Stark (CEA Saclay) and Fionn Murtagh (University of Ulster) entitled "Multiscale Image and Data Analysis: Theory, Applications and Software." The presenters reviewed the background on wavelet and other multiscale transforms and gave an overview of various multiscale image processing and analysis applications including deconvolution, compression, and feature detection.

Sunday and Wednesday afternoon also provided participants an opportunity to tour the NCSA facilities. The tour included the NCSA machine rooms which house the SGI Origin 2000, the retiring SGI Power Challenge Array, the NT Cluster, and the Mass Storage System. At the Beckman Institute, participants experienced a sampling of NCSA 3D immersive visualization environments. This included the PowerWall, a twelve by nine foot display device where Paul Woodward (University of Minnesota) displayed his hydrodynamic simulations of convection in a red giant star. Also featured on the PowerWall were 3D visualizations of the Tully Nearby Galaxies Catalogue (by Stuart Levy) and computational progress of a simulation technique called Adaptive Mesh Refinement (by John Shalf). The tour was rounded out by a visit to the NCSA Cave, a room-sized, multi-person, 3D video and audio environment. Visitors wore lightweight liquid crystal shutter glasses and interacted with the graphics environments with a hand-held device called "the wand." Paul Rajlich of NCSA demonstrated CaveVis, a toolkit for exploring multiple 3D datasets in this immersive environment.

The oral program featured eight sessions. The first session, Computational Infrastructure and Future Technologies, opened with invited speaker Larry Smarr, Director of NCSA. He shared with the audience NCSA's efforts to support distributed, collaborative research through the development of a National Technology Grid in which scientists harness the power of high-performance computing, high-bandwidth networks, and sophisticated visualization routinely in a distributed and often web-based environment. Following Dr. Smarr was invited speaker Andrew Chien (UCSD) who described his group's work assembling highly parallel supercomputers from PCs running Windows NT.

In the second session, Information Services, invited speaker Roberta Johnson (University of Michigan) reported on a ambitious web-based outreach program called Windows to the Universe[1] in which Earth and space science is presented to a K-12 audience intermixed with artistic and historical connections to the human experience. Session three featured two invited talks. First, Paul Woodward described systems and software under development at his lab to visualize Tera-byte sized datasets at high resolution using large display devices such as the PowerWall. Next, Richard Puetter (UCSD) presented a unique approach to deconvolution and image modeling which he calls the Pixon method (p. 307).

Anuradha Koratkar (STScI) and Sandra Grosvenor (Federal Data Corp.) opened Tuesday's first session, Observatory Planning, Scheduling, and Telescope Control with their joint invited talk on next generation user support tools (p. 57). Among the interesting developments in the field of observation planning they described was the recent effort by a number of observatories to collaborate on proposal preparation tools. Session five, Computational Astrophysics, opened with an invited talk from Peter Garnavich (SAO) entitled "Cosmology from High–Redshift Supernovae" (p. 33) describing the recent analysis of Type Ia supernovae that led to the surprising result of an accelerating Universe. Tuesday's last oral session featured a review of algorithms for CCD stellar photometry by invited speaker Ken Mighell (KPNO/NOAO; p. 317).

Wednesday opened with an oral session on archiving, beginning with an invited talk by Nancy Brickhouse (SAO; p. 25) who discussed the impact of supercomputing and database advances on the use of atomic data for modeling astrophysical phenomena. Leading the final session on Astrostatistics and Databases, Tom Loredo (Cornell) provided a "developer's perspective" to understanding Bayesian statistical inference as an alternative to the familiar "frequentist" approach (p. 297).

The meeting was augmented with "birds–of–a–feather" sessions, or BOFs. These informal sessions gave participants the opportunity to discuss a variety of special interests. A few BOF topics from past years returned to considerable participant interest, including Linux, IDL, and FITS. The FITS BOF reviewed the current status on several issues regarding the evolution of the standard format for astronomy. Two new BOFs were added this year which reflect some current trends in astronomical software development. The Java BOF featured a panel of developers that shared tips, resources and ideas for using Java in astronomical applications which have been summarized in a web site[2]. The other new BOF brought together two related issues (combined primarily due to scheduling constraints): Coordinating Astronomy Web Resources and The Extensible Markup Language (XML). The combination proved interesting as XML was discussed as a standard way for expressing metadata which is a major issue for Web resource interoperability. This BOF also resulted in a web site on

---

[1] `http://www.windows.umich.edu`

[2] `http://monet.astro.uiuc.edu/JavaForAstronomy`

XML for Astronomy[3] and majordomo mailing list. The final BOF, the Future of Astronomical Data–analysis Systems, or FADS, returned to the ADASS program for a fourth straight year. This BOF, led by Jan Noordam and Peter Teuben, has traditionally been a forum for discussing the issues at work in our community that so often leads to "recoding the wheel." This year's discussion examined the advantages and disadvantages of creating a repository of reusable objects for astronomical applications.

The program also included sessions for viewing poster papers and computer demonstrations. The number of posters displayed was up from previous years, but it is interesting to note the dramatic increase in computer demonstrations. In total, the program featured 10 invited talks, 32 contributed talks, 21 computer demonstrations, and 101 poster papers.

2. Proceedings Overview

Each year, the ADASS Program Organizing Committee is charged with determining—and in some cases, predicting—what will be the important themes and issues for our community of software developers and users. Often a session topic is chosen to attract contributions from fields that have been underrepresented at past ADASS conferences but are otherwise considered an important part of our community (e.g., Computational Astrophysics). With trends in technology changing so fast, it is interesting to see how these predictions bear out in the form of contributed papers. Sometimes topical sessions must be filled out with contributions that don't exactly fit the theme. On the other hand, some interesting surprises arise, such as the interest this year in XML for astronomical applications. In presenting the papers from ADASS 98, we have chosen to reorganize the contributions somewhat so as to give a more coherent picture of software development in astronomy which we hope better reflects the issues that unfolded during the meeting.

We start with a theme that cuts across our whole community: Software Development and Management. Recent ADASS conferences show that our community is beginning to pay more attention to how we develop software. Granados (p. 19) attempts to systemize this self-examination through his Software Development Methodology Survey, the results of which are expected at ADASS 99.

From Part 1, the volume travels roughly the data trail from the scientific problem (Part 2) to the end–user. Papers reporting on software to support observatories fell mainly into two groups: those that supported observation planning and scheduling (Part 3) and those that controlled instruments (Part 4). Data Compression (Part 5) emerged as a common theme even though a session was not explicitly scheduled for it.

Part 6 shows that Data Pipelines are a major area of activity in our community, and in this strong section, we see a variety of approaches to the problem. This part shows increasing importance of distributed computing. The VLT approach illustrates a distributed model using CORBA, while the Gemini Data

[3]`http://pioneer.gsfc.nasa.gov/public/xml/`

Handling System shows a client–server model. We also see the importance of *on–the–fly calibration* in which data are calibrated using the best solutions at the time a user requests it.

Part 7 shows that Archiving and Information Services to be another major area of activity. The part opens with reports on services for the end–user and progresses toward the technology behind these services. An emerging theme from this part might be interoperability and integration. Many of the contributions describe services that help bring organization to the vast amount of information available to researchers on the network.

Part 8 presents the latest work in algorithm development. Part 9 collects various general applications of statistics and databases to analysis of astronomical data. We arrive at the end of our data trail in Part 10 with the latest in user interfaces and visualization systems. We conclude the volume with another cross-cutting topic, Software Tools and Systems, which provides to both user and developer the building blocks for new applications of astronomical computing.

3. The People Behind the Conference

The ADASS 98 Program Organizing Committee (POC) was chaired by F. Rick Harden (SAO) and included Rudi Albrecht (ST-ECF/ESO), Dick Crutcher (UIUC), Brian Glendenning (NRAO), Sally Heap (GSFC), George Jacoby (NOAO), Jonathan McDowell (SAO), Glenn Miller (STScI), Jan Noordam (NFRA), Dick Shaw (STScI), Richard Simon (NRAO), Doug Tody (NOAO), and Patrick Wallace (RAL/CCLRC). Following the conference, the POC sponsored a survey of attendees in preparation for ADASS 99; the results of the survey are available on the web at `http://ra.stsci.edu/adass_survey/`.

The Local Organizing Committee (LOC) was chaired by Dick Crutcher and included Allison Clark, Dave Mehringer, Ray Plante, Harold Ravlin, and Doug Roberts. The LOC would like to thank a number of people who helped with set up including Jerry Accord, Alan Calder, Jim Farrar, Alex Farthing, José Miguel Girard, Hank Kaczmarski, and Asif Qamar. We would especially like to thank Brian Kucic, Stuart Levy, Paul Rajlich, John Shalf, and Paul Woodward for their time and enthusiasm in support of the NCSA facilities tour. We are greatly indebted to the Conferences and Institutes unit at UIUC, especially to Kendra Bair for her untiring and conscientious efforts supporting this conference. Finally, the LOC wishes to recognize Dan Briggs for his contributions as a member of the LOC. Unfortunately, Dan was killed in a skydiving accident a few months prior to the conference and didn't get see the fruits of his hard work. We at UIUC and the entire ADASS community will miss Dan.

Finally, we gratefully acknowledge the support of the conference sponsors: the National Center for Supercomputing Applications, the University of Illinois Astronomy Department, the Smithsonian Astrophysical Observatory, the National Radio Astronomy Observatory, the National Optical Astronomy Observatory, the European Southern Observatory, the Space Telescope Science Institute, the Space Telescope European Coordinating Facility, Sun Microsystems, Research Systems, Inc., the National Aeronautics and Space Administration, and the National Science Foundation. Their support in the form of hardware, software, and financing was critical to the success of ADASS 98.

4. ADASS Information

ADASS 99 will be located on the Island of Hawaii, October 3–6, 1999 and will be hosted by the Canada–France–Hawaii Telescope. For further information, visit the ADASS 99 web site (`http://www.cfht.hawaii.edu/ADASS/`) or send mail to `adass@cfht.hawaii.edu`.

Details about ADASS 98 including program abstracts, results from the participant survey, and links to past ADASS conferences are available from the ADASS 98 web site (`http://www.ncsa.uiuc.edu/ADASS98`). This year will see the establishment of a permanent ADASS web site (www.adass.org) which will serve as a centralized warehouse for ADASS conference information past and future.

David M. Mehringer
UIUC Astronomy Department

Raymond L. Plante and Douglas A. Roberts
National Center for Supercomputing Applications

April 1999

Cover Illustration: The Visual Target Tuner (VTT) from the paper by Koratkar and Grosvenor (p. 57). The VTT is an interactive field-of-view visualization tool that allows the user to import images from several sources (including the NASA Extraterrestrial Database—NED, HST's Guide Star Catalogs, and the Digitized Sky Survey). Once imported the user can overlay an instrument's field-of-view positioning and orienting the target exposure visually. Overlays of multiple apertures can also be done visually. The VTT tool was developed as part of the Next Generation Space Telescope's (NGST) Scientist's Expert Assistant (SEA) program in a joint effort between Space Telescope Science Institute (ST ScI) and the Advanced Architectures and Automation Branch of Goddard Space Flight Center's (GSFC) Information Systems Center.

Participant List

Mark Abernathy, Space Telescope Science Institute, 3700 San Martin Drive, Baltimore, MD 21218, USA (abernathy@stsci.edu)
Alberto Accomazzi, Smithsonian Astrophysical Observatory, 60 Garden Street, Cambridge, MA 02138, USA (aaccomazzi@cfa.harvard.edu)
Miguel Albrecht, European Southern Observatory, Karl-Schwarzschild-Strasse 2, Garching 85748, Germany (malbrech@eso.org)
Rudolf Albrecht, Space Telescope European Coordinating Facility, Karl Schwarzschild-Strasse 2, Garching D-85748, Germany (ralbrech@eso.org)
Jae Allen, National Center for Supercomputing Applications, 152 CAB, 605 East Springfield Avenue (MC-476), Champaign, IL 61820, USA (jallen@ncsa.uiuc.edu)
Troy Ames, Goddard Space Flight Center, Code 588, Greenbelt, MD 20771, USA (troy.ames@gsfc.nasa.gov)
Elisabetta Angeloni, European Southern Observatory, Karl-Schwarzschild-Strasse 2, Garching 85748, Germany (eangelon@eso.org)
Peter Anninos, National Center for Supercomputing Applications, 605 East Springfield Avenue (MC-476), Champaign, IL 61820, USA
Hajime Baba, Kyoto University, Japan (baba@kusastro.kyoto-u.ac.jp)
Charles Backus, Jet Propulsion Laboratories, 4800 Oak Grove Drive, Pasadena, CA 91109-8099, USA (backus@heta.jpl.nasa.gov)
Pascal Ballester, European Southern Observatory, Karl Schwarzschild-Strasse 2, Garching bei München D-85748, Germany (pballest@eso.org)
Raymond Bambery, Jet Propulsion Laboratory, 4800 Oak Grove Drive, M/S 168-414, Pasadena, CA 91109-8099, USA (raymond.j.bambery@jpl.nasa.gov)
Klaus Banse, European Southern Observatory, Karl-Schwarzschild-Strasse 2, Garching 85748, Germany (kbanse@eso.org)
Jeannette Barnes, National Optical Astronomy Observatories, P.O. Box 26732, Tucson, AZ 85726, USA (jbarnes@noao.edu)
Paul Barrett, Goddard Space Flight Center, Code 660.2, Greenbelt, MD 20771, USA (barrett@compass.gsfc.nasa.gov)
Andrea Baruffolo, Osservatorio Astronomico di Padova, vicolo Osservatorio, 5, Padova PD 35122, Italy (baruffolo@pd.astro.it)
Tony Beasley, National Radio Astronomy Observatory, 520 Edgemont Road, Charlottesville, VA 22903, USA (tbeasley@nrao.edu)
Dave Bell, National Optical Astronomy Observatories, 950 N. Cherry Avenue, Tucson, AZ 85726-6732, USA (dbell@noao.edu)
Lucio Benfante, Astronomical Observatory of Padua, via Monteriondo, 14, Monteforte d'Alpone VR 37032, Italy (benfante@dei.unipd.it)
Martin Bly, Rutherford Appleton Laboratory, Chilton Didcot, Oxfordshire OX11 OQX, United Kingdom (bly@star.rl.ac.ule)
Nancy Brickhouse, Smithsonian Astrophysical Observatory, 60 Garden Street, Mail Stop 6, Cambridge, MA 02138, USA (nbrickhouse@cfa.harvard.edu)
Alan Bridger, UK Astronomy Technology Centre, Royal Observatory, Blackford Hill, Edinburgh EH9 3HJ, United Kingdom (ab@roe.ac.uk)

Christa Brown, National Optical Astronomy Observatories, Central Computer Services, 950 N. Cherry Avenue, Tucson, AZ 85719, USA (cbrown@noao.edu)
E. Lawrence Brown, Raytheon, Goddard Space Flight Center, Code 664, Greenbelt, MD 20771, USA (elwin@redshift.gsfc.nasa.gov)
Howard Bushouse, Space Telescope Science Institute, 3700 San Martin Drive, Baltimore, MD 21218, USA (bushouse@stsci.edu)
Ivo Busko, Space Telescope Science Institute, 3700 San Martin Drive, Baltimore, MD 21218, USA (busko@stsci.edu)
Oleg Bykov, Pulkovo Observatory, St. Petersburg, Pulkovskoe shosse 65, Pulkovo Observatory, St.Petersburg 196140, Russia (bykov@venvi.usr.pu.ru)
Nicola Caon, Instituto de Astrofisica de Canarias, C/ Via Lactea s/n, La Laguna - Tenerife E-38200, Spain (ncaon@ll.iac.es)
Josephine Chan, University of Cambridge, 1 Cutter Ferry Lane, Cambridge CB4 1JR, United Kingdom (sjchan@ast.cam.ac.uk)
Alberto Maurizio Chavan, European Southern Observatory, Karl-Schwarzschild-Strasse 2, Garching D-85748, Germany (amchavan@eso.org)
Anton Chernenko, Space Research Institute, Profsojuznaja 84/32, Moscow 117810, Russia (anton@cgrsmx.iki.rssi.ru)
Matthew Cheselka, National Optical Astronomy Observatories, P.O. Box 26732, Tucson, AZ 85726, USA (cheselka@noao.edu)
Cynthia Cheung, Goddard Space Flight Center, Astrophysics Data Facility, Code 631, Greenbelt, MD 20771, USA (cynthia.cheung@gsfc.nasa.gov)
Andrew Chien, University of California, San Diego and National Center for Supercomputing Applications, 9500 Gilman Drive, Dept. 0114, La Jolla, CA 92093-0114, USA (achien@cs.ucsd.edu)
Damian Christian, Space Telescope Science Institute, 3700 San Martin Drive, Baltimore, MD 21218, USA (dchrist@stsci.edu)
Mark Clark, National Radio Astronomy Observatory, P.O. Box 2, Green Bank, WV 24944, USA (mclark@nrao.edu)
Thomas M. Comeau, Jr., Space Telescope Science Institute, 3700 San Martin Drive, Baltimore, MD 21218, USA (tcomeau@stsci.edu)
Tim Cornwell, National Radio Astronomy Observatory, P.O. Box 0, Socorro, NM 87801, USA (tcornwel@nrao.edu)
Dennis Crabtree, Canada-France-Hawaii Telescope Corporation, P.O. Box 1597, 65-1238 Mamalahoa Highway, Kamuela, HI 96743, USA (crabtree@cfht.hawaii.edu)
Ray Creager, National Radio Astronomy Observatory, P.O. Box 2, Green Bank, WV 24944, USA (rcreager@nrao.edu)
Richard Crutcher, University of Illinois, 1002 W. Green Street, Urbana, IL 61801, USA (crutcher@astro.uiuc.edu)
Chris Currie, National Center for Supercomputing Applications, 605 East Springfield Avenue (MC-476), Champaign, IL 61820, USA
Malcolm Currie, UK Astronomy Technology Centre, Royal Observatory, Blackford Hill, Edinburgh EH9 3HJ, United Kingdom (mjc@roe.ac.uk)
Greg Daues, National Center for Supercomputing Applications, 605 East Springfield Avenue (MC-476), Champaign, IL 61820, USA

Lindsey Davis, National Optical Astronomy Observatories, P.O. Box 26732, Tucson, AZ 85726, USA (davis@noao.edu)
Jean-Pierre De Cuyper, Royal Observatory of Belgium, Ringlaan 3, Ukkel B1180, Belgium (Jean-Pierre.DeCuyper@ksb-orb.oma.be)
Michele De La Pena, Space Telescope Science Institute, 3700 San Martin Drive, Baltimore, MD 21218, USA (delapena@stsci.edu)
Cor de Vries, Space Research Organization of the Netherlands, Sorbonnelaan 2, Utrecht 3584CA , Netherlands (C.deVries@sron.nl)
Erik Deul, Sterewacht Leiden, P.O. Box 9513, 2300 RA Leiden, Leiden, Netherlands (deul@strw.leidenuniv.nl)
Nicolas Devillard, European Southern Observatory, Karl Schwarzschild-Strasse 2, Garching D-85748, Germany (ndevilla@eso.org)
Markus Dolensky, Space Telescope - European Coordinating Facility, Karl-Schwarzschild - Strasse 2, Garching bei München 85748, Germany (mdolensk@eso.org)
Jennifer Dunn, National Research Council, 5071 West Saanich Road, Victoria, BC V8X 4M6, Canada (dunn@dao.nrc.ca)
Daniel Durand, National Research Council, 5071 W. Saanich Road, Victoria, BC V8X 4M6, Canada (daniel.durand@hia.nrc.ca)
Frossie Economou, UKIRT (JAC), Joint Astronomy Centre, 660 North A'ohoku Place, Hilo, HI 96720, USA (frossie@jach.hawaii.edu)
Guenther Eichhorn, Smithsonian Astrophysical Observatory, 60 Garden Street, Cambridge, MA 02138, USA (geichhorn@cfa.harvard.edu)
Christian Erd, Astrophysics Division, European Space Agency, ESTEC/SAX, Postbus 299, Noordwijk 2200 AG, Netherlands (cerd@astro.estec.esa.nl)
Allen Ferris, Space Telescope Science Institute, 3700 San Martin Drive, Baltimore, MD 21218, USA (farris@stsci.edu)
Anthony Ferro, University of Arizona, Steward Observatory/NICMOS Project, 933 N. Cherry Avenue, Tucson AZ, 85721-0065, USA (tferro@as.arizona.edu)
Mike Fitzpatrick, National Optical Astronomy Observatories/IRAF Group, 950 N. Cherry Avenue, Tucson, AZ 85726-6732, USA (fitz@noao.edu)
Michael Folk, National Center for Supercomputing Applications, 605 E. Springfield Avenue (MC-476), Champaign, IL 61820, USA (mfolk@ncsa.uiuc.edu)
James R. Fowler, Hobby & Eberly Telescope/McDonald Observatory, P.O. Box 1337, Fort Davis, TX 79734, USA (jrf@astro.as.utexas.edu)
Marian L. Frueh, McDonald Observatory, P.O. Box 1337, Fort Davis, TX 79734, USA (frueh@astro.as.utexas.edu)
Joe Futrelle, National Center for Supercomputing Applications, 152 Computing Applications Bldg., 605 E. Springfield Avenue (MC-476), Champaign, IL 61820, USA (futrelle@ncsa.uiuc.edu)
Carlos Gabriel, European Space Agency - VILSPA, Villafranca del Castillo, Satellite Tracking Station, Madrid 28080, Spain (cgabriel@iso.vilspa.esa.es)
Peter Garnavich, Center for Astrophysics, Mail Stop 20, 60 Garden Street, Cambridge, MA 02138, USA (pgarnavich@cfa.harvard.edu)
Robert Garwood, National Radio Astronomy Observatory, 520 Edgemont Road, Charlottesville, VA 22911, USA (bgarwood@nrao.edu)

Rene Gastaud, CEA Cen Saclay, Gif Sur Yvette 91 191, France
Severin Gaudet, National Research Council, 5071 W. Saanich Road, Victoria BC V8X 4M6, Canada (severin.gaudet@hia.nrc.ca)
Matt Giamporcaro, Boston University, 677 Beacon Street, Boston, MA 02215, USA
Mark Giuliano, Space Telescope Science Institute, 3700 San Martin Drive, Baltimore, MD 21218, USA (giuliano@stsci.edu)
John Glaspey, Kitt Peak National Observatory, P.O. Box 26732, Tucson, AZ 85726, USA (glaspey@noao.edu)
Brian Glendenning, National Radio Astronomy Observatory, P.O. Box 0, Socorro, NM 87801, USA (bglenden@nrao.edu)
Arno Granados, National Radio Astronomy Observatory, P.O. Box 2, Green Bank, WV 24944, USA (agranado@nrao.edu)
Victor Grechnev, Institute of Solar-Terrestrial Physics, Lermontov St. 126, Irkutsk 664033, Russia (grechnev@iszf.irk.ru)
Gretchen Greene, Space Telescope Science Institute, 3700 San Martin Drive, Baltimore, MD 21218, USA (greene@stsci.edu)
Perry Greenfield, Space Telescope Science Institute, 3700 San Martin Drive, Baltimore, MD 21218, USA (perry@stsci.edu)
Eric W. Greisen, National Radio Astronomy Observatory, 520 Edgemont Road, Charlottesville, VA 22903-2475, USA (egreisen@nrao.edu)
Preben Grosbøl, European Southern Observatory, Karl-Schwarzschild-Strasse 2, Garching D-85748, Germany (pgrosbol@eso.org)
Sandy Grosvenor, Federal Data Corporation, Goddard Space Flight Center, Mail Stop 588, Greenbelt, MD 20771, USA (Sandy.Grosvenor@gsfc.nasa.gov)
David Grumm, Smithsonian Astrophysical Observatory, 60 Garden Street, Cambridge, MA 02138, USA
Bernt Grundseth, Canada-France-Hawaii Telescope Corporation, P.O. Box 1597, Kamuela, HI 96743, USA
Damien Guillaume, University Louis-Pasteur, Strasbourg, 25 rue de la paix, Louveciennes 78430, France (DG.Guillaume@ulst.ac.uk)
Sevinc Gülseçen, Istanbul University, I.U. Observatory, Astronomy and Space Science Dept., Istanbul 34452, Turkey (sevincg@yahoo.com)
F. Rick Harnden, Jr., Smithsonian Astrophysical Observatory, 60 Garden St. MS-2, Cambridge, MA 02138, USA (frh@cfa.harvard.edu)
R. Lee Hawkins, Whitin Observatory, Wellesley College, Wellesley, MA 02481, USA (lhawkins@wellesley.edu)
Christina Heikkila, Raytheon STX, Goddard Space Flight Center, Mailstop 664.0, Greenbelt, MD 20771, USA (cwh@lheamail.gsfc.nasa.gov)
Ingolf Heinrichsen, California Institute of Technology/IPAC, Mail Code 100-22, Pasadena, CA 91125, USA (ihh@ipac.caltech.edu)
Ron Henry, Space Telescope Science Institute, 3700 San Martin Drive, Baltimore, MD 21218, USA (henry@stsci.edu)
Norman Hill, National Research Council, 5071 W. Saanich Road, Victoria, BC V8X 4M6, Canada (norman.hill@hia.nrc.ca)
Paul Hilton, Raytheon Systems/ISAS, 3-1-1 Yoshinodai, Sagamihara Kanagawa 229, Japan (paul@astro.isas.ac.jp)

Philip Hodge, Space Telescope Science Institute, 3700 San Martin Drive, Baltimore, MD 21218, USA (hodge@stsci.edu)

Richard Hook, European Southern Observatory/ST-ECF, Karl Schwarzschild-Strasse 2, Germany (rhook@eso.org)

Toshihiro Horaguchi, National Science Museum, 3-23-1, Hyakunin-cho, Shinjuku, Tokyo 169-0073, Japan (horaguti@kahaku.go.jp)

Allan Hornstrup, Danish Space Research Institute, Juliane Maries Vej 30, Copenhagen 2100, Denmark (allan@dsri.dk)

John M. Hughes, Steward Observatory, 933 North Cherry Avenue, Tucson, AZ 85721, USA (jmh@as.arizona.edu)

Edwin Huizinga, Space Telescope Science Institute, 3700 San Martin Drive, Baltimore, MD 21218, USA (huizinga@stsci.edu)

Gareth Hunt, National Radio Astronomy Observatory, P.O. Box 2, Green Bank, WV 24944, USA (ghunt@nrao.edu)

Yasuhide Ishihara, Fujitsu Co. Ltd., 9-3, Nakase 1 Chome, Mihama-ku, China 261, Japan (ishi@ssd.se.fujitsu.co.jp)

Shannon Jaeger, National Research Council, 5071 W. Saanich Road, Victoria BC V8X 4M6, Canada (shannon.jaeger@hia.nrc.ca)

Fred Jansen, European Space Agency-ESTEC, P.O. Box 299, Noordwijk 2200 AG, Netherlands (fjansen@astro.estec.esa.nl)

Tim Jenness, Joint Astronomy Center, 660 N. A'ohoku Pl, University Park, Hilo, HI 96720, USA (t.jenness@jach.hawaii.edu)

Roberta Johnson, University of Michigan, 2455 Hayward Street, 1429 Space Research Bldg., Ann Arbor, MI 48109-2143, USA (rmjohnsn@umich.edu)

Jeremy Jones, Goddard Space Flight Center, Code 588, Bldg. 23, Room E413, Greenbelt, MD 20771, USA (Jeremy.E.Jones@gsfc.nasa.gov)

William Joye, Smithsonian Astrophysical Observatory, 60 Garden Street, Cambridge, MA 02138, USA

Apu C. Kapadia, University of Illinois at Urbana-Champaign, 1010 West Green Street, 203 Daniels Hall, Urbana, IL 61801, USA (akapadia@uiuc.edu)

Vincent Kargatis, Raytheon ITSS, Goddard Space Flight Center, Code 631, Greenbelt, MD 20771, USA (vincent.kargatis@gsfc.nasa.gov)

Susan Kayser, National Science Foundation, MPS/AST Room 1045, 4201 Wilson Blvd., Arlington, VA 22230, USA (skayser@nsf.gov)

Athol Kemball, National Radio Astronomy Observatory, P.O. Box 0, Socorro, NM 87801, USA (akemball@nrao.edu)

Bridget Kennedy, Space Telescope Science Institute, 3700 San Martin Drive, Baltimore, MD 21218, USA (bkennedy@stsci.edu)

Steven Kleiner, Smithsonian Astrophysical Observatory, Mail Stop 66, 60 Garden Street, Cambridge, MA 02138, USA (skleiner@cfa.harvard.edu)

Anuradha Koratkar, Space Telescope Science Institute, 3700 San Martin Drive, Baltimore, MD 21218, USA (koratkar@stsci.edu)

Reinhold Kroll, Instituto de Astrofisica de Canarias, c/ Via Lactea s/n, E-38200 La Laguna TF, Spain (kroll@iac.es)

Leslie Kuchinski, California Institute of Technology, IPAC, MS 100-22, Pasadena, CA 91125, USA (leslie@ipac.caltech.edu)

Wayne Landsman, Raytheon ITSS, Goddard Space Flight Center, Code 681, Greenbelt, MD 20771, USA (landsman@mpb.gsfc.nasa.gov)

Jonghoon Lee, University of Illinois at Urbana Champaign, Beckman Institute, 363 Psychology Bldg. (MC-716), Urbana, IL 61821, USA

James R. Lewis, Royal Greenwich Observatory, Madingley Road, Cambridge CB3 OHJ, United Kingdom (jrl@agt.cam.ac.uk)

John Lightfoot, U.K. Astronomy Technology Centre, Royal Observatory, Blackford Hill, Edinburgh EH9 3HJ, United Kingdom (jfl@roe.ac.uk)

Rick Lively, National Radio Astronomy Observatory, P.O. Box 0, Socorro, NM 87801, USA (rlively@nrao.edu)

Antoine Llebaria, LAS (CNRS), Traverse du Siphon, PB8 13376 Marseille Cedex 12, Marseille 13, France (antoine@astrsp-mrs.fr)

Thomas Loredo, Cornell University, Department of Astronomy, 604 Space Sciences Building, Ithaca, NY 14853-6801, USA (loredo@spacenet.tn.cornell.edu)

Stephen Lubow, Space Telescope Science Institute, 3700 San Martin Drive, Baltimore, MD 21218, USA (lubow@stsci.edu)

Barry Madore, California Institute of Technology, NED 100-22, IPAC, Pasadena, CA 91125, USA (barry@ipac.caltech.edu)

Oleg Malkov, Institute of Astronomy, Russian Academy of Science, 48 Pyatnitskaya Str., Moscow 109017, Russia (malkov@inasan.rssi.ru)

Michele Maris, Trieste Astronomical Observatory (OAT), Via G.B. Tiepolo, 11, Trieste TS I-34131, Italy (maris@oat.ts.astro.it)

Jonathan McDowell, Smithsonian Astrophysical Observatory, 60 Garden Street, Cambridge, MA 02138, USA (jmcdowell@cfa.harvard.edu)

Thomas A. McGlynn, Goddard Space Flight Center, NASA/GSFC/USRA, Code 660, Greenbelt, MD 20771, USA (tam@silk.gsfc.nasa.gov)

Robert McGrath, National Center for Supercomputing Applications, University of Illinois at Urbana-Champaign, 605 E. Springfield Avenue (MC-476), Champaign, IL 61820, USA (mcgrath@ncsa.uiuc.edu)

Nuria McKay, University of Manchester - NRAL, Jodrell Bank, Macclesfield, Cheshire SK11 9DL, United Kingdom (nm@jb.man.ac.uk)

Brian McLean, Space Telescope Science Institute, 3700 San Martin Drive, Baltimore, MD 21218, USA (mclean@stsci.edu)

Joseph McMullin, National Radio Astronomy Observatory, 520 Edgemont Road, Charlottesville, VA 22903, USA (jmcmulli@nrao.edu)

David Mehringer, University of Illinois at Urbana-Champaign, Department of Astronomy, 1002 West Green Street, Urbana, IL 61801, USA (dmehring@astro.uiuc.edu)

Alberto Micol, Europoean Space Agency/ST-ECF, c/o European Southern Observatory, Karl Schwartzschild-Strasse 2, Garching b. München D85478, Germany (amicol@eso.org)

Scott Midler, Research Systems, 2995 Wilderness Place, Boulder, CO 80301, USA (smidler@rsinc.com)

Kenneth Mighell, Kitt Peak National Observatory, P.O. Box 26732, Tucson, AZ 85726-6732, USA (mighell@noao.edu)

Glenn Miller, Space Telescope Science Institute, 3700 San Martin Drive, Baltimore, MD 21218, USA (miller@stcsi.edu)

Walter Warren Miller III, Space Telescope Science Institute, 3700 San Martin Drive, Baltimore, MD 21218, USA (wmiller@stsci.edu)

Dave Mills, National Optical Astronomy Observatories, 950 North Cherry Avenue, Tucson, AZ 85719, USA (dmills@noao.edu)
Douglas Mink, Smithsonian Astrophysical Observatory, 60 Garden Street, Cambridge, MA 02138, USA (dmink@cfa.harvard.edu)
Yoshihiko Mizumoto, National Astronomical Observatory of Japan, Osawa Mitaka Tokyo 181-8588, Japan (mizumoto@optik.mtk.nao.ac.jp)
Fionn Murtagh, University of Ulster, Faculty of Informatics, Londonderry BT48 7JL, Northern Ireland, United Kingdom (fd.murtagh@ulst.ac.uk)
Tim Naylor, Keele University, Department of Physics, Keele University, Staffordshire ST5 5BG, United Kingdom (timn@astro.keele.ac.uk)
Maria Nieto-Santisteban, Space Telescope Science Institute, 3700 San Martin Drive, Baltimore, MD 21218, USA (nieto@stsci.edu)
Jan Noordam, Netherlands Foundation for Research in Astronomy, P.O. Box 2, Dwingeloo 7990 AA, Netherlands (jnoordam@nfra.nl)
Mike Norman, National Center for Supercomputing Applications, University of Illinois at Urbana-Champaign, 605 East Springfield Avenue (MC-476), Champaign, IL 61820, USA
Joel D. Offenberg, Raytheon STX, Goddard Space Flight Center, Code 681, Greenbelt, MD 20771, USA (Joel.D.Offenberg.1@gsfc.nasa.gov)
Patricio Ortiz, Centre de Donnees de Strasbourg,, 11 Rue de l'Universite, Strasbourg 67000, France (portiz@astro.u-strasbg.fr)
Stephan Ott, European Space Agency, Villafranca del Castillo Satellite Tracking Station, Madrid E28080, Spain (sott@iso.vilspa.esa.es)
Cindy Pantalone, Harvard University, 46 Oxford Street, Cambridge, MA 02138, USA
Harry Payne, Space Telescope Science Institute, 3700 San Martin Drive, Baltimore, MD 21218, USA (payne@stsci.edu)
James Peachey, Raytheon STX, Goddard Space Flight Center, Greenbelt, MD 20771, USA
William Pence, Goddard Space Flight Center, HEASARC/NASA/GSFC, Mail Code 662, Greenbelt, MD 20771, USA (pence@tetra.gsfc.nasa.gov)
Benoit Pirenne, ESO/ST-ECF, Karl-Schwarzschild-Strasse 2, D-85748 Garching bei München, Germany (bpirenne@eso.org)
Norbert Pirzkal, European Southern Observatory/ST-ECF, Karl-Schwarzschild-Strasse 2, Garching 85748, Germany (npirzkal@eso.org)
Raymond Plante, National Center for Supercomputing Applications, 103 Astronomy Bldg., 1002 West Green Street, Urbana, IL 61801, USA (rplante@ncsa.uiuc.edu)
Joel Plutchak, University of Illinois, National Center for Supercomputing Applications, 152 CAB, MC-476, Champaign, IL 61820, USA (plutchak@uiuc.edu)
Joseph Pollizzi, Space Telescope Science Institute, 3700 San Martin Drive, Baltimore, MD 21218, USA (pollizzi@stsci.edu)
Marc Pound, University of Maryland, Department of Astronomy, Computer & Space Science Building, College Park, MD 20742, USA (mpound@astro.umd.edu)
Alexei Pozanenko, Space Research Institute, Profsojuznaja str. 84/32, Moscow 117810, Russia (apozanen@iki.rssi.ru)

Richard Puetter, Center for Astrophysics & Space Sciences, University of California, San Diego, 9500 Gilman Drive, La Jolla, CA 92093-0424, USA (rpuetter@ucsd.edu)

Harold Ravlin, University of Illinois; Department of Astronomy, 1002 West Green Street, Urbana, IL 61801, USA (hr@astro.uiuc.edu)

Nicholas Rees, Joint Astronomy Center, 660 N. Aohoku Place, Hilo, HI 96720, USA (n.rees@jach.hawaii.edu)

Joel Richon, Space Telescope Science Institute, 3700 San Martin Drive, Baltimore, MD 21218, USA (richon@stsci.edu)

Doug Roberts, National Center for Supercomputing Applications, 103 Astronomy Building, 1002 West Green Street, Urbana, IL 61801, USA (dougr@ncsa.uiuc.edu)

William Roberts, Mount Stromlo and Siding Spring Observatories, Private Bag, Weston Creek Post Office, Canberra ACT 2611, Australia (bill.roberts@anu.edu.au)

Michael R. Rosa, Space Telescope - European Coordinating Facility, European Southern Observatory, Karl-Schwarzschild-Strasse 2, D-85748 Garching, Germany (MROSA@ESO.ORG)

Arnold Rots, ASC/SAO, Smithsonian Astrophysical Observatory, 60 Garden Street, Cambridge, MA 02138, USA (arots@head-cfa.harvard.edu)

George Rudenko, Institute of Solar Terrestrial Physics, Lermontov St. 126, Irkutsk 664033, Russia (rud@iszf.irk.ru)

Chris Sabbey, Yale University, Astronomy Department, New Haven, CT 06520-8101, USA (sabbey@astro.yale.edu)

Kenneth B. Sall, Century Computing, Inc., 8101 Sandy Spring Road, Laurel, MD 20707, USA (ksall@cen.com)

Annette Saunders, Raytheon, Goddard Space Flight Center, Code 664, Greenbelt, MD 20771, USA (saunders@lheamail.gsfc.nasa.gov)

Isabelle Scholl, Institut d'Astrophysique Spatiale - CNRS, Batiment 121, Universite Paris XI, Orsay 91405, France (scholl@medoc-ias.u-psud.fr)

John Schultz, Space Telescope Science Institute, 3700 San Martin Drive, Baltimore, MD 21218, USA (jschultz@stsci.edu)

Stephen Scott, California Institute of Technology/OVRO, P.O. Box 968, Big Pine, CA 93513, USA (scott@ovro.caltech.edu)

Rob Seaman, National Optical Astronomy Obsevatories/IRAF, 950 N. Cherry Avenue, Tucson, AZ 85719, USA (seaman@noao.edu)

Ratnabali Sengupta, Raytheon STX, Goddard Space Flight Center, 7474 Greenway Center Drive (Code 680), Greenbelt, MD 20770, USA (sengupta.ratna@lmmail.hst.nasa.gov)

Richard A. Shaw, Space Telescope Science Institute, 3700 San Martin Drive, Baltimore, MD 21218, USA (shaw@stsci.edu)

Edward Shaya, Raytheon STX, Goddard Space Flight Center, Code 631, Greenbelt, MD 20771, USA (edward.j.shaya.1@gsfc.nasa.gov)

Hemant Shukla, California Institute of Technology/OVRO, 100 Leighton Lane, Big Pine, CA 93513, USA (hemant@ovro.caltech.edu)

Anand Sivaramakrishnan, Space Telescope Science Institute, 3700 San Martin Drive, Baltimore, MD 21218, USA (anands@stsci.edu)

Petr Škoda, Astronomical Institute Academy of Sciences, Fricova 1, Ondrejov 251 65, Czech Republic (skoda@sunstel.asu.cas.cz)

Larry Smarr, National Center for Supercomputing Applications, 605 E. Springfield Avenue, Champaign, IL 61820, USA (ls@ncsa.uiuc.edu)
Oleg Smirnov, Institute of Astronomy, Russian Academy of Science, 48 Pyatnitskaya Str., Moscow 109017, Russia (oms@inasan.rssi.ru)
Jean-Luc Starck, CEA, 91191 Gif-Sur-Yvette Cedex, France (jstarck@cea.fz)
Mercedes M. Stevens, Canada-France-Hawaii Telescope Corporation, 65-1238, Mamalahoa Highway, Kamuela, HI 96743, USA (stevens@cfht.hawaii.edu)
Elizabeth Stobie, University of Arizona, Steward Observatory, 933 N. Cherry Avenue, Tucson, AZ 85721, USA (bstobie@as.arizona.edu)
Chris Stoughton, Fermilab, MS 127, 500 Wilson Road, Batavia, IL 60510, USA (stoughto@fnal.gov)
Daryl A. Swade, Space Telescope Science Institute, 3700 San Martin Drive, Baltimore, MD 21218, USA (swade@stsci.edu)
Michael S. Swam, Space Telescope Science Institute, 3700 San Martin Drive, Baltimore, MD 21218, USA (mswam@stsci.edu)
Johannes Terlouw, Kapteyn Astronomical Institute, Postbus 800, Groningen 9700 AV, Netherlands (J.P.Terlouw@astro.rug.nl)
Peter Teuben, University of Maryland, Astronomy Department, College Park, MD 20742, USA (teuben@astro.umd.edu)
Doug Tody, National Optical Astronomy Observatories, P.O. Box 26732, Tucson, AZ 85726, USA (tody@noao.edu)
Ralph Tremmel, Max Planck Institüt für Astronomie, Koenigstuhl 17, Heidelberg 69126, Germany (tremmel@mpia-hd.mpg.de)
Francesco Tribioli, Osservatorio Astrofisico di Arcetri, l.go E. Fermi, 5, Firenze 50127, Italy (tribioli@arcetri.astro.it)
Alexander Tsvetkov, Astronomical Institute of St.Petersburg University, St. Petersburg, Petrodvorets, Bibliotechnaya pl.2, St. Petersburg 197198, Russia (tsvetkov@ast.usr.pu.ru)
Gabriella Turek, California Institute of Technology/IPAC, Mail Code 100-22, Pasadena, CA 91125, USA (gaby@ipac.caltech.edu)
Francisco Valdes, National Optical Astronomy Observatories/IRAF, P.O. Box 26732, Tucson, AZ 85726, USA (fvaldes@noao.edu)
Robert Vallance, University of Birmingham, School of Physics & Astronomy, Edgbaston Park Road, Birmingham B15 2TT, United Kingdom (rjv@star.sr.bham.ac.uk)
Ivan Valtchanov, Institute of Astronomy, 72 Tsarigradsko Chaussee Blvd., Sofia BG-1784, Bulgaria (ivan@astro.bas.bg)
Gustaaf van Moorsel, National Radio Astronomy Observatory, P.O. Box 0, Socorro, NM 87801, USA (gvanmoor@nrao.edu)
Sergei Vasilyev, National Academy of Science of Ukraine, Institute for Electromagnetic Researches, P.O. Box 30, Kharkiv 310052, Ukraine (vvs@land.kharkov.ua)
Christian Veillet, Canada-France-Hawaii Telescope Corporation, P.O. Box 1597, Kamuela, HI 96743, USA (veillet@cfht.hawaii.edu)
Harro Verkouter, Jive Institution, Oude-Hoogeveensedijk #4, Dwingeloo 7991 PD, Netherlands (verkouter@jive.nfra.nl)
Martin Vogelaar, Kapteyn Institute, Zernikegebouw, Postbus 800, Groningen 9700 AV, Netherlands (vogelaar@astro.rug.nl)

Patrick Wallace, Rutherford Appleton Laboratory, Chilton, Didcot Oxon, OX11 0QX, United Kingdom (ptw@star.rl.ac.uk)
Michael H. Ward, McDonald Observatory, P.O. Box 1337, Fort Davis, TX 79734, USA (ward@astro.as.utexas.edu)
Rein H. Warmels, European Southern Observatory, Karl-Schwarzschild-Strasse 2, Garching b. München D-85748, Germany (rwarmels@eso.org)
Donald C. Wells, National Radio Astronomy Observatory, 520 Edgemont Road, Charlottesville, VA 22903-2475, USA (dwells@nrao.edu)
William C. Wendling, National Center for Supercomputing Applications, 604 East Springfield, Champaign, IL 61820, USA (wendling@ncsa.uiuc.edu)
Richard L. White, Space Telescope Science Institute, 3700 San Martin Drive, Baltimore, MD 21218, USA (rlw@stsci.edu)
Andreas Wicenec, ESO, Karl-Schwarzschild-Strasse 2, Garching D-85748, Germany (awicenec@eso.org)
Ekkehard Wieprecht, Max Planck Institüt für Extraterrestrische Physik, Giesenbachstrasse, Garching D-85740, Germany (ewieprec@mpe.mpg.de)
Ramon L. Williamson, II, Space Telescope Science Institute, 3700 San Martin Drive, Baltimore, MD 21218, USA (ramon@stsci.edu)
James Wolfer, Andrews University, Department of Computer Science, Berrien Springs, MI 49104-0360, USA
Paul R. Woodward, University of Minnesota, Astronomy Dept./LCSE, 116 Church Street, 356 Phys., Minneapolis, MN 55455, USA (paul@lcse.umn.edu)
Masafumi Yagi, National Astronomical Observatory of Japan, 2-21-1, Ohsawa, Mitaka, Tokyo 181-8588, Japan (yagi@optik.mtk.nao.ac.jp)
Wesley Young, National Radio Astronomy Observatory, P.O. Box O, Socorro, NM 87801, USA (wyoung@aoc.nrao.edu)
Marion Zacharias, USRA, CTIO, 950 North Cherry Avenue, Tucson, AZ 85719, USA (miz@ctioj8.ctio.noao.edu)
Norbert Zacharias, U. S. Naval Observatory, 3450 Mass. Ave. N.W., Washington, DC 20392, USA (nz@ctioj8.ctio.noao.edu)
Nelson Zarate, National Optical Astronomy Observatories, 950 N. Cherry Avenue, Tucson, AZ 85719, USA (zarate@noao.edu)

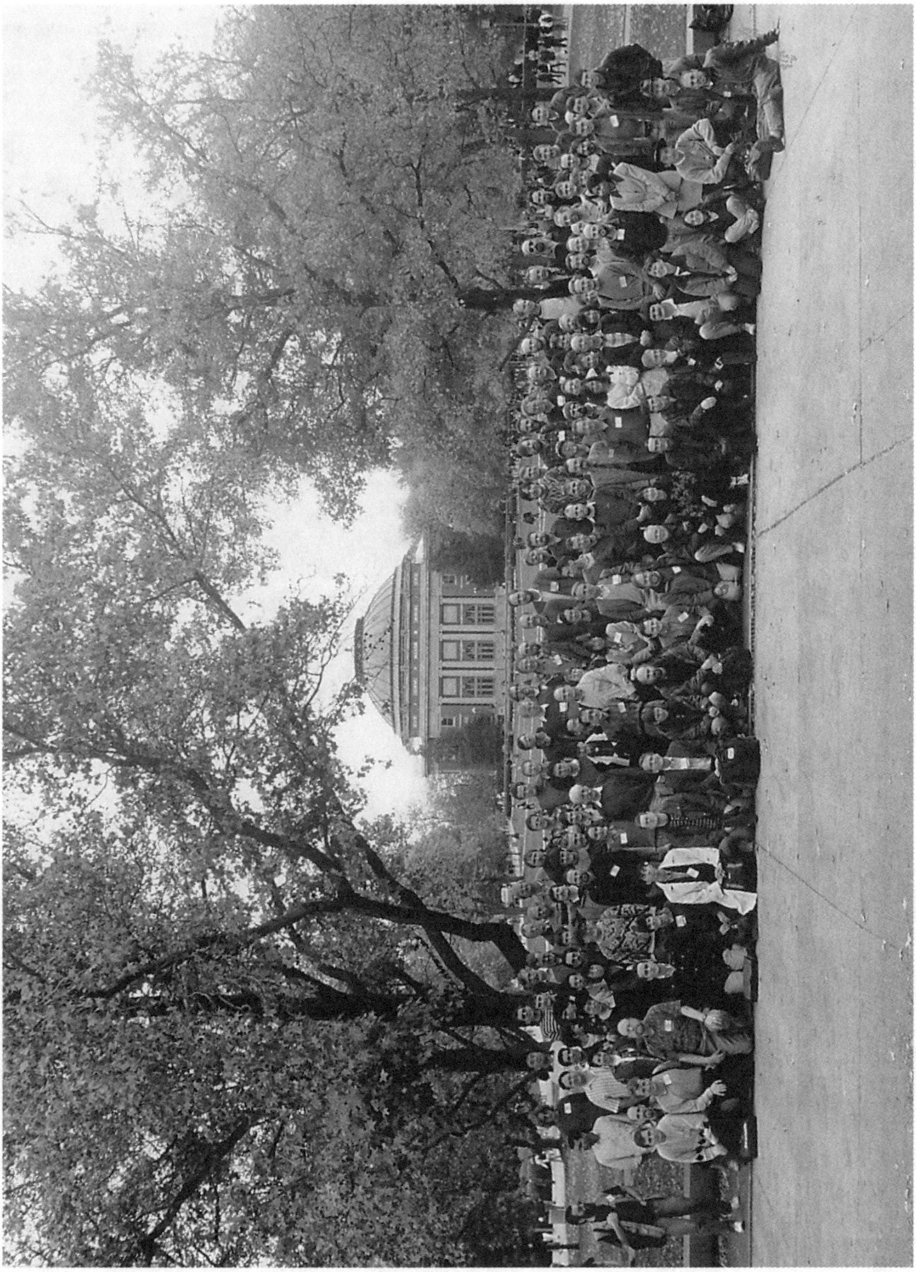

Part 1. Software Development and Management

Astronomical Data Analysis Software and Systems VIII
ASP Conference Series, Vol. 172, 1999
D. M. Mehringer, R. L. Plante, and D. A. Roberts, eds.

The Gemini Data Handling System: A Case History

Séverin Gaudet, Norman R. Hill, Jennifer Dunn, Shannon Jaeger, Steve Cockayne

National Research Council Canada, Herzberg Institute of Astrophysics, 5071 West Saanich Rd., Victoria B.C., Canada V8X 4M6

Abstract. The Gemini Data Handling System, developed by the Canadian Astronomy Data Centre for the Gemini 8m Telescopes Project, provides the data handling infrastructure for the Observatory Control System and the instrument control systems. An overview of the preliminary design was presented at ADASS '95. In August 1998, the DHS passed its acceptance tests and was released operationally to Gemini North in Hawai'i. This paper will present a case history of the project: how the requirements changed, how the design evolved to its final form, the approaches taken, the tools used, and the problems encountered.

1. Introduction

The Gemini Data Handling System (DHS), developed by the Canadian Astronomy Data Centre (CADC) for the Gemini telescopes, provides the data handling infrastructure for the Observatory Control System (OCS) and the instrument control systems. An overview of the preliminary design was presented at ADASS '95 (Gaudet 1996). This paper will not discuss the functionality, design, user interfaces, nor technologies of the DHS, as details on these topics are documented elsewhere (Cockayne et al. 1998a, 1998b; Dunn et al. 1999a, 1999b; Hill et al. 1999a, 1999b; Jaeger et al. 1999). Rather, we will present a case history of the project: how the requirements changed, how the design evolved to its final form, the approaches taken, the tools used, and the problems encountered.

2. Evolution of the Project

The Gemini software system is being produced mostly by allocated work packages to institutions in the partner countries to tap into existing skills, expertise, and experience (McGonegal 1996). An international committee worked on the high-level design of the whole control system which defined the functionality and boundaries of all work packages, including the DHS.

Early into the contract, it became obvious that the requirements for the DHS were not complete. But as the various design steps proceeded and the requirements were completed, the design evolved causing changes in both the DHS functionality and the DHS boundaries with other systems.

The design evolved from a sequenced single-threaded, single server program for each instrument into a data-driven, multiple threaded servers, each encapsulating some functionality. The DHS is also capable of accepting data from many sources simultaneously. The system also became highly configurable in terms of data structures, topology, and data dictionary to allow Gemini to adapt it to the emerging summit requirements. In addition, the system had to be distributed where subsets of the DHS would run both at the summit and at the base facility.

3. Approaches Taken

- **Object-oriented.** A team decision lead to the adoption of object-oriented methodology for both the design (Booch) and development (C++).
- **Multi-threaded distributed servers.** We used multi-threaded servers to encapsulate functionality and to localize interfaces to the rest of the Gemini software systems. For example, there is only one server handling the data put and get requests.
- **Persistent store.** The SYBASE RDBMS was used to store file status and location as well as history records.
- **Code re-use.** We have utilised many CADC data management libraries as well as *cfitsio* (NASA), the *IMP* messaging system and *SDS* data structures from AAO, ESO's *Skycat* display tool, and the Gemini Project's tcl/tk shell *OCSWish*.

4. The Environment

To allow the team to work as efficiently as possible, the proper working environment had to be created. This consisted of:

- **A well supported development environment.** The CADC already had a code management process that allowed four programmers to work simultaneously with a minimum of conflicts, allowed for configuration management, and had well defined installation and release procedures.
- **Good tools.** This included a good design tool (Rational Rose), a good set of base classes implemented early in the project (libdhs++, libGen, libdhsSta), a good debugging environment with thread support (Visual Workshop), and a good documentation tool (Framemaker).

5. Problems Encountered

- **External contract vs. partnership.** Since we were in a contractual relationship with the Gemini Project, we could have managed the evolving design with change orders, etc. However, because of our partnership status, we had a stake in the final result. We had to be driven by "We have to produce something that works, not something that meets spec!". So it became a balancing act between needs versus resources.
- **External contract vs. internal work.** The CADC has been developing software for many years, mostly to support our core archiving activity. As in most groups, internal development does not follow a formal methodology but relies on back-of-the-envelope design and ad-hoc testing. Although

not easy, the group managed the transition to a new culture of a formal methodology.
- **Not all interfaces were well defined.** As the DHS design evolved, some functions originally in the DHS were moved to other work packages and some functions initially in other packages were moved to the DHS. This necessitated the development of new interfaces, some of which took a long time to finalise.
- **Separation of UNIX vs. real-time systems.** The DHS is mostly a UNIX system, but part of the core library is to be used by instruments running under the VxWorks real-time system. Porting the UNIX library was a Gemini Project responsibility which was continually delayed by lack of resources.
- **No instrument teams ready.** The DHS team, with the help of some Gemini scientists, had to make design decisions on data structures and data processing without being able to consult with instrument teams.
- **Developers not involved in high-level design.** The high-level design of the DHS was done before the CADC became involved with the project (see §2.). We had to accept a work package with fixed costs and fixed requirements, requirements which we could only interpret based on our assumptions - some of which proved wrong!

6. Strengths

- **Good team.** This is the essential ingredient for success. The DHS team shared a vision of the final product, adapted to the methodology and the environment and worked very hard and very well together to attain the goal.
- **Solid design.** The design presented at the Preliminary Design review proved to be a solid foundation upon which detailed design of all the subsystems of the DHS could proceed.
- **Formal project methodology.** Breaking the project down into manageable, measurable tasks allowed for the trace-ability of progress and costs. Microsoft Project was used to organize this information.
- **Interfaces.** The fact that most major interfaces were fixed early in the project helped minimize the assumptions on work package boundaries.
- **Meetings.** Travel was part of the project budget and that was a wise decision. There is nothing like face-to-face meetings with other groups to resolve misunderstandings common to distributed projects.

7. Conclusion

As mentioned above, the Phase 1 delivery of the DHS is up and running on Mauna Kea in preparation for first light. Phase 2 has been negotiated and the DHS final delivery is scheduled for January 1999. Not all requirements will have been met because of both the evolving requirements and the underestimation of the cost. The cost breakdown for the DHS project to date is shown in Fig. 1.

Overall, the experience has been positive for the CADC. The formal methodology and the new technologies have been good learning experiences and have

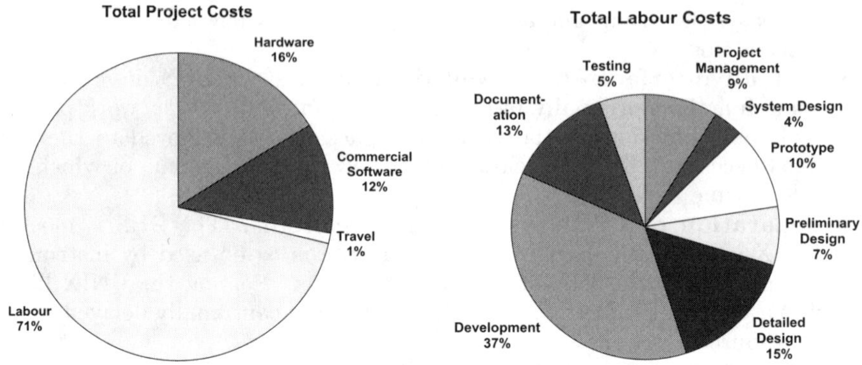

Figure 1. Total project and labour costs shown as percentages.

provided the team with intellectual challenges. This knowledge can now be applied of internal CADC projects. In addition, the CADC plans to re-use many parts of the DHS to augment the CADC archiving systems.

Acknowledgments. As in all software projects, the success of this project is a reflection of the team which included Norman Hill, Jennifer Dunn, Shannon Jaeger, Steve Cockayne, Dayle Kotturi and Séverin Gaudet. The support of the other members of the CADC, Daniel Durand, David Schade, and David Bohlender was invaluable.

References

Cockayne, S., Dunn, J., Gaudet, S., Hill, N., & Jaeger, S. 1998a, The Gemini DHS Detailed Design Document, (Gemini 8m Telescopes Project)

——, 1998b, The Gemini DHS User Manual, (Gemini 8m Telescopes Project)

Dunn, J., Cockayne, S., Gaudet, S., Jaeger, S., & Miguel, A. 1999b, this volume, 265

Dunn, J., Jaeger, S., Hill, N., Gaudet, S., & Cockayne, S. 1999a, this volume, 167

Gaudet, S. 1996, in ASP Conf. Ser., Vol. 101, Astronomical Data Analysis Software and Systems V, ed. G. H. Jacoby & J. Barnes (San Francisco: ASP), 388

Hill, N., Gaudet, S., Dunn, J., Jaeger, S., & Cockayne, S. 1999a, this volume, 163

——, 1999b, this volume, 155

Jaeger, S., Dunn, J., Cockayne, S., Gaudet, S., & Hill, N. 1999, this volume, 159

McGonegal, R. 1996, in ASP Conf. Ser., Vol. 101, Astronomical Data Analysis Software and Systems V, ed. G. H. Jacoby & J. Barnes (San Francisco: ASP), 293

Astronomical Data Analysis Software and Systems VIII
ASP Conference Series, Vol. 172, 1999
D. M. Mehringer, R. L. Plante, and D. A. Roberts, eds.

From a "Launch Readiness" System to an Astronomical Data Processing System – a Review of Four Years of CIA Development

S. Ott

ISO Data Centre, Astrophysics Division, Space Science Dept. of ESA, Villafranca, P.O. Box 50727, 28080 Madrid, Spain, Email: sott@iso.vilspa.esa.es

R. Gastaud

DAPNIA/SEI-SAP, CEA/Saclay, F-91191 Gif sur Yvette Cedex, France, Email: RGastaud@CEA.fr

S. Guest[1,2], M. Delaney[1,3], B. Altieri[1], B. Ali[4], A. Abergel[5], J.-L. Auguéres[6], H. Aussel[6], J-P. Bernard[5], A. Biviano[1,7], J. Blommaert[1], O. Boulade[6], F. Boulanger[5], C. Cesarsky[6], D. Cesarsky[5], R-R. Chary[8], V. Charmandaris[6], A. Claret[6], C. Delattre[6], F-X. Désert[5], T. Deschamps[6], P. Didelon[6], D. Elbaz[6], P. Gallais[6], K. Ganga[9], G. Helou[9], M. Kong[9], F. Lacombe[10], D. Landriu[6], O. Laurent[6], P. Le Coupanec[10], J. Li[9], L. Metcalfe[1], K. Okumura[1,4], M. Perault[5], A. Pollock[1], P. Roman[1], D. Rouan[10], M. Rupen[11], J. Sam Lone[6], M. Sauvage[6], R. Siebenmorgen[1], J-L. Starck[6], D. Tran[6], D. Van Buren[9], L. Vigroux[6], F. Vivares[5]

Abstract. The ISOCAM Interactive Analysis System (CIA) is used to calibrate and to perform the astronomical data processing of data from ISOCAM, the infrared camera on board the Infrared Space Observatory (ISO).

---

[1]ISO Data Centre, Astrophysics Division of ESA, Villafranca del Castillo, Spain

[2]Rutherford Appleton Laboratory, Chilton, Didcot, Oxon, England

[3]University College Dublin, Belfield, Dublin, Ireland

[4]University of Rochester, Rochester, USA

[5]Institut d' Astrophysique Spatiale, Orsay, France

[6]CEA, Saclay, Gif-sur-Yvette, France

[7]Osservatorio Astonomico di Trieste, Trieste, Italy

[8]University of California, Los Angeles, USA

[9]Infrared Processing and Analysis Center, JPL and Caltech, Pasadena, USA

[10]DESPA, Observatoire de Paris, Meudon, France

[11]National Radio Astronomy Observatory, Socorro, USA

CIA is generally available to the astronomical community and runs under DEC VMS Alpha, Solaris, DEC Unix, Debian (PC) Linux and HP/UX. More details, including how to obtain CIA, can be found at `http://www.iso.vilspa.esa.es/users/expl_lib/CAM_top.html`.

We discuss the challenges faced in this multi-site, multi-environment project throughout the complete software development cycle, outline the chosen approach, and review the lessons learned.

1. Introduction

Starting mid 1994, the ISOCAM Interactive Analysis System (CIA)[12] was developed to support the calibration and operation of ISOCAM, the infrared camera on board of ESA's Infrared Space Observatory (ISO)[13].

As planned, this system now includes the functionality to perform the astronomical data processing of data from ISOCAM, and is used within the ESA ISO Data Centre, by the ISOCAM consortium and by many other institutes.

The system is mainly IDL based, and contains currently 1100 IDL modules with about 220000 lines of code and comments. CPU intensive tasks are coded in C++.

2. Challenges of CIA development

2.1. Requirements

Some requirements on CIA were

- expected to change with time (from a calibration/monitoring system for operational use to an astronomical data analysis system)
- conflicting (flexibility vs. user-friendliness)
- very broad (CIA had to be prepared for the unexpected and to serve as a development bench for new algorithms, which ultimately will migrate into the general pipeline)
- unforeseen additional major requirements, e.g., CIA had also to run on Unix.

2.2. Management

The development of CIA had to be spread over multiple development sites.

Furthermore, the schedule for the first version ("launch readiness version") was extremely tight — 9 months from the start of development to acceptance testing, and the development team also had to build up IDL[14]/C++ expertise.

[12] CIA is a joint development by the ESA Astrophysics Division and the ISOCAM Consortium. The ISOCAM Consortium is led by the ISOCAM PI, C. Cesarsky, Direction des Sciences de la Matiere, C.E.A., France.

[13] ISO is an ESA project with instruments funded by ESA member states (especially the PI countries: France, Germany, the Netherlands and the United Kingdom) and with the participation of ISAS and NASA.

[14] IDL is the trademark of Research Systems, Inc.

2.3. Other challenges

As in every other software project, special attention had to be paid to software quality control and documentation. The single most difficult and time-consuming problem was to get the astrometry, e.g., the correlation of image elements (detector pixels) with astronomical (sky) coordinates, right. (ISOCAM contains two detectors, which are read-out electronically in different directions. Also its images might be re-binned to non-square pixels.)

3. Chosen approach

3.1. Requirements

In order to keep the system on track, requirements were strictly prioritized. The system is reviewed and, if necessary, pruned from time to time. In order to fulfil "expert" and "normal" user requirements, two different types of data structures were implemented. CIA was coupled with the general automatic data reduction pipeline via its architecture, not via the code. This eased development for *both* systems.

3.2. Management

Project management and control was closely entwined with software configuration control: One VMS system was chosen as the master development/configuration control system and mirrored to the other VMS development system. Unix versions are created at monthly or bi-monthly intervals from the current development version and installed at the other development sites.

The collaborative and collegial approach within the software team and among collaborators was essential to overcome all difficulties.

3.3. Other challenges

A CIA configuration control board reviews and sets the priorities of the implementation of software problem reports and software change requests. This, together with a several-month-long testing period before the release of any major new version, proved essential to achieving a high quality for the software.

For the documentation, emphasis was put on the production of an accurate and helpful User's Manual. CIA's internal help system relies on the information contained in module headers. Therefore all contributors were continuously encouraged to maintain them in case of a code change or to modify them in case of deficiencies spotted.

4. Lessons learned

1. Management
    - the effort to code properly and to maintain a stable system is normally underestimated by all participants and by their hierarchy.
    - quite often it is assumed that coding can be done as a by-product of algorithmic development. However, it was found that a proper balance in staffing between astronomers and software engineers is not only essential to produce a maintainable system but actually more efficient!
    - frequent ($\approx$ monthly) workshops between developers, and bi-annual users meetings, are essential.

2. Architecture
   - a proper definition of data structures is important. Growing data structures and flexible export tools are preferable to hard coded data structures and export tools which have to be upgraded in case of new requirements.
   - link systems with a different primary programming language only via the architecture, not via the code. Otherwise the progress of *both* systems will be slowed down.
   - if possible, avoid non-standard data structures depending on pointers. If non-standard data structures are necessary, provide an intuitive access. (NB: IDL does not provide this functionality).
   - while collaboration is the key to success, some rules have to be obeyed:
     - avoid duplication of already existing functionality.
     - enforce consistent naming conventions for routines and for the tag names within data structures.
     - avoid private ports to other operating systems.
   - avoid operating-system-dependent data files.
   - store the processing history as part of the data instead of using log-files.
3. Implementation
   - IDL permits an easy port between systems, compared to classical languages. (But the support of different operating systems remains nevertheless an overhead). However, internal IDL changes from version 3.6 to 5.0 for the widget interface led to a major recoding effort.
   - the maintenance of module headers is a Sisyphean task which needs dedicated manpower.
   - knowledge of "expert" language features like *_extra* can speed up the development significantly.
   - look for solutions available in other packages (e.g., ASTROLIB). But the use of several packages can lead to compatibility problems (e.g., name conflicts between functions and variables).
4. Configuration Control & Multiple Development Sites
   - good network connections and easy-to-use configuration control systems are essential.
   - the use of mirrors reduces synchronization effort dramatically.
   - to avoid bad surprises for users, both a floating development version and a frozen user's version are necessary.

Astronomical Data Analysis Software and Systems VIII
ASP Conference Series, Vol. 172, 1999
D. M. Mehringer, R. L. Plante, and D. A. Roberts, eds.

ORAC-DR: Pipelining With Other People's Code

Frossie Economou

Joint Astronomy Centre, 660 N. A'ohōkū Place, Hilo, HI 96720

Alan Bridger, Gillian S. Wright

UKATC, Royal Observatory, Blackford Hill, Edinburgh, UK

Tim Jenness

Joint Astronomy Centre, 660 N. A'ohōkū Place, Hilo, HI 96720

Malcolm J. Currie

UKATC, Royal Observatory, Blackford Hill, Edinburgh, UK

Andy Adamson

Joint Astronomy Centre, 660 N. A'ohōkū Place, Hilo, HI 96720

Abstract. As part of the UKIRT ORAC project, we have developed a pipeline (ORAC-DR) for driving on-line data reduction using existing astronomical packages as algorithm engines and display tools. The design is modular and extensible on several levels, allowing it to be easily adapted to a wide variety of instruments. Here we briefly review the design, discuss the robustness and speed of execution issues inherent in such pipelines, and address what constitutes a desirable (in terms of "buy-in" effort) engine or tool.

1. Throwing Away Software

UKIRT is a 4-meter class infrared telescope that has recently celebrated its 20th anniversary. ORAC (Observatory Reduction and Acquisition Control, Bridger, Wright, & Economou 1998) is a project to re-implement all software with which an observer interacts, from observation preparation through data acquisition to data reduction.

One might well ask why it is necessary to throw away perfectly good software. The answer lies in a variety of reasons, including the following:

- Most of the existing software is tied in to a particular (VAX/VMS) platform and is non-portable. We wish to have software that runs under a Unix platform but is also portable to other operating systems.
- We would like to capitalize on modern technologies and techniques.

- Some of the existing software is not only arcane, but has grown in such a fashion as to be difficult to support and extend.

More specifically, let's consider data reduction. In the past, UKIRT has provided on-line data reduction for its near-IR spectrometer (the instrument CGS4) using the purpose built programme CGS4DR. CGS4DR provided in many cases publication quality data while running automatically and on-line – a remarkable feat when one considers the complexity of infrared spectroscopic data analysis.

However, because UKIRT is on the brink of receiving two new instruments (the mid-IR spectrometer MICHELLE and the imaging spectrometer UIST), we are forced to contemplate dedicated instrument-specific software with some misgivings:

- Software effort to cater for what are increasingly more (and more complex) instruments on an individual basis is lacking.
- Existing software re-invent many wheels (e.g., image display) that greatly increase the body of code that must be supported by observatory staff.
- Instrument-specific packages often require great effort to adapt to new instruments, even if these have fundamental scientific similarities with existing ones.

2. ORAC-DR: A Framework for Software Re-use

We have developed a data reduction pipeline (ORAC-DR) with the particular aim of addressing the shortcomings of instrument specific software outlined above. ORAC-DR is a pipeline that drives external packages (algorithm engines) via a messaging system using data reduction recipes. This has the following advantages:

- The use of external algorithm engines for data analysis and display uses software supported by the community, thus restricting observatory effort and support to instrument-specific aspects of data reduction.
- The use of recipes (a series of instructions) and primitives (instructions for a single reduction step) to direct the pipeline provides a modular system of components, many of which can be shared amongst different instruments.
- The software effort to cater to new instruments is restricted to the scientific aspects of data reduction rather than the mechanics of it (file manipulation, display, etc)

The detailed implementation of the software design and its implementation in Perl is partly described elsewhere in these proceedings (Jenness & Economou 1999). An outline is presented in Fig. 1.

3. Re-using Other People's Software

In last year's proceedings, Mandel & Murray (1998) address the issue of software buy-in (the effort required to use somebody else's software), and discuss the concept of minimal buy-in – software that's too easy not to try out. Trying out software, however, is only a beginning. When one is prepared to put other people's code on their telescope's mission-critical machines (not to mention in front

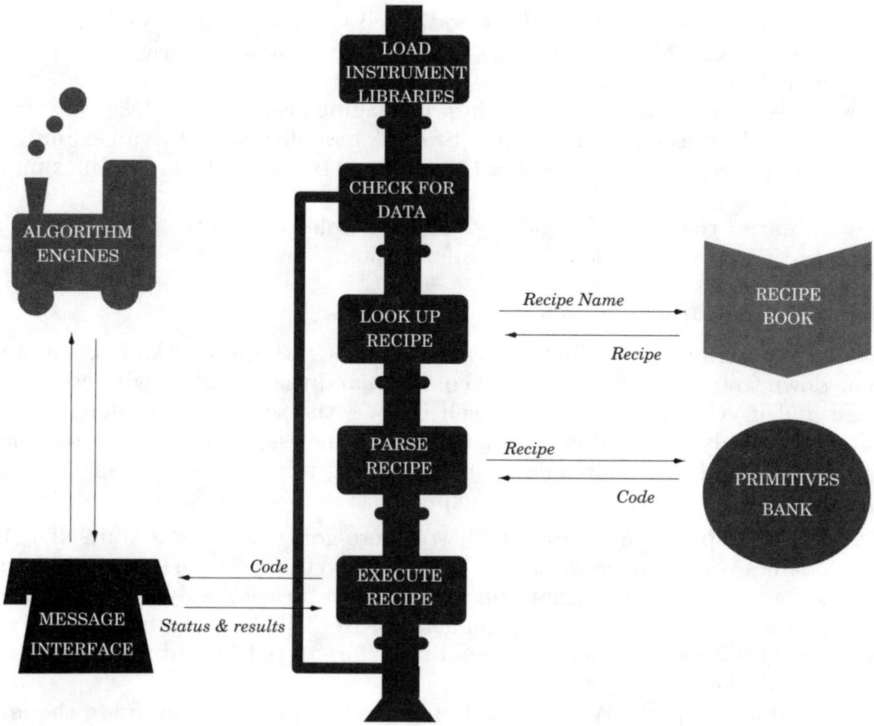

Figure 1. The components of the ORAC-DR pipeline. In our operational model, the pipeline itself is supported by the ORAC-DR development team, the recipes by an instrument scientist, the primitives by an instrument software specialist, and the messaging and algorithm engines by the community.

of funding-critical observers) a variety of issues, technical and non-technical, achieve paramount importance.

Here we discuss each in turn, providing by way of illustration some of the choices we made for the ORAC-DR pipeline. Note that we are specifically referring to software use by software developers in, or in lieu of, in-house software development, rather than end-user software consumption.

3.1. Technical Issues

By far the most important issue is robustness. When using external packages and tools, one has to not only issue a command, but also receive back parameters and a meaningful status that can differentiate between predictable errors (such as data being too noisy for an algorithm to provide a reliable result) to catastrophic errors (missing files, crashes, etc.).

The key to this is the ability to use an interface that allows reliable two-way communication between the pipeline and its external components. In practice this means:

- Algorithm engines that can be addressed via a messaging system, e.g., we use the ADAM messaging systems to control a wide variety of Starlink monoliths,
- Tools that can be addressed via a messaging system or a socket interface. ORAC-DR uses for its display the Starlink monolith KAPPA as an example of the former, and the tool GAIA (based on RTD aka SKYCAT) as an example of the latter.
- Libraries that have C-bindings. For example, we use the NDF library to directly access files for header information.

3.2. Non-Technical Issues

Mandel & Murray suggest that software re-use, or rather the lack of it, may well come down to to "money, power and control", and that "dealing with economic, social and psychological factors is well beyond the scope of this short paper" and we can only echo their sentiments. Nevertheless, there are non-technical issues that should be addressed by those willing to adopt an existing body of code. Here is some advice from our experience:

- Design: Spec your product as if you were going to write it yourself. This enables you to understand what you want (so that you can evaluate existing software against it), what you have to give up (probably not as much as you may fear) and what do you get "for free" (bells and whistles that you would not have wanted to expend the effort to put in, but that your users would like).
- Product support: When considering a certain package, evaluate the level of support – and remember that support for you the developer (who might require code changes to support greater flexibility) is a different matter than end-user support (help and bug fixing). Also beware of single points of failure – a tool supported by a single person, no matter how expert and responsive, may be left unsupported if that person changes jobs. Also beware of commercial packages and support who may raise their price tag above the level you can afford.
- Consumer savvy: Also consider local expertise. It is preferable to have a local expert in the technology that you are adopting. Furthermore, it is preferable to adopt a technology that is not arcane so that your local expert may be replaced if they leave. Using technologies that are in use outside the immediate astronomical community is one way to ensure this.

References

Bridger, A. B., Wright, G. S., & Economou, F. 1998, in ASP Conf. Ser., Vol. 145, Astronomical Data Analysis Software and Systems VII, ed. R. Albrecht, R. N. Hook, & H. A. Bushouse (San Francisco: ASP), 292

Jenness, T. & Economou, F. 1999, this volume, 171

Mandel, E. & Murray, S. S. 1998, in ASP Conf. Ser., Vol. 145, Astronomical Data Analysis Software and Systems VII, ed. R. Albrecht, R. N. Hook, & H. A. Bushouse (San Francisco: ASP), 142

Astronomical Data Analysis Software and Systems VIII
ASP Conference Series, Vol. 172, 1999
D. M. Mehringer, R. L. Plante, and D. A. Roberts, eds.

Status and Future Plans for Parallelization of AIPS++

Douglas A. Roberts, Richard Crutcher

National Center for Supercomputing Applications, University of Illinois Urbana-Champaign, Urbana, IL 61801

Wes Young, Athol J. Kemball

National Radio Astronomy Observatory, P.O. Box O, Socorro, NM, 87801

Abstract. During the past year, the parallel group of AIPS++ has been actively incorporating parallel processing into the most computationally intensive aspects of AIPS++. We report here the status of the AIPS++ parallelization effort and plans for future development. The biggest success of the past year of AIPS++ parallelization has been the implementation of the parallel algorithm applicator class. This class implements parallelism using the Message Passing Interface (MPI). MPI is a portable system that allows data and instructions to be sent to remote processors (either on the same machine or on different machines). New classes derived from this parallel base class will carry out parallelization with a minimal effort from application programmers. This solution addresses most embarrassingly parallel problems, notably spectral line processing. We are also investigating tuned libraries (starting with FFT's), notably the Sgi/Cray Scientific Library (SCSL).

1. Goals of AIPS++ Parallelization

The goal of the AIPS++ parallelization project is to provide an easy to use, high performance processing system that allows astronomers to processes large datasets. Specifically our project has the following goals:

1. Give astronomers at least an order-of-magnitude increase in computational power.
    - This processing power can be used to process data sets larger than currently possible.
    - The increased capability can also be used on moderate-sized data sets to explore the use of a variety of algorithms (some of which may be computationally expensive).
2. Use a familiar interface.

- Since we are building the parallel system within the AIPS++ system[1] we should use the same interface.
- Astronomers who use the parallel system for some portion of their processing should notice almost no difference in the interface (even though some of their processes are being carried out on a batch system).

3. Get as much code parallelized in a short time as possible.
   - This priority motivates us to implement a parallel infrastructure into the existing code.
   - Also we should make it easy for a programmer to use the parallel infrastructure.

2. State of AIPS++ Parallelization

An Algorithm-Applicator scheme has been chosen to deal with embarrassingly parallel (e.g., spectral line) problems. The Applicator is the controller, which sets up the problem, transports data to the Algorithm processes using the Message-Passing-Interface (MPI)[2]. MPI is a portable system that allows data and instructions to be sent to remote processors (either on the same machine or on different machines). Wes Young (see page 506) presented a demonstration of our first implementation of the algorithm applicator class to carry out spectral line deconvolution.

A graphical interface to the parallel system and batch processing has been implemented (see Fig. 1). The user only has to tell the application the number of processors to use. Submission to the batch queue is carried out with only one additional step.

Figure 2 shows the speed up of a deconvolution as a function of processors. Currently the parallel system shows good scaling up to 16 processors.

- Lower degree of speed up beyond 32 processors may be due to I/O inefficiencies.
- Collaboration with NCSA enabling technologies team is important to obtain I/O performance statistics of the AIPS++ code.

3. Uses of Parallel AIPS++ in the Short Term

Power users can begin to use the parallel system on the NCSA computers after the first of the year. The parallel system will include parallelization of various spectral line processing: gridding, imaging, deconvolution (various flavors of CLEAN and MEM). The parallel infrastructure should work on a generic multiple processor machines.

---

[1] `http://aips2.nrao.edu/aips++/docs/html/aips++.html`

[2] `http://www-unix.mcs.anl.gov/mpi/index.html`

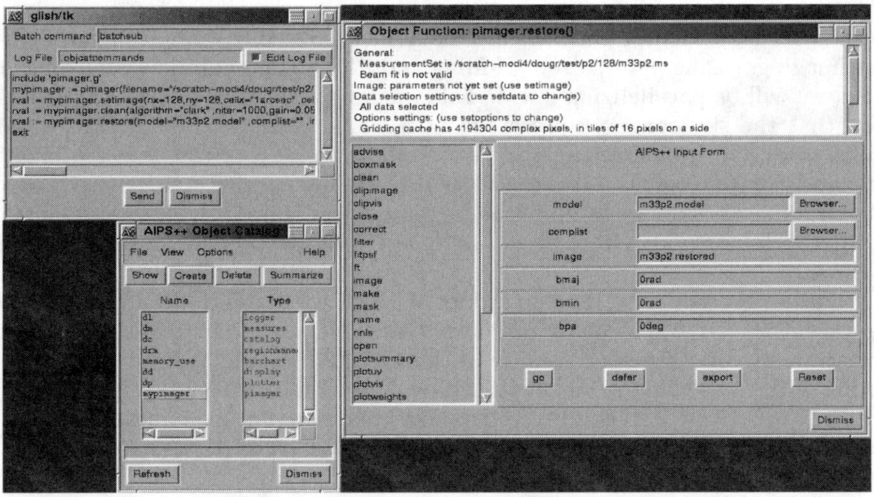

Figure 1. GUI for batch submission.

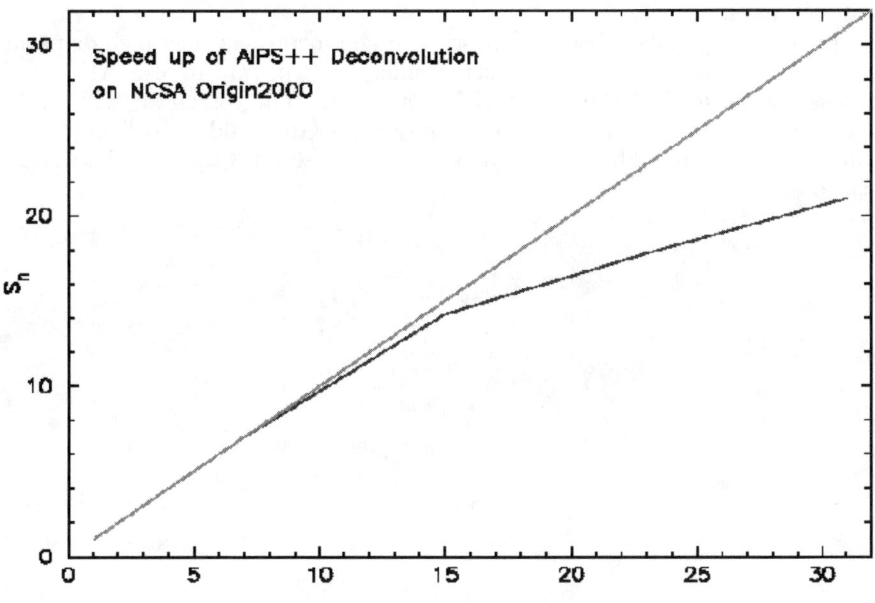

Figure 2. Clark CLEAN speed up of M33 HI 512 × 512 pixels by 100 channels using AIPS++ Applicator/Algorithm classes.

4. Longer Term Goals

Our future goals include parallelizing the AIPS++ IMAGER application to carry out imaging, (using parallelized gridding and FFT's) and parallel deconvolution. Also, we will be parallelizing a wide-field imaging algorithm (where the assumption that the sky can be represented at a single tangent plane breaks down). Finally, we will be increasing the user support to help astronomers, who need the large computational resources of NCSA, use the parallel AIPS++ system on the Origin2000. The first step of the increased user support has been to increase the network bandwidth from the VLA to NCSA by putting NCSA on the NRAO intranet.

Future projects also include a port of AIPS++ to NT, in order to take advantage of the large NT cluster now available at NCSA and increasingly available to astronomy departments because of their modest prices. We are also collaborating with the Pablo group[3] at the University of Illinois to investigate the I/O patterns of AIPS++. Their group has instrumented our code and is now working to identify I/O patterns and statistics. This will be important to show where better performance due to increased I/O would be possible. The next generation of the MPI standard (MPI-2) includes a standard for MPI I/O. The MPI I/O standard is finalized and implementations are now available. We intend to explore its use as a complement to the parallel processing development.

We also are looking into integration of batch communication into the Glish IPC. Continue to use the parallel algorithm applicator to implement more algorithms in parallel.

In addition to post-observation processing, our project may include parallel on-line processing. The JCMT has decided in principle to use AIPS++ to process the backend of the new 1 GHz correlator. The correlator will create a data rate of about 10 MB/sec (up to about 3.8 GB could be collected in a 6 minute observation). They will have a network of about 8 to 16 workstations to carry out the processing.

---

[3]`http://www-pablo.cs.uiuc.edu/`

Astronomical Data Analysis Software and Systems VIII
ASP Conference Series, Vol. 172, 1999
D. M. Mehringer, R. L. Plante, and D. A. Roberts, eds.

A "Scientific" Approach to Software Project Management Part I: A Survey of Development Methodologies in Scientific Computing

Arno F. Granados

National Radio Astronomy Observatory[1], *Green Bank, WV 24944-0002, USA*

Abstract. The author presents the motivation for creating a survey of software project management practices in scientific computing. The survey was distributed to attendees of ADASS'98, and the results and analysis will be presented at ADASS'99.

1. Motivation

I am a software engineer. I was an astronomer. My metamorphosis from astronomer to engineer occurred during a two year period when I worked for a small high-tech start up company creating commercial image processing software for remote sensing applications. Witnessing firsthand the differences between the process of creating "scientific" software, vs. writing "code-for-sale", made me question many aspects of software engineering, ranging from how many spaces to indent *if ... else* blocks, to project scheduling and workplace ethics. I have recently re-entered the world of astronomy, however this time I am working as an engineer, not a scientist. I have found that my background provides me with a unique perspective on the process of writing scientific software, and what I see from my perspective alarms me. Scientific software projects are becoming large and complex, even by industrial standards, and while scientists seem eager to borrow bleeding edge technology from industry (witness the proliferation of Java based tools at ADASS'98), they appear much less eager to borrow the methods used to manage these software projects.

I have been told that science and commercial programming practices don't mix, or that the situation is not as bad as I think it is. After hearing more than a modicum of contradictory information, I decided that it would be interesting and informative to ask scientists, scientific programmers, and software engineers working in science environments, just what exactly is going on in scientific software development. The amount of information was potentially overwhelming, so I decided to focus on the issues of software project management.

In the commercial-industrial world, Project Management (or lack thereof) and formal software development procedures are often cited for the success or failure of a software project (Whitten 1995). In contrast, scientific software has

[1]The National Radio Astronomy Observatory is a facility of the National Science Foundation operated under cooperative agreement by Associated Universities, Inc.

developed a reputation for being undisciplined, and formal development procedures are rare for even large projects–as an example, it wasn't until 1997 that the Science Software group at Space Telescope Science Institute implemented a formal program of design review and source code inspections (Shaw & Greenfield 1997).

Moreover, astronomical software is often "developed" by "Domain Experts", not software engineers. Historical inertia often binds institutions to a "code-and-fix" mind-set, amplified by the oft spoken observation that this is "research." Requirements are ill defined, if they exist at all. Yet, we all borrow software programs from colleagues, commission graduate students to write and maintain applications and scripts, utilize source code libraries (such as the IDL Astronomical Library), and invest resources in pseudo-commercial analysis packages such as AIPS, AIPS++, IRAF, and OPUS (Mandel & Murray 1998). While the computer industry has capitalized on the buzzword of "re-use," astronomers have been re-using code for decades. Unfortunately much of that code was originally written for single purpose applications, e.g., a single user, or as "throw-away" code, e.g., a single use. Such code is often associated with the words "unmaintainable" and "spaghetti-code." The ubiquitous computer has created the impetus for scientific software to be portable, robust, and maintainable (Conroy, Mandel, & Roll 1998): traits of software that has been designed and developed, not just coded and fixed.

2. The Survey

In order to analyze the current state of scientific software development in a more quantifiable way, I created a "Software Development Methodology Survey" which was distributed at the 1998 ADASS meeting. The purpose of the survey was to create a formal baseline of understanding of the software development methodologies in use in astronomical computing–what works and what doesn't. The survey was comprised of 33 questions covering the following topics:

- Are software groups creating formal requirements documents?
- How are developers obtaining requirements?
- Are software projects being tracked by project managers?
- Are developers using commercial tools (e.g., defect tracking, CASE tools, GUI tools), or building tools in-house?
- Do coding groups use techniques such as walk-throughs and code-reviews to limit code defects and improve reliability and maintainability?
- Are formal development procedures helpful or just too much overhead?
- What percentage of software groups have formal training in programming or computer science?

The survey was printed on multiple choice ScanTron forms. Statistics and analysis were not available by the December 15, 1998 deadline for proceedings submissions. The results and analysis of the survey will be presented at the 1999 meeting of ADASS.

Acknowledgments. I wish to thank the ADASS Organizing committee for their support and cooperation in distributing the survey. I also wish to thank

Kendra Bair at the University of Illinois for her assistance in preparing and processing the survey.

References

Conroy, M., Mandel, E., & Roll, J. 1998, in ASP Conf. Ser., Vol. 145, Astronomical Data Analysis Software and Systems VII, ed. R. Albrecht, R. N. Hook, & H. A. Bushouse (San Francisco: ASP), 150

Mandel, E. & Murray, S. S. 1998, in ASP Conf. Ser., Vol. 145, Astronomical Data Analysis Software and Systems VII, ed. R. Albrecht, R. N. Hook, & H. A. Bushouse (San Francisco: ASP), 142

Shaw, R. A. & Greenfield, P. 1997, in ASP Conf. Ser., Vol. 125, Astronomical Data Analysis Software and Systems VI, ed. G. Hunt & H. E. Payne (San Francisco: ASP), 26

Whitten, N. 1995, Managing Software Development Projects, (New York: Wiley)

Part 2. Applications to Problems in Astronomy

Astronomical Data Analysis Software and Systems VIII
ASP Conference Series, Vol. 172, 1999
D. M. Mehringer, R. L. Plante, and D. A. Roberts, eds.

Systemization and Use of Atomic Data for Astrophysical Modeling

Nancy S. Brickhouse

Smithsonian Astrophysical Observatory, Cambridge, MA 02138

Abstract. The growth of supercomputing capabilities over the past decade has led to tremendous opportunities to produce theoretical atomic data for use in modeling astrophysical plasmas; however, not all data are created equally. Both accurate and complete atomic data are important to astrophysical modeling via plasma codes. Critical theoretical evaluations, experimental benchmarks, and even astrophysical observations themselves are all useful in assessing the temperatures, densities, opacities, abundances, and other physical quantities determined from spectroscopy. The database challenge for the future is to move toward increasing both flexibility and standardization of atomic data formats so that the quality of the data can be taken into account in modeling. Problems in X-ray spectroscopy will illustrate the importance of opening up what has often been a black box to many astronomers.

1. Introduction

Spectroscopists in astrophysics may be heard to remark, "If one picture is worth a thousand words, then one spectrum must be worth a thousand pictures." While ultimately the goal of spectroscopy is to determine physical properties and discern among physical processes, the levels of inference involved in "diagnosing" the plasma tend to make the art of spectroscopy less than easily accessible to the general astronomy community. A spectrum's value lies in the quality of the underlying models.

The interpretation of spectra is built upon models of the astrophysical source (e.g., the magnetic confinement of hot plasma in stellar coronae), of the plasma processes (e.g., the equilibration of electron velocities through collisions to achieve a Maxwellian distribution), and of the atomic physics (e.g., the representation of the microphysics of atomic rates). Thus, determining the average density of an emitting source from a measured diagnostic line ratio may appear to be straightforward, but the reliability of such a density is difficult to assess without accounting for all the levels of modeling. Furthermore, the reliability of a diagnostic also depends on its observational context, not only signal-to-noise and calibration, but also the degree of contamination by line blending. The isolation of a diagnostic feature from other spectral components thus becomes an additional modeling concern.

The application of astronomy to fundamental spectroscopic studies has its origins with the Fraunhofer lines in the Sun in the early 1800's, about the same time as the beginning of quantitative laboratory work. Basic spectroscopy continues to be important to astronomy as we probe the universe with increasing spectral coverage, spectral resolution, throughput, and calibration. Atomic data that result from fundamental theoretical and experimental work then become critical components for astrophysical modeling.

Calculation, compilation, and distribution of atomic data are increasingly becoming the responsibility of the astrophysics community, as our needs diverge from the primary interests and activities of atomic and laboratory plasma physicists. Systemization will enhance the value of atomic data for astrophysical modeling. While each field of astrophysical spectroscopy has a diverse set of needs and expectations for spectral analysis, we expect that some general design principles are likely to apply. This paper discusses some of these design issues, using our modeling efforts in high energy astrophysics as an example. The ongoing development project for the Astrophysical Plasma Emission Code (APEC) is implementing these ideas (Smith et al. 1998).

2. Approaches to Spectral Analysis (with a Bias toward High Energy Spectroscopy)

Two complementary approaches are used routinely to analyze spectra. One approach is to choose one or two lines from which to derive a physical quantity, such as density or temperature. This approach has the advantage that bright isolated lines with good theoretical models may be selected. On the other hand, this approach can lead to errors if the plasma is not isothermal or isochoric. Furthermore, the theoretical models may not be as reliable as one might hope. Finally, to test for consistency among derived quantities, one needs to consider other information, often in the same spectrum.

A second approach is to construct a model with sufficient numbers of variable parameters that a good fit to the entire spectrum can be found. This "global fitting" approach is still widely used in X-ray astronomy because only moderate resolution ($\lambda/\Delta\lambda \sim 50$) has been obtained beyond the Sun. Even the strongest emission lines are significantly blended with continuum and other lines at this resolution, as shown in Fig. 1. Although this approach utilizes the entire spectrum and is easily automated, disadvantages are inherent. The information content of the spectrum is not necessarily reflected in bin-based fitting since the model parameters are not independent. Furthermore, methods of this type place a huge burden on the accuracy and completeness of the atomic data needed.

X-ray spectra which are to be obtained with the *Chandra X-ray Observatory* gratings will cover a broad band, providing enormous information content in each observation. An integration of both approaches is needed: to derive initial models from the strong, well understood lines and to test these models against the entire spectrum for consistency or problems. To make the atomic data and spectral emission models available for both purposes, we need to open up the black box of spectral analysis.

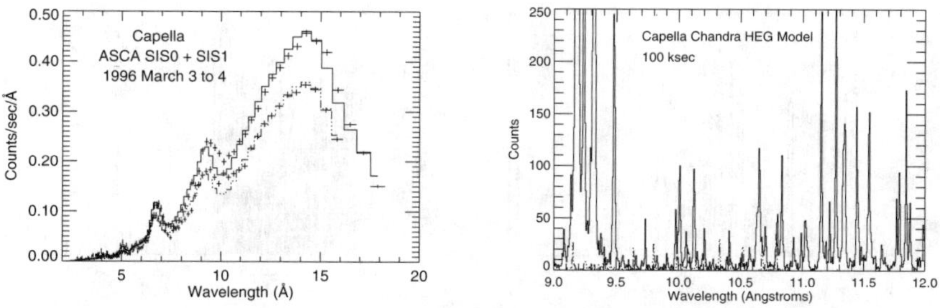

Figure 1. Global fitting methods are used to analyze moderate resolution spectral data, placing high demands on the accuracy and completeness of the plasma emission model. *Left*: Models of the ASCA spectrum of Capella show a deficit of flux around 10 Å (Brickhouse et al. 1999). New calculations of Liedahl & Brickhouse (1999) show that the flux deficit is removed if additional lines from high principal quantum number $n > 5$ of Fe XVII to XIX are added to the model. *Right*: Simulations of the *Chandra* High Energy Transmission Grating observation show a wealth of emission lines blended at CCD resolution. The new lines of Liedahl & Brickhouse are shown separately as dash-dotted lines.

3. Opening Up the Black Box

The spectral codes developed in the 1970's for hot, collisionally ionized plasmas compute the fundamental rates of ionization/recombination and excitation/decay processes (Raymond & Smith 1977; Raymond 1998; Mewe 1991 and references therein). While the original codes model most of the important processes, increased computing capacity over the past two decades has led to improvement in the accuracy of many atomic rate coefficients and to an increase in the numbers and detail of available rates, as illustrated in Fig. 2.

Modular design allows the atomic data, formerly buried in Fortran DATA statements, now to be stored in a stand-alone database. Accommodating simple changes to atomic data should be much easier than in the past. The increase in functionality, discussed in more detail below, is enormous.

We are currently compiling the Astrophysical Plasma Emission Database (APED) for high energy spectroscopy as a set of FITS files. These files are easily viewed, accessed, and modified with standard software. Multiple sources of data are supported, as are multiple functional forms for data. APED will facilitate identification of spectral lines in complex spectra, comparison of atomic data, access to reference material associated with the data, and other spectral analysis functions.

3.1. Spectral Synthesis Codes

Spectral synthesis codes calculate the relevant atomic processes leading to line and continuum emission, under assumptions about the physical conditions. Examples include codes to model stellar atmospheres, photoionized plasmas, and

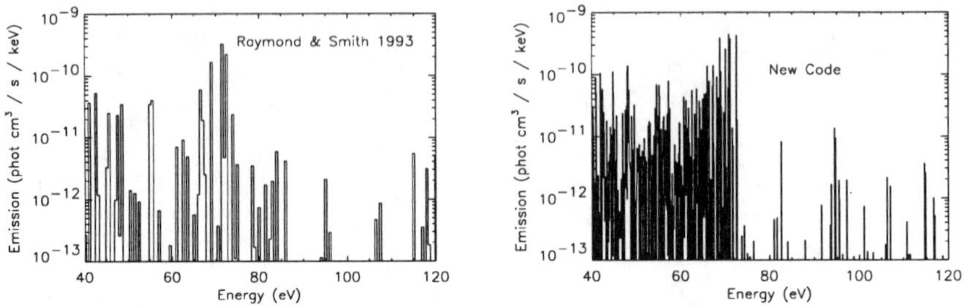

Figure 2. The number of emission lines currently included in new models greatly exceeds the numbers from previous work. *Left*: A Raymond-Smith model using the 1983 version of the code. *Right*: APEC model (Smith et al. 1998), using atomic data compiled by the Chianti collaboration (Dere et al. 1998) for the same spectral band. Not only does the new model show more emission lines, but strong features do not coincide because the older model uses multiplet level averages rather than individual fine-structure levels.

collisionally ionized plasmas. The predicted spectra are compared to observations using a set of physical parameters, which are then adjusted to obtain the best fit. The relevant observations can either be a select set of diagnostics or the entire spectrum (and even beyond). The physical parameters obtained in this way information may be temperature, density, ionization state, structure, abundances or other parameters constructed from these characteristics.

Typically the model parameters may be both underdetermined and overdetermined by the observed spectrum. Using the example of a collisionally ionized plasma, let us take the simplest case of a "coronal model." The coronal model assumes that the plasma is in ionization equilibrium, is optically thin, and is of relatively low density. If we also assume that the observed emission comes from a single temperature, then the "best fit" temperature and its emission measure (i.e., the amount of emitting material) are easily determined from the observed photon flux. Stellar coronal plasmas are not generally isothermal, however, and thus an emission measure distribution (emission measure as a function of temperature) is used to parameterize the coronal structure. Emission lines may come from a range of temperatures in the model, leading to an overdetermined model, or interesting temperatures may simply not be represented by features in the spectral range under observation. Moreover, observed features may simply not be represented by the plasma model. For example, the He II UV and EUV lines observed in stellar transition regions are almost certainly optically thick, and possibly populated by photoionization followed by recombination cascades, rather than by the collisional ionization and excitation processes which dominate most of a coronal spectrum.

The challenges of using a spectral synthesis code to obtain physical information about the source are not only to try to get as much detail as possible correct, but also to understand the limitations of the model. These limitations are generally lumped into the concept of "systematic" or model uncertainties and can be difficult to quantify. Such uncertainties include errors associated with the input atomic rates. For example, collisional data are almost entirely compiled from theoretical sources, so that one must consider the type of theoretical calculation to estimate the uncertainty. The model may be incomplete, by not including all the relevant excitation processes (leaving out a recombination cascade that contributes 20% of the level population), or in its treatment of the model ion (not enough energy levels or lines). Of course, one must also consider the possibility that the model does not apply at all to a particular feature, as in the example of He II above. Finally, of course, there are observational uncertainties as well, which can be either statistical (e.g., counts) or systematic (e.g., calibration; line blending) in nature.

The scientist judges that a model is acceptable if it does a pretty good job of fitting the spectrum. Unfortunately, quantifying what "pretty good" means is only mathematically straightforward in the statistical sense (e.g., by using χ^2). One builds confidence by getting familiar spectral features correct, while one worries that problem areas may either be astrophysically interesting (meaning that the model might not be entirely correct and that some exciting new phenomena must be invoked) or boring (just another inaccurate atomic rate). If one wants to rely on only a few diagnostic lines or line ratios, the problem of distinguishing between oversimplification in the treatment of the emissivity calculation, inaccuracies in the atomic rates, complexity at the source (such as multiple densities when trying to use a density diagnostic), must all be taken into account before the diagnostic is judged reliable. Usually this involves finding consistency with a large set of other diagnostics. Improving scientific judgment on the quality of the spectral fit requires opening up the atomic data and plasma emission models to scrutiny.

3.2. Atomic Database

Improvements in the atomic data that are input to a spectral synthesis code include improvements in both accuracy and completeness. Atomic rate uncertainties range from a few percent to factors of 2 or even larger, depending on the theoretical atomic model approximations used. Quantities such as wavelengths and atomic transition probabilities (or oscillator strengths) are available electronically for many transitions, although lack of standardization makes accessing the information time consuming. Publicly available collisional data tend to be sparse and poorly organized. Standardization might be yet another improvement, but the astrophysics community will need to negotiate standardization with the producers of atomic data and other users.

Converting collisional data into a standard format is a daunting task. As an example, electron impact (or "collisional") excitation data exist in the literature as cross sections, rate coefficients, collision strengths, and Maxwellian-integrated "effective" collision strengths. These quantities are not simple multiplicative factors of each other but can require extensive numerical computation to convert from one form to another. The situation is made still more complex because the data may be stored as a grid of data values or as fits to physically-based functions. Older data do not cover a sufficient energy grid, and thus need to be extrapolated with physical models. The most accurate collisional rate data include the calculation of resonance structure near threshold, requiring cross sec-

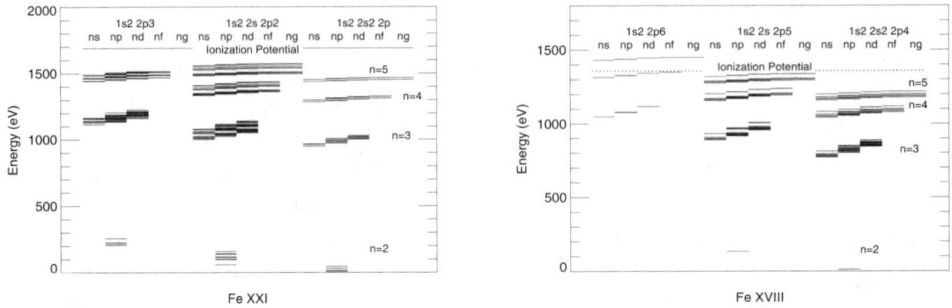

Figure 3. Energy level diagrams from APED illustrate the issue of database size. *Left*: The Fe XVIII model includes 501 energy levels up through $n = 5$ and $\sim$ 13,000 transitions (sparse relative to a possible $501 \times 501 = 250,000$), compared with Raymond-Smith, which lists 11 multiplet lines. *Right*: The Fe XXI model includes 848 energy levels up through $n = 5$ and $\sim$ 16,000 emission lines, compared with 12 lines in a Raymond-Smith model.

tions to be calculated at thousands of energies. These data are then integrated over the electron energy distribution (usually taken to be Maxwellian) to produce effective collision strengths. Current computing power only allows a small number of transitions to be calculated with resonances. Large numbers of lines are obtained by building bigger, but less accurate, atomic models. By combining these two types of theoretical data, we can incorporate the best information for a given ion. With large numbers and diversity of collisional data sets today, an approach which stores data in as original a form as possible, along with keys to the rate calculation method is less labor-intensive and more reliable.

The APED project is compiling data now for $\sim 10^6$ lines of use for highly ionized plasmas, as illustrated in Fig. 3. Even with this number of lines, we now know that we are missing important bands of radiation (e.g., $n > 5$), which, when theoretically calculated and incorporated into the database, will almost certainly be multiplet or quantum-number averaged, rather than fine-structure lines. While the number of lines included now in stellar atmosphere models far exceeds this (e.g., Hauschildt et al. 1997), APED will probably of comparable size. Oscillator strengths constituting the atomic data needed for opacity calculations are constants intrinsic to the spectral line, whereas collisionally-dominated spectral line emissivities vary with temperature and density.

The critical evaluation of atomic data remains an important area of research. Laboratory measurements provide key inputs to the assessment; however, the large numbers of data needed prohibit the exclusive use of experiments to test collisional cross section theory. In X-ray spectroscopy, astrophysical observations are compared with models to determine problem areas, and set priorities for laboratory work (Fig. 3). The *Chandra* X-ray Observatory Center (CXC) is sponsoring a collaborative effort known as the Emission Line Project to produce X-ray catalogs of three bright stellar coronal sources. These catalogs will provide guidance as to the overall quality of the coronal models, identify problem spectral features, and help to set priorities for further theoretical and experimental work.

4. Improvements in Functionality

With an atomic database (including error estimates) that stands alone, the functionality of the spectral synthesis code can be greatly increased. The most accurate theoretical data can be selectively used for line and line ratio diagnostics. Whereas high accuracy for an individual line might be obtained by patching together disparate sources of atomic data, there may be an advantage to using a complete set of atomic data calculated self-consistently (even with lower accuracy) for a global fit model.

For radiative cooling calculations called by hydrodynamics codes, or in non-equilibrium ionization calculations for which the number of unknown parameters is large, more efficient computations are required. Rather than modifying code, one can manipulate the atomic database, producing a smaller averaged database in the same format. Bundling fine structure energy levels following standard atomic physics "sum rules" is straightforward. The accuracy of new models may then be benchmarked against models using the original data.

By tracking accuracy estimates in the atomic database, uncertainty analysis can be folded into the modeling at three levels: by comparing different data sets in the database, by sensitivity testing (varying database values "by hand" to see how sensitive the line emission is), and by Monte Carlo modeling of the entire set. While it is difficult to assess systematic errors in the data (such as might occur, for example, if resonance contributions have been left out), at least data analyzers can be aware of the pitfalls.

Finally, the atomic database as well as models derived from spectral synthesis codes can be used directly in data analysis as table models. The atomic database alone is useful for line identification, for understanding line formation through access to an energy level diagram, and for choosing diagnostics based on accuracy.

5. Conclusions

Not all atomic data are created equally; fortunately, modern storage and computational methods allow us to account for this diversity. Both accuracy and completeness are necessary for spectral modeling. Database formats which allow for different types and qualities of data will improve the integration of data into a critically evaluated set, while flexibility in processing allows for intercomparison of different data within the context of the plasma models.

Since spectral synthesis codes are called on to perform a wide variety of tasks, ranging from calculating diagnostic line ratios to calculating radiative cooling rates needed by hydrodynamics codes, a flexible database can be tailored to the needs of the user.

A range of tools makes possible better assessment of the reliability of astrophysical interpretations, accounting for atomic data, atomic processes, plasma processes, and the quality of observations. Collaboration among people involved in rates, codes, and astrophysics is the best way to address needs in these areas.

Acknowledgments. My colleagues on the development of the Astrophysical Plasma Emission Code (APEC) have contributed enormously to the improvements discussed here. I gratefully acknowledge John Raymond, Duane Liedahl, and Randall Smith for sharing their knowledge, insights, and opinions with me on all levels of modeling. I also acknowledge the support of the Chandra X-ray Observatory Center (formerly the AXAF Science Center) and NASA Grant NAG5-3559.

References

Brickhouse, N. S., Dupree, A. K., Edgar, R. J., Liedahl, D. A., Drake, S. A., White, N. E., & Singh, K. P. 1999, in preparation

Dere, K. P., Landi, E., Mason, H. E., Monsignori Fossi, B. C., & Young, P. R. 1998, in ASP Conf. Ser., Vol. 143, The Scientific Impact of the Goddard High Resolution Spectrograph, ed. J. C. Brandt, T. B. Ake, & C. C. Petersen (San Francisco: ASP), 390

Hauschildt, P. H., Shore, S. N., Schwarz, G. J., Baron, E., Starrfield, S., & Allard, F. 1997, ApJ, 490, 803

Liedahl, D. A. & Brickhouse, N. S. 1999, in preparation

Mewe, R. 1991, Astron. Astrophys. Rev., 3, 127

Raymond, J. C. 1988, in Hot Thin Plasmas in Astrophysics, ed. R. Pallavicini (Dordrecht: Kluwer), 3

Raymond, J. C., & Smith, B. W. 1977, ApJS, 35, 419

Smith, R., Brickhouse, N., Raymond, J., & Liedahl, D. 1998, "High Resolution X-Ray Spectroscopy: Issues for Creating an Atomic Database for a Collisional Plasma Spectral Emission Code", European Space Research and Technology Centre,
`http://astro.estec.esa.nl/XMM/news/ws1/procs/smithrk.ps.gz`

Astronomical Data Analysis Software and Systems VIII
ASP Conference Series, Vol. 172, 1999
D. M. Mehringer, R. L. Plante, and D. A. Roberts, eds.

Cosmology from High Redshift Supernovae

Peter Garnavich

Harvard-Smithsonian Center for Astrophysics

Abstract. The discovery of a correlation between the light curve shape and intrinsic brightness has made Type Ia supernovae exceptionally accurate distance indicators out to cosmologically interesting redshifts. Ground-based searches and follow-up as well as Hubble Space Telescope observations of Type Ia supernovae have produced a significant number of objects with redshifts between 0.3 and 1.0. The distant SNe, when combined with a local sample analyzed in the same way, provide reliable constraints on the deceleration and age of the Universe. Early this year, an analysis of a handful of Type Ia events indicated that the deceleration was too small for gravitating matter alone to make a flat Universe. A larger sample of supernovae gives the surprising result that the Universe is accelerating, implying the existence of a cosmological constant or some other exotic form of energy.

The success of this research has depended on the development of algorithms and software to register, scale and subtract CCD images taken weeks apart and to search for variable objects. A good fraction of the point-sources identified are asteroids, variable stars, or AGN, so spectra are needed to confirm the identification as a Type Ia supernova and obtain a redshift. The best candidates are followed photometrically to construct light curves. The steps to transform the observed light curves into cosmologically interesting results will also be described.

1. Introduction

Supernovae are the ultimate variable stars. Type Ia explosions are thermonuclear detonations of lowly white dwarfs. Over two weeks their brightness can rise by more than 20 magnitudes, a factor of 100 million. These exploding stars are easily seen in in small telescopes as far out as the Virgo cluster of galaxies and if we were lucky enough to have one go off in our Milky Way, the supernova would be easily visible during the day.

But supernovae are relatively rare events. On average, a Type Ia supernova explodes in a galaxy the size of ours once every 500 years. To be guaranteed a discovery on any given night requires searching tens of thousands of galaxies. Such wholesale searching has been made possible by the development of large area charge-coupled device (CCD) detectors and mosaics of CCDs used at the focus of 4-meter aperture telescopes.

Two groups, the High-Z Supernova Search Team led by Brian Schmidt (Schmidt et al. 1998) and the Supernova Cosmology Project led by Saul Perlmutter (Perlmutter et al. 1999) have begun very successful searches for distant supernovae using the 4-m telescope at the Cerro-Tololo Inter-American Observatory. Between them, the two teams have nearly tripled the supernova discovery rate (Fig. 1). The purpose of these searches is to find supernovae at cosmologically interesting distances. Type Ia supernovae make excellent distance indicators, so these events at high redshift can be used to map out the history of the expanding Universe.

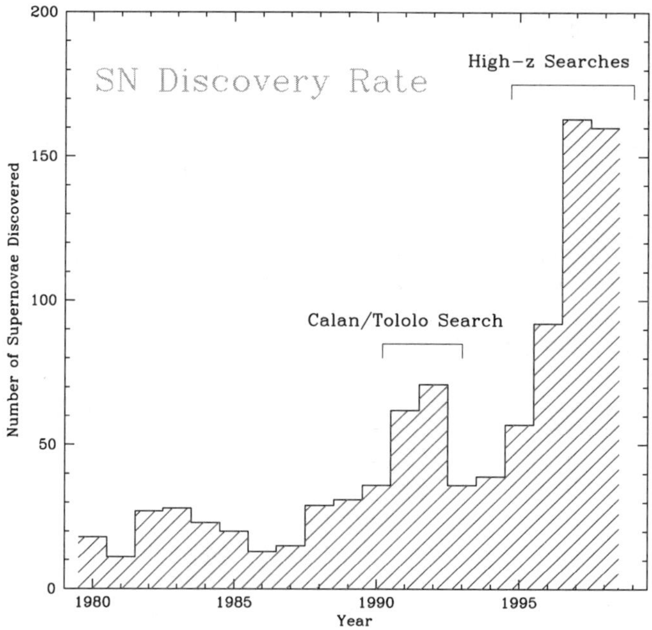

Figure 1. The discovery rate of all supernovae between 1980 and 1998. The Calan/Tololo Survey (Hamuy et al. 1996) was a collaboration of Chilean astronomers that was very successful in finding supernovae in the early 1990's. The sharp rise in the number of discovered supernovae beginning in 1996 is due to the success of two groups searching for supernovae at high redshift.

2. High-Redshift Searches

At very faint magnitude limits, $m \sim 22$, supernova searches become efficient because so many galaxies can be seen in a single deep CCD exposure on a large telescope. Five to ten supernovae per night can be found using a 4-m aperture telescope and a large area detector with a $0\overset{\circ}{.}5$ field of view.

The high-z searches require observing the same fields more than 3 weeks apart and subtracting the images using sophisticated software that matches seeing, background, and flux levels between the epochs. What should remain after this process are the objects that have changed in some way, asteroids, variable

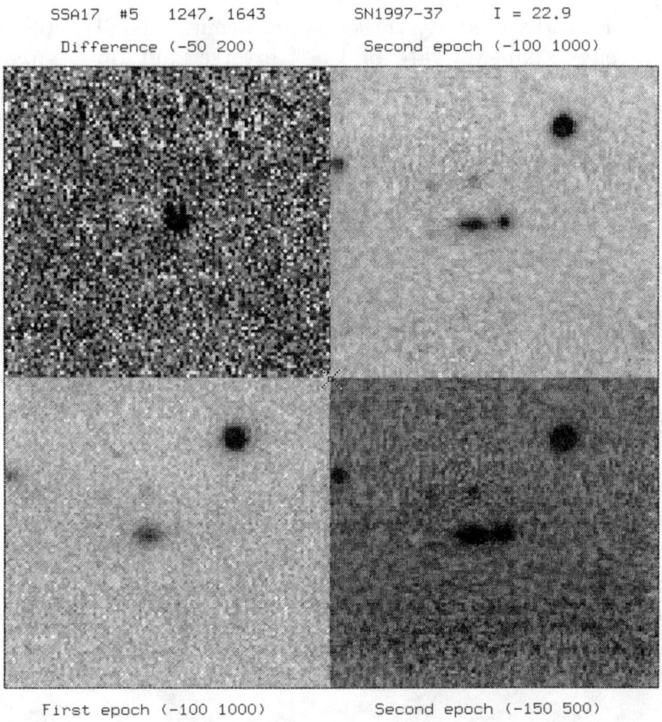

Figure 2. The discovery images of SN 1997cd at a redshift of 0.51. The lower-left panel shows the 'template' image taken three weeks before discovery. The two panels on the right show the discovery image at two contrast levels. The upper-left panel displays the difference between the two epochs and clearly shows the galaxy and foreground stars subtracted away and the presence of the supernova.

stars, active galactic nuclei (AGN) and supernovae (Fig. 2). Asteroids are very numerous at these magnitude limits but can be identified by their motion if multiple images of each field are taken on the same night. Distinguishing supernovae from AGNs and other variables is more difficult. Spectra are generally taken to confirm the identification as well as measure a redshift and attempt to determine the supernova type.

The High-Z Search Team currently uses a Perl script to combine existing programs, such as IRAF, DoPHOT and DAOPHOT, to manipulate the images and extract the photometric information. This approach has the advantage of reducing the amount of new code that had to be created. But it does complicate the process of transporting the search, as many of the versions of the existing programs differ from machine to machine.

3. Results

Why work so hard to find and study these distant events? The Hubble constant describes the current expansion rate of the Universe, but by comparing the brightnesses and redshifts of nearby and distant Type Ia supernovae, it is possible to see if the rate of expansion has changed over cosmic time. At $z \sim 1$, the Universe was half its current age and between then and now the deceleration of the expansion should be detectable from accurate measurements of the distances and redshifts of supernovae. The deceleration rate is a direct measure of the total matter density of the Universe since it is the gravitational pull of the matter which is slowing the expansion (Fig. 3).

The initial results suggested very little deceleration over the past 7 billion years and therefore, insufficient matter in the Universe to reverse the expansion (Garnavich et al. 1998a). Now, with a total of more than fifty supernovae analyzed by the two groups, it is clear that the Universe is currently accelerating. This surprising result suggests that the vacuum may have a non-zero energy density which currently dominates over the matter density. Of course, there are other, even more exotic, forms of energy which would cause the observed acceleration (Garnavich et al. 1998b).

There is some danger involved. If dust or properties of Type Ia supernovae have evolved between $z = 1$ and the present, then the distances measured from the supernovae may be systematically affected. Currently there is no evidence for evolution in the supernovae or in the dust extinction. But more work on understanding both the high$-z$ supernovae and nearby events are needed to place the cosmological observations on firm ground.

4. Conclusions

It is currently possible to discover and study supernovae out to redshifts of ≈ 1. These searches require two epochs of imaging more than one square degree of sky to $R > 22$ in order to be assured of finding supernovae. Two groups have now discovered over 100 supernovae and have analyzed the light curves of more than half of these events. The results are surprising. The Universe appears to be accelerating at the current epoch, suggesting that the Universe is dominated by an unknown energy component with a large negative pressure.

Acknowledgments. Thanks to Brian Schmidt, Robert Kirshner, the High-z Supernova Search Team.

References

Garnavich, P.M., et al. 1998a, ApJ, 493, L53

Garnavich, P.M., et al. 1998b, ApJ, 509, 74

Hamuy, M., et al. 1996, AJ, 112, 2408

Perlmutter, S., et al. 1999, ApJ, in press

Riess, A., et al. 1998, AJ, 116, 1009

Schmidt, B., et al. 1998, ApJ, 507, 46

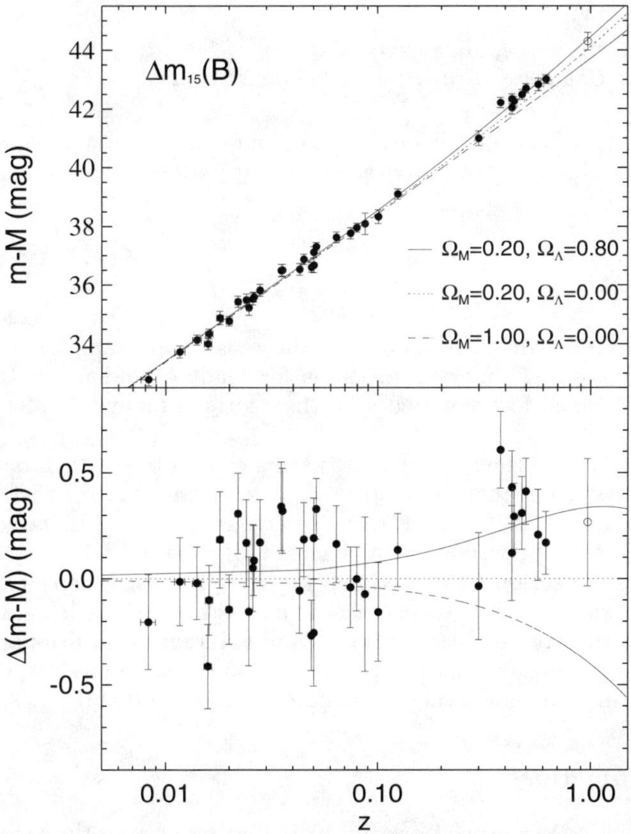

Figure 3. The supernova Ia Hubble diagram (distance as $m - M$ versus redshift z) extending to nearly $z = 1.0$ from Riess et al. (1998). The many low-redshift supernovae define the zeropoint (Hubble constant and Type Ia absolute magnitude), so that measuring the brightness of the high-z events provides a direct estimate of how the expansion rate has changed. The lower panel shows the deviation of the supernovae magnitudes from a model Universe with a low matter density (Ω_m) and no vacuum energy (Ω_Λ).

Astronomical Data Analysis Software and Systems VIII
ASP Conference Series, Vol. 172, 1999
D. M. Mehringer, R. L. Plante, and D. A. Roberts, eds.

Large Scale Clustering in the Universe – A Package of Codes

Ivan Valtchanov

Institute of Astronomy, Bulgarian Academy of Sciences, 72 Tsarigradsko Chaussee Blvd, BG-1784 Sofia, Bulgaria

Abstract. A collection of codes is compiled for studying some aspects of the large scale clustering of galaxies, clusters and superclusters of galaxies.

1. Introduction

Measuring the degree of clustering is one of the most important tasks in the contemporary cosmology. Numerous methods for analysis and many theories were developed in the past forty years due to the efforts of many scientists (e.g., Peebles 1980, 1993).

Because the analysis of the large scale structure cannot be made into routine work, most of the steps for analysis require specific methods and construction of specific programs as well. That is why a package of codes for large scale structure analysis cannot be written as a general tool.

Nevertheless, a collection of procedures for studying some aspects of the clustering in the Universe can be constructed; inevitably any such package or collection of codes will only be a starting point and will require additional development. All the codes constituting the package are applicable to a large range of problems concerning the clustering of galaxies, clusters and superclusters.

2. Correlation functions

The correlation functions are very powerful instrument to study the large scale clustering of galaxies and clusters (e.g., Peebles 1980; Sicotte 1995). All the procedures implemented in the package require random catalog generations and these mock or Monte Carlo catalogs must contains all the selection effects presented in the real catalog.

The following selection effects are analyzed – distance, galactic latitude, galactic longitude, declination, supergalactic latitude.

2.1. Incorporating the selections

Each selection can be represented by a functional form and then it can easily be incorporated in the coordinate generation. Other possible variants where no assumption for the functional form for the selection were also implemented:

- Smoothing the cumulative distribution function (CDF) of the coordinate in consideration and using it as a template for random coordinate generation.
- Bootstrap resampling (Efron 1979).

We have used extensively all the methods for random catalogs creation and even combinations of them is possible. For example, we take uniform distribution of the galactic longitude, for the galactic latitude we implement the functional cosecant law and for the distance we use the bootstrap resampling.

2.2. Correlation function estimators and uncertainties

Depending on the dimensions of interest, we have developed the one dimensional (1D), 2D and 3D two-point correlation function, the cross-correlation function (Peebles 1980) and the correlation function for the line-of-sight and transverse direction pair separations $\xi(r_p, \pi)$ (e.g., Guzzo et al. 1997). An extensive use of the procedures can be found in Kalinkov, Valtchinov, & Kuneva (1998a, 1998b).

The following estimators were implemented in all the codes – RR estimator, DP estimator (Davis & Peebles 1983), H estimator (Hamilton 1993) and LS estimator (Landy & Szalay 1993).

We regard the following uncertainties for each estimator – Poissonian error, bootstrap error (Ling, Frenk, & Barrows 1986), bias-corrected bootstrap uncertainty (Efron & Tibshirani 1986).

3. Substructures searching

Many different methods are developed to search for substructures and comparisons to computer simulations are used for assessment of their performance (see for example Pinkney, Roettiger, & Burns 1996). In the package we have implemented a qualitative approach through the use of different mapping techniques – multiscale wavelet analysis (e.g., Slezak, Bijaoui, & Mars 1990; Starck, Murtagh, & Bijaoui 1998) and Gaussian adaptive and non-adaptive smoothing and filtering (e.g., Silverman 1986; Pisani 1993, 1996).

For the sake of quantifying the observed peaks in the density distribution, a method based on the work of Escalera & Mazure (1992) is implemented.

An application and results with the use of our procedures can be found in Kalinkov et al. (1998a).

4. Finding groups of objects

We have developed procedures based on the variants of the friends-of-friends (FOF) algorithm. Such procedures are used mainly for construction of catalogs of groups or superclusters of galaxies.

Huchra & Geller (1982), Nolthenius & White (1987) and Nolthenius (1993) procedures are implemented. A comprehensive comparison for the methods has been made by Frederic (1995).

A different approach for finding groups is developed by Kalinkov & Kuneva (1993). It is based on searching for groups of objects with chosen density contrast over the local density in given local volume. In a sense, this method is adaptive with respect to the local density and is also FOF method. The result from the application of this procedure is the construction of the largest catalog of superclusters of galaxies (Kalinkov & Kuneva 1993).

5. Void probability function

The void probability function (VPF) depends on the correlation function of all orders and in that aspect it is complementary to the correlation function; in some cases it also provides informative statistics (White 1979). An example of using the procedures described here can be found in Kalinkov et al. (1998a).

VPF in 2D/3D is the probability that a randomly thrown disk/sphere, with given area/volume in the distribution with mean surface/volume density, is empty of objects. The procedures implement the VPF definition with the including edge and selection effects.

6. Utilities

In this section various utility procedures are described. They are independent and can be used as a standalone procedures.

- An Interactive tool for robust estimation of location and scale (Beers, Flynn, & Gebhardt 1990)
- Cone diagrams
- Cosmological distances

7. Future Prospects

The future development of the package will be toward searching for different non spherical structures – filaments, sheets, walls. Searching for voids among galaxies, clusters and superclusters. Procedures for 3-point correlation function and correlation functions of higher order as well as power spectrum analysis and count-in-cell techniques are on the way. Other procedures for searching of groups, clusters and superclusters will be implemented in the package too. Refining some of the codes and making them more efficient for larger samples is an everyday work.

8. Implementation

The procedures requiring much CPU time are written in C (correlation functions, friends-of-friends algorithms) and where the analysis requires much visual inspection the procedures are written in IDL[1]. A few external procedures and functions are used in the C programs and they are taken from Numerical Recipes (Press et al. 1986) but replacing them with any other publicly available procedures is not a problem.

The IDL procedures use ASTROLIB library which is part of the library with many astronomical applications for IDL and is publicly available from `http://idlastro.gsfc.nasa.gov`. All the package with extensive documentation will be available from `http://www.astro.bas.bg/~ivan/lssp/` or by request.

---

[1]IDL is the trademark of Research Systems, Inc.

Acknowledgments. I am thankful to A. Pisani for providing me with his DEDICA package, T. Beers for ROSTAT package. I would like to thank H. Sicotte for providing his invaluable Ph.D. thesis work.

This work was supported by National Research Fund of the Bulgarian Ministry of Science, Education and Technology – contract F721/1997.

References

Beers T., Flynn K., & Gebhardt K. 1990, AJ, 100, 32

Davis M. & Peebles P. J. E. 1983, ApJ, 267, 465

Efron B. 1979, Ann. Stat., 7, 1

Efron B. & Tibshirani R. 1986, Stat. Sci., 1, 54

Escalera E. & Mazure A. 1992, ApJ, 338, 23

Frederic J. 1995, ApJS, 97, 259

Guzzo, L., Strauss, M. A., Fisher, K. B., Giovanelli, R., & Haynes, M. P. 1997, ApJ, 489, 37

Hamilton A. J. S. 1993, ApJ, 417, 19

Huchra J. & Geller M. 1982, ApJ, 257, 423

Kalinkov M. & Kuneva I. 1993, A&AS, 113, 451

Kalinkov M., Valtchanov I., & Kuneva I. 1998a, A&A, 331, 838

———, 1998b, ApJ, 506, 509

Landy S. D. & Szalay A. 1993, ApJ, 412, 64

Ling E. N., Frenk C. S., & Barrows J. D. 1986, MNRAS, 223, 21

Nolthenius R. 1993, ApJS, 85, 1

Nolthenius R. & White S. 1987, MNRAS, 235, 505

Peebles P. J. E. 1980, The Large Scale Structure of the Universe (Princeton: Princeton Univ. Press)

———, 1993, Principles of Physical Cosmology, (Princeton: Princeton Univ. Press)

Pinkney J., Roettiger K., & Burns J. O. 1996, ApJS, 104, 1

Pisani A. 1993, MNRAS, 265, 706

———, 1996, MNRAS, 278, 697

Press W., Teukolsky S., Vetterling W., & Flannery B. 1992, Numerical Recipes in Fortran (2nd ed.), (Cambridge: Cambridge Univ. Press)

Sicotte, H. 1995, Ph.D. thesis, Princeton Univ.

Silverman, B. 1986, Density Estimation for Data Analysis and Statistics (London: Chapman & Hall)

Slezak E., Bijaoui A., & Mars G. 1990, A&A, 227, 301

Starck, J.-L., Murtagh, F., & Bijaoui, A. 1998, Image Processing and Data Analysis (Cambridge: Cambridge Univ. Press)

White S. D. M. 1979, MNRAS, 186, 145

Astronomical Data Analysis Software and Systems VIII
ASP Conference Series, Vol. 172, 1999
D. M. Mehringer, R. L. Plante, and D. A. Roberts, eds.

A Method to Test the Adequacy of a Model to Observations

Alexander Tsvetkov

Astronomical Institute of St. Petersburg University, 198904, St. Petersburg, Petrodvorets, Bibliotechnaya pl. 2, Russia

Abstract. It is well known that the Least Squares method returns the values of the parameters regardless the correspondence of the model to observations. If the model does not correspond to the data one obtains the formal values of the parameters, which have no physical sense. Nevertheless, many people use values thus obtained in their theories and practice.

The proposed method based on the orthogonal representation of the equations of a model is free of this fault since it tests the consistency of the model and data.

This method have been used to find out the compatibility of the standard kinematical model with the proper motions of stars in the Hipparcos catalogue. It shows that there are some difficulties with the standard model of the Galactic rotation, and, on the contrary, there is a good agreement with the model of the Sun motion.

1. Traditional Way to Study Stellar Kinematics

A common practice in astronomy is: gathering data, constructing a model, evaluating of the parameters of the model by the Least Squares technique.

Let us take an example from stellar kinematics. Usually, we have a catalogue, which contains coordinates (l, b) and proper motions $(\mu_l \cos b, \mu_b)$ of stars. General model of proper motions describes the Sun motion and the galactic rotation and may be represented as

$$\mu = \sum_{i=1}^{n} P_i \Phi_i(l, b) + \varepsilon, \qquad (1)$$

where

P_i are the parameters of the model to be determined,

ε is the noise and systematic trends, which are not included into the model.

Φ_i are the functional basis of the model.

The main condition of correct use of the LST is: "ε in the formula (1) is nothing else but the Gaussian noise". We forget this very often since the LST always does produce a result, but it may be

- Formal, non physical
- Unreliable due to large r.m.s.e
- Distorted by systematic trends

It means that we always have to remember two things:

- Does the model or its part correspond to observations?
- What is beyond the accepted model? Is the model complete?

2. Orthogonal Representation

We call equation (2) the **physical model** of proper motions where functions Φ_i form the functional basis of the model $\mathbf{\Phi}$. Let us consider the **formal model** of proper motions with another functional basis **K**:

$$\mu = \sum_{j=1}^{m} C_j K_i(l, b). \tag{2}$$

In this equation

C_j are the coefficients of expansion,

K_j are the set of functions, which are *orthogonal* on the celestial sphere.

We have to find out the dependence between C_j and P_i. In some bases **K**, we can obtain

$$P_i \sim C_{p_i} \sim C_{q_i} \sim C_{r_i} \sim \ldots \tag{3}$$

for each P_i. It means, that we may derive each parameter of the physical model more than once and, consequently, we can compare them. This is the *basic idea* of the method. The method allows to estimate whether the behavior of the observational material follows the functions of physical basis or not.

Good agreement between the values derived from several harmonics shows the correspondence of the model to observations.

Bad agreement can show:

- Adopted model does not correspond to observations
- There are unknown systematic motions which penetrate into expansion coefficients

In addition, the existence of unpredictable non-zero coefficients indicates that adopted model is not complete.

Concerning the basis **K**, we have to choose such set of functions which being applied to the physical basis, produces the relations like (3). Usually, in stellar kinematics, the *spherical functions* are good.

3. Test of the Oort-Linblad Model

The Oort-Linblad's model is widely used for the analysis of proper motions. The general view of the equation for proper motions in the galactic longitude is

$$k\mu_l \cos b = \pi V_x \sin l - \pi V_y \cos l + A \cos b \cos 2l + B \cos b \tag{4}$$

where

V_x, V_y are the components of the Sun velocity,

A, B are the Oort's parameters,

π is a trigonometric parallax,

$k = 47.4$ is a scale factor to reduce arcsec/cy into km/s kps^{-1}

We select as formal basis **K** the spherical harmonics defined as

$$K_{n^2+2k+m-1}(l,b) = \begin{cases} L_{n0}(b) & k=0, m=1 \\ L_{nk}(b)\sin kl & k\neq 0, m=0 \\ L_{nk}(b)\cos kl & k\neq 0, m=1 \end{cases} \tag{5}$$

In this basis for the Oort-Linblad's model we find that the parameters under consideration are proportional to the coefficients of the expansion:

$$\begin{array}{ll} V_x & \sim C_2 \sim C_{10} \sim \ldots \\ V_y & \sim C_3 \sim C_{11} \sim \ldots \\ A & \sim C_8 \sim C_{20} \sim \ldots \\ B & \sim C_0 \sim C_4 \sim C_{16} \sim \ldots \end{array} \tag{6}$$

The factors of proportionality may be found either theoretically for uniform distribution of stars over the celestial sphere, or numerically if the distribution is not uniform.

In some models (for example, three-dimensional rotation of the Galaxy), several parameters can hit one and same harmonics. In this case, we have to use the linear combinations of these parameters instead of individual ones.

4. Numerical Results

The universal program was written that can do all these things automatically. This program was applied to investigate the samples from the Hipparcos catalogue.

We illustrate the program by two samples for nearby (75–125 ps) and distant stars (300–500 ps) from the main sequence in H-R diagram. Also, the program was tested on artificial data[1] to show the effect in ideal case. The traditional solution using LST is shown in the table, too. The missed data in the table denote that the F-test (Broshe 1966) rejected the value in this cell.

This table shows that only the Sun motion can be detected certainly in the proper motions. The galactic rotation is polluted by unknown systematic motions which distort the 4th coefficient, though the Least Squares technique produces the solution that seems to be very reliable.

[1]The following values of parameters had been accepted for simulation: $V_x = 9$, $V_y = 15$ km/s; $A = 12$, $B = -12$ km/s kps^{-1}; standard deviation of the Gaussian noise was 2 mas/yr and the range of distances was from 300 to 500 ps.

Parameter		Simulation	75 – 125 pc	300 – 500 pc
V_x	2	7.8 ± 0.3	9.1 ± 0.5	9.1 ± 0.6
	10	8.3 ± 0.7	5.2 ± 1.3	9.8 ± 1.7
	LST	8.0 ± 0.0	8.9 ± 0.3	9.2 ± 0.3
V_y	3	14.9 ± 0.3	15.8 ± 0.5	15.1 ± 0.6
	11	15.2 ± 0.7	14.5 ± 1.3	21.0 ± 1.7
	LST	15.0 ± 0.0	15.9 ± 0.3	15.3 ± 0.3
A	8	11.7 ± 0.8	—	12.9 ± 2.1
	20	8.5 ± 4.6	—	—
	LST	12.3 ± 0.1	12.4 ± 3.8	12.9 ± 1.0
B	0	-12.3 ± 0.6	-11.9 ± 4.1	-10.4 ± 1.4
	4	-11.7 ± 1.8	$-33.0 \pm 12.$	-21.8 ± 4.3
	16	-20.0 ± 8.9	—	—
	LST	-12.1 ± 0.1	-13.7 ± 2.7	-12.9 ± 1.0

5. Conclusion

The first attepmt of the method apllication for stellar kinematics was made in two papers by Vityazev & Tsvetkov (1989a, 1989b). The basic principles of our method are described in the paper by Vityazev (1994) who proposed to use the spherical functions to derive the mutual rotation between two reference frames. Our method is an extension of this approach for kinematics. It may be easily modified for many purposes if one wishes to test a model for correspondence to observational data.

Acknowledgments. The author expresses gratitude to the LOC of the ADASS'98 conference for hospitality and financial support.

References

Brosche P. 1966, Representation of Systematic Differences in Positions and Proper Motions of Stars by Spherical Harmonics, (Heidelberg: Veroff. Astr. Rechen-Inst), N17

Vityazev, V. V. & Tsvetkov, A. S. 1989a, Vestnik LGU, 1(2), 73

———, 1989b, Vestnik LGU, 1(2), 79

Vityazev, V. V. 1994, Astron. and Astrop. Trans., 4, 195

Astronomical Data Analysis Software and Systems VIII
ASP Conference Series, Vol. 172, 1999
D. M. Mehringer, R. L. Plante, and D. A. Roberts, eds.

Time Domain Analysis of Solar Coronal Structures Through Hough Transform Techniques

A. Llebaria, P. Lamy

Laboratoire d'Astronomie Spatiale du CNRS, Marseille, FRANCE

Abstract. The Hough transform technique is applied to a series of images from the LASCO/C2 coronagraph in order to understand the temporal evolution of radial structures on the solar corona, the so-called "polar plumes". They may corotate with the solar corona and therefore change aspect with time due to projection effects on the plane of the sky. In the images these structures suddenly appear, shift, mix, and fade away on short periods of time. From a long series of images ($\sim$ 100), we determined for each image the polar intensity profile and built up the evolution of such profiles against time. The result is a Time Intensity Diagram (TID) where intensities are plotted with respect to time and position coordinates. Radial structures appear as peak intensities in each profile and therefore as bright points in the TID. The Hough transform techniques are applied to detect coherent trajectories. This technique has been applied successfully to study coronal plumes on coronal images obtained by different instruments aboard the SOHO satellite.

1. Introduction

The solar corona is composed of two components:

1. a highly structured, time variable K-corona due to Thompson scattering of solar light by coronal electrons,
2. a smooth, quasi-stable F-corona due to solar light scattering by interplanetary dust particles.

Both components exhibit an exponential decrease of their radiance with increasing distance from the Sun with the K-component having a steeper gradient than the F-corona. The K-corona structures are tightly linked to the magnetic field of the Sun and their appearance varies with the solar activity (the solar cycle has a period of 11 years). During the minima of activity, the K-corona exhibits two main stable structures, the equatorial streamers and the north and south coronal holes. These holes are filled with quasi-radial ray-like structures, the so-called polar plumes which apparently outline the polar magnetic field lines, a dipole in a first approximation. Plumes are difficult to detect as they appear as faint azimuthal modulations of the radiance of the corona. Their observations has been limited to solar eclipses which are too short to allow a study of their temporal behaviour. The LASCO-C2 coronograph aboard the SOHO satellite has been continuously observing the corona for more than 2 years and its excellent photometric performances are well suited to an in–depth analysis of the nature and temporal evolution of polar plumes and address first two main questions: are they permanent structures, and are they in rigid body rotation with the Sun? If this is so, a movie made of successive images should easily reveal the plumes as oscillating around the polar axis, a result of the projection of the rotation onto the plane of the sky. In practice, such movies are extremely difficult to interpret because of several confusing effects:

1. numerous plumes populate the coronal holes, those in the foreground moving from east to west, those in the background in the opposite direction,
2. phase function of electrons is a strong function of phase angle and heavily biases the visibility of the plumes; low latitude ones rapidly fade out

when they leave the plane of the sky. In addition, any intrinsic variations of the plumes – and we shall see below that they indeed turn on and off – will further complicate the picture.

We present below a novel approach to the problem. We first construct a Time Intensity Diagram (TID) by piling circular profiles extracted from a time-series of images at a given radial distance. Incidentally, the origin of the polar coordinates is not at the center of the Sun but closer to the limb, at the two points (either north or south) of apparent divergence of the plumes. The TID is therefore an image giving the radiance as a function of position angle (x-axis) and time (y-axis) where plumes appear as bright points. Coherent alignments, that is trajectories, are first visually detected and then established using the Hough transform technique. The dotted pseudo-sinusoidal tracks allow to conclude that plumes are recurrent structures, transitory lit, in rigid body rotation with the solar corona.

2. The Construction of the TID

The process of constructing the TID starts with a selection of a set of LASCO/C2 images covering at least half a solar rotation (14 days) with a minimum frequency of 2 images per day to insure interframe continuity. This set is rigorously photometrically equalized by comparing two successive images and correcting for any difference in radiance of more than 0.1 percent. A minimum background image is calculated from the full set after a median smoothing. Then for each equalized image of plumes the following steps are implemented:

1. Subtract the minimum background image.
2. Introduce a polar coordinate system centered at the apparent "divergence point" of the plumes (located at $\sim$ 0.843 solar radius from the center of the Sun on the polar axis).
3. Apply a transform to cartesian coordinates so that plumes now appear as a set of parallel rays.
4. Apply a 1-D median filter, 16 pixels wide and parallel to the plumes direction, to preserve the lateral spatial resolution. This filter was favored over others such as gaussian filters because it better handles impulsive noises and discontinuities which are present in the LASCO images.
5. Estimate and subtract a residual local background.
6. Obtain a mean intensity profile by averaging 128 neighbouring lines.

The resulting profiles are piled up with increasing time to construct the TID diagram. A sample TID appears in Fig. 1 with its x-axis corresponding to the

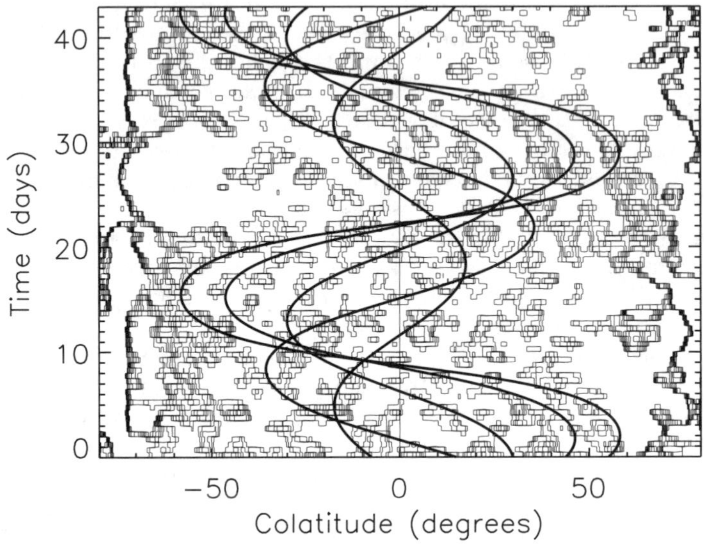

Figure 1. TID diagram for a sequence of 86 images (43 days).

position angle and the y-axis, to the time.

Under the assumption of rigid body rotation, individual plumes will follow sinusoidal curves on the TID with a common axis corresponding to the polar direction. The instantaneous position angle can be written as a function of time. The maximum elongation (or colatitude), the period and the initial phase angle are the three parameters which define a particular trajectory. The exact equation, including a corrective term for the apparent variation of the polar axis with respect to the plane of the sky can be found in Lamy et al. (1998).

3. Analysis of the trajectories using a Hough-like transform

The TIDs built from white light images do not display continuous tracks for the plumes, except at high colatitudes. Their patchy appearance immediately suggests that identified plumes are not permanent structures. However dotted sinusoidal paths are suggested and must further be explored. Their reality will indicate that plumes are enduring, recurrent structures that are transiently lit.

In order to find objective tracks, we introduce a variant of the Hough transform, a method derived from the Radon transform, which makes it possible to detect many kinds of geometric alignments in an image. The specific advantage of the Hough method is its ability to detect alignments of disjoint tracks and points as a single entity. A discussion of this method can be found in Ballester (1994, 1996) and astrophysical applications, for instance in Ragazzoni & Barbieri (1993, 1994).

In our case, alignments would coincide with the sinusoidal tracks of the plumes. The method consists in summing the TID levels along the expected trajectory for each combination of colatitude, initial phase and period. In theory, the track profile over the TID is a Dirac function, but for practical situations the Dirac kernel is replaced by a so-called influence curve. It must be positive, integrable and with a finite support. We chose a normalized truncated gaussian curve corresponding to the estimated uncertainty in the plumes position over time.

Each sum defines a point in the 3-D parameter space. The result is a 3-D intensity function of the synodic rotation period which ranges from 27 days at the equator to 35 days at the poles, with a half day uncertainty, the colatitude which varies from 0° to 80° with a reasonable resolution of 2° and the phase which covers the whole period, with a half day resolution.

For an initial estimation, this function was reduced to a more tractable 2-D function, by fixing the period to the mean synodic solar rotation. The standard technique consists in determining first the coordinates of the absolute maximum, which defines the most conspicuous track, then to subtract from the TID all the points corresponding to the track and then to restart the whole process and look for the next maximum. In practise, we do not follow this procedure because the same point of TID can belong to two distinct trajectories. Instead, we select the most prominent set of local maxima at once as indicators of most probable trajectories.

4. Results

We built the TID corresponding to a time interval of 43 days centered on December 1997, as shown in Fig. 1. We found 7 reliable trajectories with a period of 27 days and with colatitudes ranging from 14° to 54°; five of them are overplotted in the figure. This method has been effective to track the dotted trajectories of the white light polar plumes establishing for the first time their temporal evolution and their correlation with the extreme ultra-violet plumes present in the solar polar regions (Llebaria et al. 1998).

References

Ballester, P. 1994, A&A, 286, 1011

———, 1996, Vistas in Astronomy 40, 479

Lamy, P., Llebaria, A., Koutchmy, S., Reynet, P., Molodensky, M., Howard, R., Schwenn, R., & Simnett, G. 1997, in Proc. of the Fifth SOHO Workshop (ESA Publ. SP-404), ed. A. Wilson (Noordwijk: ESA), 487

Llebaria, A., Lamy, P., Deforest, C. E., & Koutchmy S. 1998, in Solar Jets and Coronal Holes (ESA Publ. SP-421), ed. P. Martens & T. D. Guyenne (Noordwijk: ESA), 87

Ragazzoni, R. & Barbieri C. 1993, in Proc. of Applications of Time Series in Astronomy and Metereology, ed. O. Lessi (Padova: Univ. of Padova), 233

———, 1994, PASP, 106, 683

Astronomical Data Analysis Software and Systems VIII
ASP Conference Series, Vol. 172, 1999
D. M. Mehringer, R. L. Plante, and D. A. Roberts, eds.

Automated Programming and Neutrino (Astro)Physics

Michele Maris

Astronomical Observatory of Trieste (OAT), Via G.B. Tiepolo 11, I34131, Trieste, Italy

Abstract. A project for the creation of a code for accurate numerical predictions for solar neutrino oscillations is presented. The project was realized through an automated program generator which scans the simulation parameter space. The software validation procedure and the production guidelines are illustrated.

1. Introduction

After the recent Super Kamiokande announcement of a convincing detection of neutrino mass (Kajita 1998), there has been increasing demand for exhaustive and accurate numerical simulations of neutrino oscillations in (astro)physical environments. Large numerical simulations are a fundamental tool when the numerical accuracy has to be compatible with the experimental one. From the point of view of computer work, two of the most computationally intensive tasks are the determination of neutrino oscillations parameters and the prediction of results of new experiments. Both require an "exploratory" scan of the full parameters space composed by: the source model (Sun, accelerator beam or cosmic rays), neutrino parameters (the neutrino mixing angle in vacuum θ_v and the difference of square masses Δm^2), the oscillation mechanism (MSW, Vacuum Oscillations, etc.), and the neutrino experimental set-up.

The COALA project[1] is designed for the creation and management of a large data base of simulation to be used in the analysis of solar neutrino data under the assumption that solar neutrinos are affected by matter enhanced oscillations or vacuum oscillations during their propagation. The project is presented as a practical example of application of automated programming, both for simulations management, results analysis and results display in a theoretical field where exploratory simulations are essential.

2. Software Development and Simulations Execution Guidelines

Formal guidelines where adopted along the software production cycle and its subsequent employment. The experience gained during this stage of the project suggested a more general, abstract and automated procedure, leading to the concept of Automated Program Generator (Maris & Staniszkis 1999).

[1] `http://www.pv.infn.it/~maris/nuesite/coala.html`

The code for the COALA project is written in standard FORTRAN 77, managed by the CERN `CMZ` software[2] and is based on a previous version written in Borland `TurboPascal 4.0` for a 486 – 33 MHz PC (Bertotti & Maris 1996). A set of coherent rules, mostly related to names composition, was applied rigorously to improve the readability and reduce the documentation. Variables and function names are composed by a set of prefixes, postfixes and keywords organized as a dictionary, in which describes a specific program feature or physical quantity.

A four stages validation chain was adopted and its first stage is the *FORTRAN* validation where formal correctness of the code and the absence of run time errors is ensured. In the *DEBUG* validation step the code is observed while working through a debugger and test messages. In the *NUMERICAL* validation step the code is checked with the results produced by a comparison code. In this step any critical functions of the `FORTRAN` code are replicated by an equivalent one written for `Mathematica`, the two functions are then quantitatively compared within a significant set of sample values of the input variables, and a validation document is written. In the *PHYSICAL* validation step the results are studied to be sure of their physical consistency. This scheme ensures the numerical correctness and stability of the final code (Maris 1997) and a `Mathematica` package for neutrino (astro)physics was produced.

The compiled code is simplified by suppressing all `FORTRAN` user interfaces. No string parsing or complex input management is performed by it. The FORTRAN code, executed in batch mode, receives all the data from the standard input, and produces only one simulation at each call. Due to such limitations all services required by the FORTRAN code are given by a hierarchy of script files which operate as a *Virtual Scanner*. From the top to the bottom of the hierarchy there are script files which either generate the script files which form the underlying level, or execute the compiled FORTRAN code or the already generated script files for a specific simulation. In this way the user interface and the services are isolated from the computation, the `FORTRAN` code is both standardized and simplified and the effort to port, maintain, update and reuse the code is reduced and many simulations can be generated in a fast and effective way.

The `FORTRAN` programs are chained through the script files hierarchy to form a simulation pipeline. The data exchange is ensured through files, stored into a database of intermediate results. Each of them carries both the data and the parameters required to generate it. In this way each simulation parameter is introduced in the pipeline only once, and the files are promptly identified by their header. The creation of an archive of intermediate results increases the disk load but it is fundamental for the *NUMERICAL* and the *PHYSICAL* validation steps, since they are used to ensure the quantitative consistency between the various simulations steps. In addition, new investigations not planned in the original research plan may be performed in a short time. Numerical results are stored in `gzip` compressed ASCII files. Up to 80% compression efficiency was achieved, which means that the final archive does not exceed the dimension of a binary archive, while the use of ASCII files ensures portability and allows

[2] `http://www.cn.cern.ch/cmz/`

scanning of the data without a special browser. To avoid the production of wrong results due to the failure of some steps in the pipeline, each program generates its output file only at the end of its task. Subsequent programs interpret the lack of a file as a failure of one of the previous steps.

Simulations descriptions, data, scripts and log files are stored into a database to improve the software reusage, the simulation documentation and to allow the reconstruction of the simulation chain when needed. For this purpose simulations are classified according to the execution date and their scopes.

Whenever possible standard and automated results presentations and analyses are applied, and once again script files generators play a central role. In this case scripts programs drive plotting packages, table formatters, or various kinds of software agents which, as an example, compile statistics about data stored in the data base.

3. The Virtual Scanner

The simplest model for a program which can explore a mesh over a given parameter space is the *Virtual Scanner* (*VuScanner*). In its simplest form a VuScanner linearizes the cartesian product of the mesh which generates all the sets of possible parameters combinations and runs the specific simulation process on them. Constraints are applied to skip forbidden combinations of parameters.

A VuScanner is composed by a set of nested loops with the same structure. Each loop is in charge of scanning the list of values associated to a given parameter, the loops are nested according to the parameters space hierarchy. The most external loop scans the first parameter in the hierarchy, the next loop the second parameter, and so on. The user links his own routines in the loops structures only at predefined points.

At the beginning of the project for any application a specific VuScanner was written, but a general purpose VuScanner based on MatLab 4.0 is presently in use. Here, a recursive algorithm is applied to scan a table which reports the list of parameters to be scanned, their values, the constraints and the sequence of user routines to be linked to the VuScanner. This VuScanner may be used both to generate a script file which will be subsequently executed (code generation mode) or to drive the execution of codes directly (simulation driving mode). The more recent MatLab VuScanner runs directly code on a PC with a Pentium II processor and generates script files for VMS-DCL, MS-DOS batch language, MatLab and in the future for UNIX. Moreover such version of VuScanner is at the root of the IDL VuScanner described in (Maris et al. 1999).

4. Results

A FORTRAN77 package for solar neutrino (astro)physics was developed, tested in an accurate way and successfully used to explore the behaviour of neutrinos for several hypothesis of neutrino masses. The code works on VMS-ALPHA machines and on PC Pentium II without any relevant modification. During the validation phase a `Mathematica` package for the neutrino physics was created. A series of different prototypes of Virtual Scanners were created to drive such simulations and to accomplish the analysis of results. A large database of simulations (up to 2 GBytes) was generated and studied. The main scientific achievements obtained from this code are described in Bertotti & Maris (1996), Liu, Maris, & Petcov (1997), Maris & Petcov (1997, 1998), Maris (1997), Chizhov, Maris & Petcov (1998), and Dolfini et al. (1998). The most relevant of them is an accurate and exhaustive study of the Day/Night asymmetry expected for the event rates in the Super Kamiokande solar neutrino detector. This study leads to the quantitative study of the enhancement of such effect through the selection of those solar neutrinos which cross the Earth Core, which is now applied by the Super Kamiokande collaboration (Nakahata 1998).

5. Conclusions

A Virtual Scanner may be practically applied to studies in the field of solar neutrino (astro)physics. Its use reduces the manpower required to manage such complex simulations, and allows a deepest systematic study of the subject through quantitative predictions.

Acknowledgments. The author is deeply indebted with S.T. Petcov of SISSA – Trieste for fruitful discussions, with E. Calligarich, director of the Pavia INFN Section who allowed the use of the ICARUS computing facilities at Pavia University, with F. Pasian of OAT – Trieste for support and suggestions about the formalization of automated programming, with G. Schiulaz of OAT – Trieste for improvements in the english style.

References

Bertotti B. & Maris M. 1996, Ph.D. thesis of M. Maris, Pavia, DFNT

Chizhov, M., Maris, M., & Petcov, S. T. 1998, Phys. Rev. D, submitted

Dolfini, D., et al. 1998, in preparation

Kajita, T. 1998, in Proc. of the XVIII Intl. Conf. on Neutrino Physics and Astrophysics, (North-Holland: Elsevier), in press

Liu, Q. Y., Maris, M., & Petcov, S. T. 1997, Phys. Rev. D, 56, 26

Maris, M. 1997, in Proc. of the Fourth Solar Neutrino Conference, ed. W. Hampel (Heidelberg: Springer-Verlag), 342

Maris, M., Pasian, F., Smareglia, R., Maino, D., Burigana, C., Staniszkis, M., & Barriga, J. 1999, this volume, 145

Maris, M. & Petcov, S. T. 1997, Phys. Rev. D, 56, 7444

———, 1998, Phys. Rev. D, 58, 113008

Maris, M. & Staniszkis M. 1999, this volume, 470

Nakahata, M. 1998, in Proc. of the XVIII Intl. Conf. on Neutrino Physics and Astrophysics, (North-Holland: Elsevier), in press

Part 3. Observatory Planning and Scheduling

Astronomical Data Analysis Software and Systems VIII
ASP Conference Series, Vol. 172, 1999
D. M. Mehringer, R. L. Plante, and D. A. Roberts, eds.

The Next Generation User Support Tools

Anuradha Koratkar

Space Telescope Science Institute

Sandy Grosvenor

Federal Data Systems

Abstract. Technological advances such as widespread use of the Internet, multi-platform visual development tools such as Java and CORBA, and overall increases in the power of desktop hardware are allowing for significant improvements in user support tools. Consequently, many of us are expending considerable effort developing or rewriting observing tools to take advantage of these advances. We will present what we think are the key challenges and issues facing the developers of the Next Generation Support Tools, and describe how some groups are approaching these challenges.

1. The Need for User Support Tools

Observing, be it ground-based or space-based, classical or service mode, is a pipeline from start to finish. It starts with the submission of a proposal and ends with properly documented data sets in the archive. In the present era, it is often the case that observing instruments are developed by one group, and commonly used by another group of people. Further, the instrumentation can be very complex with large capital investment both in instrumentation and observatory infrastructure. In such an environment, for any type of observing to maximize scientific returns, information has to be passed back and forth between instrument developer, observatory staff and the user community. Observational conditions, instrument capabilities, and scheduling complexities must be thoroughly understood, and this information has to be provided to the user community in a concise, timely, effective and efficient manner. Thus, to disseminate information effectively to the user community observatory staff depend on:

- documentation
- software tools
- human user support

The overriding goal of maximizing the scientific returns becomes challenging when it has to be achieved with limited resources for user support. In such an environment, user support provided by a person has to be minimized without losing the human aspect of support. This implies that the support provided via documentation and tools has to be friendly, up to date, and easily accessible to users of varying levels of expertise. Therefore, we need to move towards electronic documentation that is well indexed and is well suited for frequent and dynamic updating. An important implication of electronic documentation is that software tools can effectively access information to provide context sensitive help. This integration of documentation and tools must be exploited to reduce manual effort.

2. The Proposal Preparation Process

The proposal preparation process is normally divided into two phases:

- *Phase I:* In this phase the proposer describes a project, its scientific goals and objectives and defends the science. A proposal in this stage is the basis for the initial planning and scheduling by the observatory staff.
- *Phase II:* In this phase the successful proposer provides the observatory staff with a precise executable observing.

From the user's perspective during the the proposal preparation process he/she needs to understand:

- what is the instrumentation available at the observatory,
- what are the characteristics of the observatory/instrumentation,
- how much observing time will be required for a given science project, and finally
- how the observatory plans and schedules observing programs.

From the observatory staff perspective all the above information has to be provided to the users, so that at the end of the proposal preparation process the observatory has obtained a pool of well-defined, accurate, unambiguous, flexible, feasible and schedulable programs. This pool of "good" observing programs then has to be organized into an efficient observing time line. It is essential that the observing pool have good programs, since every problem in the beginning of the pipeline has a tendency to propagate down the observing pipeline and make the whole system inefficient. Thus, it is crucial to have an effective proposal preparation process.

Various steps in the proposal preparation and scheduling processes are governed by a set of rules, may be complex rules, conducive to some form of automation. In this paper we will concentrate on the proposal preparation process, i.e., the Phase I and Phase II process from the users perspective. Note that we will not be differentiating between the Phase I and Phase II, since many of the software tools can be used in both phases.

3. What are Good User Support Tools?

"Good" proposal preparation software tools are essential not only because they help decrease the amount of human user support, but because of the complexity of both instruments and observing programs. Also, the increase in competition for observing time implies that users want to explore their parameter space to justify their scientific objectives properly. Further, with the rise in service mode observing information must be passed precisely and correctly from the user to the observatory staff. Thus, let us define what we consider a "good" software tool. A good software tool has the following properties:

- easy to use for users of varying degrees of expertise.
- provides reduction in manual support from observatory staff. This implies that the tool decreases complexity of instruments/process, provides easy access up-to-date reference information, is flexible and helps guide user towards completion of that tool's process.
- allows exploration of observing parameter space.
- allows visualization and is as interactive as possible.
- is aesthetically pleasing.
- is common for both observatory staff and observer.

4. New Technologies

It is an understatement to say that there has been an explosion of new technologies in the past decade. Major advances continue to occur in software, hardware, and telecommunications. The Internet revolution has fostered the development of distributed tools and languages. Java, while still maturing and developing, shows promise at succeeding in becoming a truly platform independent language. CORBA (Common Object Request Broker Architecture) is emerging as a single standard to allow objects on different systems to truly become distributed and to interact with each other, regardless of the host system and language.

On the hardware side, the exponential growth in computing speed, memory, and storage shows no sign of slowing. What was impossible on personal machines just a few years ago is now virtually instantaneous. And with the growth in computing power comes new possibilities in the capabilities of visual and interactive tools and interfaces.

Similarly, expert systems tools are no longer relegated to mainframe systems and unfriendly interfaces. The new generation tools are visual, fast, and much cheaper that their predecessors.

Finally, a whole new technology arena is emerging: data mining. Data mining is a combination of all of the above technologies to provide new ways of organizing, studying, analyzing, and managing the vast sources of information now at astronomer's fingertips.

5. Emerging Strategies for User Support Tools

There are two efforts that are aggressively using these new technologies. The tools generated by both these efforts are globally attempting to reduce complexity in the proposal preparation process. They use interactive and dynamic visualization to help prepare an observing program. Java is the preferred language for both these efforts.

5.1. The Gemini Effort

The Gemini Observatory has proposal preparation tools both for Phase I and Phase II. In the Phase I tool there is a Navigator that guides the user through the process and leads them to completion. The check list is a very useful visual aid (see Fig. 1). Although the administrative information is gathered through a form based tool, this tool is effective because of two features. (1) the tool simultaneously shows a concise form of the information; and (2) if previewing or printing the formatted Phase I proposal information, the tool launches an HTML browser window in which the current proposal information is displayed. The user can use the HTML window in the manual mode or the automatic mode to update the proposal information.

The Phase II tool allows the observer to completely define an observation or observing program. The definition of an observation includes components such as target position, scheduling constraints, instrument configuration, guide star selection, sequencing, calibration and configuration of the data processing pipeline. The visualization and interactive capabilities of the tool allow the user a lot of freedom to explore the parameter space. This effort is now entering the final phases of development. For further information see the Gemini web site at `http://www.gemini.edu`.

5.2. The Next Generation Space Telescope's Scientist's Expert Assistant (SEA)

The SEA is a research and development effort, investigating new approaches for proposal development with the goal of significantly reducing the costs of supporting the general observer program. It is a joint effort between Space Telescope Science Institute (ST ScI) and the Advanced Architectures and Automation Branch of Goddard Space Flight Center's (GSFC) Information Systems Center. SEA's prototyping effort priorities are visualization and expert systems. Because of its prototyping nature, we are able to explore riskier and more radical approaches than is possible conventionally.

The SEA has an interactive exposure time calculator (ETC) that allows exploration of the instrument, target and observatory parameter space. It can access reference material dynamically to immediately see the effects of changes in the exposure's parameters (see Fig. 2). It can also handle multiple instruments and even multiple observatories. The Visual Target Tuner (VTT) is another main feature of SEA. The VTT is an interactive field-of-view visualization tool that allows the user to import images from several sources (including the NASA Extraterrestrial Database—NED, HST's Guide Star Catalogs, and the Digitized Sky Survey). Once imported the user can overlay an instrument's field-of-view positioning and orienting the target exposure visually. Overlays of multiple exposures can also be done visually.

The various SEA tools are designed to have a common "look and feel". We anticipate that most of the tools will be valuable during all phases of proposal preparation: exploratory, Phase I, and Phase II. And information will pass between tools and phases seamlessly with no manual transfer of information. This alone will avoid many unnecessary mistakes. Although the SEA tools are integrated they can be used individually.

SEA is about halfway through the development cycle. Currently, the expert system capabilities are limited and have not yet been fully explored. Additional features expected over the next year include spectroscopic support for the (ETC), substantial expansion of expert systems capabilities especially in the areas of visit planning, and context sensitive help. SEA has also not yet undergone significant testing for its effectiveness. For further information see the SEA web site at `http://aaaprod.gsfc.nasa.gov/SEA/`.

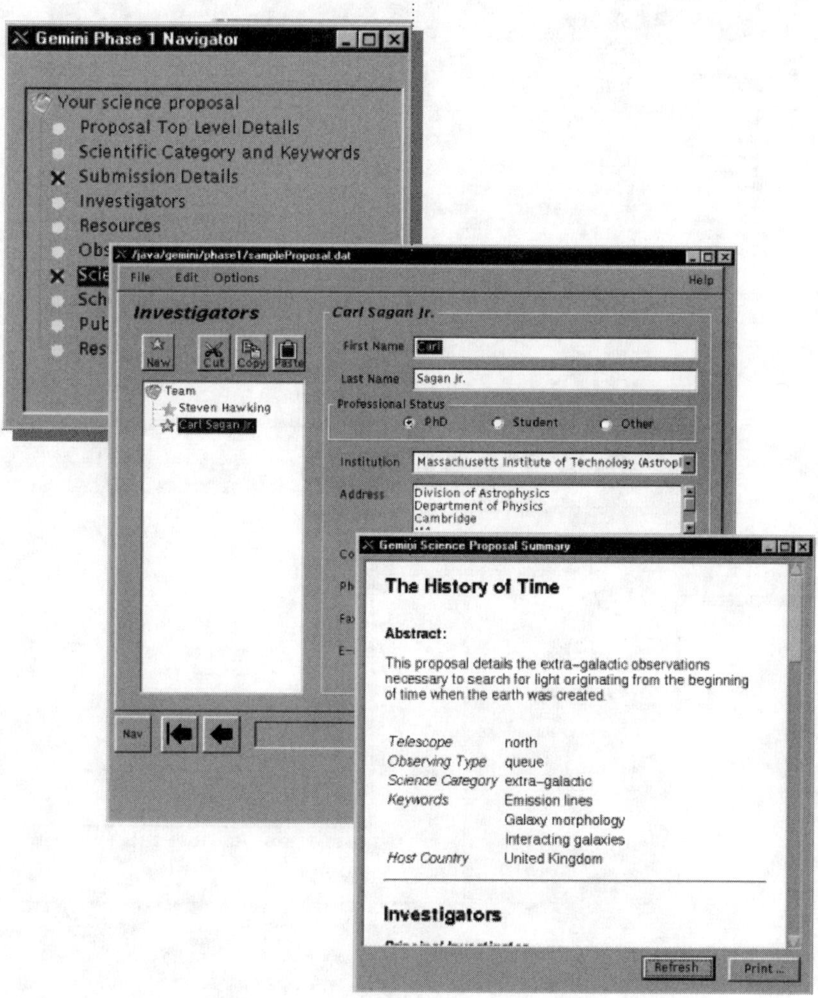

Figure 1. Gemini Phase I Tool highlights.

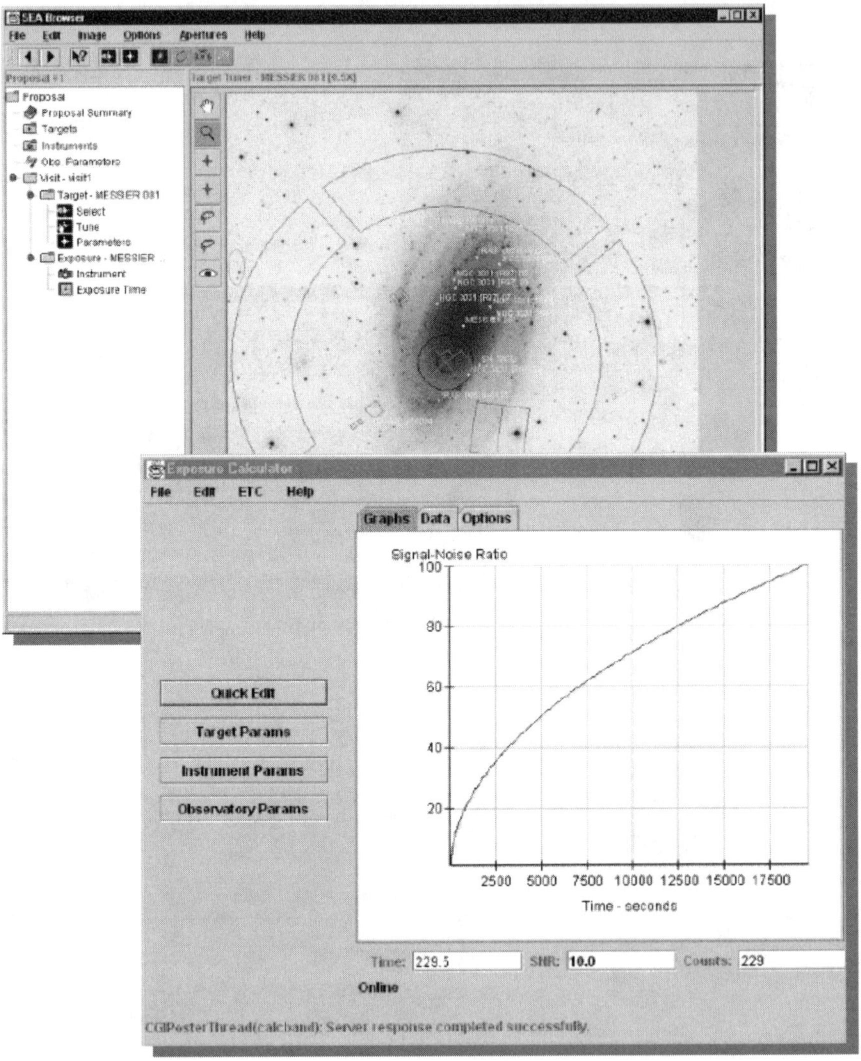

Figure 2. The SEA Exposure Time Calculator and Visual Target Tuner.

6. Workshop on Observing Tools

At the SPIE meeting on "Observatory Operations to Optimize Scientific Returns," there was consensus that we explore the mutual concerns and constraints of the various observatories, so that we can understand where joint, shared, or collaborative effort might be beneficial. Such a collaborative effort across observatories can be considered because:

- the fundamental details of preparing a proposal are common to most observatories whether they be ground-based or space-based.
- from the users' perspective it would be extremely useful if one did not have to be intimately familiar with a given observatory to be able to use it effectively. Common proposal preparation and user support tools that are similar across observatories reduce the learning curve and help in achieving familiarity.
- also from the users' perspective, there should be ways to effectively compare the various observatory/instrument setups within an observatory and across observatories.
- finally, many of us are developing or rewriting our proposal preparation tools using the latest software technologies, which is a unique opportunity for an effective and efficient collaboration.

We therefore organized the "Workshop on Observing Tools" on the 28th and 29th of October 1998. Over fifty representatives, a mixture of astronomers and software developers, representing a variety of observatories (ground and space-based, optical, x-ray, infrared) attended this workshop.

The minimal goal of the workshop was to get to know each other, our areas of expertise and priorities. We were to share ideas and encourage cooperation, and if possible come to a consensus on

- *basic shared requirements:* Understand the requirements and procedures at each observatory to determine what are the various software tools that can be included in a common user support tools package.
- *a common tool framework:* Determine if tools can be shared/made compatible with each other. Then define a common set of libraries that allow different software modules to be easily added/reused. Establish broad guidelines that will maximize compatibility of the various tools.
- *shared tools:* Define a software framework that will allow seamless integration of tools across observatories, and also determine what technologies are best suited for such a purpose.
- Develop consensus of where/how to apply specialization.

At the end of the workshop we formed six working groups that will report back in six months. These working groups have very simple short-term goals and will explore the concept of collaboration on their own terms. Broadly, the short-term goals are to develop a common definition of what each tool should contain. The long-term goals are to define interfaces that will make tools inter-operable, share code and pool efforts in the development of the next generation tools' capabilities. The true measure of success of this workshop will be the tangible results that these working groups will produce. For details on the workshop and the working groups, please see the workshop web site at `http://aaaprod.gsfc.nasa.gov/Workshop`.

Acknowledgments. Both AK and SG thank ST ScI and GSFC for their support and personally thank the SEA team member J. Jones, T. Brooks, M. Fisher, L. Ruley, C. Burkhardt and K. Wolf for the lively discussions and the behind the scenes effort.

Astronomical Data Analysis Software and Systems VIII
ASP Conference Series, Vol. 172, 1999
D. M. Mehringer, R. L. Plante, and D. A. Roberts, eds.

Results from the NOAO Telescope Proposal Process Workshop

Caty Pilachowski, Jeannette Barnes

National Optical Astronomy Observatories[1], *P.O. Box 26732, Tucson, AZ 85726*

Abstract. The National Optical Astronomy Observatories hosted a Telescope Proposal Process Workshop in Tucson in mid-August in which representatives of several observatories met to discuss common issues and goals related to proposals for astronomical telescopes. The goals of the Proposal Process Workshop were to develop a shared understanding among observatories, including the National Gemini Offices and the Gemini Project, of the requirements and procedures for telescope proposals, and to encourage cooperation among the national observatories of the Gemini partner countries, STScI, and other institutions faced with similar issues and concerns related to the telescope proposal process. This paper will summarize the issues discussed at the workshop and the actions resulting from these discussions.

1. Introduction

Representatives of several observatories, including NOAO, Gemini, STScI, McDonald Observatory and the HET, the MMT Observatory, the CFHT, the National Research Council of Canada, the AAO, and JACH, met together in Tucson to discuss common issues and goals related to telescope proposals for astronomical telescopes. The goals of the Proposal Process Workshop were (1) to develop a shared understanding among observatories, including the National Gemini Offices and the Gemini Project, of the requirements and procedures for telescope proposals, and (2) to encourage cooperation among the national observatories of the partner countries, STScI, and other institutions faced with similar issues and concerns related to the telescope proposal process.

The primary motivation for the workshop arose from NOAO's plan to unify its proposal process which includes proposals and procedures for all its nighttime facilities (KPNO, CTIO, HET, MMT, South Pole, and Gemini) over the next year. We realized that many of the Gemini partner countries would be facing the same issues as NOAO—how to present a unified proposal submission system to its user community for Gemini and their other nationally accessible telescopes. The Kona SPIE meeting sparked an interest in a collaborative effort among other institutions as well.

[1]Operated by the Association of Universities for Research in Astronomy, Inc. (AURA) under cooperative agreement with the National Science Foundation

2. The Meeting Agenda

Broad issues and goals were explored during the first morning of the 2-day meeting, and the first afternoon was used to investigate the content of proposals in order to identify the complete set of information necessary to include in a telescope proposal. On the second morning we discussed the proposal process to identify in some detail the scientific and procedural requirements for the different phases, from preparation through submission, merit review, technical review, and scheduling. On the final afternoon we looked more deeply into specific implementation issues, with the goal of identifying a clear path to providing a set of tools for investigators and observatory staff to manage the proposal process.

3. Discussions

Prior to the workshop, participants were asked to provide URLs to proposal information at their respective institutions. Upon reviewing the posted URLs it was clear that most of the participants favored a LaTeX-based proposal form (not so surprising since LaTeX is the "publishing" language for astronomy!). Most proposal forms asked investigators for similar information although perhaps not in the same way (LaTeX keywords varied from form to form as well as descriptive information). The actual processing of proposals at the individual institutions varied as did the database systems used to track the proposal information.

During the discussions several key items were clearly identified as important issues in the proposal process.

- Some pre-proposal preparation tools should be available to the investigators to help them plan their proposals. Some tools that would be useful during this "Phase 0" are 1) exposure time calculators for the different instruments at each site, 2) a unit translator that would allow investigators to convert unfamiliar units, and 3) image quality estimators for determining the type of sky needed for queue observing.

- Observatories should use as similar proposal forms as possible. The participants felt that the proposal submission process should include an option for Web-based preparation and submission that does not require the distribution of software, similar to the system used at NOAO. The NOAO Web-based proposal form transcribes the information entered into the Web form into the standard NOAO LaTeX template for final submission.

- The proposal form should only ask for information that is truly needed to evaluate the proposal request. The process should be kept as simple as possible for users while allowing them to present their scientific case in a clear and concise manner. Information that is not needed for the initial submission (Phase I) should be left for the Phase II step (target stars, guide stars, etc.).

- Items on the proposal form should be requested in terminology common in the astronomical community. This improves the chances that what is requested is what you will get.

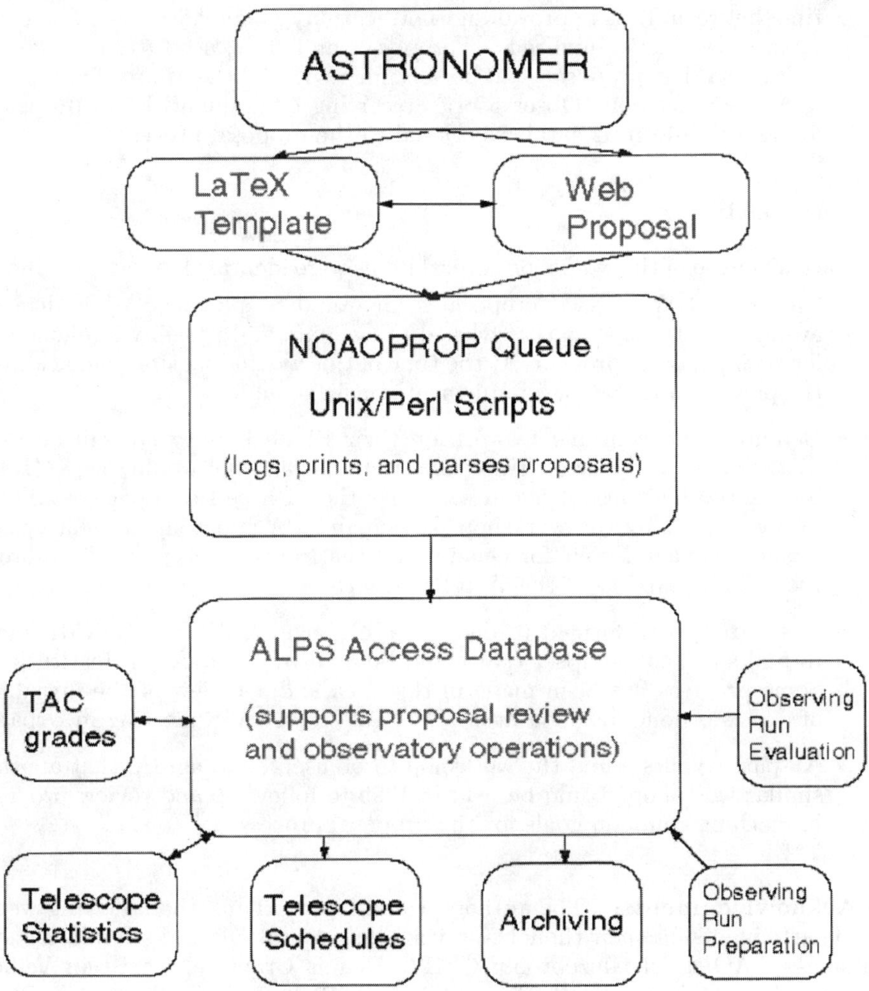

Figure 1. The NOAO Proposal Process.

- Proposals need to appear "professional" for the review process. The Telescope Allocation Committees prefer to review proposals on "paper" and it is important that the formatting is consistent from proposal to proposal. It is not enough to have a Web-based submission process, but the submitted proposals need to print in a consistent and readable format.
- Local procedures for processing need to be fast and efficient to minimize time between TAC approval and notification of time allocated. Some observatories receive hundreds of proposals and these need to be processed quickly so that proposals can be sent for review to the appropriate panels in a timely fashion. Observatories receiving large quantities of proposals found a well-formed database critical to the proposal process.

4. Action Items

At the conclusion of the workshop several items were identified as "action" items.

- The NOAO Web-based proposal form would be packaged up and made available to those sites that wished to evaluate and possibly implement it for their proposal process. At the time of this writing six sites had acquired the package and two had implemented it in some form.
- Gemini would continue to work on their Phase I tool that will be used by the AAO for their next proposal round. This tool would use XML for its markup language. Gemini will make their Phase I tool pages available for evaluation by the workshop participants. A subgroup was formed to review the page layout for Gemini and the NOAO Web pages to improve the overall usability of the Web-based form.
- A subgroup was formed to define LaTeX tags (and possibly XML tags) to find some common set that could be used by the different institutions. Some progress has been made in this area and a review of the initial set of tags was done during a brief session at the ADASS meeting in Urbana.
- All participants found the workshop to be useful and agreed that another similar workshop should be held in 1999 to follow up and review progress in reaching common goals for the proposal process.

Acknowledgments. The authors would like to thank the following workshop participants for their thought-provoking presentations and stimulating discussions: AAO: Karl Glazebrook; CFHT: Dennis Crabtree, Christian Veillet; Gemini: Kim Gillies, Phil Puxley; JAC: Nick Rees; NOAO: Dave Bell, Todd Boroson, Christa Brown, Buell Jannuzi; NRC: Tim Davidge; STScI: Brett Blacker, Glenn Miller; University of Arizona: Craig Foltz; University of Texas: Mark Cornell.

Additional information on the workshop can be found on its Web page at `http://www.noao.edu/scope/tpp_workshop/`.

Astronomical Data Analysis Software and Systems VIII
ASP Conference Series, Vol. 172, 1999
D. M. Mehringer, R. L. Plante, and D. A. Roberts, eds.

A VBA Desktop Database for Proposal Processing at National Optical Astronomy Observatories

Christa L. Brown

National Optical Astronomy Observatories, Tucson, AZ 85719, Email: cbrown@noao.edu

Abstract. National Optical Astronomy Observatories (NOAO) has developed a relational Microsoft Windows desktop database using Microsoft Access and the Microsoft Office programming language, Visual Basic for Applications (VBA). The database is used to track data relating to observing proposals from original receipt through the review process, scheduling, observing, and final statistical reporting. The database has automated proposal processing and distribution of information. It allows NOAO to collect and archive data so as to query and analyze information about our science programs in new ways.

1. Design Goals

ALPS is a custom database and application designed using Microsoft Access as a development tool. ALPS is designed to meet the Observatories' needs in handling proposals for telescope time in an environment accessible to both the administrative and the technical staff. NOAO chose Access as the development platform for a number of reasons. On the surface, Access offers a user-friendly approach for database neophytes. Its user-friendly features include:

- a well organized Database window,
- wizards for constructing database objects,
- a variety of built-in properties to define each object,
- and a simplified macro scripting language.

However, beneath the surface Access provides a powerful environment for the developer.

- Automation facilitates the exchange of information between Microsoft Office suite applications, providing both the developer and the end user with a wide range of tools best suited for various needs.
- The Visual Basic for Applications (VBA) programming language allows the developer to create a robust application and to automate complex tasks.
- The Access relational data model and Structured Query Language (SQL) foundation allow the developer to make uncomplicated representations of complex data.
- The Jet Database Engine exposes programmable Data Access Objects that allow program code direct access to database data and structures.

2. Application Architecture

The term "application architecture" describes the conceptual and physical arrangement of objects and processes. Within application architecture are these object-related subsets: data architecture, code architecture, and interface architecture. Data and interface architecture are discussed below.

2.1. Data Architecture

Most Access solutions are constructed in two pieces: the front-end application database file with form, report, and program objects, and the back-end data file with tables that are linked to the front-end. The front-end portion of the ALPS application resides on the user's workstation. The back-end is stored apart from the front-end in a Jet Database Engine on a file server. This model creates a file server database application, detaching the shared data (located on a shared drive for multiuser support) from the application objects (placed on a local hard drive for better performance). The separation of data from the interface allows for multiuser data access.

Data enter ALPS via two mechanisms, data entry and imports. The standard mechanism for entering records into tables is for users to key data into forms. Data are also pulled into ALPS by code routines that fetch data from external files submitted electronically, such as observing proposals, Telescope Allocation Committee (TAC) grades, and observing logs. ALPS buffers external data's entry into the application by loading them to a temporary table where they are validated before they are posted to transaction tables.

2.2. Interface Architecture

Almost all of a user's interaction with ALPS is through forms. Because forms are the primary device for allowing user interaction with data, they are the most visible and heavily used application element. Forms are the gateway to data records, provide the mechanism for displaying switchboard menus and selection dialogs, and are used to provide programmable class objects. At least 50 percent of the ALPS application's development budget and time has been spent on forms.

ALPS employs a Single Document Interface (SDI) paradigm. ALPS presents the user with one form at a time and a clear path forward and backward between related records and forms. We have not found a compelling need for users to compare different sets of data at the same time on a single screen. In addition, since many of our users are still limited by 15-inch monitors, screen real-estate is limited.

ALPS does not provide the everyday user with access to the Database window. Instead, users navigate from form to form, occupying a very restrictive and well-defined application universe. Our preferred user-interface model directs users through forms one by one, or at least limits them to some small, manageable group of related forms at the same time. The penalties for not following this model include the following:

- Users could enter dependent data out of sequence. For example, they might try to enter child records before parent records.

- Users could open multiple forms that point to the same record, introducing possible record-locking challenges.
- Users could open too many forms at the same time, impacting application performance or losing track of their workflow.
- Users could keep multiple unsaved records open, usually by minimizing forms on the desktop without saving their edits. This situation would prevent other users from seeing current data.
- Users could lock records, usually by minimizing forms that have edited records in an unsaved state (depending on the form's locking model). This scenario would potentially lock other users out of specific records for an unreasonable amount of time.

By enforcing only permitted behavior and events, users are guided through the application in a friendly and logical manner. The ALPS user-interface model includes the following components:

- A main switchboard menu form enables users to navigate to the various features of the application.
- Toolbars and menu options aid navigation, but do not include any built-in Access options that are dangerous to the specific application.
- Forms that open in a single-instance modal state create a simplified environment in which the user performs one task at a time.
- A controlled data mode provides a filtered recordset with the ability to add, edit, browse, and delete selectively within the recordset, as appropriate to the context.
- Keystroke trapping provides form and record navigation. ALPS doesn't assume that users always navigate with a mouse.
- A limited feature set, with features disabled when they are not needed in a specific context, gives users access only to options that they can use effectively to execute the task at hand.
- A clear path to navigate enables users to move forward and backward between application elements. The application establishes and enforces order in the workflow.

3. Switchboard Menus

At the top of the ALPS navigation pyramid is a switchboard menu. The switchboard menu is a single page form with a button for each main functional area in the application. Each menu button contains attached code that opens a form, prints a report, runs a data process, or opens another switchboard.

The ALPS switchboard interface model uses a column of small buttons, each with a text descriptor next to it. The strength of this design style is the capability of conveying varying lengths of explanatory text to the user for each application option without creating wide command buttons.

Compared to table-based menu forms, however, this option is very inflexible. To add a feature to the menu, a developer must add a button and its related code to the form. On the positive side, this switchboard loads and executes faster than a form tied to a menu options table.

Another positive aspect of button-based menus is the capability to enable and disable features selectively on the fly by toggling the Enabled property of each button. For example, the Import Proposals button on the Proposal Submission switchboard is enabled only if a proposal has been submitted and resides in the directory reserved for these files. Thus, a user can see if there are any proposals to import by simply looking at the button status on the Proposal Submission switchboard. This very simple programming technique conveys immediate status information to the users via the menu form. In addition, the technique produces highly visible results for the small amount of development time it takes to create a smart menu. Achieving this result is much more difficult in a menu based on a list box, in which you must change a flag value in the underlying table and requery the menu. Consequently, you must also keep the menu list in the local database so that table changes do not affect other users, which may adversely affect your application maintenance strategy.

Another weakness of the button-based menu is that only a limited number of options can be displayed on this menu. Using a list box instead of a button-based metaphor would enable the list of menu options to be essentially infinite. In addition, a list-box menu could be driven by a table containing its options. Modifying the application's capabilities would become simply a matter of adding or editing the table's records. A list-box menu could also allow a layer of programmatic security to be implemented by only listing options in the list box that are permitted for the current user.

Options in the menu list are sorted based on a process-flow model. The order of menu options follows the orders that tasks are executed. For example, the proposal submission and Telescope Allocation Committee (TAC) preparation processes have a logical flow; steps must be performed in sequence.

4. Conclusion

The ALPS database and application have been in operation at NOAO for two years, during which time they have grown in complexity and functionality. ALPS is presently being modified to allow support for multiple observing sites and for replication among sites to meet NOAO's changing needs for support of observing proposals.

Astronomical Data Analysis Software and Systems VIII
ASP Conference Series, Vol. 172, 1999
D. M. Mehringer, R. L. Plante, and D. A. Roberts, eds.

TransVERSE: An Architecture for Configuring Astronomical Observations

Gary Curtis, Tom Donaldson, Rob Douglas, Andy Gerb, Mark Giuliano, Natalie Nigro

Space Telescope Science Institute

Abstract. Typically, an observation specified by an observer cannot be directly executed by an observatory. Observation specifications need to be configured so that they can be used by other components of the observatory. This paper motivates the need for configuration tools and describes a software architecture which supports configuring astronomical observations.

1. Introduction

Many software systems have been written for scheduling astronomical observations (Chavan 1998; Johnston & Miller 1994; Kleiner 1999). These systems optimize the scheduling of multiple observations. In contrast, comparatively little effort has been placed in planning and optimizing individual observations. This paper describes a system for preparing and configuring individual astronomical observations. The rest of the paper reads as follows: §2 motivates the need for configuring astronomical observations, §3 outlines a software architecture for configuring astronomical observations, and §4 gives a brief summary of the main results.

2. Motivation

Typically, an observation specified by an observer as a phase 1 or 2 program cannot be directly executed by an observatory. Observation specifications need to be configured so that they can be used by other components of the observatory (e.g., schedulers, commanding, archiving). As described below, the configuration process is more than a simple translation. The process adds information and optimizes the input specification.

Many required activities such as calibrations, data dumps, instrument configurations, and slews are not explicitly specified by the observer. The configuration process expands the user specification, creating an observation plan which contains all the necessary activities to execute the observation. Figure 1 gives an example of adding activities to an observation specification. The example shows a specification with three exposures. Each exposure specifies a filter, a target, and the number of copies of the exposure to be made. The observation plan shows how the configuration process adds activities to the specification. In particular, the process adds a filter move between exposures 1 and 2, creates

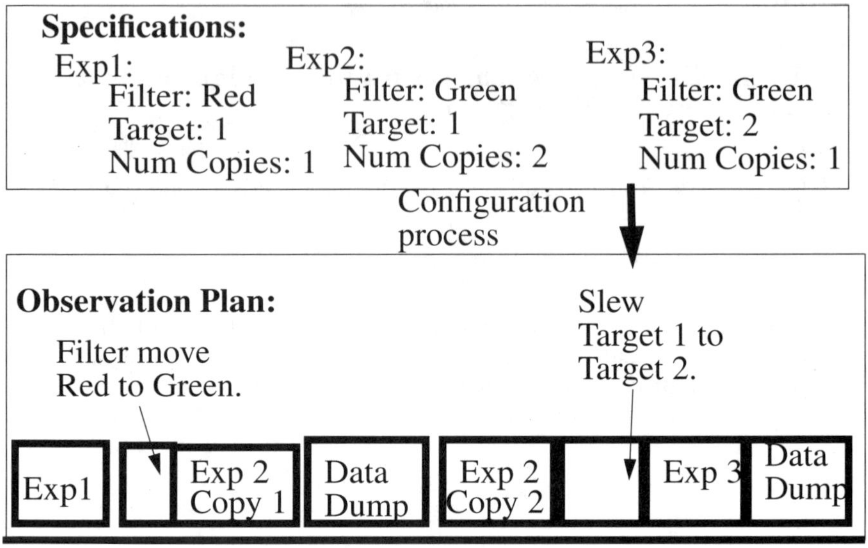

Figure 1. Adding Activities.

two copies of exposure 2, adds a slew between exposures 2 and 3, and adds data dumps.

The configuration system embeds detailed knowledge about how the observatory works so the observer does not need to know these details. As a result, the observer can concentrate on his scientific goals and not the details of how to run an observatory.

In some cases an observation specification only gives ranges of values for input parameters. For example, observations can give ranges of legal times for exposure durations. The choice of actual exposing durations is left up to the observatory and should optimize the use of the telescope. Figure 2 shows a specification with three exposures each with a range of exposure durations. In the example, there are 60 minutes of visibility available for the exposures and the configuration system adjusts the actual exposure durations to fill the orbit.

This approach supports more efficient usage of the observatory. An observation specification defines a set of potential observation plans. The actual observation plan for a specification is selected based on the current conditions available for executing the observation. For low-earth orbiting telescopes, such as the HST, the current conditions include the available orbital visibility. For ground based telescopes, the current conditions include the current weather and seeing conditions.

The configuration process requires the ability to optimize multiple competing objectives. This may require making trade-offs between competing objectives. For example, an observatory might want to:

1. Minimize the duration of observations (i.e., leave more time for other observations).
2. Maximize the time on target (i.e., optimize the current observation).

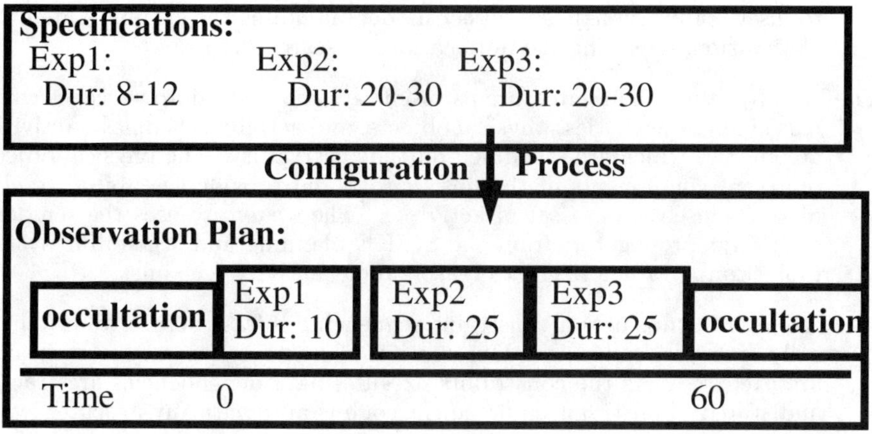

Figure 2. Adjusting Exposure Durations.

3. Minimize overheads (i.e., slews, data dumps, instrument configurations).
4. Place overheads during dead times (i.e., during occultations).

In general, the configuration process requires knowledge of the instrument and ground system in order to optimize multiple competing criteria.

As the use of service mode observing increases, there will be an increased demand for flexible tools for configuring astronomical observations. In classical observing, the investigator actually performs the observations and is free to adjust his plan to meet the current conditions. In service mode observing, the investigator is not available to adjust observations. The investigator must specify any contingencies as part of the specification. A configuration system allows the contingencies to be automatically taken into account.

3. System Architecture

The TRANS system has provided support for configuring HST observations since 1989 (Gerb 1991). The system is being re-engineered under the TransVERSE project. The resulting system will provide a flexible yet powerful tool for configuring astronomical observations.

The TransVERSE architecture supports configuring observations through the following features:

I. The architecture separates input/output functionalities from the process of managing the structure of an observation. Structure management defines the semantics of the system. The input/output modules provide different views of the core semantics. This feature is useful as down-stream systems often require different views of the data.

II. The architecture provides an explicit search model which supports optimizing interacting criteria. Structure management is divided into two components. A search engine generates decisions about the structure (e.g., where to insert calibrations). An object model maintains the current state of an observation, including the implications of search decisions.

III. The object model architecture partitions its knowledge into packages which reason about activities, physical objects, and activity schedules. Activities are entities which are scheduled on the observatory. The physical objects represent components of the observatory and provide the ability to simulate themselves on a set of activities. The system re-uses the temporal constraint propagator, from the SPIKE planning and scheduling system (Johnston & Miller 1994), to reason about activity schedules.

IV. The architecture uses a constraint sequencing infrastructure to automatically track dependencies. On creation of an object, a set of constraints are executed. As the constraints execute, data dependencies are tracked and stored. Constraint enforcement code is automatically executed when a dependency is triggered. This infrastructure frees the developer from having to explicitly sequence the execution of constraints. The constraint sequencer is like a spreadsheet for constraints. A spreadsheet automatically updates the values for cells when dependent cells are modified. Likewise, the constraint sequencing infrastructure automatically executes constraints when dependent values change.

4. Conclusions

The functionality provided by the TRANS system, observation configuration and optimization, has proved invaluable to the efficient and correct use of the Hubble Space Telescope. The job of TRANS is complex as it requires detailed knowledge of the observatory and the ability to optimize competing priorities. The architecture outlined above provides an infrastructure for configuring and optimizing observations for the Hubble Space Telescope and other missions.

References

Chavan, A. M., Giannone, G., Silva, D., Krueger, T., & Miller, G. 1998, in ASP Conf. Ser., Vol. 145, Astronomical Data Analysis Software and Systems VII, ed. R. Albrecht, R. N. Hook, & H. A. Bushouse (San Francisco: ASP), 255

Gerb, A. 1991, Goddard Conference on Space Applications of Artificial Intelligence (NASA Conference Publication 3110), 45, 58

Johnston, M., & Miller, G. 1994, in Intelligent Scheduling, ed. M. Zweben & M. Fox, (San Fransico: Morgan Kaufmann), 391, 422

Kleiner, S. 1999, this volume, 77

Astronomical Data Analysis Software and Systems VIII
ASP Conference Series, Vol. 172, 1999
D. M. Mehringer, R. L. Plante, and D. A. Roberts, eds.

Small, Fast and Reusable: A Satellite Planning and Scheduling System

Steven Kleiner
Smithsonian Astrophysical Observatory, Cambridge MA 02176, Email: skleiner@cfa.harvard.edu

Abstract. The SWAS planning and scheduling system provides all functions from ephemeris processing to guide star selection to timeline generation. It embodies a dynamic scheduling algorithm which allows schedules to be rebuilt or slipped while conserving the scheduling effort already invested. A novel graphical interface provides interactive scheduling. The system uses a component design for efficient reuse. The software is available[1] for use by other missions.

1. Introduction

The Submillimeter Wave Astronomy Satellite is 600 lb NASA Small Explorer satellite launched into a 650 km orbit on 12/5/98. It is a radio telescope which observes the O_2, CI, H_2O, ^{13}CO and $H_2\,^{18}O$ spectral lines in the 500 GHz (0.6 mm) regime from molecular clouds in our Galaxy. These observations will help us to understand the chemistry, energetics and kinematics of star-forming regions.

The spacecraft was built by the NASA Goddard Space Flight Center. The mission is under the scientific direction of the Smithsonian Astrophysical Observatory. The PI is Gary Melnick. The NASA Astrophysical Data Processing program supports the distribution of this package for other missions. At the time of this writing, SWAS has completed its launch and early orbit testing and is taking beautiful data. The planning and scheduling system has worked faultlessly.

Noteworthy features of the SWAS planning and scheduling system include:

- full range of capabilities, speed and relative simplicity;
- efficient planning algorithms;
- dynamic scheduling algorithm;
- novel graphical user interface;
- a component design for convenient reuse;

[1] `http://www.quaternion.harvard.edu`

2. The SWAS Planning and Scheduling System

The planning and scheduling system provides all the functions required to plan, schedule and command SWAS. The system starts with a predictive ephemeris describing the orbit and produces a timeline which is then converted to a binary load and uplinked to the spacecraft. Between these two points the system must compute Keplerian orbital elements, planetary positions, Sun, Moon and Earth limb pointing constraints, target rise and set times, gyro calibration and waypoint targets, line-of-sight velocities for LO commanding, guide stars for the Ball CT601 star tracker, slew paths between targets, mapping offsets and attitude quaternions. The system must then generate commanding for the science instrument and the spacecraft data handling and attitude control subsystems.

The system is very fast. We can generate a week's worth of observation timelines, from ephemeris ingestion to guide star selection and command generation, in about three minutes on a SUN Ultra-10. The code is in C, uses Motif for the widgets, and consists of 18.5K lines of code.

3. Planning and Scheduling Algorithms

We have adopted a geometrical approach to planning calculations. We convert the Cartesian state vectors in the ephemeris file to Keplerian elements which describe the geometry of the orbit. We calculate such things as target rise and set times, shadow entry times, nominal roll positions, Sun avoidance satisfaction, pointing offsets, guide star selection and gyro calibration and waypoint target generation by clever use of coordinate transformations. We have entirely avoided the need to step through an orbit. We consistently use an R_{313} rotation corresponding to the classical Euler angles to do our transformations.

There is a paradox in scheduling. The better the schedule, the more fragile it generally is. A highly efficient schedule, for example, will rarely allow one to swap targets in and out or to slip the schedule for a week without degrading it. But the fact of real life is that schedules must be rebuilt or replanned all the time, and usually on very short notice. Too often a carefully crafted schedule must be thrown out in favor of a jury-rigged replan schedule. One goal of the SWAS planning and scheduling design was to build dynamic and conservative schedules – schedules which could be modified or slipped without degrading their efficiency while still retaining the scheduling effort already invested in them.

We have been able to realize this goal by redirecting our effort from building good schedules to building good order forms. An order form is a specification of the activities to be placed on a schedule, without reference to the time at which the activities are to be scheduled. The order form is a simple abstraction of the schedule, and a schedule is a concrete instance of the order form. This is an almost trivially simple differentiation, but it has powerful practical consequences. First, given a good order form, it is very simple to build a good schedule from that order form. Secondly, order forms are quite tolerant of changes. The schedules are not tolerant of change, but given that it is easy to generate a new schedule from an order form, we simply throw away the old broken schedules and build new good schedules when we change the order form. It works! The rule enforced in the SWAS planning and scheduling system is that the planner

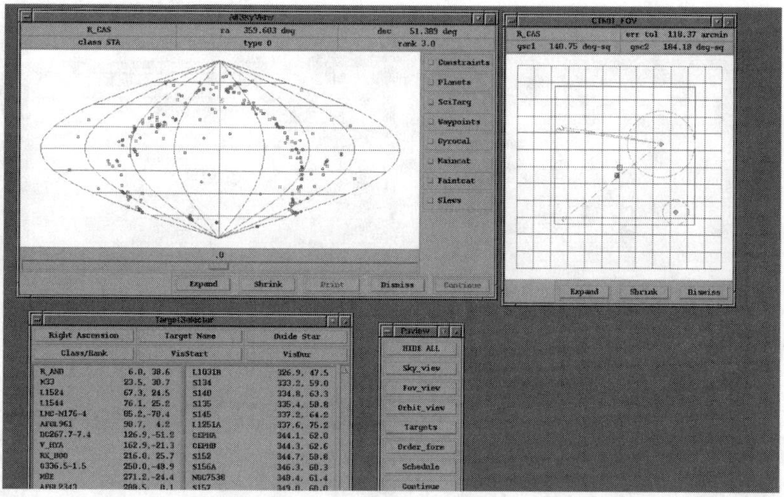

Figure 1. SWAS planning displays such as an all-sky map, the star tracker field of view with guide stars and a target browser which can select targets based on different criteria.

never interacts with the schedule directly. All changes are made to the order form and a new schedule is immediately run off from the order form.

4. Graphical User Interface

Some SWAS planning displays are shown in Fig. 1 The all-sky display shows how the SWAS target catalog is largely confined to the galactic plane. The field of view display to the right shows the sky as seen by the Ball CT601 star tracker and shows the guide stars for a given target. The target browser allows ones to examine targets based on their location in the sky, their visibility, the quality of their guide stars or other criteria.

Figure 2 shows the SWAS scheduling displays. The schedule to the left shows eight days of science observations. The waypoints which provide safe telescope pointings between science observations are not shown. The mixer to the right allows the planner to dynamically adjust the composition of the order form. The schedule display is then updated immediately to show the resulting new schedule.

Clicking on any gap in the schedule display highlights those activities on the order form which could be used to fill the gap. It is then simple to adjust the order form to fill the gap. Conversely, clicking on any scheduled activity highlights that activity on the order form, allowing one to eliminate the activity from the schedule. One can build or modify schedules in a very fluid and intuitive manner.

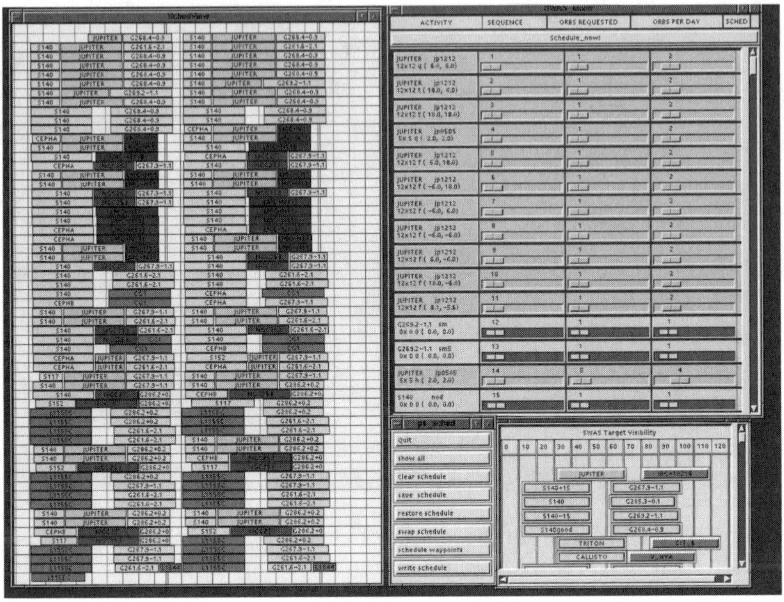

Figure 2. The SWAS scheduling displays showing an eight day schedule, the mixer which controls the composition of the schedule and a visibility display showing target rise and set times.

5. Design for Reuse

The SWAS planning and scheduling system is a set of components which can be assembled in different configurations to do different tasks. The design follows that of the CAMAC crates in nuclear physics which provide a chassis, a power supply and a bus. Different modular components can then be plugged into the crate to provide any needed functionality. We have one single, small, well-defined global PSTask data structure which acts as a bus and allows different components to communicate with each other. The data structure also carries strobe signals which synchronize the activities of the different components. We build a tools such as the SWAS scheduler or the SWAS timeline generator by selecting different components and plugging them into the core data structure. Components include a Target object which manages the SWAS target list and waypoint and gyrocal and planetary targets, an Orbit object which can either read an existing ephemeris or generate a model orbit, an ACS object which embodies the state of the spacecraft Attitude Control System and a Receiver object which manages the science instrument.

A web site, `http://www.quaternion.harvard.edu`, is under construction to describe the SWAS planning and scheduling code in greater detail and to make it available for use by other teams.

Astronomical Data Analysis Software and Systems VIII
ASP Conference Series, Vol. 172, 1999
D. M. Mehringer, R. L. Plante, and D. A. Roberts, eds.

Recent Improvements to HST Parallel Scheduling

Ronald Henry, Mike Butschky

Space Telescope Science Institute, 3700 San Martin Drive, Baltimore, MD 21218

Abstract. The Hubble Space Telescope (HST) has several scientific instruments (SIs) that may be used at any given time. Most primary visits submitted by HST observers only use one SI, leaving the other SIs free to be requested by "pure parallel" observing programs. In order to accomplish this, separate scheduling units (SUs) for each parallel SI must be created and then scheduled by the Science Planning and Scheduling System (SPSS), taking into account numerous orbital and scientific constraints. The Parallel Observation Matching System (POMS) has the task of matching parallel visits to primary observations and "crafting" appropriate parallel SUs at each opportunity, taking scientific criteria and orbital constraints into account.

The process for planning and scheduling parallel observations is thus quite different from the process for primary science. In the past, custom crafting rules for each parallel program were necessary, requiring full-time support from a software developer. In addition, because POMS ran as a standalone system, its ability to model how long parallel SUs would take was limited, especially with the flexible buffer-management schemes used for the second-generation SIs.

A new version of POMS was developed in 1997. This version uses a formal proposal syntax (the same used for primary observations) for parallels, so that different proposals can be handled uniformly and without the need for customized "crafting rules." In addition, POMS is integrated with the Transformation (TRANS) planning system in order to give it full knowledge of overheads within an SU, eliminating the need for ad hoc modeling. The power and versatility of this approach has paid off in improved utilization of parallel opportunities, greatly reduced maintenance costs, and an ability to gracefully handle new parallel proposals and new SIs with minimal software effort.

This paper discusses the requirements, design, and operational results of the new POMS.

1. POMS Requirements

POMS has two main sources of input: a description of available parallel proposals, and a description of the primary calendar (this description is generated **after** primary observations are scheduled). Parallel proposals are ranked by priority and structured into visits, which are ranked by priority within each proposal. The task of POMS is to match parallel visits to appropriate intervals on the calendar.

The first step in this process is to structure the calendar into a set of **slots** where parallel alignments (uninterrupted pointings) can go. Oversimplifying somewhat, a slot for a given parallel SI is an interval on the calendar during which there is target visibility, no use of that SI by primary alignments, no spacecraft maneuvers or SAA passages, no loss of pointing control, and no primary alignments marked as non-parallel-able.

For each parallel visit, POMS then structures the slots into a collection of **parallel fields**. Whereas a slot is a place where a parallel alignment can go, a field is a place where an entire parallel SU can go. An SU consists of a set of alignments, but there may be gaps between the alignments (to allow for occultations, for instance), so a parallel SU can extend over many orbits. A field is a set of adjacent slots that (a) all fit within the same primary SU (no major slews within a field), (b) are all scientifically compatible with the parallel visit under consideration (the primary pointing must satisfy any pointing constraints of the parallel), and (c) do not differ in pointing by more than the tolerance specified in the parallel proposal. In order to allow for internal alignments (to do "clean-up" commanding and dump the SI buffers) at the end of a parallel SU, a field is allowed to extend past the end of the last slot in it, until an event is encountered that violates one of the above constraints.

Field sizes provide an upper bound on the size of SUs that can be crafted in them. (In POMS, unlike in primary HST scheduling, there is a difference between the term **visit** and **SU:** a visit refers to the set of exposures submitted by the observer, while the SU is the entity actually scheduled from that visit. Individual visits can be instantiated into parallel SUs many times, whenever POMS finds a suitable opportunity.) Depending on the calendar structure, there will be a distribution of fields of various sizes. The longest opportunities (fields) are considered most valuable scientifically. Normally, parallel proposers are advised to provide a range of visits of various sizes, with the longer ones given higher priority so that POMS will consider them first. If it proves infeasible to craft an SU for a long visit in a field, a shorter visit may be tried in the same place later. If an SU only fills part of a field, the field may be reused for the same visit, as well as for lower-priority visits. Thus the visit sizes act as a sieve, which produces smaller and smaller holes as the shorter visits are encountered. For each SI, at least one low-priority "sponge" visit is provided to fill all the remaining fields that are large enough to allow science that might be conceivably useful.

In order to determine whether crafting is feasible for a particular visit in a field, POMS must compare the field size to the duration of the SU the visit would produce. The duration of an SU includes overheads such as filter wheel motions, cleanup commanding, and buffer dumps. In addition, exposure durations are adjustable within limits provided by the observer. Therefore POMS faces an **overhead inversion** problem. For efficient crafting, it must determine the maximum exposure durations which, after overheads are added, would produce an SU that comes as close as possible to filling the field without exceeding it. However, POMS itself has no knowledge of SI overheads.

This problem was one of the main reasons POMS was redesigned for greater integration with TRANS, the planning subsystem that, among other things, computes SI overheads and also has an algorithm for inverting them to optimize primary observations. The old system attempted to "model" what TRANS would do, an approach that became untenable with the complexities of buffer management that were added to TRANS to support NICMOS and STIS. Instead, POMS now calls TRANS directly each time it attempts crafting. TRANS returns either the longest suitable visit that may be crafted in the field (taking advantage of its ability to adjust exposure durations) or an indication of failure, meaning that even with the minimum exposure durations, the field was still too short for the visit.

One major problem with the crafting approach is performance. Crafting attempts to take advantage of late information – the structure of the primary calendar – in order to optimize efficiency for parallel SUs. The disadvantage is that this creates a real-time process. Adding parallels is only one small step in the larger process of building a calendar for a week's worth of HST science, and normally only a few hours are allotted for it. TRANS was designed as a batch process that does planning for an entire visit, and had never been run in a real-time environment before. Calls to TRANS are computationally expensive. In the POMS environment, a typical TRANS call takes anywhere from 10 seconds to several minutes, depending on the size and complexity of the visit. To cut down on the number of futile TRANS calls, POMS uses a heuristic which identifies cases where crafting is bound to fail and gives up immediately, but there are still several hundred TRANS calls in the course of crafting for a typical calendar. POMS provides an **escape hatch** whereby crafting can be terminated after a specified time limit, even if not yet complete. Performance considerations have limited the number of visits, especially long visits, that can be specified per parallel SI. There are plans to address this problem by partitioning the calendar and distributing the load among many machines.

2. POMS Software Design

POMS is written in Common Lisp using the Common Lisp Object System (CLOS). The main classes in POMS are as follows:

Timeline: the calendar after primary observations are scheduled.

Parallel-Proposal-Set: The set of all parallel proposals POMS is attempting to craft.

Parallel-Proposal: A single parallel proposal in the Parallel-Proposal-Set.

Parallel-Visit: A single parallel visit in a Parallel-Proposal.

SI-Slot-Set: Set of slots computed by POMS for a single parallel SI.

Field-Set: Set of fields computed by POMS for a single parallel visit.

Trans: Interface to the TRANS system.

Crafted-Parallel-Visit-Set: Set of POMS crafted parallel visits.

Crafted-Parallel-Visit: A single POMS crafted parallel visit

The algorithm POMS uses to craft parallels is as follows:

1. POMS-Main creates and initializes the Timeline using the Calendar File.
2. POMS-Main creates and initializes the Parallel-Proposal-Set. Parallel-Proposal-Set creates and initializes all the Parallel-Proposals. Parallel-Proposal creates and initializes Parallel-Visits for each visit in the Parallel-Proposal.
3. POMS-Main creates an empty Crafted-Parallel-Visit-Set.
4. POMS-Main calls Parallel-Proposal-Set's Schedule method to schedule all Parallel-Proposals on the Timeline. Parallel-Proposal-Set calls the Get-Slot-Set method to get the SI-Slot-Set for each parallel-able instrument. Parallel-Proposal-Set calls each Parallel-Proposal's Schedule method in priority order.
5. Parallel-Proposal calls each Parallel-Visit's Schedule method in priority order. Parallel-Visit calls the SI-Slot-Set's Get-Fields method to get the Field-Set from the SI-Slot-Set for this Parallel-Visit's instrument. Parallel-Visit determines if each Field in the Field-Set looks valid for crafting, and if it is, calls Trans's Run method for further verification and crafting.
6. POMS-Main calls Crafted-Parallel-Visit's Write-Report to write the crafting report.

3. Operational Results

The redesigned POMS was installed for operational use at the start of 1998. Prior to this installation, old POMS had been used only for WFPC2 parallels and a set of operational tools based on static (rather than crafted) parallel SUs had been used for NICMOS and STIS. An analysis of parallel performance based on the first four months of operations showed an improvement in total parallel time of about 16%, attributable both to improvements in POMS and to operational improvements during this period. Working under very tight real-time constraints, POMS has crafted about 100 parallel SUs per week, including a substantial number of long SUs (three orbits or more), and is averaging about 48 hours of parallel alignment time per week over the three parallel-able SIs. (Analysis shows that this is still only about 43% of the total potential for parallel time, due to various inefficiencies in the system.) After a few startup glitches, operational performance of POMS has been reliable. An average of 5-10% of parallel SUs produced by POMS have failed to schedule, for reasons that are complex and rooted in the imperfect model of the scheduling system used by POMS and TRANS. Holding the run-time down has been a persistent challenge for the operators.

The main objective of the redesign, however, was not to improve performance but to support the new SIs and reduce maintenance costs. By these measures POMS has been an outstanding success. The integration with TRANS has enabled POMS to incorporate the complex overhead models with the newer SIs, and the formal proposal syntax has greatly reduced the effort needed to support new proposals. POMS no longer requires the support of a full-time developer. Nearly all SI-specific code has been removed from POMS, so that few if any changes will be necessary to support future SIs.

Part 4. Telescope and Instrument Control Systems

Astronomical Data Analysis Software and Systems VIII
ASP Conference Series, Vol. 172, 1999
D. M. Mehringer, R. L. Plante, and D. A. Roberts, eds.

A Radical Approach to Software Modularity in Telescope Control Software

M. H. Clark
National Radio Astronomy Observatory, Green Bank, WV 24944-0002, USA

Abstract. A difficulty in writing control software for research telescopes is designing sufficient flexibility to handle continually changing requirements. The primary goal of the control software architecture for NRAO's new 100-meter radio telescope in Green Bank, West Virginia is flexibility through the construction of a radically modular software system.

1. Introduction

The original motivation for the design used in the monitor and control system for the Green Bank Telescope (GBT) (Lockman 1998) was to generate a radically modular "straw man" in order to better understand the limits of modularity[1] when implementing a control system for a general purpose telescope such as the GBT. However, to our surprise, repeated scenarios did not break the model, so it was decided to adopt the design for the GBT. There were two other influences affecting the choice of design. We were given the requirement to think in terms of a virtual telescope, i.e., to design a system that was not telescope specific. This was important not so much because there were plans to use the system on additional telescopes, but rather the fact that a telescope does, in fact, become a different telescope over time because of changes in instrumentation. The second influence was that, at the time, I was first becoming impressed with object-oriented methods, and my inclination was to design everything as objects. No regrets or apologies.

The basic design[2] (Clark 1998) is simple. A radio telescope is defined as a laboratory rather than an instrument, i.e., it is a set of devices (of which the antenna is only one) which needs to be configured in novel ways to accomplish observations (or scans). Each device is an autonomous subsystem which can be fully configured prior to a scan through a finite set of control parameters in order to run the scan in coordination with other devices. For purposes of configuration and observing, each device requires no more than four interfaces (control, monitor, message/alarm, and data) which are identical for all devices. The same control interface for user programs is also used to recursively build a tree of control modules whose root is a generic "scan coordinator". Any

[1] See Meyer (1988) for a definition of software modularity.

[2] `http://info.gb.nrao.edu/~mclark/article/article.html`

control module of the tree may be used to initiate a scan for the sub-tree under its control because each control module is derived from the same base class (called a Manager) and therefore inherits all of the control protocols required to coordinate and initiate a scan.

2. Advantages

In the GBT, a device is implemented using two sources of code: generic code from base classes which is used across all devices, and code written specifically for the device either by writing virtual methods inherited from a base class or by implementing new classes. Because the four device interfaces are identical for all devices, a larger proportion of the functionality can be designed into the base classes, and even much of the device-specific code is implemented as virtual methods. This approach allows many features to be handled wholly in the base classes.

An example of the use of base classes is the integration of the message/alarm system into the control system. A Manager's scan parameters can generate two possible errors: one is the result of user input (illegal value) and the other is the result of failures in digital interfaces (activate fault). Since some Managers may require a large numbers of scan parameters in order to define the operation of the device, specifying individual messages for illegal values and activate failures can be a non-trivial effort. In addition, there is the possibility of inconsistency of implementation across devices or even in the wording of the messages. It would be an advantage if these messages could be wholly implemented in the base Manager. Though the methods to check and activate scan parameter values are written for specific parameters in the derived Manager, these methods are called in the base class. The base Manager class uses the return value of the scan parameter virtual methods to generate and report the appropriate messages for all parameters.

Another example of integration of the control and message/alarm systems is the generation of a summary status for each device. Each message is assigned a severity level (i.e., information, notice, warning, error, fault, or fatal). The base Manager class computes a status based on its most severe message currently active. The control hierarchy of Managers that make up the control system recursively passes these summary statuses up the hierarchy so that the status of each device and subsystem reflects the most severe level of all of its messages. From this it becomes trivial to create a display for indicating statuses for the entire system.

Each individual monitor value and scan parameter is referenced by an object of the class DataDescriptor which contains a complete description of the value including name, explanation, data type, size, units, and array count (if a vector). These DataDescriptors allow more functionality to be handled by the base classes than otherwise would be possible. For example, the user programs do not have to be rewritten (or even recompiled) to handle new devices and their associated scan parameters and monitor values since the reading, writing, communicating, and displaying of their values depends on code based on the DataDescriptors, not on compiled data types. Also, the displays can be embellished with explanations and units since the information is readily available. In fact, the DataDescriptor is complete enough that headers for FITS binary tables (Cotton, Tody, & Pence 1995) can be generated automatically. This allows the engineers to select any monitor point, whose values are sent to a logging program which generates files in FITS format. Likewise, the observer can select any monitor point's values to be stored as part of the observational data along with traditionally sampled values such as antenna positions and weather.

3. Pitfalls Avoided

The project's highly modular approach has allowed us to avoid many problems which are traditionally encountered while developing a telescope control system.

The GBT control system is a distributed system. Rarely anymore does one find a large telescope control system where all devices are controlled and monitored from one computer. Even if one tries to avoid building a distributed system, the continuing introduction of new devices often with their own control computers will create a range of computers that need to communicate with each other. If one designs a control system so that each device acts as an autonomous unit, then one can remove the need for real-time dependencies between computers. Because each device in the GBT is almost wholly independent of other devices, the GBT monitor and control system is able to use "vanilla" ONC RPCs to handle those communications which are needed. So, though it is a distributed system, the development effort is greatly reduced.

As in any system, the devil is in the details, and we have expended a significant amount of effort in debugging code as devices undergo increasing amounts of use. However, integration per se, i.e., problems resulting from adding devices to the system as they come on line, has been almost painless. The interfaces between devices are so few and small that system integration has simply not been an issue.

Related to system integration is device testing and maintenance. In early reviews of the requirements and design for the system, a recurrent request from engineers was the ability to run and test a specific piece of equipment while the rest of the telescope was being used by other engineers, was shut down, or was being used for observing. Specifically, they wanted to know what they had to do to get the device to work, e.g., how to run a backend with no antenna which traditionally initiates the "go" signal and generates antenna positions. Because of the autonomy of each device, the answer is "nothing."

Some software engineers like to stress that they are "real-time" programmers, almost as if it were a badge of honor. And it is certainly true that the constraints of deadlines in code adds an entirely new dimension to every aspect of a problem. In large real-time systems the problems become very complicated very quickly. Many large real-time systems are tours de force and the brain children of some extremely talented people. However, by defining a telescope control system not as a real-time system, but as a computer system that contains small pieces that are real-time, the problem becomes greatly simplified. Because each device is an autonomous unit, one can design devices so that real-time loops are closed within the same processor. And perhaps most importantly, because real-time dependencies are local to a device, the implementation of these real-time programs can be handled by a single engineer.

One of the most gratifying aspects of dealing with a system where modularity has allowed a single set of interface to be developed for all devices on the telescope, is the ease of integrating this interface into interpreters which are used both as a user-interface and a tool to generate graphical user interfaces. We currently have two interfaces for the system: Glish (Schiebel & Paxson 1998) and Tcl/Tk (Ousterhout 1994). The original implementation of a Glish client for the antenna was a two day effort, and the current client which works for all devices took three weeks. The enhanced Tcl interpreter for communicating with all telescope devices took less than a month. As a device comes on line, we expend no effort toward interfacing the new device to our interpreters, though the appearance of the graphical user interface per se for a specific device does take significant thought and planning.

Acknowledgments. My original concepts (Clark 1992)[3] underwent extensive change and enhancement because of the experiences gained and the close-knit team effort during development. I especially want to thank J. Richard Fisher and Geoff Croes, without whose confidence and invaluable insights, the project would have fallen far short of its goals.

The National Radio Astronomy Observatory is a facility of the National Science Foundation operated under cooperative agreement by Associated Universities, Inc.

References

Clark, M. H. 1992, GBT M&C Requirements Analysis Model (GBT Memo Ser. 88), (Green Bank: NRAO)

——, 1998, in SPIE Proc., Vol. 3351, Telescope Control Systems III, ed. H. Lewis (Bellingham: SPIE), 287

Cotton, W. D., Tody, D. & Pence, W. D. 1995, A&AS, 113, 159

Lockman, F. J. 1998, in SPIE Proc., Vol. 3357, Advanced Technology MMW, Radio, and Terahertz Telescopes, ed. T. G. Phillips, 656

Meyer, B. 1988, Object-oriented Software Construction, (New York: Prentice Hall)

Ousterhout, J. K. 1994, Tcl and the Tk Toolkit, (Reading: Addison-Wesley)

Schiebel, D. & Paxson, V. 1998, "The Glish 2.7 User Manual", National Astronomy Radio Observatory, `http://aips2.nrao.edu/aips++/docs/reference/Glish/Glish.html`

[3] `http://info.gb.nrao.edu/GBT/memos/GBTmemos.html#1992`

Astronomical Data Analysis Software and Systems VIII
ASP Conference Series, Vol. 172, 1999
D. M. Mehringer, R. L. Plante, and D. A. Roberts, eds.

The Green Bank Telescope Laser Metrology Computer Control System

Ramón E. Creager

National Radio Astronomy Observatory, Green Bank, WV 24944-0002, USA

Abstract. To use the 100-meter Green Bank Telescope at millimeter wavelengths, the antenna surface must be continuously adjusted to compensate for environmental effects. The metrology system, comprising a network of infrared laser rangers, has been developed to measure the position of the surface to an accuracy of 50 microns. The metrology computer control system is described here. The embedded systems which point the 18 lasers and take range measurements are based on Intel x86 computers. The central control of these computers is done using a Pentium PC running Windows NT. Engineering displays of the data are produced by piping the data in real-time to Microsoft Excel.

1. Introduction

The GBT is a fully steerable, offset paraboloid radio telescope currently under construction at the National Radio Astronomy Observatory's Green Bank, WV site[1]. It has a main reflector that is 100 by 110 meters in aperture, with a 60 meter focal length. As this massive telescope is moved in elevation, the main dish backup structure will deform. To bring the main reflector surface back to an acceptable parabolic shape, the GBT will feature an active surface, which consists of 2209 panels (each with a surface accuracy of 100 μm RMS) joined at the corners by actuators designed to provide approximately $\pm$ 1 inch of travel. In addition, the massive feedarm assembly will bend, causing the receivers to move relative to the telescope's prime focus. To compensate for this, and other factors such as thermal expansion of the structure, the GBT will have an ellipsoidal secondary reflector that can be driven to maintain the Gregorian focus on the desired receiver.

The telescope will feature a laser metrology system (Payne, Parker, & Bradley 1992) that will be used to quantify these deformations, and also to locate cardinal points on the structure to enable high precision pointing of the telescope. The laser metrology system consists of 12 individual laser rangefinder units located on the ground in a 240 meter diameter circle centered on the telescope's pintle bearing, and six more mounted on the telescope's feed arm. The ground based rangefinders will measure the cardinal points on the telescope, allowing the position of these points to be fixed with great accuracy. The feed

[1] `http://www.gb.nrao.edu`

arm based rangers will measure the position of 2209 corner cube retro-reflectors mounted on each panel near the actuators.

2. Metrology Computer System

The metrology system is controlled by several computers arranged as a distributed system. The basic unit of the metrology system is the laser rangefinder itself, known as the ZY unit (Parker 1991). This unit is mounted at a known location either on the ground or on the telescope and can measure ranges to pre-determined targets. These units are controlled over an ethernet connection (using TCP) by a master process known as the ZIY, whose job it is to unify the various rangefinders to provide a complete system. This system in turn provides ONC RPC interfaces to the rest of the telescope system.

Early in the development of the metrology system, the decision was made to use commercial off-the-shelf (COTS) components whenever possible, as overall cost and development time were important considerations. Because the system did not have any real-time requirements, and because the data throughput for this system would be quite low, x86 ISA bus components were selected to control the hardware and DOS and Windows NT were used to run the software.

2.1. Laser Ranger Computer Hardware

The laser ranger system's mission is to measure a range from a reference point located on the instrument to the selected target retroreflector. The laser itself is a modulated semiconductor laser whose beam is placed in the instrument's optical axis by a system of mirrors and prisms. A 2 axis mirror then redirects the beam to the chosen retroreflector target. The return beam is redirected to the detector optics by the same mirror. The detected return signal is demodulated and mixed, and the resulting signal is then sampled by a 100 kHz 16 bit A/D converter. The ZY software then converts these data into raw phase and amplitude, and from there these data are further refined into a range measurement.

The following components accomplish this task:

- A 2 axis motion control board from Octagon Systems is used to control the 2 axis mirror that the ZY uses to aim the laser at targets. This controller is based on the National Semiconductor LM628 Precision Motion Controller.
- A Quatech DAQ-16 100 kHz, 16-bit A/D converter is used to generate reference signals and to digitize the detected and converted phase signal.
- A Bancomm bc630AT Real Time Clock, connected to an IRIG-B time signal, provides accurate time stamps and allows the scheduling of operations.
- A 24-bit digital I/O adapter is used to read switch positions and operate the RF Oscillator and servo amplifier power supplies.
- A 130 MHz AMD 5x86 based Single Board Computer (SBC) from Industrial Computer Source provides ample processing power and control over the peripherals.

All of these components are mounted in an industrial ISA card cage and enclosed in a weather proof enclosure.

3. ZY Software

When the decision was made to use x86 hardware for the ZY prototypes, no real-time requirements were identified for the software and therefore DOS was initially used for the ZY system's OS. Initially, this decision was driven by the desire to keep costs down. At the time, the ZY system was experimental, its ultimate success very much an open question. As the system evolved from an experimental, proof-of-concept system to near production prototype status, DOS has continued to serve well.

To avoid losing hardware events, all hardware devices are supported by interrupt service routines. Other time critical tasks are also performed in interrupt service routines, and assembly language is used in areas where high performance was critical.

This x86/DOS system has proven to have certain advantages:

- Excellent, inexpensive software development tools were readily available.
- The ZY program code has proven remarkably immune to processor/SBC upgrades. The ZY software has run on processors ranging from a 4.77 MHz 8086 all the way to an 133 MHz AMD 5x86 based SBC with only minor changes to the sources.
- The GBT Metrology Group relies a great deal on temporary employees, such as co-op and summer students, to meet some of its manpower needs. The simplicity of DOS systems allows these team members to get up to speed quickly.

4. ZIY Software

The ZIY software currently consists of two main components. The main process, known simply as the ZIY, acts as the brains of the metrology system. It is in charge of orchestrating the actions of all of the ZYs that it connects to, and it provides the ONC RPC[2] interface to the metrology system through which any other GBT process must access this system. The second component, the ZIY Client, is an engineering client to the ZIY process. It is used for system tests and to conduct the pre-commissioning measurement program (Hall et al. 1998). The ZIY Client communicates with the ZIY through the interface provided by the ZIY. (This ZIY client will eventually be replaced by a GBT M&C client.)

Microsoft Windows NT was chosen to run both of these programs. In addition to being readily available and relatively inexpensive, NT has qualities that are well suited for both of these very different programs. For the ZIY, NT offers the following services:

- A capable threading API
- 32-bit flat address space
- Native TCP/IP support
- Service API
- Remote administration of service processes

[2] `http://set.gmd.de/~mfg/oncrpc.html`

In addition, NT provides the following services useful in the implementation of the ZIY client:

- A rich GUI API
- DCOM (Distributed Component Object Model) support

This last item has proven most useful for the Metrology Group. It has allowed us to automate data gathering and processing, using widely available and understood data analysis tools such as Microsoft Excel. Using DCOM, the ZIY client can automatically start Excel, and pipe data directly to the spreadsheet as they become available. The Excel spreadsheet itself is programmed (in Visual Basic for Applications) to place the data in appropriate worksheets and cells, and display real-time stripcharts of the data as they are received. Once all of the data have been received by the spreadsheet, our co-op student operators can run pre-written Excel VBA routines to further process the data and make it ready for initial presentation and analysis.

5. Conclusion

As the Green Bank metrology system shows, the use of inexpensive embedded PC components for computer control is a viable alternative to the more traditional solutions (VMEbus, StdBus), provided that the requirements are not extraordinary. The wide availability of ISA, PC-104 and CompactPCI components means that there is a large array of COTS components to choose from, allowing the embedded systems engineer to concentrate on the problem at hand rather than designing and constructing custom components. This abundance in hardware components is now complemented by the availability of solid, capable operating systems and tools, including commercial and freely available RTOSes for these systems. The result of this is the ability to create a computer control system quickly and inexpensively.

Acknowledgments. I would like to thank John M. Payne, who launched the Metrology project and boosted it through its early stages with his contagious energy and enthusiasm. I would also like to thank David H. Parker, whose wisdom and patience has helped guide this project from the experimental stages to the more difficult production stage.

References

Hall, R., Goldman, M. A., Parker, D. H., & Payne, J. M. 1998, in SPIE Proc., Vol. 3357, Advanced Technology MMW, Radio, and Terahertz Telescopes, ed. T. G. Phillips, (Bellingham: SPIE), 265

Parker, D. H. 1991, GBT Instrument Identification, (GBT Archive M0071), (Green Bank: NRAO)

Payne, J. M., Parker, D. H., & Bradley, R. 1992, A Rangefinder With Fast Multiple Range Capability, (GBT Memo Ser. 73), (Green Bank: NRAO)

Astronomical Data Analysis Software and Systems VIII
ASP Conference Series, Vol. 172, 1999
D. M. Mehringer, R. L. Plante, and D. A. Roberts, eds.

Remote Observing With Java

Hemant Shukla, Stephen Scott

Owens Valley Radio Observatory, PO Box 968, Big Pine, CA 93513

Simon Weaver

Principia College, Elsah, IL 62028

Abstract. We demonstrate the next generation of monitor and control software for Caltech's Millimeter Array at the Owens Valley Radio Observatory. The software is based on a client/server architecture and provides remote access for controlling, monitoring and troubleshooting the array. The client is written in Java and thus provides a Web based interface that is supported on many machine architectures and operating systems. The server software is developed in C++ and runs on a UNIX host machine. Data from the microcomputers embedded in the array are transported to the UNIX host shared memory area where they are then available to the server programs. Security algorithms are implemented to authenticate remote access for control of the array. The design of the software makes it portable to other observatories with little modification.

1. Introduction

This computer demonstration consists of remote observing with the Caltech Millimeter Array located at Owens Valley in California to show the feasibility and features of this software package. The demo with realtime status of the array is available on the web[1]. Many of the details of the implementation have been presented in the paper submitted in ADASS VII (Scott & Finch 1998) and this is complementary to that paper. However, in the last year there have been various additions to the program that are further discussed.

Any remote access request to control the array has to be authenticated at the server end. The authentication is made against the UNIX username/password pair in the OVRO server, thereby requiring the user to have a UNIX account on the server beforehand. To avoid password sniffing over the net, the username/password pair is encrypted. Only the monitoring of the array (browsing) is allowed for the non-authenticated users. The modular nature of the authentication function makes it easier to implement other restriction/permission rules based upon machine address or domain name if desired.

[1] `http:/www.ovro.caltech.edu/cma`

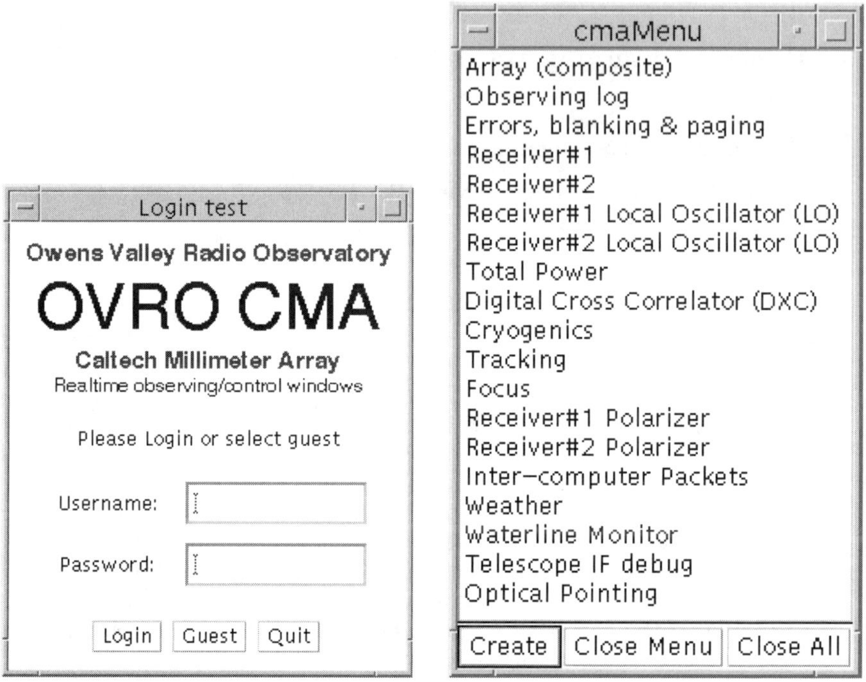

Figure 1. The Login window required for remote authentication (*left*). Unauthorized users may click on the Guest button and provide courtesy user name for only monitoring mode. Menu window (*right*).

The flexibility of the basic realtime window has been increased by adding a pulldown menu that allows other windows to be launched separately or overlay the existing window. Space has been conserved by eliminating the buttons controlling plotting and printing and migrating the functionality to pulldown menus.

The package has proved to be very portable if realtime data is available in shared memory. The observatory specific features have been segregated to allow for easy customization.

Figures 1 through 3 show some sample windows running on a UNIX based machine.

References

Scott, S. & Finch, R. 1998, in ASP Conf. Ser., Vol. 145, Astronomical Data Analysis Software and Systems VII, ed. R. Albrecht, R. N. Hook, & H. A. Bushouse (San Francisco: ASP), 49

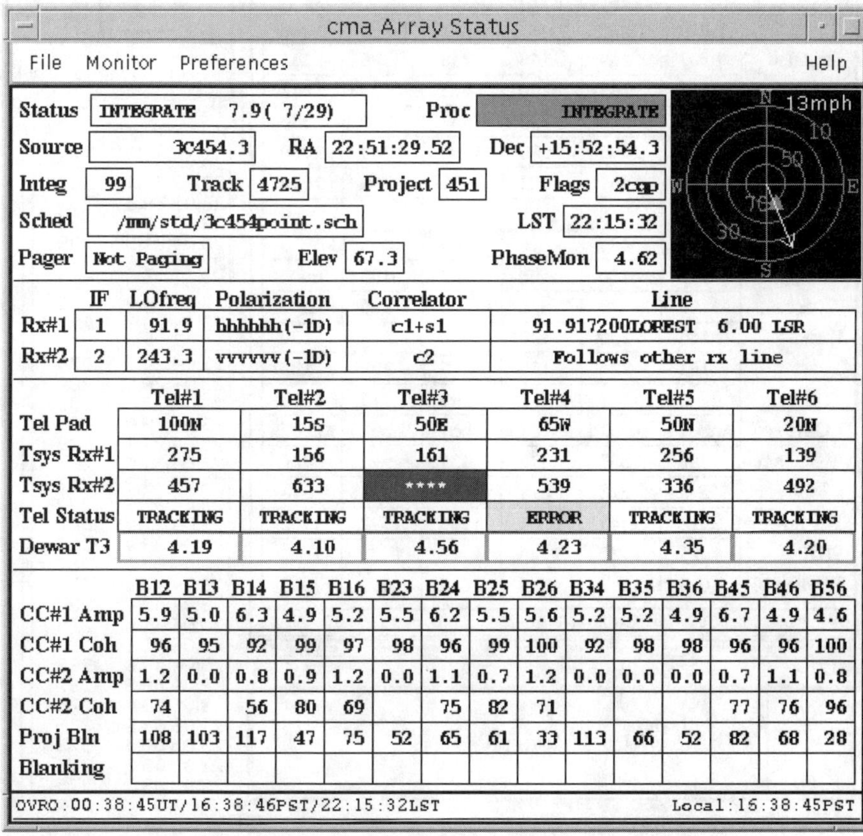

Figure 2. Array Status window.

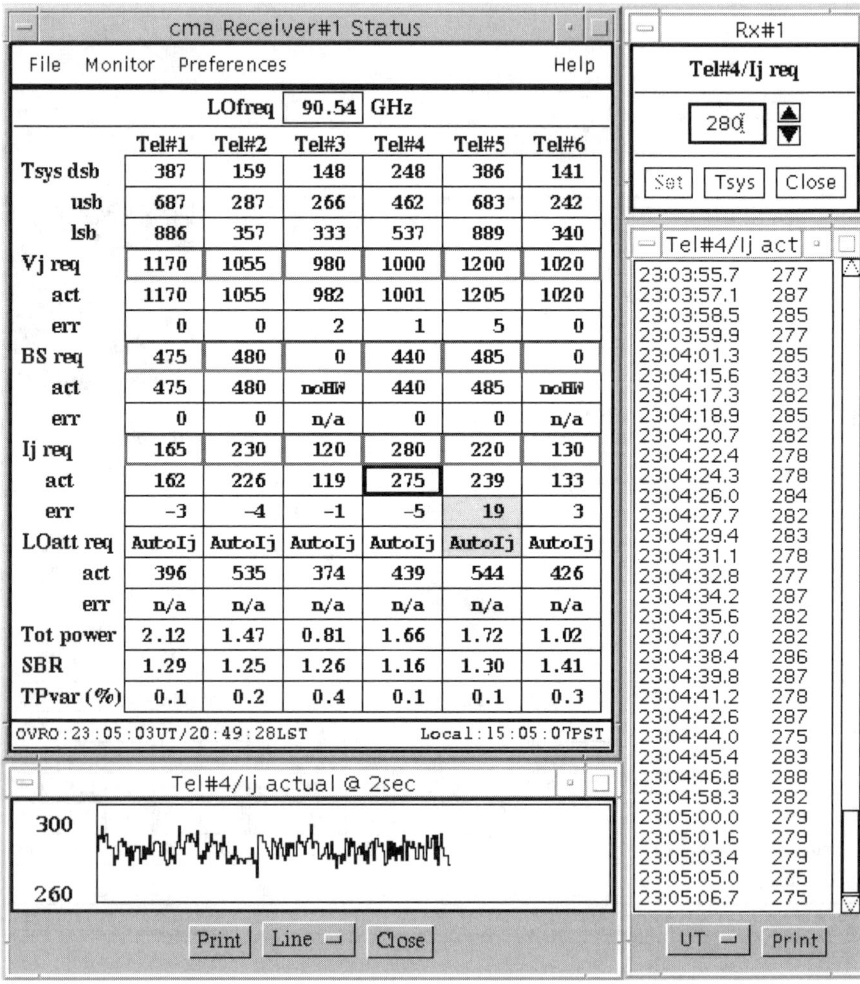

Figure 3. Clockwise from upper left — receiver window, control widget for changing the values, list of the cell values, and the plot of the values.

Astronomical Data Analysis Software and Systems VIII
ASP Conference Series, Vol. 172, 1999
D. M. Mehringer, R. L. Plante, and D. A. Roberts, eds.

Real-Time Linux Driving a Spectrometer

Peter Teuben, Andrew Harris, Kate Isaak[1], James Morgan[2]

Astronomy Department, University of Maryland, College Park, MD

Jonas Zmuidzinas

California Institute of Technology, Pasadena, CA

Abstract. We have been upgrading the software front-end to WASP (Wideband Autocorrelating SPectrometer, operating from 800 to 4000 MHz) from Lynx-OS to Real-Time Linux and will discuss Real-Time software aspects. We are using a standard parallel "PIO" card to receive data from the spectrometer and standard GPIB interfaces that drive a frequency synthesizer for bandpass calibration.

WASP and its front-end computer are fully mobile and can be taken to any suitable radio telescope. The front-end is typically connected via Ethernet to the actual observing computer at the telescope. We will also discuss a typical observing setup in such a loosely coupled environment.

1. Experimental setup

A ruggedized Pentium-133 was acquired (replacing the previous 486-33 Lynx-OS-based setup) to which a standard NI GPIB card and Metrabyte PIO-12 DAQ card were added. RedHat 5.1 Linux was installed and modularized device drivers for the PIO-12 and GPIB cards were obtained from the Linux Lab project.[3]

The WASP[4] (Harris, Isaak, & Zmuidzinas 1998) signal feeds directly into the PIO-12 and is driven at 4.603 kHz by WASP. WASP is also controlled via the PIO-12 card as the i8255 based PIO-12 card contains 3 bytes, configurable in many I/O modes. We use PA to read, PB to write, PC(low nibble) to write and PC(high nibble) to read.

The GPIB card is connected to an HP8648C synthesizer for bandpass calibration, which in turn drives WASP in calibration mode, (see also Fig. 1).

---

[1] now at: MRAO, Cambridge, UK

[2] now at: Edge Technologies, Inc., Laurel, MD

[3] `http://www.llp.fu-berlin.de/`

[4] `http://www.astro.umd.edu/~teuben/wasp/`

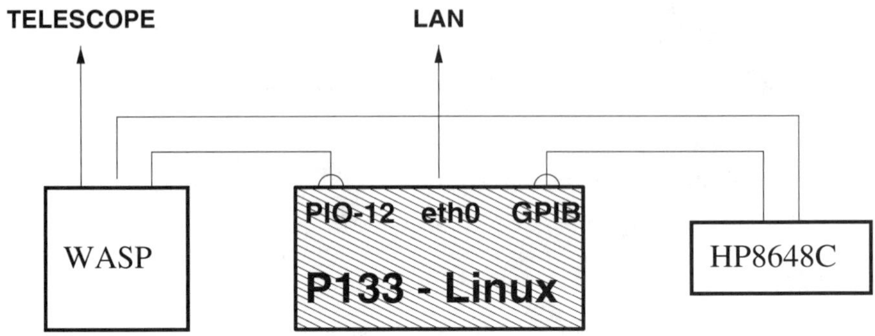

Figure 1. Schematic setup. The Pentium-133 computer is connected to a LAN via Ethernet, through which telescope control will occur. The GPIB card is connected to an HP8648C synthesizer for calibration and the PIO-12 data acquisition card is connected to WASP

2. Real-Time Alternatives (cheating)

Several cheap alternatives to increase performance and/or responsiveness are available on Linux:

`nice` Re-prioritizing of the CPU usage that hooks into the standard scheduling algorithms the kernel uses. An obvious choice, but rarely sufficient, especially for instrument control.

`swapoff` Turning off the swap can sometimes be an effective way to avoid paging in the system and avoid dreaded latencies. Of course this does not solve task switching delays.

`irqtune` This is a simple interrupt optimizer, with the advantage that it works in user-space, without the need to recompile the kernel. For serial lines (e.g., modems) this can make a big performance difference. It is generally available for Linux.

`QNX scheduling` This needs a true patch to the Linux kernel[5], supplied by Adam McKee, and replaces the scheduling algorithm with one that was freely modeled after the commercially available QNX Real-Time operating system scheduling algorithms. Using a supplied program, `qsched`, users can then prepend commands that re-schedules the commands in QNX style, e.g.,

```
qsched -t 10000 rmsobs
```

would run `rmsobs` in guaranteed timeslices of 10 seconds. This can of course (like many in these situations) hang the machine if not carefully

[5] `ftp://sunsite.unc.edu/pub/Linux/kernel/patches/misc/QNXsched-1.21.tar.gz`

used. We tried QNX scheduling and found it to be reliable for short observations, but lack of low level scheduling made it impossible to adopt it for complex tasks like the bandpass calibration.

3. Real-Time Linux

Victor Yodaiken and collaborators at New Mexico Tech (Socorro) have recently made patches available to the Linux kernel which turns it into a good Real-Time operating system[6]. This works by a user supplied kernel module, which takes over control of the computer. Only when this user module has determined there is time left will the kernel get time allotted and can do its usual things. Communications between these modules and user program occurs via special RT FIFO buffers. As usual in this business, programmers need to take special care not to produce situations that can lock up the computer.

We wrote an interrupt driven Real-Time module to guarantee that no data will be lost. Data are then read until a full "lag-readout" has been accumulated (192 bytes at a rate of 4.6kHz) and sent to a user program via a Real-Time FIFO. This turned out to work very reliably. The periodicity and nature of this signal also made it possible to accurately determine if data were missed.

4. Observing

Since WASP was designed to be mobile, a typical observing mode might consist of shipping the WASP box (the current prototype is a 1 cubic meter) together with the HP synthesizer and Linux box to a remote observatory where the IF mixed signal can be spectrum analyzed.

The Linux box will only be controlling the spectrometer itself and is normally part of a local network. The observing telescope computer will drive the telescope and send synchronized commands (in the form of remote shell scripts) to the Linux box to perform integrations, calibrate the signal, set the chopper, etc.

5. Future

WASP2 is being planned, taking data at 50kHz instead of the current 5kHz rate. Our current P133 should be able to handle around 30-50kHz, a Pentium-II-266 should be able to handle a 150kHz interrupt rate.

References

Harris, A. I., Isaak, K. G., & Zmuidzinas, J. 1998, in SPIE Proc., Vol. 3357, Advanced Technology MMW, Radio, and Terahertz Telescopes, ed. T. G. Phillips (Bellingham: SPIE), 384

[6] `http://www.rtlinux.org`

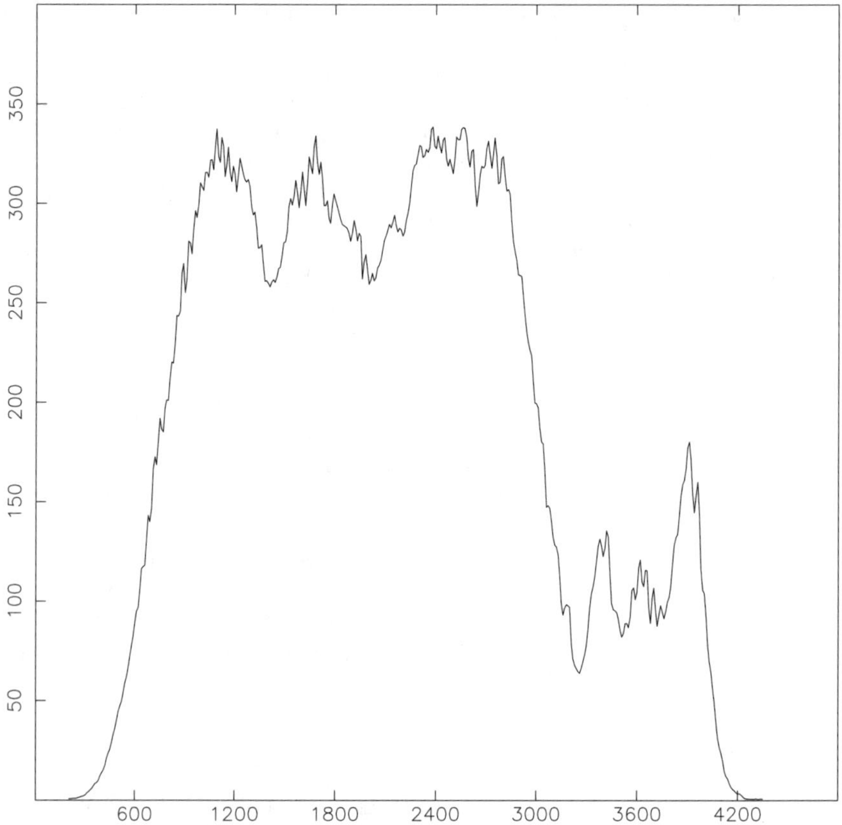

Figure 2. Bandpass filter shape (actually a one-dimensional projection) as determined with the current experimental setup. Horizontal units are MHz, vertical units arbitrary.

Astronomical Data Analysis Software and Systems VIII
ASP Conference Series, Vol. 172, 1999
D. M. Mehringer, R. L. Plante, and D. A. Roberts, eds.

NASA's Instrument Control Markup Language (ICML)

Troy J. Ames

NASA/Goddard Space Flight Center

Kenneth B. Sall, Craig E. Warsaw

Century Computing, Division of AppNet, Inc.

Abstract. This paper describes our experience in using XML to aid in collaborative scientific research involving infrared astronomy. We describe our current efforts for SOFIA in developing an XML vocabulary for instrument description, communication and control, pipeline algorithms, metadata, and documentation. Out focus has been infrared astronomy, although most of the techniques employed have much wider applicability.

1. Instrument Remote Control

Instrument Remote Control (IRC) is a joint effort between NASA/GSFC's Advanced Architectures and Automation branch[1] (Code 588) and Century Computing; it will enable trusted infrared astronomers around the world to easily access infrared astronomical instruments located in remote, inhospitable environments.

1.1. Project Goals and Methodology

The long-term goal of the Instrument Remote Control (IRC) project is to develop a distributed framework from science user to instrument which will provide robust interactive and reconfigurable control and monitoring of remote instrumentation. It is crucial that the next generation of astronomical instruments be largely free of the constraints imposed by specific operating systems and rigid computer hardware configurations that are likely to need to be changed many times during the lifetime of the instruments. IRC promotes the idea of distributed environment with processes assigned to the CPU and hardware best suited to the task. Our assumption is that better equipment will appear soon after commissioning; we must therefore be able to accommodate changes easily. Our software methodology is characterized by several key goals and principles:

- It must be relatively easy to develop, enhance, and maintain the software.
- Addition of a new instrument should be an incremental not monumental.
- Object-oriented design techniques facilitate the development of a general solution that can be applied to a diverse range of instruments.
- The methodology must support instrument integration, testing, control.
- IRC explicitly promotes reuse – it is part of our design, not an afterthought.

[1] `http://aaaprod.gsfc.nasa.gov/website/`

To accomplish our goals, we are developing a comprehensive instrument description based on Extensible Markup Language (XML)[2,3], a World Wide Web Consortium standard. This description will drive nearly all of our software development efforts. We will make use of automated tools to assist in the definition of this description, as well as in incorporating changes to the description. Our ultimate system will interpret the descriptions at runtime, where performance allows. In other cases, we will generate code from the latest instrument description.

1.2. Interested Parties

Our initial prototype was a Java application developed for the Center for Astrophysical Research in Antarctica (CARA) that ran on a Mac Quadra to control heaters for SPIREX at the South Pole in early 1998. IRC is currently focusing on the design and implementation of the control software for HAWC, the High-resolution Airborne Wideband Camera[4], for SOFIA[5] (Stratospheric Observatory For Infrared Astronomy). Our second SOFIA instrument will be the Submillimeter And Far InfraRed Experiment (SAFIRE). Both HAWC and SAFIRE are SOFIA First Light Instruments.[6]

We believe our techniques will also be applicable to ESA's SPIRE (Spectral and Photometric Imaging REceiver for FIRST – Far InfraRed and Submillimetre Telescope) as well. These three instruments have some features in common (e.g., their detectors) which make them good candidates. However, we emphasize that our techniques can be extended to any astronomical instruments. In fact, our solution can only become truly general if we have a diverse set of data points; we encourage members of the astronomical community who share our concerns to contact the authors to contribute in some manner to this development effort.

2. Astronomical Instrument Markup Language (AIML)

Our long-term focus is to develop an extensible framework to which new instruments can be added with relative ease. This will eventually be accomplished by implementing our own *AIML*[7] based on a custom Document Type Definition (DTD) which will evolve over time, as a direct effort of NASA/GSFC and Century Computing. Since AIML is an XML vocabulary, it will be well suited to describing hierarchical, structured information. AIML will be used to describe control capabilities, data streams, message formats, and communication mechanisms, as well as for online documentation and the association of housekeeping

---

[2] `http://WDVL.com/Authoring/Languages/XML/`

[3] `http://www.w3.org/XML/Activity.html`

[4] `http://pioneer.gsfc.nasa.gov/public/hawc`

[5] `http://sofia.arc.nasa.gov/`

[6] `http://www.sofia.usra.edu/observatory/instruments/first_light/announce.html`

[7] The original name for our language was *Instrument Control Markup Language (ICML)*. However, since control is no longer our only focus, we decided to rename our language *Astronomical Instrument Markup Language (AIML)*.

metadata with acquired images. Some of these aspects of instrument control will be reflected in Java graphical user interfaces, generated from the instrument descriptions. Other sections of the instrument description will be applied to data capture and data pipeline algorithms, as well as to other instrument subsystems. Different parts of the description can be used by different components of the software architecture, and even by separate applications. Furthermore, once we collect a number of instrument descriptions from a variety of observatories, astronomers will be able to specify what characteristics they need for a particular proposal and generate a query that will return a list of suitable instruments.

2.1. Applicability of XML to IRC

The hierarchical, structured markup of XML facilitates many aspects of instrument description, algorithm description, and documentation.

- Control Capabilities – what the instrument can do
- Data Stream Descriptions
  - images – data collection, archiving, indexing, and retrieval
  - housekeeping – instrument status, health and safety
  - command responses
- Pipeline algorithms – inputs and outputs of algorithms; connecting algorithms; defining visualizations as consumers of the pipeline outputs
- Message Formats – separate formats for Control and Data Streams
- Communication Mechanisms – ports, TCP, RS232
- Documentation
  - template with specific styles applied
  - novice vs. expert views of documentation
  - role-dependent view – investigator, operator, or astronomer role
  - short descriptions – displayed as tool tips in the GUI
  - long descriptions – multiple sentences or paragraphs for on-line help
- Style and Presentation
- Metadata / Housekeeping

Down the road, we expect to expand AIML to encompass:

- Planning Observations (proposals)
- Scheduling Instruments for Observations

The following examples are meant to be illustrative of our approach.

2.2. Metadata Example

The AIML example below is a simple description of an element we call an *Image* which consists of two parts: (1) the data itself, stored in a separate file; and (2) the metadata which defines the circumstances under which the data was collected. Although the example specifies FITS, our notion of an Image is independent of any particular binary representation. A key advantage of this approach is that the metadata portion can be processed by any XML-aware tool and is fully searchable, regardless of the binary format which the metadata describes.

```
<Image>
<Data href="m42th1a.fits" type="FITS" series="1247"/>
<Metadata>
  <ObjectName>M42, Orion Nebula, Theta-1</ObjectName>
```

```
  <TimeStamp>
<Date month="09" date="16" year="2001" />
<Time hrs="17" min="24" sec="13" msec="145" />
  </TimeStamp>
  <Integration duration="20" filter="300micron" />
  <FlightInfo ra="5:33" dec="5:25" />
          <Notebook>This stellar image was taken on our maiden voyage.
  Weather conditions were...
  </Notebook>
</Metadata>
</Image>
```

We recently learned of Astronomical Markup Language[8] by Guillaume & Murtagh, which describes astronomical objects, articles, tables, set of tables, images, and authors. We hope to coordinate our efforts with those of Guillaume & Murtagh, however, our AIML description is aimed at the wider domain of instrument description and control, as well as data collection and pipeline algorithms.

2.3. ADR Subsystem Example

The next example describes two Commands, HeatSwitch and HeatSwitchCurrent, both of which accept exactly one argument. Note that the "current" argument of the HeatSwitchCurrent command has Valids (constraints) defined. This results in a GUI component which will guarantee input only within the constrained range.

```
<Control id="HAWC.ADR">
  <Command id="HeatSwitch">
    <Argument id="state" type="boolean" required="true"/>
    <!-- other Arguments would go here -->
  </Command>
  <Command id="HeatSwitchCurrent">
    <Argument id="current" type="float" required="true">
      <Valids>
        <Range low="0.0" high="200.0" units="amps" />
      </Valids>
    </Argument>
  </Command>
  <!-- other Commands would go here -->
</Control>
```

3. Summary

We have attempted to convey the significant advantages gained by describing astronomical instruments and data pipeline algorithms using AIML, our XML vocabulary. Separation of GUI descriptions from application logic has long been a favored practice in the software industry. It is time for the astronomical community to favor an approach that similarly separates out as much of the instrument details from the code as possible. By so doing, we can develop and modify instruments far more quickly than those created by conventional software development techniques. We will also produce a database of instrument descriptions that can greatly simplify the task of locating the right instrument for achieving specific proposal goals. To join our low-volume XML for astronomy mailing list, send the one-line message `subscribe xml your_email_address` to `majordomo@pioneer.gsfc.nasa.gov`.

---

[8] http://www.infm.ulst.ac.uk/%7Edamien/these/

Astronomical Data Analysis Software and Systems VIII
ASP Conference Series, Vol. 172, 1999
D. M. Mehringer, R. L. Plante, and D. A. Roberts, eds.

The Scientific Uplink System for SIRTF, a Design Study

Ingolf Heinrichsen, Booth Hartley, Bob Narron, Su Potts, Gabriella Turek, Emmanuel Valjavec, Xiuqin Wu

SIRTF Science Center, IPAC, CALTECH, MS 100-22, Pasadena, CA 91125, USA

Abstract. The Space Infrared Telescope Facility (*SIRTF*) is one of NASA's Great Observatory missions, scheduled for launch in 2001. As such its ground segment design is driven by the requirement to provide strong support for the entire astronomical community in the form of Guaranteed Time, Legacy Program and Open Time Observers.

In this contribution we present the current status of the software design of the Scientific Uplink Software for *SIRTF*. The astronomical observing modes of the scientific instruments on board of *SIRTF* are captured in the form of Astronomical Observation Templates, which will be deployed to the community as Java based WWW applications. In addition web-based tools will be provided to the observers in order to support the preparation of their mid- and far-infrared observations. These tools will allow the observer to estimate observing times, sensitivities and expected data volumes by making calls to the Interpreter and Resource Estimator Subsystem. At a later stage the same software will be used to derive the detailed command sequences needed by the *SIRTF* instruments to execute a given observation. All information is captured in the central Science Operations Database, which will support the entire *SIRTF* science operations. The fundamental need to support a large astronomical community lead to an object oriented Java and WWW based design approach for this entire up-link software system.

1. Observing with SIRTF

1.1. Instrumentation

The Space InfraRed Telescope Facility *SIRTF*,[1] (Fanson et al. 1998), is the fourth and final element in NASA's family of "Great Observatories." It consists of a 0.85m telescope and three cryogenically-cooled science instruments.

IRAC (Fazio et al. 1998a, 1998b) is a four-channel camera that provides simultaneous $5\overset{'}{.}12\times5\overset{'}{.}12$ images at 3.6, 4.5, 5.8, and 8μm. The pixel size is $1\overset{''}{.}2$ in all bands. All four detector arrays in the camera are 256×256 pixels in size, with the two short wavelength channels using In:Sb and the two longer wavelength channels using Si:As IBC detectors. The IRAC point-source sensitivity requirements (5σ, 200s) at 3.6, 4.5, 5.8, and 8.0μm are 6, 7, 36, and 54 μJy, respectively.

---

[1] `http://ssc.ipac.caltech.edu/sirtf`

IRS (Roellig et al. 1998) will provide *SIRTF* with low and moderate-spectral resolution spectroscopic capabilities from 4–40μm. The IRS is composed of four separate modules, with two of the modules providing R$\approx$50 spectral resolution over 4–40μm and two modules providing R$\approx$600 spectral resolution over 10–37μm.

The **MIPS** (Heim et al. 1998; Rieke et al. 1998) instrument will be comprised of three detector arrays and attendant optics and calibrators. The arrays are a 128$\times$128 arsenic-doped silicon for imaging at a wavelength of 24 m, a 32$\times$32 gallium-doped germanium at 70 μm, and a 2$\times$20 stressed Ge:Ga array at 160μm. The 32$\times$32 array also takes very low-resolution spectra from 50–100μm.

1.2. Astronomical Observing Templates

The science capabilities of the *SIRTF* instruments are represented by seven "observing modes" or Astronomical Observation Templates (AOTs). The AOT is a central design concept to *SIRTF* Science Operations. Observers design experiments by selecting an AOT and choosing its parameters. They fill out WWW-based forms and submit "Astronomical Observation Requests" (or AORs). The completed AORs are deposited into the *SIRTF* Science Operations Database (SODB). Specially designed software expands each AOR into activities and an uplink sequence for transmission to the spacecraft. AORs are the basic building blocks for observatory scheduling, pipeline data processing and science archiving.

From an observer's viewpoint, the AOT "Front End" is the electronic form filled out to define an experiment to be executed by *SIRTF*. An Astronomical Observation Request (AOR) is the result of filling out an AOT form. AORs are the fundamental scheduling unit for the *SIRTF* Observatory, they can be thought of as a list of parameters that completely describe an observation. In the proposal containing a set of AORs the observers can influence the scheduling process by providing information about the timing and relationship of the individual observations. In this way the observatory can be asked to perform observations as a block at a specific time or in a specific order. There is generally no need for observers to modify AORs after submission with the observing proposal. Modification of AORs can be made, however, if it is required by recommendations of the Time Allocation Committees or because of changes in Observatory performance, i.e., *SIRTF* will have a single phase proposal submission process.

For repetitive engineering or calibration functions the *SIRTF* uplink software will provide IETs and IERs, which are the Instrument Engineering analogs of AOTs and AORs.

2. SIRTF Uplink Software

2.1. Design Drivers

The mayor design drivers for the *SIRTF* Uplink software are:
Data Volume: It is estimated, that the *SIRTF* Science Center will solicit, evaluate and select from $\approx$1200 proposals per proposal cycle. It will have to support observation planning and modification for $\approx$400 of these proposals per cycle and thus prepare $\approx$20000 observations per year, including special calibration and engineering activities.

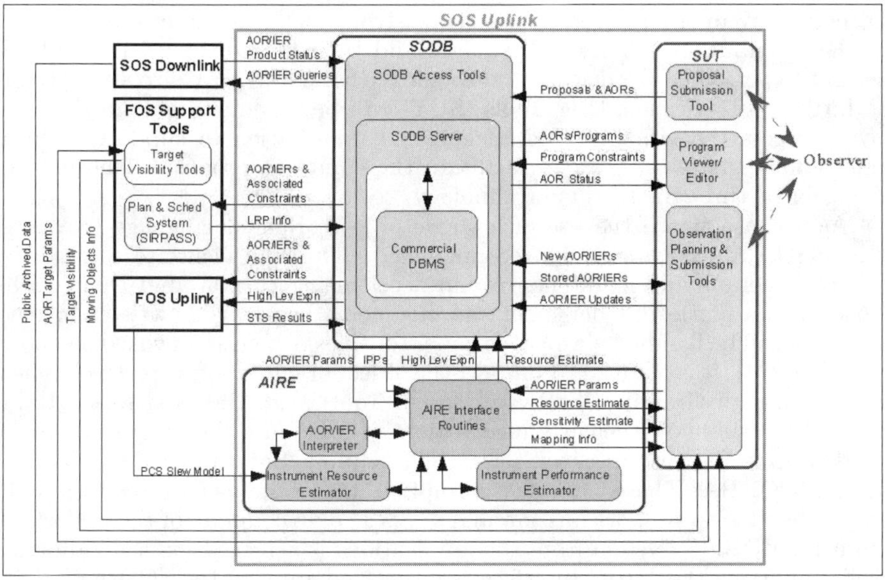

Figure 1. Overview of the SIRTF Uplink Software Design

User Community: The *SIRTF* Science Center must be able to support the entire astronomical community, providing tools for fast and easy preparation of scientific proposals and tracking of AORs.

Easy and Fast Software Maintenance: The software shall allow quick adaptation to changing instrument performance or operating conditions. It is the goal to achieve a turn-around time of less than 4 weeks for the software life cycle and to be able to react to emergencies within one week during the actual *SIRTF* operations.

Independence from Commercial Supplier: The Database design shall allow to easily change the supplier of the underlying Data-Base-Management System (DBMS) to be able to adapt to the evolving software market in order minimize risk and cost of ownership.

Software Re-Use: The software shall be designed in such a way that major portions can be re-used for future missions and legacy code from previous NASA missions can be incorporated.

Resources and SIRTF Autonomy: The software must be able to estimate resources like Execution Times/Uplink- and Down-link Data Volumes with sufficient accuracy to support the planning of the satellite and Deep Space Network (DSN) operations. This is important as it is planned to routinely have contact with the spacecraft only once a day for $\approx$30min and to uplink command sequences approximately once a week.

2.2. Design Overview and Guidelines

The high level requirements outlined above lead to a set of design guidelines for the SIRTF Uplink Software System. The User Interface needed to be deployed via the WWW in order to reach the entire astronomical community, while avoiding the distribution of software packages and their maintenance. Examples of the prototype AOTs are reachable through the SIRTF WWW site at IPAC[2]. In the future tools (Science User Tools (SUT) will be made available in the same way to access the vital infrared background information to support proposal preparations. It is the idea to facilitate the complete proposal handling and management through the same technology.

As a consequence the use of a single object oriented language (Java) for entire SSC Uplink System was recommended. This will facilitate software reuse, while simplifying maintenance and accelerating turn-around times. This also allows for platform independent development. Legacy code written in other languages will be handled via Java wrapper routines, which are invoked on a local server via RMI. It is planned to support the object oriented software development process through Graphical Software Design Tools, e.g., Rational Rose. Rapid prototyping will allow the early involvement of users into the design process.

The external interfaces of the SIRTF Uplink Software will be handled through CORBA. These interfaces comprise, e.g., large software systems developed by JPL to support mission planning. The de-coupling of the underlying commercial DBMS System from the applications is achieved via a middle-ware layer supporting the JDBC Interface Routines, comprising the SODB.

This entire concept should allow as far as possible the re-use for future missions by separating the instrument specific models from generic parts as like the expansion engine (AIRE) used to generate the command sequences for a given observation. An overview of the results of these design considerations can be seen in Fig. 1.

References

Fanson, J., Fazio, G. G., Houck, J. R., Rieke, G. H., Tenerelli, D. J., & Whitten, M. 1998, in SPIE Proc., Vol. 3356, Space Telescopes and Instruments V, ed. P. Y. Bely & J. B. Breckinridge (Bellingham: SPIE), 478

Fazio G. G., et al. 1998a, in SPIE Proc., Vol. 3354, Infrared Astronomical Instrumentation, ed. A. M. Fowler (Bellingham: SPIE), 1024

Fazio G. G., et al. 1998b, in Proc. of Astrophysics with Infrared Surveys: A Prelude to SIRTF, ed. M. Bicay, R. M. Cutri, & B. F. Madore (San Francisco: ASP), in press

Heim, G. B., et al. 1998, in SPIE Proc., Vol. 3356, Space Telescopes and Instruments V, ed. P. Y. Bely & J. B. Breckinridge (Bellingham: SPIE), 985

Rieke, G. H., Young, E. T., Rivlis, G., & Gautier, T. N. 1996, BAAS, 189, 3.09

Roellig T. L., Rakowski, A., Taudien, G. E., Huisjen, M., Bolton, M., Seib, D. H. 1998, in SPIE Proc., Vol. 3354, Infrared Astronomical Instrumentation, ed. A. M. Fowler (Bellingham: SPIE), 1192

---

[2]`http://ssc.ipac.caltech.edu/sirtf/`

Astronomical Data Analysis Software and Systems VIII
ASP Conference Series, Vol. 172, 1999
D. M. Mehringer, R. L. Plante, and D. A. Roberts, eds.

Distribution of the OPUS Data Processing System

Daryl A. Swade, James F. Rose

Computer Sciences Corporation, Space Telescope Science Institute, 3700 San Martin Drive, Baltimore, MD 21218

Abstract. The OPUS data processing software system was developed for converting raw telemetry from the HST into standard FITS format data files. The OPUS process control and monitoring components, which are applicable to many pipeline projects, are being distributed on CD-ROM along with associated documentation.

1. Introduction

The Space Telescope Science Institute (STScI) has developed OPUS as a data processing software system for converting raw telemetry into standard Flexible Image Transport System (FITS) format data files. OPUS is a dynamic, event-driven, distributed processing system that provides an environment designed to handle a large number of observations processed through many steps across a network of computers (Rose 1998). OPUS is also an automated system that monitors processing and provides facilities for error identification and repair.

The Hubble Space Telescope (HST) OPUS pipeline has been operational at STScI since December 1995, and the OPUS baseline system has now been packaged so that other spacecraft missions and observatories can take advantage of this flexible system for data processing projects. OPUS has been recently ported to the UNIX operating system and can currently run on Sparc/Solaris, ALPHA/DIGITAL UNIX, PC/LINUX, VAX/VMS, and ALPHA/VMS platforms. OPUS also supports a pipeline running on any mix of these platforms.

2. OPUS Components

OPUS can be considered as having three components. First, the baseline OPUS system is a distributed processing platform, independent of the applications that it runs and monitors. Second, the Application Programming Interface (API) is a library of software packages that supports the blackboard architecture for the distributed platform as well as for the standard applications. In addition, OPUS consists of a collection of instrument specific pipeline applications that, until recently, have been developed primarily for the instruments aboard the HST. STScI is currently also developing applications for the Far-Ultraviolet Spectroscopic Explorer (FUSE) mission (Rose et al. 1998).

2.1. Baseline System

OPUS has adopted a blackboard architecture using the standard file system of the native operating system for interprocess communication (Rose et al. 1995). In this model processes do not communicate with one another, but simply read and write to a common 'blackboard' instead of having a single 'controller' process which must be continuously aware of the activity of other processes in the system (Nii 1989). This technique effectively decouples the interprocess communication from the individual processes that comprise the data processing system.

Within the OPUS distributed processing system, a variety of independent processes are run sequentially as processing steps in the pipeline. The system also allows multiple instances of any single process to be run simultaneously without interfering with each another. Multiple pipelines, or independent paths, are also supported. Any pipeline can be configured to run its processes across multiple nodes on a network of computers.

In addition to several copies of pipelines with identical processing steps, OPUS supports any number of different pipelines all running on the same network. Thus, in addition to the science pipelines, OPUS can, for example, accommodate an engineering data pipeline.

The OPUS environment is configured through a set of simple ASCII text resource files that describe the command line arguments, pipeline triggers, how steps get triggered, and other control information (Boyer & Choo 1997). The pipeline path file defines a set of cluster-visible directories on the available disks.

In order to monitor the system, OPUS provides two Motif pipeline managers (Rose, Choo, & Rose 1996). The process manager assists with the task of configuring the system and monitors the status of each process. The observation manager views the pipeline activities, monitoring the progress of datasets through the pipeline, and flagging observations that are unable to complete pipeline processing. Multiple managers can each monitor separate pipelines without interference from one another.

OPUS provides facilities for handling data processing problems such as missing data, absent calibration files, or other unexpected situations. The OPUS pipeline provides convenient ways to investigate problems: examine process log files, list data file headers, view observation processing history (trailer) files, and finally restart the troubled exposure at any step in the pipeline.

2.2. API

An OPUS Application Programming Interface is currently under development. See Miller (1999) for details.

2.3. Pipeline Applications

The OPUS applications are programs or scripts that tend to be specific to individual missions or science instruments. Processes (or scripts) can be triggered in three ways: the most common is to allow the completion of one or more previous pipeline steps to act as the process trigger. Another useful technique is to use the existence of a file of a certain class as a trigger. Alternatively, one can use a timing device to trigger an OPUS process (e.g., wake up once an hour).

With OPUS, mission software developers can write specific applications that use the OPUS architecture and API. Experienced STScI OPUS developers are available for consultation and help (opushelp@stsci.edu). For examples of OPUS pipeline applications see Rose (1997), Schultz et al. (1999) and Swam & Swade (1999).

3. Other Missions Using OPUS

While the OPUS system was developed at the STScI, the blackboard system and the OPUS API are independent of the HST mission. The following missions are now using OPUS as their science data processing pipeline system: FUSE (Far Ultraviolet Spectrographic Explorer), INTEGRAL (International Gamma-Ray Astrophysics Laboratory), SIRTF (Space Infra-Red Telescope Facility), AXAF (Advanced X-Ray Astrophysics Facility), and the MSSSO (Mount Stromlo and Siding Spring Observatories) MOSAIC project.

In addition, many groups are considering the OPUS platform for their data processing projects. Over 40 copies of OPUS CD-ROM and FAQ have been distributed to potential users.

4. OPUS CD-ROM

The OPUS baseline system is currently distributed on CD-ROM to help other institutions with their own pipeline management. The OPUS CD-ROM comes complete with the process manager, the observation manager, a set of sample applications, and all the resource files required to get a sample pipeline running.

The sample pipeline is distributed with the OPUS environment to demonstrate some of the major capabilities of the OPUS system. This example simply converts a stack of public GIF images from the HST archives into standard FITS files. In the sample pipeline, the GIF images provide the 'raw telemetry' files that would normally feed a real data reduction pipeline.

More importantly, the sample pipeline is a working illustration of how to put together your own production pipeline. Each of the tasks in the pipeline demonstrates a different variety of 'trigger' that activates the task. For each task there is a separate text resource file showing the variety of switches and parameters available to the system. In addition, the pipeline resource files and path files for the sample pipeline are provided in simple text files that you can both examine and modify.

5. OPUS FAQ

The OPUS CD-ROM and the sample pipeline are fully documented in the OPUS Frequently Asked Questions (FAQ). The FAQ is available on the OPUS CD-ROM or at `http://www.stsci.edu/software/OPUS/opusfaq.html`. This document explains how to install the system on your computers, how to run the sample pipeline, how each of the managers works, and what the different resource files are about. As more projects get experience with the OPUS environment, this document expands with further clarifications, and is revised with each new release of the OPUS CD-ROM.

6. Summary

While the OPUS system was developed at the STScI, the blackboard system and the API are independent of the HST mission. HST mission specific applications are not portable, but the experience of the OPUS team in developing complete pipelines for the HST, FUSE, and other potential missions is available. Packages such as OPUS provide a true resource for NASA and ESA projects in a cost-conscious era where the software development cycle can and should be better controlled. The OPUS platform and the OPUS software libraries can be reused, forming the basis for the rapid development of robust data processing applications.

For further information see the OPUS web page at `http://www.stsci.edu/software/OPUS` or send e-mail to opushelp@stsci.edu.

References

Boyer, C. & Choo, T. H. 1997, in ASP Conf. Ser., Vol. 125, Astronomical Data Analysis Software and Systems VI, ed. G. Hunt & H. E. Payne (San Francisco: ASP), 42

Miller, W. W., III 1999, this volume, 195

Nii, H. P. 1989, in Blackboard Architectures and Applications, ed. V. Jagannathan, R. Dodhiawala, & L. Baum (San Diego: Academic Press), xix

Rose, J. 1998, in ASP Conf. Ser., Vol. 145, Astronomical Data Analysis Software and Systems VII, ed. R. Albrecht, R. N. Hook, & H. A. Bushouse (San Francisco: ASP), 344

Rose, J., Heller-Boyer, C., Rose, M. A., Swam, M., Miller, W. W., III, Kriss G. A., & Oegerle, W. 1998, in SPIE Proc., Vol. 3349, Observatory Operations to Optimize Scientific Return, (Bellingham: SPIE), 410

Rose, J., et al. 1995, in ASP Conf. Ser., Vol. 77, Astronomical Data Analysis Software and Systems IV, ed. R. A. Shaw, H. E. Payne, & J. J. E. Hayes (San Francisco: ASP), 429

Rose, J. F. 1997, in ASP Conf. Ser., Vol. 125, Astronomical Data Analysis Software and Systems VI, ed. G. Hunt & H. E. Payne (San Francisco: ASP), 38

Rose, J. F., Choo, T. H., & Rose M. A. 1996, in ASP Conf. Ser., Vol. 101, Astronomical Data Analysis Software and Systems V, ed. G. H. Jacoby & J. Barnes (San Francisco: ASP), 311

Schultz, J. J., Goldstein, P., Hyde, P., Rose, M. A., Steuerman, K., Baum, J., Perrine, R., & Swade, D. A. 1999, this volume, 199

Swam, M. S. & Swade, D. A. 1999, this volume, 203

Astronomical Data Analysis Software and Systems VIII
ASP Conference Series, Vol. 172, 1999
D. M. Mehringer, R. L. Plante, and D. A. Roberts, eds.

CICADA – Configurable Instrument Control and Data Acquisition

Peter J. Young, William H. Roberts, Kim M. Sebo

Mount Stromlo and Siding Spring Observatories, Private Bag, Weston Creek, ACT, 2611, AUSTRALIA

Abstract. CICADA (Young et al. 1997) is a multi-process, distributed application for the control of astronomical data acquisition systems. It comprises elements that control the operation of, and data flow from CCD camera systems; and the operation of telescope instrument control systems. CICADA can be used to dynamically configure support for astronomical instruments that can be made up of multiple cameras and multiple instrument controllers. Each camera is described by a hierarchy of parts that are each individually configured and linked together. Most of CICADA is written in C++ and much of the configurability of CICADA comes from the use of inheritance and polymorphism. An example of a multiple part instrument configuration – a wide field imager (WFI) – is described here. WFI, presently under construction, is made up of eight 2k $\times$ 4k CCDs with dual SDSU II controllers and will be used at Siding Spring's ANU 40in and AAO 3.9m telescopes.

1. Introduction

Early releases of CICADA[1] had many built-in site-specific sections of code which meant that it was very difficult to "export" to other observatories. One of the aims of CICADA v2 was to replace these sections of the code with "self-configuring" code based around simple ASCII tables. Another aim of release 2 of CICADA was to provide support for multi-controller, multi-CCD instruments. This requirement also impacted on the design of the Configuration Tables. The table design now used by CICADA is a linked list of table items and each item is a hash table of fields which describe the item. There are tables describing Instruments, Cameras, Controllers, CCDs, Temperature Sensors, Focalplanes, Dewars, etc.

2. Building a Configuration – the Configuration GUI

Figure 1 shows the CICADA Hardware Configuration windows for the Controller Table. It is by using these and other configuration windows that the tables are built which are eventually used to dynamically configure the CICADA software.

[1] `http://msowww.anu.edu.au/computing/cicada`

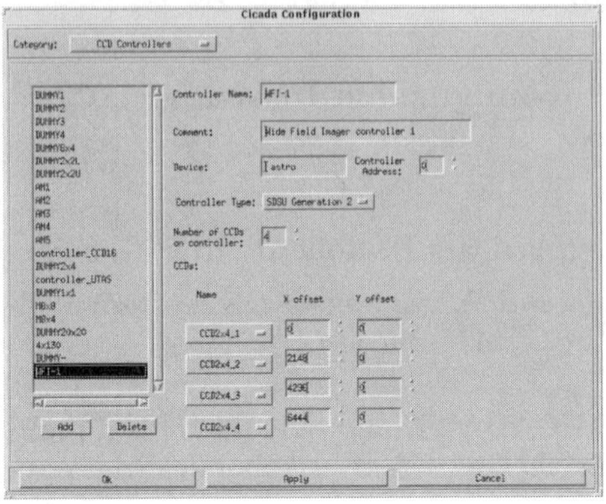

Figure 1. The Controller Configuration GUI.

3. Dynamic Implementation of a Configuration

The Cicada_Master process is the communication and operations control centre for CICADA, functioning as the request interpreter, the activity status monitor and the traffic director. It is started at the beginning of a CICADA run and is responsible for interpreting the instrument configuration.

This is done by instantiating a Cicada Instrument object that represents the hierarchical instrument description by using a set of individual Instrument_Part objects. Each of these Instrument_Part objects starts a set of instrument control processes on configured instrument hosts and handles activity and status requests passed to it by the Cicada_Master process. Each activity request runs as a separate POSIX thread, so simultaneous control of each part is maintained. In addition separate threads are started to fetch status information from the instrument parts. Only one activity can be running on an individual part at one time, attempts to start new activities are blocked.

The use of threads allows for full control of activities running on an instrument part, including monitoring, pausing/resuming and aborting.

The Cicada_Master process also starts a TCL interpreter and handles the execution of embedded CICADA commands in TCL scripts. These commands are passed to the Instrument_Part objects in the same way as the interactive command stream which could be from a GUI or from a command line interface.

4. Handling Arbitrary Multi-Amp Detectors

An important component of the configurability of CICADA is the ability to handle the geometry of CCD mosaic layouts in a generalised way.

At the GUI level (see Fig. 2, left), the astronomer selects one (or several) regions of interest, spanning some part, or all of the mosaic and possibly including overscan and/or prescan pixels. However, at the readout level, a simple rectangular region in mosaic coordinates may actually span multiple CCDs and amplifiers of the mosaic.

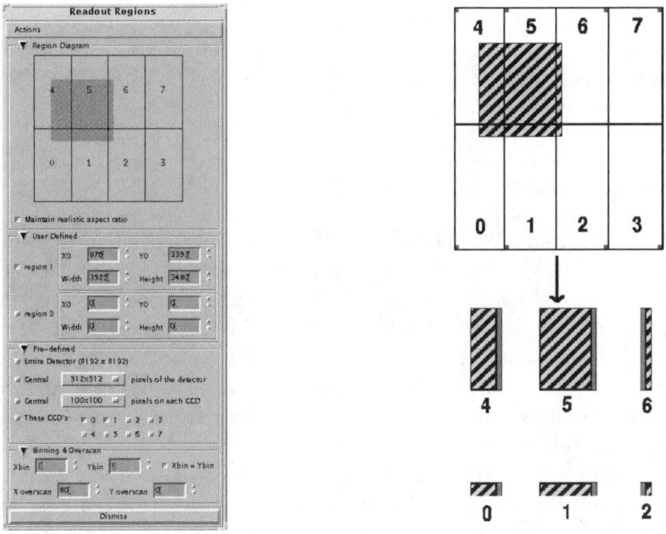

Figure 2. The regions GUI (*left*) and output from the CICADA Regions Class (*right*).

The CICADA Regions class is a general solution to the problem of converting the specified rectangular readout subregion (in mosaic coordinates) into the actual pixels clocked out by the CCD controller(s) (in CCD/amplifier coordinates).

The geometry of the mosaic camera is specified in the CICADA hardware configuration window, where the sizes, positions and readout directions of the CCDs in the mosaic are enumerated.

The CICADA Regions class is capable of dealing with almost any practical mosaic geometry. It handles any (reasonable) number of CCDs of any size, with arbitrary multiple amplifier positions and directions, including support for overscan rows and columns for each amplifier, and independent pixel binning in either axis. Up to two possibly overlapping (mosaic-space) rectangular regions are supported, however this limit is arbitrary and could be increased without difficulty. The only assumption made is that the size of the full readout of each amplifier must be the same for all amplifiers. This constraint is likely to be true for all conventional CCD mosaic systems.

In general, a single mosaic sub-region spanning multiple amplifiers is converted to multiple smaller readout regions, one region per readout amplifier. Each of these smaller regions are ultimately written into separate FITS extensions of a multi-extension FITS file. The FITS extension headers contain sufficient geometry information so that the original mosaic-level geometry can be reconstructed.

The returned information from the Regions class is used by CICADA to construct the output FITS file from the interleaved steam of pixels coming from the controller(s), and also to display the resultant image reassembled into the same rectangular mosaic-space region selected by the astronomer.

Figure 2 at right shows at the top an example readout region selected by the astronomer in mosaic coordinates, spanning multiple CCDs. Below are shown the multiple sub-regions for each amplifier, returned to CICADA by the Regions class including overscan from each amplifier.

5. Site-specific Plug-Ins

Currently CICADA can be extended to support telescopes and filter controllers for sites other than MSSSO by user-written "plug-in" code. This is done by writing code that conforms to a CICADA defined API, building a shared library using a supplied Makefile and then replacing the default shared library. A base level CICADA C++ class implements the API as pure virtual methods that need overriding by the user derived class. Template telescope and filter control definitions are provided for a user to easily follow the required steps.

The GUI for telescope and filter control is then dynamically built based on descriptions provided in the CICADA configuration tables. This method has been used successfully at the UNSW APT telescope, the University of Tasmania and work is now progressing toward code to control the Anglo Australian Telescope.

6. Current Applications

CICADA is in use at MSSSO, controlling instruments at several telescopes. The same software is used at each telescope, only the configuration files differ to cater for different controller types and CCD configurations. Currently, CICADA supports the SDSU Generation I and II controllers as well as the Astromed 3200 controller. MSSSO is currently building a Wide Field Imager (WFI)[2] in collaboration with the Anglo-Australian Observatory and the University of Melbourne. This new Instrument is an 8 CCD mosaic using dual SDSU Generation II controllers and will make full use of the configurability of CICADA as well as its multi-process, distributed design. The next major project for CICADA is control of MSSSOs existing Double Beam Spectrograph which is a dual-camera configuration (i.e., two shutters, two controllers, two CCDs, two image displays).

References

Young, P. J., Brooks, M., Meatheringham, S. J., & Roberts, W. H. 1997, in ASP Conf. Ser., Vol. 125, Astronomical Data Analysis Software and Systems VI, ed. G. Hunt & H. E. Payne (San Francisco: ASP), 385

[2] `http://msowww.anu.edu.au/observing/wfi/`

Astronomical Data Analysis Software and Systems VIII
ASP Conference Series, Vol. 172, 1999
D. M. Mehringer, R. L. Plante, and D. A. Roberts, eds.

Novel Use of a Simulator for the Calibrations of Scientific Instruments

Christian Erd, Fred Jansen, Jon Brumfitt, Marco Beijersbergen, Philippe Gondoin, Hassan Siddiqui, Jorgo Bakker, Peter Videler, Lalit Jalota

Astrophysics Division, European Space Agency, Space Science Department

Abstract. The XMM Science Operations Team developed a detailed simulator to describe the performance of the scientific payload on-board European Space Agency's X-Ray Mission XMM. This simulator was written in modern programming languages (C++ and F90). Its application is not limited to simulations of instrumental data, but the simulator also plays a key role in calibrating the instruments, and as such it will become an integral part of the XMM analysis system.

1. Introduction

The European Space Agency's X-ray observatory mission XMM[1] (Lumb et al. 1996 and references therein) consists of three telescopes, each containing 58 highly polished mirror shells. About 50% of the light of two of these mirror modules is intercepted by two reflection grating arrays, which disperse the X-rays to a secondary focus. The dispersed rays are detected by a strip of nine CCDs per instrument. This comprises the Reflection Grating Spectrometer (RGS), which has a bandwidth from 0.3–2.4 keV (5–35 Å) and a resolving power ($\lambda/\Delta\lambda$) from 300 to 800.

Each of the three primary foci is equipped with a European Photon Imaging Camera (EPIC), with each containing assemblies of CCDs. The EPIC detectors provide imaging and medium spectroscopic information over a band from 0.1–12 keV (1–123 Å).

In order to complement X-ray observations, an Optical Monitor (OM) is included in the payload of XMM to monitor sources simultaneously in the visible/UV band from 70–600 nm.

All instruments are co-aligned, operate in parallel, and cover the same field of the sky. Photon counting of all instruments provides spectral and imaging information at the same time.

---

[1] `http://astro.estec.esa.nl/XMM`

2. Instrumental Calibration and the Role of Simulations

The scientific calibration of the instruments depends in a complex way on many variables, such as operating parameters of the instruments, and also on the characteristics of the celestial target and possible background. This large variety prohibits full calibration of all aspects of the instruments during the pre-launch phase. Therefore, detailed models of each instrument were developed in the form of simulation programs, which were subsequently calibrated with a selected set of critical test cases. Simulations are then carried out to generate further data such that the remainder of the calibration parameter space is filled and the transfer of the ground calibrations to in-orbit conditions can be performed. A major effort was spent in building the XMM simulator SciSim[2].

SciSim will also be used to establish the dependencies of the calibrations on the exact circumstances of each observation (e.g., precise position of the source in the field of view, background, stray-light, etc.). With the help of SciSim, calibration data can be generated every observation for each specific set of modified parameters, and the optimum calibration accuracy can be obtained.

3. Summary of Additional Applications of the XMM Simulator

Due to the detailed instrument specific properties included in SciSim, it is is also used for the generation of test data to aid the development of instrument health and monitoring software, which is part of the XMM ground station.

During the development of the scientific analysis software (as discussed at this conference by F. Jansen), SciSim provides input not only in the form of test data, but also as an aid to investigate analysis strategies and to optimize the pipeline data production for the continuous build up of the XMM archive.

General observers are making use of SciSim during the preparations of observation proposals to optimize exposure times and instrument configuration details. Simulations will be performed during data analysis for the definition of data selection and filtering strategies, allowing for feed-back on observed features. Instrumental and calibration effects will be separable from true source characteristics enabling ultimate verification of observed data.

In summary, SciSim represents a novel use of a simulator, because of its key position to conduct X-ray observations with XMM and to analyze data from XMM. It represents an integral part of XMM data.

4. Software Implementation

The simulations of the instruments of XMM are implemented in the form of several free standing modules, each module simulating only one part of the instrumentation. Communication between modules is performed via rays of a uniform structure by using pipes. Each ray represents the probability of one incident photon. Therefore, the simulation process is very flexible and different groupings of the individual simulation modules are possible for the purpose

---

[2]`http://astro.estec.esa.nl/XMM/scisim`

of addition of analysis and filtering tasks, or for efficiently bypassing unused instruments.

Modern programming languages such as FORTRAN 90 and C++ were used in a combined way. Great care was taken to only select libraries from the public domain so that distribution of SciSim is not subject to limitations due to software licensing issues.

Detailed documentation is also made available, which allows efficient use of SciSim. Furthermore, documentation of the source code is also publicly available and should be at a level that is adequate for future projects to re-use SciSim. Adaptation not only of configuration files, but also of code, can be performed in a controlled manner where necessary.

Because several separate processes are used for one simulation, the tasks can be split among several processors and executed in parallel. The performance of SciSim easily allows simulations of exposures faster than real time.

5. User Interface and Setup of Simulations

The user interface is provided by a set of GUIs which control the simulation. Separate widgets are started for access to the most commonly used parameters (additional flexibility is provided via ASCII configuration files). Figure 1 shows the main GUI which summarizes and controls the status of the simulation. The large window shows a celestial display. Source positions are indicated by small dots, and the field of views of the three basic instruments (EPIC, RGS, OM) are represented by colored lines. Positions of sources may either be provided by the user via a dedicated widget, where the spectral parameters are also accessible, or may be obtained from a catalogue. Several sources can be defined for the same simulation with different emission spectra.

To help the user with the identification of the necessary modules, a schematic layout of the data flow is provided (see Fig. 2). Through this interface, the control of each simulation component is achieved by separate widgets.

To support the user with the input for each field, two levels of help are available. When moving with the mouse to any field, tool-tips provide additional description. Secondly, a help button is available for all widgets. When pressed, this button displays the locally available HTML formatted on-line documentation using an internet browser.

References

Lumb, D. H., Eggel, H., Lainé, R., & Peacock, A. 1996, in SPIE Proc., Vol. 2808, EUV, X-Ray, and Gamma-Ray Instrumentation for Astronomy VII, ed. O. H. Siegmund & M. A. Gummin, (Bellingham: SPIE), 326

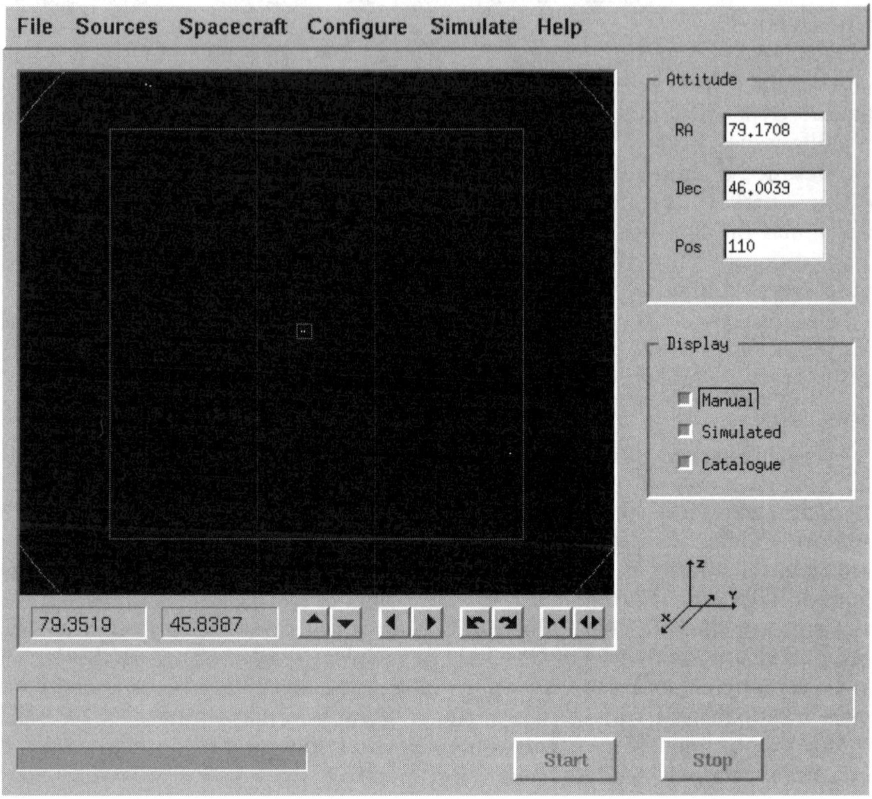

Figure 1. The main SciSim GUI.

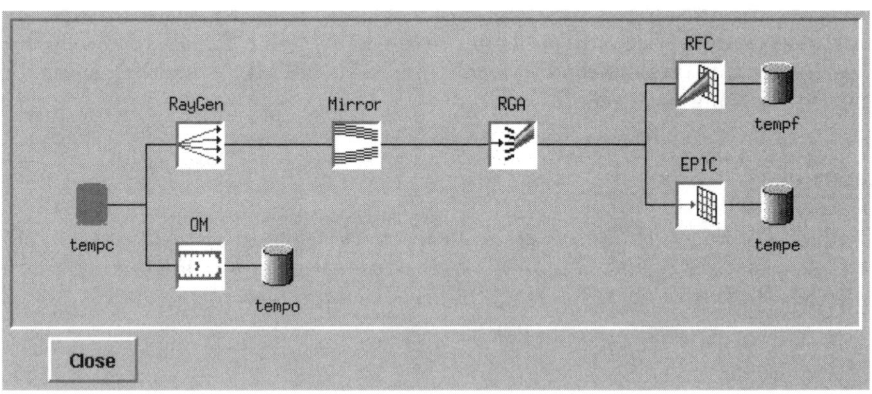

Figure 2. Schematic layout of the data flow that is also used for the control of instrument simulation parameters.

Part 5. Data Compression

Astronomical Data Analysis Software and Systems VIII
ASP Conference Series, Vol. 172, 1999
D. M. Mehringer, R. L. Plante, and D. A. Roberts, eds.

A Scheme for Compressing Floating-Point Images

Richard L. White, Perry Greenfield

Space Telescope Science Institute, 3700 San Martin Dr., Baltimore, MD 21218

Abstract. While many techniques have been used to compress integer data, compressing floating-point data presents a number of additional problems. We have implemented a scheme for compressing floating-point images that is fast, robust, and automatic, that allows random access to pixels without decompressing the whole image, and that generally has a scientifically negligible effect on the noise present in the image. The compressed data are stored in an FITS binary table. Most astronomical images can be compressed by approximately a factor of 3, using conservative settings for the permitted level of changes in the data.

We intend to work with NOAO to incorporate this compression method into the IRAF image kernel, so that FITS images compressed using this scheme can be accessed transparently from IRAF applications without any explicit decompression steps. The scheme is simple, and it should be possible to include it in other FITS libraries as well.

1. Introduction

Our goal is to develop a compressed image format that is suitable for use as a working format — images will be stored and used directly in the compressed format. This is in contrast to most compressed file formats (e.g., gzip), where the entire file must be uncompressed before the data can be used. We also want the compressed data to be of archival quality, so that access to the original uncompressed data is not necessary for any scientific purposes. With these goals in mind, there are a number of desirable characteristics for the compression method:

Fast in execution time. The time to read and decompress the data and should be comparable to (or faster than) the time to read the uncompressed data.

Effective in compressing the data. It should produce compression factors competitive with the best available methods.

Lossless (exactly reversible) for integer data. Lossy compression methods can produce substantially higher compression ratios. However, for most modern astronomical detectors the noise in the raw data is sufficiently low that it must be compressed losslessly for archival purposes.

Small memory requirements. Any memory needed for the compression or decompression should be much smaller than the full image size.

Random access to pixels. It must be possible to read a small image section located anywhere within the image without having to decompress the entire image.

For nearly all floating point astronomical images, the least significant bits of the float mantissa are essentially random and so are incompressible. Zeroing out some of these random bits corresponds to quantizing the floating point values. This leads to some additional considerations:
Automatic setting of quantization parameters. The appropriate quantization parameters should be determined directly from the data.
Nearly lossless quantization. The quantization must not significantly increase the noise in the data for scientific purposes.
Unbiased noise properties. Features smaller than the quantization should be recoverable from the mean of many data sets.
Easily understandable effect on noise. The quantization scheme should not introduce correlations in the noise in adjacent pixels. It must therefore operate on each pixel independently, which eliminates, e.g., wavelet-transform based methods like `hcompress` (White, Postman, & Lattanzi 1992; White & Percival 1994).

2. Compression Overview

We quantize and compress the data one row at a time. This allows random access to different parts of the image; no more than one row need be decompressed to access any pixel, and when a pixel value changes, it affects only a single row in the compressed image. It also helps avoid bias and dynamic range problems, which can result from a global quantization of the image. Finally, it makes the quantization more spatially adaptive so that it can respond to noise characteristics that change across the image.

For each row, there are 3 basic steps in the algorithm: (1) Estimate the noise level in the row. (2) Convert float pixel values to integers by linearly quantizing them at levels that are a specified fraction of the noise. (3) Losslessly compress the integer pixel values using the Rice algorithm. (Note that for integer images, the first 2 steps are not necessary.) The steps are described in more detail below.
Noise Estimation The rms noise σ in each row is robustly estimated as 1.483 times the median of the absolute value of the differences of successive pixels. If the median absolute difference is zero, then we fall back on a direct computation of the rms using a sigma-clipping algorithm to reject large outliers.
Quantization The separation of quantization levels is $2^{-B}\sigma$, where B is the user-specified number of bits of accuracy in the quantization of the noise. By specifying the quantization level in terms of σ, we ensure that a given value of B will produce comparable quality images regardless of the noise level. B is directly related to the compressed file size: increasing B by 1 will increase the file size by 1 bit/pixel.
Rice Coding We use the Rice algorithm (Rice, Yeh, & Miller 1993) to compress the integer data values. The Rice algorithm is simple and very fast, compressing or decompressing $\sim 5 \times 10^6$ pixels/sec on modern workstations. It requires only enough memory to hold a single block of 16 or 32 pixels at a time. It codes the pixels in small blocks and so is able to adapt very quickly to changes in the input image statistics (e.g., Rice has no problem handling cosmic rays, bright stars, saturated pixels, etc.).

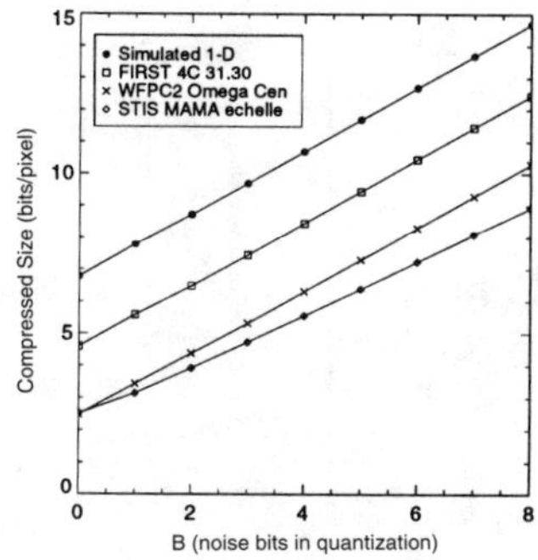

Figure 1. Compression as a function of quantization parameter B for floating point images.

For lossless compression schemes, the compressed file size depends on the nature of the image. For typical images the compressed file size is about 4 to 8 bits/pixel for $B = 1$; one can thus keep 4–6 bits in the noise (which is plenty, as we show below) and still achieve a compression factor of 3 from the original 32 bit/pixel floating point image.

Data Format We store the compressed data in FITS binary tables. Each row of the input image is associated with a row of the table. The compressed byte stream is stored as a variable length array in the table heap. The quantization scale factors are stored as additional columns.

3. Performance on Float Images

For floating point images, one can effectively choose the compression factor for the data by varying the target number of quantization bits B. In this case the compressed image has slightly different pixel values than the original image, so we are concerned both with the compression factor and the degradation of the data. We assess the data quality by applying realistic data analysis procedures to the data before and after compression.

Compression Performance The size of the compressed data (in bits/pixel) is linearly related to the number of noise bits B used in quantization. The compression as a function of B for various images is shown in Fig. 1.

WFPC2 Astrometry & Photometry Test We tested the effects of quantization on astrometry and photometry using Omega Cen observations and the WFPC2 photometry script (Whitmore & Heyer 1995). The effect of quantization is much smaller than the noise (Fig. 2).

Bias Test To test for bias, we constructed 1-D simulated data with a weak signal buried in the noise. The "truth" spectrum includes a strong source at pixel 75 and a weak source at pixel 25. Each simulated realization of the spectrum has Gaussian noise of rms 1 count added (Fig. 3a). We averaged 1000 such spectra to recover the faint signal; the mean spectrum is shown in Fig. 3b, and the mean of coarsely quantized ($B = 1$) spectra in Fig. 3c. The signal is smaller than the quantization but is recovered very accurately. The quantization method is quite unbiased and introduces almost no additional noise into the data.

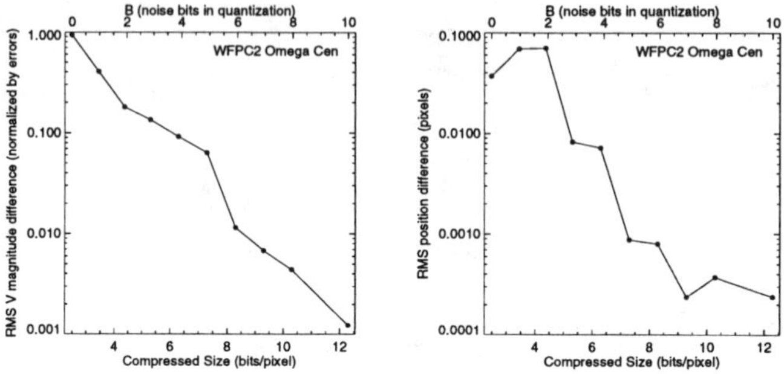

(a) (b)

Figure 2. (a) RMS difference in magnitudes (normalized by uncertainty in the magnitudes) as a function of quantization scale factor B (*top axis*) and the compressed size (*bottom axis*). (b) RMS difference in positions as a function of quantization scale factor B.

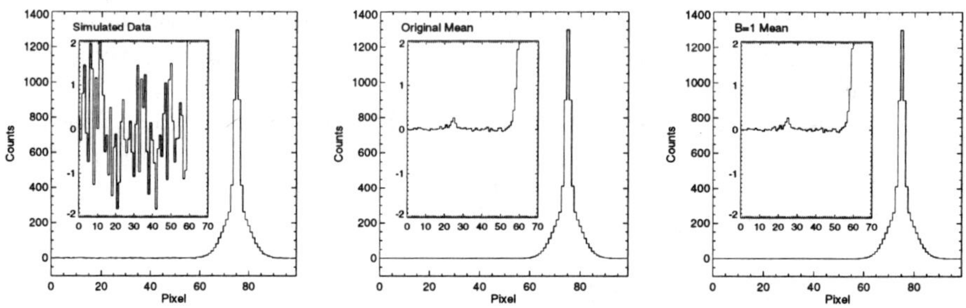

(a) (b) (c)

Figure 3. (a) Sample realization of simulated data. (b) Mean spectrum. (c) Mean of $B = 1$ quantized spectra.

Acknowledgments. Thanks to Eric Wyckoff for his assistance in testing.

References

Rice, R. F., Yeh, P.-S., & Miller, W. H. 1993, in Proc. of the 9th AIAA Computing in Aerospace Conf., (AIAA-93-4541-CP), American Institute of Aeronautics and Astronautics

White, R. L. & Percival, J. W. 1994, in SPIE Proc., Vol. 2199, Advanced Technology Optical Telescopes V, ed. L. M. Stepp, (Bellingham: SPIE), 703

White, R. L., Postman, M., & Lattanzi, M. G. 1992, in Digitised Optical Sky Surveys, ed. H. T. MacGillivray & E. B. Thomson (Amsterdam: Kluwer), 167

Whitmore, B. & Heyer, I. 1995, "A Demonstration Analysis Script for Performing Aperture Photometry, WFPC2 ISR 95-04," `http://www.stsci.edu/instrument-news/isr/wfpc2/9504/9504_1.html`

Astronomical Data Analysis Software and Systems VIII
ASP Conference Series, Vol. 172, 1999
D. M. Mehringer, R. L. Plante, and D. A. Roberts, eds.

Adaptive, Lossless Compression of 2-D Astronomical Images

C. N. Sabbey

Department of Astronomy, Yale University, New Haven, CT 06520-8101

Abstract. We describe a lossless compression program developed for incorporation into the data acquisition system of a 16 CCD drift scan survey. The program is based on switched linear prediction with Rice coding of the high order bits of the mapped prediction residuals. Several notable features of the program are its speed (≈ 5 MB/sec on a 300 MHz Pentium II), adaptivity to a variety of image types without any user-tuned parameters, and suitability for online work (it is a one-pass algorithm with minimal buffering requirements and a steady output stream).

The compression ratios and run times of the program are compared to other well-known compression programs using a variety of test images. A stand-alone version of the **encode** program is available in C from `ftp://www.astro.yale.edu/pub/sabbey/encode.tar.gz`.

1. Introduction

Although numerous programs are available for the lossy compression of astronomical images (e.g., White & Percival 1994; Starck et al. 1996), much less work has been published on lossless compression, in particular for online applications (although see Veran & Wright 1994). As a result, in producing the online software for the QUEST (QUasar Equatorial Survey Team) 16 CCD drift scan survey (Sabbey, Coppi, & Oemler 1998), we designed a custom, lossless compression program. The real-time compression provided has proved crucial for reliably acquiring and storing the nightly 30 GB of drift scan image data with the existing hardware system.

The program was required to be fast (i.e., compress at a rate comparable to the disk I/O), work in one-pass with minimal buffering, be easy to implement and incorporate into a complex system, allow streaming of the data to disk or tape, adapt to a variety of image types, and provide compression ratios comparable to more sophisticated offline programs without requiring any user-tuned parameters.

To accomplish this, we started with the well-known technique (Rice 1991) of Rice coding (Rice & Plaunt 1971) the difference between each pixel value and its linear prediction (a linear combination of the already-transmitted, neighboring pixel values). More specifically, the prediction residuals are partitioned into the N low order "noise" bits, which cannot be compressed, and the remaining (16-N) high order bits, which form a sharply peaked (Laplacian) distribution, and can be efficiently transmitted as Rice codes. Rice coding is a form of entropy coding

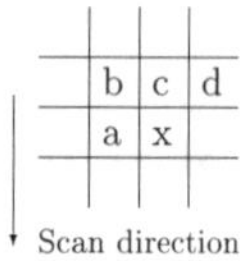

Graham's Rule:

$$\hat{x} = \begin{cases} a & \text{if } \delta_x \leq \delta_y \\ b & \text{otherwise} \end{cases}$$

Modified Rule:

$$\hat{x} = \begin{cases} (a+b+c+d)/4, & \text{if } \delta_x < \epsilon,\ \delta_y < \epsilon \\ a, & \text{if } \delta_x < \epsilon,\ \delta_y \geq \epsilon \\ c, & \text{if } \delta_x \geq \epsilon,\ \delta_y < \epsilon \\ a + (c-b)/2, & \text{if } \delta_x \geq \epsilon,\ \delta_y \geq \epsilon,\ \delta_x < \delta_y \\ c + (a-b)/2, & \text{if } \delta_x \geq \epsilon,\ \delta_y \geq \epsilon,\ \delta_x \geq \delta_y \end{cases}$$

Figure 1. A prediction, $\hat{x}$, is formed for the current pixel, x, using one of several possible linear combinations of the neighboring pixels (a, b, c, and d). A modification of Graham's rule is used to select one of the linear combinations based on a comparison of the local gradient approximations ($\delta_x = |c - b|$ and $\delta_y = |a - b|$) to a noise threshold ϵ.

(in which more frequently occuring values are assigned shorter codewords) that is less optimal but faster than Huffman coding.

To extend this technique of Rice coding the most significant bits of the differential signal, we investigated the selection of the linear predictor, the calculation of the partition size N, and how to make both adaptive at each pixel of the image in a fast and automated way without transmitting overhead information. The adaptivity is crucial for obtaining consistently high compression ratios and eliminating parameters that require user tuning. Some examples of image variation that the program should adapt to are changes in the background level, the noise level, the object surface density and shape, the image type (e.g., photometry or slitless spectroscopy), and the existence of bad CCD columns and other defects.

To adaptively select the predictor we modified Graham's rule (see Netravali & Lamb 1980, and Fig. 1) for use with astronomical images. In relatively flat image regions, the mean of several neighboring pixels will be selected as the predictor; otherwise the direction and order of a predictor will be selected based on the local correlation. For example, pixel values in a bad CCD column will be predicted using the adjacent previous row value, while a higher order predictor in the dispersion direction will be used to predict bright spectrum pixels in a slitless spectroscopy image.

An alternative approach to making the prediction adaptive is to subdivide the image into blocks (e.g., 8×8). Then for each block all the predictors can be tested and the most effective one used. However, for our image set the optimal block size and shape varied among the images and within a given image. Also, the blocks were not aligned with the image features, and there was a conflict between using small blocks to allow rapid adaptivity but needing large blocks to limit the overhead of indicating the predictor used. Due to these and other obstacles (e.g., the increased computation required), we resorted to the modified Graham's rule described above.

The partition size (number of "noise" bits $N = N_s + N_r$) is adaptively selected at each pixel from the sum of a slowly varying term, N_s, which accounts for the background noise level, and a rapidly varying term, N_r, which accounts for structure remaining in the prediction residual image. N_s for a given image block is obtained using the number of 1-bit Rice codes transmitted in the previous image block (N_s is adjusted to keep the probability of Rice codeword 0 near

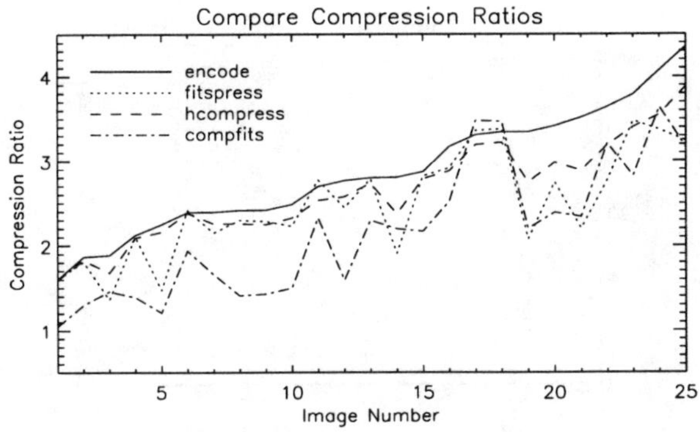

Figure 2. The performance of **encode** is evaluated by comparison to the lossless output of three well-known compression programs. The **encode** program often provides significantly improved compression ratios (the average percent difference is 22% improvement).

optimal). N_r is obtained at each pixel from a table lookup using the prediction residual from the previous pixel as an index.

The compression ratios and run times of the **encode** program were evaluated by comparison to three well-known compression programs: **hcompress**, **fitspress**, and **compfits+compact**. The test images used are listed in Table 1, and the results are presented in Figs. 2 and 3. Although the three comparison programs have important uses beyond lossless compression (e.g., **hcompress** and **fitspress** are often used for lossy compression), they nonetheless represent the state of the art and provide a relevant standard. Further details of the **encode** algorithm can be found in Sabbey et al. (1998).

References

Netravali, A. N. & Limb, J. O. 1980, Proc. IEEE, 68, 366

Rice, R. F. 1991, Some Practical Universal Noiseless Coding Techniques, Part III, Module PSI14,K+ (JPL Publ. 91-3), (Pasadena: JPL)

Rice, R. F. & Plaunt, J. R. 1971, IEEE Trans. Comm. Tech., 19, 889

Sabbey, C. N., Coppi, P., & Oemler, A. 1998, PASP, 110, 1067

Starck J.-L., Murtagh, F., Pirenne, B., & Albrecht, M. 1996, PASP, 108, 446

Veran, J. P. & Wright J. R. 1994, in ASP Conf. Ser., Vol. 61, Astronomical Data Analysis Software and Systems III, ed. D. R. Crabtree, R. J. Hanisch, & J. Barnes (San Francisco: ASP), 519

White, R. L. & Percival, J. W. 1994, in SPIE Proc., Vol. 2199, Advanced Technology Optical Telescopes V, ed. L. M. Stepp (Bellingham: SPIE), 703

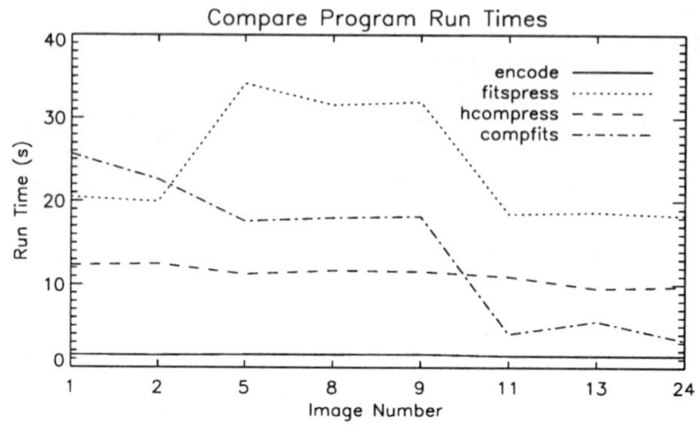

Figure 3. The speed of **encode** is evaluated by comparison to three well-known compression programs (using a 300 MHz Pentium II). Only the large ($\approx$ 8 MB) images were used to avoid measuring times $\ll 1$ s and to use a consistent image size. The **encode** program is found to be an order of magnitude faster in many cases.

Table 1. 25 test images used to evaluate the **encode** program.

No.	Name	Size	Description	Bkg Noise
1	simn200	2048×2048	IRAF simulated stars/galaxies	200.0
2	prism	2048×2048	objective prism slitless spectroscopy	70.0
3	ngc0001	321×507	NGC 3210 in V, globular cluster	45.5
4	com0001	500×500	halo of NGC 4874, strong image gradients	50.0
5	flat	2176×2048	Flatfield image, tapered edges	40.0
6	noise	1024×1024	Simulated noise image, flat background	20.0
7	gal0002	264×427	galaxy cluster (coadded)	24.7
8	grism2	2176×2048	grism slitless spectroscopy	26.0
9	grism	2176×2048	grism slitless spectroscopy	25.0
10	m57	1080×1080	M57 Ring Nebula, strong image gradients	10.0
11	simusamp	2048×2048	IRAF simulated undersampled stars/galaxies	10.0
12	fiber	100×512	IRAF simulated multifiber spectra image	3.0
13	simosamp	2048×2048	IRAF simulated oversampled stars/galaxies	10.0
14	ngc0002	322×508	NGC 3210 in V, shorter exposure	7.5
15	gal0001	320×503	galaxy cluster 0637-53	9.7
16	sgp0001	234×441	deep SGP in J	7.7
17	gal0004	1000×1000	SA 68	5.9
18	gal0003	1000×1000	SA 68	6.0
19	tuc0003	420×246	47 Tuc off-center in V	8.6
20	for0001	305×512	Fornax dwarf in V, crowded star field	8.6
21	tuc0004	420×246	47 Tuc off-center in B	6.3
22	hst	800×800	HST image of globular cluster, crowded	2.0
23	sgp0002	234×441	deep SGP in R	4.0
24	simn2	2048×2048	IRAF simulated stars/galaxies	2.0
25	for0002	305×497	Fornax dwarf in B, crowded star field	4.8

Of the 25 test images used to evaluate the **encode** program, 13 comprise the "standard test images" suite used at the Two-Dimensional Photometry Systems and Test Images Workshop (ESO/ECF 1989). The remaining 12 images include simulated and observed photometry and spectroscopy image data with varying noise levels, crowding, etc.

Astronomical Data Analysis Software and Systems VIII
ASP Conference Series, Vol. 172, 1999
D. M. Mehringer, R. L. Plante, and D. A. Roberts, eds.

Contour-Based Image Compression for Fast Real-Time Coding

Sergei Vasilyev

SOLERC, Institute for Electromagnetic Researches, P.O. Box 30, Kharkiv, 310052, Ukraine

Abstract. A new method based on simultaneous contouring the image content with subsequent converting of the contours to a compact chained bit-flow, thus providing efficient spatial image compression, is proposed. It is computationally inexpensive and can be directly applied to compressing the high-resolution bitonal imagery, allowing to approach the ultimate speed performance. Combining the method with other compression schemes, for example, Huffman-type or arithmetic encoding, provides better lossless compression to the current telecommunication compression standards. The problems of method application to compressing the color images for remote sensing and mapping applications, as well as lossy method implementation, are discussed.

1. Introduction

The current trends in astronomy require essential improving the resolution and broadening the spectral range of imagery acquired by space borne instruments. Better productivity of space observations yields enormous growths of data volumes that causes a serious technical problem of reducing the increase in data storage and space transmission costs. As it is not uncommon for multispectral/hyperspectral remote sensing, a typical data set for a single frame of imagery may occupy a few gigabytes of space. The onboard compression can allow to essentially unload the data transmission channels, as well as enable more efficient data storage and processing.

Most of the existing effective image encoding techniques designated to still image compression are not suitable to parallel operation and often fail with large real images, as they assume loading the entire image or its significant portion into the memory and usually have fairly high demand to the onboard computer performance and storage media capacity. The state of the art transform-based techniques, including wavelet and fractals, implicate complex mathematical transformations involving every pixel of the image, thus resulting in very active CPU use. High speed readout processing, which is crucial for very dense inputs, can also be hardly attainable with these techniques. The existing real-time encoding schemes, allowing incremental image compression, like those implemented in CCITT/ITU-T telecommunication coding standards, are usually behind the leading still image compression techniques. They rarely give a compression factor of more than 10 in the lossless mode, and still do not give satisfactory speed performance with the high-resolution imagery.

In this paper we describe an attempt to implement a simple and robust compression algorithm, which would combine reasonable compression rate and high coding speed to be suitable for real-time processing the large images.

2. Implementation

The discussed technique is designed to be most efficient for compressing the bitonal imagery, however, like many other bitonal compression schemes, its application range can be extended to other image types. Unlike most leading bitonal image compression techniques being currently in use that implement direct bit map compression across the raster lines with run-length, arithmetic or other coding schemes, the contour-based technique embodies another approach. It exploits spatial information redundancy, which resides in most real images, and takes advantage of replacing the regions of adjacent pixels with the same either color or classification characteristics by their contours, thus eliminating this redundancy. The approach allows efficient and fast one- stage compression through image decomposition into a set of contours and binding the contours in the chain link bit-flow, thus storing only those bits belonging to the contours.

Contour recording is performed through detection of each contour and sequential coding the relative coordinates of the contour pixels exploiting their continuity and logical linking. Obviously, every successive move along the contour line can be completely defined by only three bits. To achieve better compression, some additional logical restrictions to the possible moves can be applied, so that only two bits per pixel may be used providing virtually lossless compression. In this case a minimum buffering is required to keep the previous running coordinate for each contour pixel. Figure 1 illustrates contour recording with restrictions based on the previous contour pixel coordinates and clockwise contour bypass rule. As a result of encoding, the contours represented in the original bitmap become stored in the one-dimensional bit sequence. The contour entry coordinates can be stored either separately or in the resulting bit flow.

Actual implementation of the contour-based compression algorithm implicates simultaneous edge detection and storage without the need of preliminary image contouring. Our algorithm compresses images line by line and handles only current relative positions on particular edges at a time, so that the resulting bit flow is shaped by sliding the current bypass point along the successive edge pixels of the image regions with same attribute pixels. Therefore we also call this algorithm and general approach as Running Edge Image Compression (REIC).

The REIC implementation for compressing the grayscale and color images assumes performing the Gray coding (Gray 1953) and bit plane separation first and, then, passing each plane to a separate instance of the algorithm (see, e.g., Abdat & Bellanger 1994 for an example of Gray/bilevel coding combination). An alternative approach assumes assigning the classification codes based on image color mode and bit- depth to each contour object with subsequent storage of the codes in the compressed bit sequence.

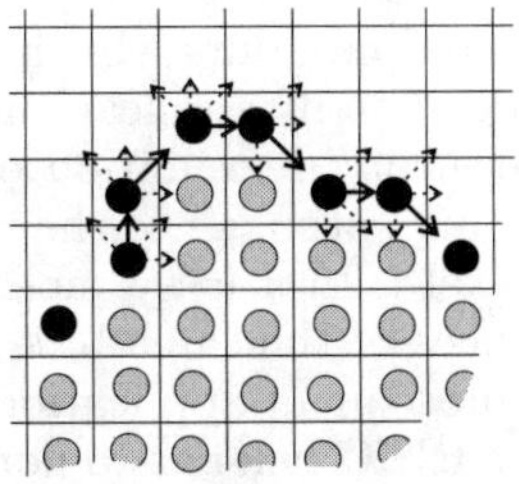

Figure 1. Clockwise contour following and simultaneous two-bit restricted coding sequence determination using the previous contour pixel position. The area being contoured is shown as light-tone circles. Dashed arrows show the allowable moves to the adjacent positions for each contour pixel along the contour line (solid circles), and the solid arrow shows the actually selected move.

3. Compression rates and performance

We performed the preliminary comparative tests of the contour-based compression technique with the high resolution bitonal images. Our tests have shown that, in the two bits per pixel mode, the contour-based encoding combined with the generic GZIP encoding gives about 30 per cent better compression for still 600 dots per inch images to the widely used CCITT/ITU-T Group 4 and about three times to Group 3 compression standards. Application of the REIC technique alone gives near-equal compression to Group 4, however, results in several times higher encoding speeds. The intrinsic nature of the contour-based compression allows to utilize mathematical logic for edge detection and contour encoding rather than arithmetic operations whenever possible, thus reducing the CPU use and approaching the ultimate speed performance. Higher resolution images allow for increasingly better compression.

4. Discussion

While implementation of the REIC technique steadily exhibits good compression in our tests with a variety of bitonal images, its efficiency for other kinds of imagery is essentially determined by image type and nature. Among the factors affecting the contour-based compression, most important are image noisiness and complexity. When compressing color imagery, the color depth puts additional limits to REIC efficiency. Our tests gave satisfactory results with high-resolution noise-free images of up to 8 bits per pixel.

Scanning line operation of the REIC allows it to be conjoined well with the spectral compression technique proposed by the author earlier (Vasilyev 1998), and these two techniques can be concurrently implemented for efficient onboard multispectral/hyperspectral image compression.

Lossy version of the REIC technique assumes sparse representation of contours first, thus providing additional compression, and then coding the contours in an ordinary way. In this scheme, the image decompression is performed beginning with interpolation of the contours. Such a lossy-type scheme can be customized and optimized for different applications to achieve maximum compression ratio with negligible visual degradation of the image. It also should not significantly affect the coding speed, since the required transformations will be applied to only contour pixels. In a lossy mode, the resulting bit flow will have longer continuous sequences of the same bits, thus giving some additional compression opportunity for subsequent application of other coding schemes. It should be noted that the lossy REIC belongs to nondegrading methods, and, as once applied, its repeated application to the same image does not lead to any further change to the image.

5. Conclusion

The operation principles of the contour-based compression allow to encode images in the scanning-line mode that makes the technique applicable to the real time image processing systems and on board compression. The approach allows sequential encoding of images of virtually any size including multispectral and hyperspectral imagery, while consuming small amount of RAM memory and operating in a computationally inexpensive way with no need for special hardware. The method can be promisingly efficient for automatic-mode high-resolution color image recording and mapping when combined with the pattern recognition and edge detection techniques. It can also be suitable for document imaging and sending imagery applications.

Acknowledgments. The author is grateful to the ADASS VIII Organizing Committee and European Southern Observatory for offering him the financial assistance to attend the Conference.

References

Abdat, M. & Bellanger, M. G. 1994, in Proc. IEEE First Intl. Conf. on Image Processing, 3, 851

Gray, F. 1953, U.S. Patent 2632058

Vasilyev, S. 1998, in ASP Conf. Ser., Vol. 145, Astronomical Data Analysis Software and Systems VII, ed. R. Albrecht, R. N. Hook, & H. A. Bushouse (San Francisco: ASP), 504

Astronomical Data Analysis Software and Systems VIII
ASP Conference Series, Vol. 172, 1999
D. M. Mehringer, R. L. Plante, and D. A. Roberts, eds.

Data Compression for NGST

María A. Nieto-Santisteban, Dale J. Fixsen[1], Joel D. Offenberg[1], Robert J. Hanisch, H. S. (Peter) Stockman

Space Telescope Science Institute, 3700 San Martin Drive, MD 21218
Email: nieto, fixsen, offenbrg, hanisch, stockman@stsci.edu
URL: `http://ngst.gsfc.nasa.gov/cgi-bin/iptsprodpage?Id=14`

Abstract. The immense amount of data that the Next Generation Space Telescope (NGST) will produce and its distant orbit from Earth make it mandatory to do some amount of on-board image processing and data compression. This paper gives a summary of the performance of several lossless compression methods. We also show results of prescaling the image prior to compression using a square-root function. This imposes a slightly lossy compression, but the scaling can be adjusted so as to retain the desired number of noise bits.

1. Introduction

The Next Generation Space Telescope (NGST) will produce about 600 GB/day, assuming we use the NASA Yardstick 8k × 8k NIR camera (16 bits/pixel), save and transmit 64 non-destructive read-outs per image, and the camera is in continuous use (about 80 observations/day, 10^3 s each). However, with an L2 halo orbit, the NASA NGST study estimates a downlink rate of 5.35 GB/day using X-band. Clearly the volume of data to downlink must be reduced by at least a factor of 100.

Astronomical images are noisy. This fact makes them difficult to compress by lossless compression algorithms such as Huffman, Lempel-Ziv, run-length, or arithmetic code. However, they also have the virtue of showing similar values among adjacent pixels. Techniques such as Rice's algorithm (Rice, Yeh, & Miller 1993) and derivatives (White & Becker 1998; Stiavelli & White 1997) can take advantage of this. In this paper, we present how some of these compression techniques would work with NGST images. Unfortunately, these lossless algorithms give us compression ratios that still exceed the telemetry guidelines. We have also looked into the feasibility of doing lossy compression by scaling the original image prior to the lossless compression. Under this scheme, we find substantial data reduction with a negligible effect on the data quality.

2. The Data Compression Process: Ratios

The first and more important compression ratio, **64:1**, is obtained after applying a cosmic ray rejection process and fitting 64 readouts into one single image

[1]Raytheon STX Corporation, 4400 Forbes Blvd, Lanham, MD 20706

(Offenberg et al. 1999). (In fact, 65 readouts involve the cosmic ray rejection and fitting process, however, the first readout [dark frame] should be rarely downlinked.) An additional compression factor close to **3:1** is achieved by using predictive compression techniques such as the Lossless JPEG and Rice's algorithm. Dictionary-based lossless compression programs such as "gzip" and "compress" present lower compression ratios (see Table 1).

We can further reduce the data volume by using a prescaled image as input to the lossless encoder. The prescaling process, based on the square-root function, can be adjusted so as to retain as many bits of noise as desired. Similar results are obtained independently of the lossless compression technique in use, with overall compression ratios of **4:1** (keeping 4 noise bits), up to as much as **8:1** (keeping 1 noise bit).

Table 1. Performance Summary.

Compressor	Time (s)	Memory (kB)	Lossless (bpp)	Lossy (bpp) $N_b = 1$	$N_b = 2$	$N_b = 3$	$N_b = 4$
rcomp	0.253	2904	5.473	2.104	3.087	3.612	3.997
uses	0.737	4040	5.477	2.107	2.988	3.529	3.952
LLjpeg	2.539	4152	5.341	2.097	2.933	3.431	3.877
gzip	4.871	1112	6.369	2.125	3.233	3.886	4.425
compress	0.747	1600	6.036	1.691	2.762	3.474	4.015
RMS Error				0.510	0.252	0.186	0.123
Mean Diff.				–0.079	–0.024	–0.078	0.007

This table give the summary in time, memory and compression rates (lossless and lossy) achieved for five **lossless compressors**. As test input data, a simulated NGST deep image was used (1024 $\times$ 1024, 2 bytes per pixel, DN units) obtained in a 10^3s exposure after the cosmic ray removal and slope fitting algorithm. A readout noise of 15 electrons and a gain of 4 are assumed. Scaling function applied for lossy compression, $N_B\sqrt{G}\sqrt{D+Y'}$. Values of the Normalized Root-Mean-Square Error and Mean Difference are also shown. *Rcomp* and *uses* are implementations of the Rice algorithm. *Rcomp* was developed at STScI by Rick White. *Uses* was developed at University of New Mexico Microelectronics Research Center. *LLjpeg* is a lossless JPEG developed by the PVRG-JPEG. *gzip* and *compress* are the well known general purpose compression programs based on dictionary techniques. The tests were run on a Sun Ultra 10.

3. Lossy Compression: Square-Root Prescaling

Pure noise by its very nature is impossible to compress. In order to compress the data and retain as much information as possible, it is useful to eliminate the low order data bits (i.e., the noise). Truncating the bits at some level is one possibility, but the noise level differs across the picture. We describe a simple approximation to the noise which is carried in the data itself.

There are two important sources of non-systematic noise: the Poisson distribution of the photons themselves and the readout noise in the readout electronics. From these two sources, a reasonable approximation to the standard deviation is $\sigma = \sqrt{S+R}$ where S is the signal in units of photons (or equivalently electrons) and R is the readout variance in the same units.

In order to simplify both the encoding and the decoding we make an approximation to $\sigma \simeq \sqrt{S+Y}$ where S is the readout (in electrons) and Y is a number which we will derive from the readout variance R. First note an absolute offset does not incur a significant penalty in data transmission since virtually any lossless data compression scheme will use very little bandwidth to transmit the offset in a large block of data. The key then is to get a good approximation to the derivative. Note if we use $\sqrt{S+Y}$ for $S/\sqrt{S+R}$ for large S the derivatives already match. By taking the derivatives with respect to S and setting them equal to each other at $S = 0$ we have $Y = R/4$. Thus we can use $\sqrt{S + R/4}$ and truncation as a lossy compression to reduce the data dynamic range in an approximately uniform way. Since our NGST simulation is in data numbers (DN), we must rescale this formula to $\sqrt{G}\sqrt{D+Y'}$, where G is the gain, D is the signal in units of DN and $Y' = R/4G$. Therefore, assuming a 15 electrons readout noise and a gain of 4, $Y' = 14$. (Note $R = 15^2$).

The remaining issue is where to do the truncation. We multiply by a value N_B and round to the nearest integer. The value N_B is then the number of bits into the noise that we save. Setting N_B to 1 has the effect of severely restricting the noise at each pixel but it also may affect averages and fits that attempt to pull a signal out of the noise. Larger values of N_B allow the compressed data to more closely match the original values, but at less effective compression ratios (see RMS Error in Table 1). We suggest using $N_B \sim 2$, a value that gets us close to but under the $1/\sqrt{12}$ that we could expect from the digitization noise. Figure 1 shows how this lossy scheme does not introduce any pattern or systematic error into the residual images. The small bias shown is due to the digitization process. If this were a problem in practise, we could add a random bias which will (on average) remove it.

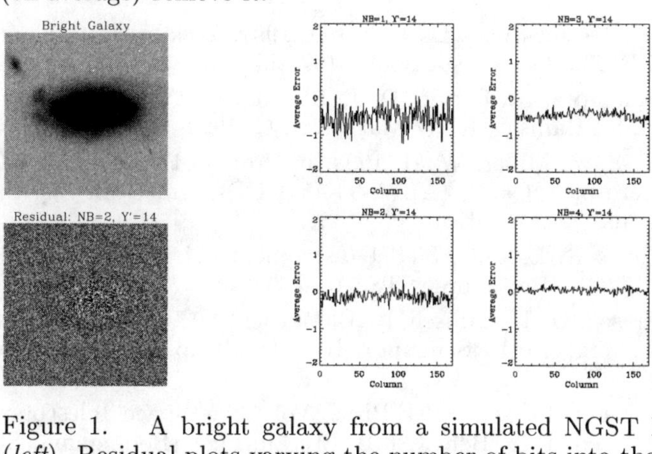

Figure 1. A bright galaxy from a simulated NGST long exposure (*left*). Residual plots varying the number of bits into the noise (*right*).

4. Compression Benefits from Cosmic Ray Removal

For the the maximum exposure time we have adopted, 10^3 s, it is expected that 10% of the image is affected by cosmic rays hits (Stockman et al. 1998). Most of the cosmic rays produce high values in one single pixel unrelated to other values in the vicinity. All tested compression programs benefit from prior cosmic ray removal. The Rice implementation used in this paper utilizes a linear first-order unit-delay predictor whose output is equal to the difference between the input data value and the preceding data value (CCSDS 1997). This algorithm particularly benefits from CR removal.

5. Conclusions

- It is possible to reduce by more than two orders of magnitude the anticipated volume of data.
- Our lossy method does not introduce any important distortion into the data. The prescaling may be very useful for scenes with high dynamic range, such as 2-D spectra or images of dense star clusters.
- Compression benefits from prior cosmic ray rejection.
- According to the Performance Summary (Table 1) Rice would be our choice of compression. It is the fastest algorithm. The bit rates given by Rice are among the best. Current implementations require the most memory, but this can be improved. The Consultative Committee for Space Data Systems (CCSDS) recommends Rice for Lossless Data Compression.

Acknowledgments. The authors thank Rick White for his advice and letting us use his Rice algorithm implementation. These studies are supported by the NASA Remote Exploration and Experimentation Project (REE), which is administered at JPL under Dr. Robert Ferraro, Project Manager.

References

Consultative Committee for Space Data System 1997, Lossless Data Compression, (CCSDS120.0-G-1 Green Book), (Washington: NASA)

Offenberg, J. D., Sengupta, R., Fixsen, D. J., Stockman, P., Nieto-Santisteban, M., Stallcup, S., Hanisch, R., & Mather, J. C. 1999, this volume, 141

Rice, R. F., Yeh, P.-S., & Miller, W. H. 1993, in Proc. of the 9th AIAA Computing in Aerospace Conf., (AIAA-93-4541-CP), American Institute of Aeronautics and Astronautics

Stiavelli, M. & White, R. L. 1997, STScI Instrument Science Report, (STScI Publ. ACS-97-02), (Baltimore: STScI)

Stockman, H. S., Fixsen, D. J., Hanisch, R. J., Mather, J. C., Nieto-Santisteban, M. A., Offenberg, J. D., Sengupta, R. & Stallcup, S. 1998, astro-ph/9808051

White, R. L. & Becker, I. 1998, in SPIE Proc., Vol. 3356, Space Telescopes and Instruments V, ed. P. Y. Bely & J. B. Breckinridge, (Bellingham: SPIE), 823

Astronomical Data Analysis Software and Systems VIII
ASP Conference Series, Vol. 172, 1999
D. M. Mehringer, R. L. Plante, and D. A. Roberts, eds.

Cosmic Ray Rejection with NGST

Joel D. Offenberg, Ratnabali Sengupta, Dale J. Fixsen

Raytheon/ITSS, Space Telescope Science Institute, 3700 San Martin Drive Baltimore, MD 21818

Peter Stockman, Maria Nieto-Santisteban, Scott Stallcup, Robert Hanisch

Space Telescope Science Institute

John C. Mather

Code 680, NASA Goddard Space Flight Center

Abstract. The Next Generation Space Telescope (NGST) will perform wide-field imaging in the high-radiation environment of deep space. Outside any protective magnetic fields, cosmic rays will produce many charged-particle events on the imaging detector, affecting as much as 10% of the field-of-view during a baseline 1000-second obervation. We present an algorithm that identifies cosmic ray events in a series of non-destructive readouts during an observation. Test results, in which over 99% of of the cosmic rays are identified and removed without significant degradation of the accumulated image data, are also presented.

1. Introduction—Why Reject Comic Rays?

The Next Generation Space Telescope (NGST), planned for launch in 2007, will be a large aperture, passively-cooled observatory concentrating on the infrared (IR) part of the spectrum. One of the primary missions of the NGST is to provide infrared deep images of distant (high-redshift) galaxies. As of this writing, the project team's goal is to push the NGST observation bandpass to wavelengths as long as 10 microns and beyond. To do this, the telescope's mirror and optical assembly must be kept extremely cold—on the order of 40 K. In order to achieve this low temperature without expensive and massive active cooling equipment, the telescope must be located in deep space. Therefore, the NGST is planned to be launched into an L2 halo orbit, about 1 million miles from the Earth.

This puts the NGST beyond the protective influence of any planetary magnetic field, exposing it to a large number of cosmic rays. When a cosmic ray passes through the detector, its charge will falsely trigger the detector, ruining the data in that part of the detector. During a baseline 1000-second exposure, we anticipate that 10% of the detector will be hit by a cosmic ray. This high level of data loss will significantly impact NGST's science return.

2. Rejecting Cosmic Rays

By "rejecting" cosmic rays, we are referring to techniques to digitally analyze image data and identify and discard cosmic ray events in the detector, preserving only clean data. We can accomplish this by reading the detector frequently and then identifying cosmic rays as unusual events on the detector.

The NGST detectors will have non-destructive read-out capability. Therefore, we can read the detector multiple times during an observation while keeping the integrated observation data. The detector is assumed to have a 16-bit dynamic range and an intrinsic read noise uncertainty of ± 15 electron units. We unrealistically assume, for now, that there are no systematic errors and that we know the dark field and flat field completely. (We plan to add non-trivial dark field and flat field components in future tests.)

Our cosmic ray identification algorithm involves performing 65 non-destructive reads on the detector. We designate these values S_0, S_1 and so on up to S_{64}. S_0 is performed at the start of the sequence (t=0) and the remaining reads are distributed evenly throughout the 1000-second observation (i.e. $t_i = i * 1000/64$). For the purpose of identifying cosmic rays, we compute the differences $D_0 ... D_{63}$, where $D_i = S_{i+1} - S_i$. We then identify cosmic rays events as being instances where $Abs(D_i - Median(D_i)) > 5 * AbsDev(D_i)$. We use the median and absolute deviation (instead of the more traditional mean and standard deviation) because the former are more robust when the data sample contains outliers (Press et al. 1986). In particular, we discovered that the median and standard deviation failed when a data sample was impacted by multiple cosmic rays (0.6% of the detector will be impacted by multiple cosmic rays). We use the absolute value of the difference to avoid biasing the data; since a few data reads (> 1 in 10^4) will randomly lie more than 5 deviations from the median, we must be careful to discard all of the "natural" outliers in both directions from the mean as well as the cosmic rays.

If a cosmic ray is detected in interval j, we repeat this algorithm for $D_0 ... D_{j-1}$ and for $D_{j+1} ... D_{63}$, and so on until no cosmic ray candidates are found. Since we are looking for events at the 5-deviation level, we expect that fewer than 1 "good" data value in 10^4 data points will be rejected falsely.

3. Finding Flux Values

After rejecting cosmic rays, we have a series of data read values $S_0 ... S_{64}$ and a list of zero or more outliers from the mean for the series, the vast majority of which should be cosmic ray events on the detector. To calculate the value of the flux of the object, we apply the optimum slope-fitting routine to the up-the-ramp data, discarding the outliers. (We are concerned only with the slope—the increase in detector counts over time. The zero-point of the line is not needed for this calculation.)

We use a variant on the linear least-squares fit for each segment of consecutive data reads that are not impacted by cosmic rays. We compute a covariance matrix for the data values $M_{i,j}$, which is a tridiagonal matrix where $M_{i,i} = S_i + 2 * \sigma^2$, ($\sigma$ is the readout noise, ± 15 electron units), $M_{i,i+1} = M_{i,i-1} = -\sigma^2$

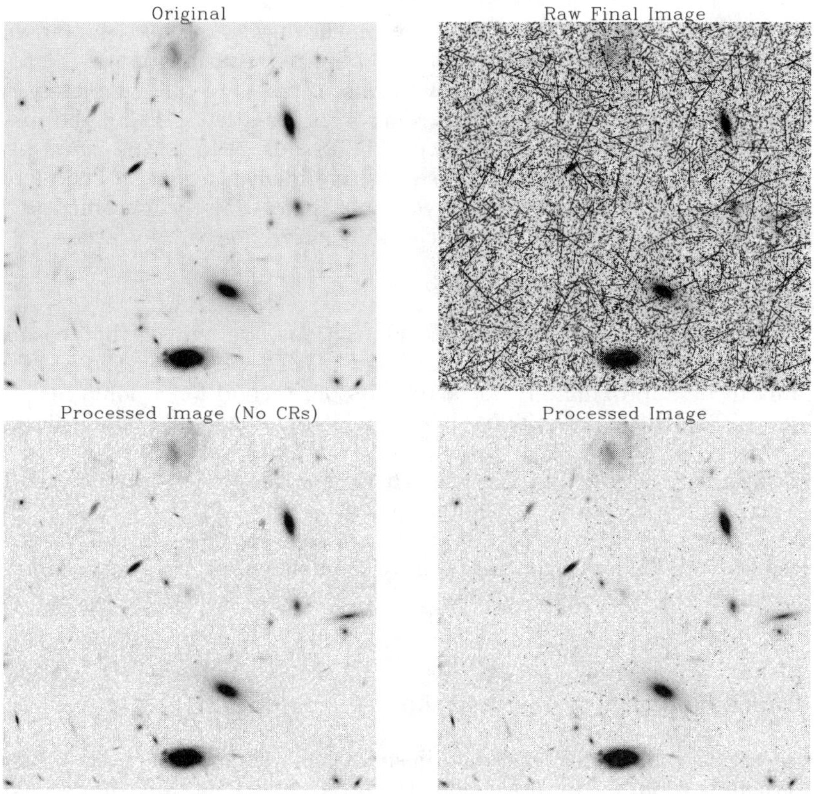

Figure 1. Sample Results using simulated NGST data. Clockwise from upper left: Simulated NGST field of view (Im & Stockman 1998); Simulated FOV as read directly from hypothetical NGST detector (includes noise & detector cosmic ray events, 1000-second exposure); Processed image including cosmic ray removal; Processed image generated from observations with cosmic rays turned off in data simulator.

and $M_{i,j} = 0$ where $j \neq \{i-1, i, i+1\}$. We then compute $C_{i,j} = M_{i,j}^{-1}$, and find the slope of the line $A = \Sigma_{i=0}^{n-1}(S_i * \Sigma_{j=0}^{n-1} C_{i,j})$.

The slope A is computed for each line segment that is not interrupted by a cosmic ray event. Multiple slopes are then combined into a single value using $S_{out} = \frac{\Sigma_{j=1}^{\#segs}[A_j*(N_j)(N_{j-1})(N_{j+1})]}{\Sigma_{j=1}^{\#segs}(N_j)(N_{j-1})(N_{j+1})}$, where A_j is the slope of line segment j and N_j is the number of data reads making up segment j.

4. Results

As seen from Fig. 1, the cosmic ray rejection algorithm removes most of the cosmic ray events from the detector. 114319 cosmic ray events occured on the detector during the 1000-second observation shown here. 1669 cosmic rays, just over 1% of the total, survived the detection and removal process.

If left unremoved, the cosmic ray events in the detector completely ruined 10.3% of the image. The cosmic ray removal process left 1.4% of the image as completely lost, and, because it threw out data reads, reduced the signal-to-noise in 10.2% of the image (including false-positive identifications). The algorithm is still in need of refinement; there were 606 pixels falsely identified as being impacted by cosmic rays along with the 1669 surviving cosmic rays.

4.1. When to reject Cosmic Rays?

Due to financial and communications restrictions, we expect that cosmic ray rejection may have to be done on-board the NGST. We expect the NGST data downlink to be approximately 1.6×10^6 bits per second for 8 hours per day for a total of 5.35GB per day. The near-IR detector alone will contain 64 million 16-bit pixels, for a raw data content of 128MB per data read. For 80 1000-second observations per day, the near-IR camera alone will produce 10GB per day. This requires a data compression ratio of a factor of 2 (Nieto-Santisteban et al. 1999). However, if we wished to perform cosmic ray rejection after downlink, the data from all 65 data reads would have to be downlinked. This would require a communication rate of more than 600 GB per day, requiring a compression ratio of over 100.

5. Conclusion

We can identify and remove 99% of cosmic ray events on the NGST detector. While there is room for improvement, the data quality and information content are largely preserved by this algorithm. This cosmic ray rejection method provides a way to combine 65 images into one, and contributes a significant amount of data compression. This, in turn, loosens one limiting factor on the NIR camera size due to downlink capability. However, this will require significant computer resources to properly handle a full 8k $\times$ 8k pixel detector.

Acknowledgments. These studies are supported by the NASA Remote Exploration and Experimentation Project (REE), which is administered at the Jet Propulsion Laboratory under Dr. Robert Ferraro, Project Manager.

References

Im, M. & Stockman, H. S. 1998, in ASP Conf. Ser., Vol. 133, Science With The Next Generation of Space Telescope, ed. E. P. Smith & A. Koratkar (San Fransisco: ASP), 263

Nieto-Santisteban, M. A., Fixsen, D. J., Offenberg, J. D., Hanisch, R. J., & Stockman, H. S. 1999, this volume, 137

Press, W. H., Flannery, B. P., Teukolsky, S. A., & Vetterling, W. T. 1986, Numerical Recipes, (Cambridge: Cambridge Univ. Press)

Astronomical Data Analysis Software and Systems VIII
ASP Conference Series, Vol. 172, 1999
D. M. Mehringer, R. L. Plante, and D. A. Roberts, eds.

Automated Programming and Simulations of Signal Compression from the Cosmic Microwave Background for the Planck Low Frequency Instrument

M. Maris, F. Pasian, R. Smareglia

Astronomical Observatory of Trieste (OAT), Via G.B.Tiepolo, 11, I-34131, Trieste, Italy

D. Maino

SISSA, Via Beirut, 2/4, I-34131, Trieste, Italy

C. Burigana

Istituto TeSRE / CNR, Via Gobetti 101, I-20129, Bologna, Italy

M. Staniszkis

Mathematics and Informatics Department Udine University, Via delle Scienze 206, I-33100, Udine, Italy

J. Barriga

IEEC, Gran Capitá 2-4, Barcelona, Spain

Abstract. The concept of Automatic Programming and Managing of simulations is applied to the compression of the data stream expected from the Low Frequency Instrument (LFI) aboard Planck Surveyor, the next millennium European Space Agency (ESA) satellite for imaging the Cosmic Microwave Background (CMB). A first analysis of the efficiency of several compression algorithms for different combinations of cosmological and astrophysical signals and for instrumental noise is presented.

1. Introduction

The Planck satellite (formerly COBRAS/SAMBA, Bersanelli et al. 1996), which is planned to be launched in 2007, will produce full sky CMB maps with high accuracy and resolution over a wide range of frequencies. Table 1 summarizes the basic properties of LFI aboard Planck (the reported sensitivities per resolution element, in terms of thermodynamic temperature, represents the goals of LFI for a one-year mission).

The limited bandwidth reserved to the downlink of scientific data calls for huge lossless compression, the theoretical upper limit being a bit less than eight. Careful simulations are demanded to quantify the capability of true compressors for "realistic" synthetic data including CMB signal, foregrounds and instru-

Table 1. Summary of LFI characteristics (Mandolesi et al. 1998a).

Characteristic				
Center frequency [GHz]	30	44	70	100
Number of detectors	4	6	12	34
Angular resolutions, FWHM [$'$]	33	23	14	10
Bandwidth [$\Delta\nu/\nu$]	0.2	0.2	0.2	0.2
$10^6 \Delta T/T$	1.6	2.4	3.6	4.3
ΔT_{rms} [μK]	4	7	10	12
Both polarizations	yes	yes	yes	yes

mental noise. An Automated Program Generator was applied to perform this exploratory process by changing the signal components, the dynamic range of detectors and the compression algorithms, which enables to save time, reduce errors and simplify the simulations management.

2. Planck Operations Data Rate and Compression

During the data acquisition phase, the Planck satellite will rotate at a rate of one circle per minute around a given spin axis that changes its direction every hour (of $2\overset{'}{.}5$ on the ecliptic plane in the case of simple scanning strategy), thus observing the same circle on the sky for 60 consecutive times (Mandolesi et al. 1998a,b). The LFI instrument will produce continuous data streams of temperature differences between the microwave sky and a set of onboard reference sources; both differential measurements and reference source temperatures must be recorded.

The data rate will depend on the sampling time, the present LFI Proposal is $\sim$ 7 msec for each detector (Mandolesi et al. 1998a), thus calling for a typical data rate of $\sim$ 260 kB/sec, while the allocated bandwidth for the data is on average only 20 kB/sec. The data flow have to be downloaded on ground without information losses, and by minimizing scientific elaboration on board. The solution of this problem asks for an efficient and lossless compression of LFI data before they are sent to Earth.

As shown by Gaztñaga et al. (1998), theoretical upper limits for the lossless compression factor of noise data are located between 7.75 and 7.82. On the other hand, such a level of efficiency can be reached under certain optimization hypotheses, it is crucial to study the efficiency of real compression algorithms applied to "real" LFI data (as obtained by simulations) in the form of time series. This may lead, as a second step, to the definition of some level of simple "a priori" data coding or preprocessing to be performed, aimed at increasing the compression factor by allowing the simulated data to better satisfy the optimization hypotheses. The whole work is aimed at the optimization of the transmission bandwidth dedicated to the downlink of LFI data from the Planck spacecraft to the FIRST/Planck Ground Segment.

3. Experimental Strategy

For the successful development of the project, the following steps have been identified:

- select different compression algorithms
- run simulations producing data which contain different types of signal sources, instrumental characteristics and detection electronics:
  1. noise spectrum: *i*) white noise; *ii*) white noise coupled to $1/f$ noise with different knee frequencies
  2. signal sources subsequently added to the noise: *i*) + CMB fluctuations; *ii*) + dipole; *iii*) + Galaxy; *iv*) + extragalactic point sources
  3. different time lengths
  4. reference load temperature (20 K for present text)
  5. different frequency channels with the corresponding beam sizes (we have considered the 30 GHz and 100 GHz channels with $30'$ and $10'$ FWHM beam sizes respectively)
  6. different signal offset and scaling
- apply compression algorithms to simulated data
- analysis of results and discussion on possible data coding or preprocessing.

The combination of the above items yields a large number of combinations to be systematically explored. Automatic simulation generation and management through one simplified Automated Program Generator (APG; Maris 1998; Maris & Staniszkis 1999) implementing a Virtual Scanner (VuScanner) under `IDL` were considered both for analysis and realization. This version of the Virtual Scanner is similar to the `MatLab` version used for neutrino studies (Maris 1999) with some modifications to take into account the reduced flexibility of the `IDL` language if compared to `MatLab`.

4. Results and Conclusions

The simulated cosmological and astrophysical components are generated according to the methods described in Burigana et al. (1998) and the data stream and noise generation as in Burigana et al. (1997) and Seiffert et al. (1997).

Signal detection is simulated through the formula (M. Bersanelli, J. Herrera, & M. Seiffert 1998, private communication) $V = AFO + VOT \cdot \Delta T$, where V is the detection chain output in Volts ($-10\text{V} \leq V \leq 10\text{V}$), AFO is the detection chain offset ($-5\text{V} \leq AFO \leq +5\text{V}$), VOT is the antenna temperature / voltage conversion factor ($-0.5\text{V/K} \leq VOT \leq +1.5$ V/K), ΔT is the difference between antenna temperatures due to the sky signal and a reference load. Sixteen bits quantization is applied to the signal before compression and before writing into a binary file.

The output of the compression chain is a set of compression evaluation tables for the components of the incremental cosmological signal, one for each set of conditions. The compression efficiency η_c is evaluated as the ratio $\eta_c = L_u/L_c$, where L_u is the uncompressed signal length in bytes and L_c is the compressed signal size.

We have analyzed the efficiency of four kinds of compression algorithms (`compress`, `compact`, `szip` and `gzip`) for a limited sample of the full set of simulations. For these tests, the full range found was 1.9 – 6.5. Our preliminary conclusions are summarized as follows:

1. the compression efficiency η_c is influenced by: VOT, AFO, and by the signal length L_u
2. compressing more circles at a time (i.e. increasing L_u) slightly improves η_c
3. the compression efficiency depends on the frequency and on the components included in the sky signal
4. in particular, removing the cosmological dipole through a reversible fit improves the η_c
5. the best compressor tested is `gzip`, but `szip` is comparable to it and is NASA space qualified
6. the `gzip` compression efficiency for the "central" detector values ($VOT = 1$ V/K, $AFO = 0.0$ V) for 1 min data stream at 30 GHz (100 GHz) is $\eta_c^{\mathtt{gzip}} \sim 4.9 - 5.1$ ($\eta_c^{\mathtt{gzip}} \sim 4.4 - 4.5$).

In conclusion these realistic compression factors, albeit encouraging as a first result, are about a factor two worse than the theoretical upper limit. Further investigations are required to reach the target compression factor without any on-board data averaging at all. In addition, the choice of a given compressor can not be based only on its efficiency, but also on the on-board available CPU and on the official ESA space qualification: tests with this hardware platform and other compressors will be made. In the near future, long–duration flight balloon experiments will provide a solid base to test and improve compression algorithms.

Acknowledgments. We warmly acknowledge a number of people which actively support this work, in particular F. Argüeso, M. Bersanelli, L. Danese, G. De Zotti, E. Gaztñaga, J. Herrera, N. Mandolesi, P. Platania, A. Romeo, M. Seiffert and L. Toffolatti for fruitful discussions and K. Gorski, E. Hivon and B. Wandelt for providing us with Healpix pixelisation tools. J.B. acknowledges an INTA grant spent at the OAT in July – August 1998.

References

Bersanelli, M., et al. 1996, COBRAS/SAMBA, (Rep. Phase A Study, ESA D/SCI(96)3), (Noordwijk: ESA)

Burigana, C., et al. 1997, (TeSRE/CNR Internal Rep., 198/1997)

Burigana, C., et al. 1998, Ap. Lett. Comm., submitted

Gaztñaga, E., et al. 1998, Ap. Lett. Comm. in press

Mandolesi, N., et al. 1998a, "Low Frequency Instrument for Planck", Proposal submitted to ESA

———, 1998b, Ap. Lett. Comm., submitted

Maris, M. (ed.) 1998, "Automated Program Generation at PLANCK/LFI", PLANCK/LFI Project, `http://www.pv.infn.it/~maris/apgatpl.html`

Maris, M. 1999, this volume, 50

Maris, M. & Staniszkis, M. 1999, this volume, 470

Seiffert, M., et al. 1997, Rev. of Sci. Instr., submitted

Part 6. Data Pipelines

Astronomical Data Analysis Software and Systems VIII
ASP Conference Series, Vol. 172, 1999
D. M. Mehringer, R. L. Plante, and D. A. Roberts, eds.

Pipeline Processing of Data in the VLT Era

Preben Grosbøl, Klaus Banse, Pascal Ballester

European Southern Observatory,
Karl-Schwarzschild-Str. 2, D-85748 Garching, Germany

Abstract. The VLT Data Flow System pipeline and quality control subsystem provides a general infrastructure for standard reduction and quality assessment of data obtained at the VLT facility. The main design challenge is to support a wide range of instruments in a distributed environment. The pipeline system can be configured, through a set of ASCII files, to handle several instrument simultaneously. It was designed using the object-oriented methodology and major parts of the baseline version will be written in Java using OMG/CORBA technology to support distributed objects.

1. Introduction

The VLT Data Flow System (DFS) provides a single, homogeneous, end-to-end system for handling science data from the VLT facility (see Grosbøl & Peron 1997). It can be divided into three main parts, namely: a) pre–observation tasks which include preparation of observing proposals, detailed specification of observations and tentative scheduling, b) observation support containing archiving services, and c) post–observation processing including pipeline reduction and quality control of data acquired.

This paper focuses on the post-observation modules with emphasis on the standard pipeline reduction of VLT data. A discussion of the quality control aspects was given by Ballester et al. (1998).

2. Requirements and Challenges

The vast amount of data produced by the VLT and the multitude of instruments demand that raw data can be reduced very efficiently and with a minimum of manual intervention. The DFS pipeline has been designed for this purpose and will be used in four main scenarios:

- near real time reduction of data at the VLT observatory to provide a first assessment of data quality,
- off-line reduction of data at the ESO headquarters to offer users standard reduced data products for service-mode observations,
- data reduction at the users home institute to make a more customized processing possible, and
- re–processing of data from the VLT Science Archive.

Whereas it is trivial to make an explicit pipeline to reduce data from a given instrument, the main challenge for the DFS pipeline is to create a unique infrastructure which can serve all of the more than 15 different instruments on the four 8.2 m telescopes plus ancillary units for interferometry and wide field imaging. The long expected life time of the facility makes it mandatory to rely on a single concept to ease operation and reduce maintenance costs while employing a modular design and thereby enable a gradual replacement of components in the course of time.

There must be a clear separation between the pipeline infrastructure and data processing tasks to ensure that any suitable data reduction system (DRS) can be used. Some reduction tasks for a specific instrument may already be available in a particular system (e.g., AIPS++ could provide interferometric reduction procedures). It is also prudent to assume that not all current DRS's will be fully updated or supported over the next decades. A smooth migration of reduction tasks from one to another system would therefore be important. Further, the large data volume also makes it essential to support parallel processing to take advantage of multi–processor or loosely coupled computers e.g. Beowulf type of systems.

3. Pipeline Model and Assumptions

The pipeline processing model assumes that all raw frames can be uniquely associated to a specific instrument and the full description of its setup can be obtained from their FITS headers. It must also be possible to determine the observational and operational context of a raw frame relative to others by analyzing its FITS keywords. This provides a hierarchical grouping of frames e.g. based on their relation to Observation Blocks and Templates (see Grosbøl & Peron 1997). Each instrument must define a unique classification of all frames used by the pipeline including raw frames, calibration data and products generated.

Actual pipeline tasks are triggered by different events such as *arrival of new raw frame* or *end of observing template*. The frames associated to the event are identified and the necessary set of reduction recipes are obtained following rules defined for the instrument. The recipes specify the algorithms and the required parameter including calibration data such as CCD flat fields or wavelength tables. Calibration files are either stored in a local database or file directory structure. The appropriate calibration frames for a given input data set are found through their classification and by matching a primary key defined as a set of FITS keywords.

4. Architectural Design

The architecture of the pipeline system is based on a distributed, object oriented design. The system can either be driven by events or activated through graphical user interfaces. Five main applications are defined to support the basic pipeline functionality:

- **Pipeline:** is responsible for the standard reduction of data. It creates the explicit reduction tasks (*ReductionBlocks*) to be executed following a given event.

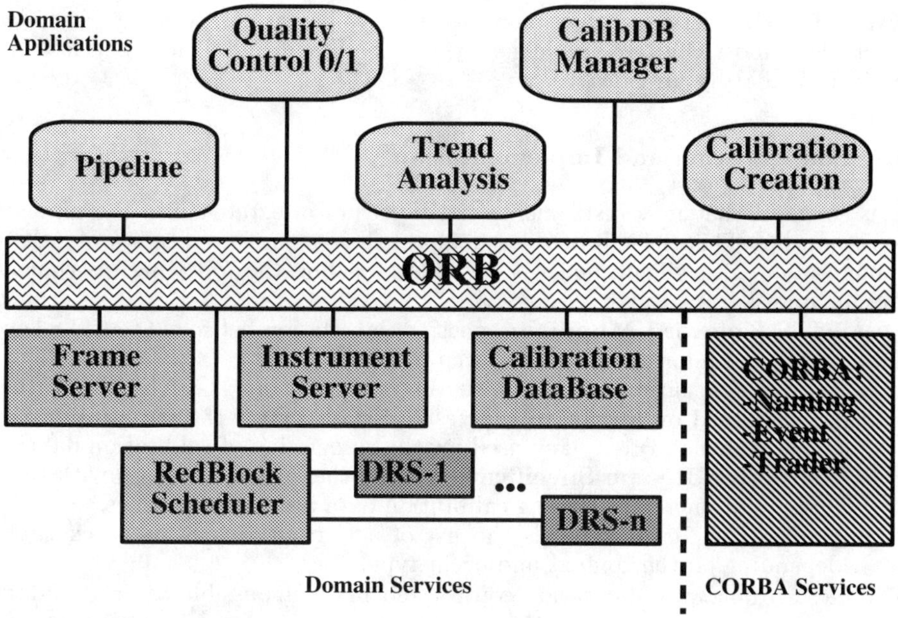

Figure 1. Main applications and services of the DFS Pipeline subsystem

- **Quality Control:** checks the quality of both raw and reduced data by comparing them with requirements and models.
- **Trend Analysis:** monitors the calibration solutions to identify possible variations.
- **Calibration Database Manager:** controls and maintains the calibration data used for the pipeline reductions.
- **Calibration Creation:** creates new calibration data which after a certification process may be included in the database.

The first implementation is aimed mainly at an automatic, batch type environment but it is expected that more clients will be added to provide better interactive control. The applications use a set of general services:

- **Frame Server:** determines context and classification of new frames and groups them.
- **Instrument Server:** provides several instrument specific facilities e.g. FITS keyword mapping, classification, reduction recipe descriptions and rules for required reduction steps.
- **Calibration Database:** is the depository for calibration data and support search methods to locate them.
- **ReductionBlock Scheduler:** receives *ReductionBlocks* and schedules them for execution by one of the Data Reductions Systems it has access to.

The communication between clients and servers is based on the OMG/CORBA distributed object model. The components are shown in Fig. 1 together with several CORBA services which may be used. Whereas the Naming Service is available in most CORBA implementations, the Event and Trader Services are not yet standardly provided.

5. Configuration and Implementation

It is essential that new instruments can be easily integrated into the pipeline environment. This is facilitated by defining their behaviors in a set of ASCII configuration files. They make it possible to define the following instrument specific items:

- **FITS Keyword Mapping:** specifies the relation between the FITS keywords and information used by the pipeline.
- **Frame Classification:** defines the criterion for classifying individual frames based on a boolean expression. It also gives the structure of the frame and the primary key used for the association of calibration data.
- **Reduction Recipe Specification:** lists the available recipes with their formal parameters including calibration data and default values.
- **Reduction Rule:** defines the list of action or recipes to be executed depending on the context and event type.

The files are under configuration control and make it possible to process data from the Science Archive using the appropriate versions for the instrument definitions.

It is foreseen that the DFS pipeline will be employed at computer systems distributed over the ESO sites and possibly exported to external institutes. Web based interfaces are also expected to play an important role in operating and monitoring the pipelines. The major parts of the DFS pipeline will be implemented in Java which provides excellent support for distributed object systems and user interfaces. It is expected that OMG/CORBA based tools will be used for the object bus and general services.

References

Grosbøl, P. & Peron, M. 1997, in ASP Conf. Ser., Vol. 125, Astronomical Data Analysis Software and Systems VI, ed. G. Hunt & H. E. Payne (San Francisco: ASP), 23

Ballester, P., Kalicharan, V., Banse, K., Grosbøl, P., Peron, M., & Wiedmer, M. 1998, in ASP Conf. Ser., Vol. 145, Astronomical Data Analysis Software and Systems VII, ed. R. Albrecht, R. N. Hook, & H. A. Bushouse (San Francisco: ASP), 259

Astronomical Data Analysis Software and Systems VIII
ASP Conference Series, Vol. 172, 1999
D. M. Mehringer, R. L. Plante, and D. A. Roberts, eds.

The Client Server Design of the Gemini Data Handling System

Norman Hill, Séverin Gaudet, Jennifer Dunn, Shannon Jaeger, Steve Cockayne

National Research Council Canada/Herzberg Institute of Astrophysics, 5071 West Saanich Rd. Victoria B.C. Canada V8X 4M6

Abstract. The Gemini Telescopes Data Handling System (DHS) developed by the Canadian Astronomy Data Centre (CADC) has diverse requirements to support the operation of the Gemini telescopes. The DHS is implemented as a group of servers, where each performs separate functions. The servers use a client server model to communicate between themselves and with other Gemini software systems. This paper describes the client server model of the Gemini Data Handling System.

1. Introduction

The role of the DHS in the Gemini Telescopes systems is to a provide data handling infrastructure for the other Gemini systems and to isolate the other principal systems as much as possible from the details of how and where the data is stored. In addition the DHS provides an infrastructure for data quality assessment, which includes processing data to remove instrument signatures, and displaying data to telescope users. The DHS runs on Sun workstations which use the Solaris operating system.

The DHS servers communicate with each other, and with Gemini systems outside the DHS, using a client server model. The reasons for designing the DHS as separate servers are:

- To allow parallel development of the various components of the DHS.
- To reduce the complexity of the DHS overall.
- To reduce the complexity of the tasks under development at any given time.
- To allow each of the servers to be designed and implemented as a separate development task, allowing better management of the project.

2. The DHS Library

The DHS library (Hill, Gaudet, & Kotturi 1998; Hill & Gaudet 1998) is a C library which was developed to provide an interface between the DHS and its client programs. The library sends messages between programs, which consist of commands and attributes (parameters) or bulk data (image data and descriptive attributes produced by Gemini instrument, Dunn et al. 1999). Both commands and bulk data are DRAMA SDS structures (Baily 1993) and are sent with the DRAMA IMP (Shortridge 1997) messaging system (see the DRAMA home

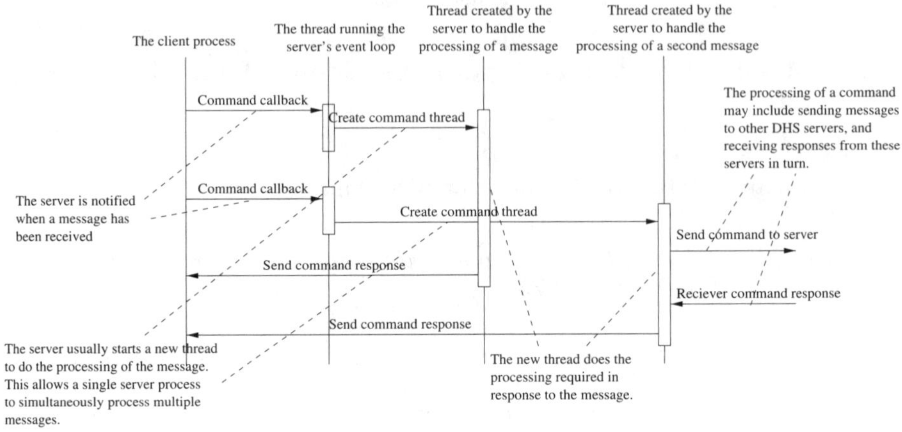

Figure 1. Server event processing message diagram.

page[1]). Responses to commands or bulk data can be received asynchronously via a callback from the DHS library, or synchronously by explicitly waiting for a response.

The DHS library was ported to the VxWorks real-time operating system by the International Gemini Project Office to allow instruments to directly transfer data to the DHS. The Gemini Observatory Control System (OCS) development group incorporated the DHS library into OcsWish, the Tcl/Tk interpreter used by the OCS (Walker & Gillies 1998). The Tcl/Tk interface is used by the DHS consoles (Jaeger et al. 1999), and by the Quick Look System (Hill et al. 1999).

IMP was selected as the message passing mechanism for the interface to software outside of the DHS because it provides an efficient interface for sending both large and small data messages between systems, and because it is available for both the VxWorks and Sun/Solaris systems to be used by the Gemini project. IMP was also used for internal communications within the DHS because it fulfilled the communications needs of the DHS; it was very efficient for message passing between tasks on the same computer, and it avoided problems associated with having different interfaces for internal and external communications.

3. Client Server Communications

The DHS requires a very flexible communications architecture. A client may need to communicate with more than one server, a server may receive commands and data from more than one client, and servers may communicate with several other servers.

Figure 1 is a Booch event trace diagram that depicts the events which occur when two commands are received by a server. The two commands may have been initiated by the same client, or by two separate clients. Vertical lines on the diagram represent threads or processes in the interaction, the vertical

[1] `http://www.aao.gov.au/drama/html/dramaintro.html`

bars overlaying the vertical lines indicate when the thread or process is active. Horizontal lines represent messages or function calls between the threads or processes.

Servers use the DHS library to specify callback functions which will executed when commands or bulk data requests are received. When a callback function is executed, the server is responsible for processing the command or bulk data request. The DHS library also provides functions to allow responses to be returned for commands or bulk data requests.

4. The DHS Servers

The following servers comprise the Gemini DHS:

The Command Server provides a common command interface to all of the DHS servers, and eliminates the need for external clients of the DHS to be aware of which DHS servers are actually executing commands.

The Status Server provides a status interface between the DHS servers and Gemini systems outside of the DHS. The Status Server allows status information to be distributed to the Gemini system, without each server being required to know how or to whom the status is being reported. The Status Server also monitors and reports DHS disk and database space usage.

The Data Server accepts bulk data requests from other Gemini systems, supplies data labels for other Gemini systems, creates FITS files from the data, and informs other DHS Servers when new data is available.

The Quick Look Server is responsible for distributing data to appropriate Quick Look Tools (Hill et al. 1999).

The Storage Server creates storage media for the Gemini archive and for delivery to Gemini Telescope users.

The History Server monitors history log records in the Gemini EPICS (Experimental Physics and Industrial Control System) databases and creates a permanent record of all logged messages.

The On-Line Data Processing Server processes data collected by the Gemini Telescopes. The processed data is then archived and/or sent to the Quick Look Server for display.

The Synchronous Data Processing Server performs data processing tasks for other Gemini systems. This server returns processed results to the Gemini system which initiated the data processing.

5. Conclusions

The modular design of the DHS has several advantages over a monolithic design:

- Interactions between modules are minimized. This means even major modifications to a module are unlikely to effect other modules.
- It is possible to run a subset of the DHS. This is desirable because some of the servers require considerable setup and configuration.

- It is possible to distribute the DHS components over more than one computer or even over more than one site. This was useful when it was discovered that writing CD-ROMs at high altitude wasn't recommended. The design allowed the DHS Storage Server to be located at the Gemini base facility in Hilo, without any modification to the DHS.
- Development of each of the servers could proceed independently and in parallel.
- Components of the DHS which had not yet been developed, or which were not complete were simulated using a generic server simulator. This allowed individual servers to be tested without running the entire DHS.
- Potential problem areas were isolated in specific servers. (For example, the EPICS channel access event loop was only required in the History and Status Servers.)
- The threaded nature of the servers allows a single copy of each server to service multiple clients.

The ability to run a subset of the DHS has already been exploited by the CADC to set up experimental systems to transfer data directly from the Canada France Hawai'i Telescope (CFHT) and the James Clark Maxwell Telescope (JCMT) on Mauna Kea to the archives in Victoria.

There were also disadvantages to the client server design.

- Debugging the system was sometimes difficult due to the asynchronous nature of the interaction between servers.
- Each server has its own configuration file, and information is often repeated in several configuration files. This repetition makes the system more difficult to configure than it would be with a monolithic program.

However, overall the advantages of the client server design far outweighed the disadvantages.

References

Baily, J. 1993, "Self-Defining Data System (SDS) Version 1.4", Anglo-Australian Observatory, `http://www.aao.gov.au/drama/doc/ps/sds.ps`

Dunn, J., Jaeger, S., Hill, N., Gaudet, S., & Cockayne, S. 1999, this volume, 167

Hill, N. & Gaudet, S. 1998, ICD 3 - Bulk Data Transfer, (Gemini Data Handling System Rep. No. dhs_pdr_icd3/23, Gemini Project)

Hill, N., Gaudet, S., Dunn, J., Jaeger, S., & Cockayne, S. 1999, this volume, 163

Hill, N., Gaudet, S., & Kotturi, D. 1998, ICD 1c - Baseline DHS Interface (Gemini Data Handling System Rep. No. dhs_pdr_icd1c/05, Gemini Project)

Jaeger, S., Dunn, J., Cockayne, S., Gaudet, S., & Hill, N. 1999, this volume, 159

Shortridge, K. 1997, Interprocess Message Passing (IMP) System, (AAO/IMP_MANUAL_8, Drama Software Report 8) (Sydney: Anglo-Australian Observatory)

Walker, S. & Gillies, K. 1998, OcsWish Technical Manual, (ocs.ocs.015-OcsWishTechnical/02, Gemini Project) (Tucson: Gemini Observatory)

Astronomical Data Analysis Software and Systems VIII
ASP Conference Series, Vol. 172, 1999
D. M. Mehringer, R. L. Plante, and D. A. Roberts, eds.

Interfacing with the Gemini Data Handling System

Shannon Jaeger, Jennifer Dunn, Steve Cockayne, Séverin Gaudet, Norman Hill

National Research Council Canada/Herzberg Institute of Astrophysics, 5071 West Saanich Rd. Victoria, B.C. Canada V8X 4M6

Abstract. The Gemini Telescopes Data Handling System (DHS), developed by the Canadian Astronomy Data Centre, consists of eight server programs, a graphical user interface (GUI) and a tool for displaying images. The DHS GUI controls and monitors the entire DHS by communicating with only two of the eight DHS server programs. This paper describes the communication model used by the DHS GUI.

1. DHS Overview

The role the DHS plays within the Gemini systems is that of the the data controller. DHS is data-driven, acting on any data it receives from any Gemini principal system. DHS is responsible for maintaining a permanent data store, creation of archivable media, data processing, and displaying data for quality control. These tasks are performed by the eight server programs described below and in more detail in Hill et al. (1999b).

Command Server is the command interface between DHS and external software, including the DHS GUI.

Status Server is the interface between DHS and the outside world for outgoing status information. It also monitors resource usage, such as disk and database space.

Data Server accepts bulk data from other Gemini systems, and informs other DHS servers when data is available. It is the data manager for the DHS (Dunn et al. 1999a).

Quick Look Server (Hill et al. 1999a) and Quick Look Tool together are responsible for displaying selected data within the DHS in a timely and useful manner.

Storage Server is responsible for the creation of media. This server shares source code and concepts with the European Southern Observatory's ASTO system, (Dunn et al. 1999b).

History Server collects time-stamped history records from all Gemini systems and stores them in a database.

On-line Data Processing Server processes data collected by the Gemini Telescopes. The processed data is then archived and/or sent to the Quick Look Server for display.

Synchronous Data Processing Server performs data processing tasks for other Gemini Systems. The server returns processed results to the Gemini system which initiated the task.

The DHS will run at both the base and summit facilities for both Gemini Telescope locations: Cerra Pachon, Chile and Mauna Kea, Hawaii. It is expected that some subset of the set of DHS servers will run at each facility.

2. DHS GUI Overview

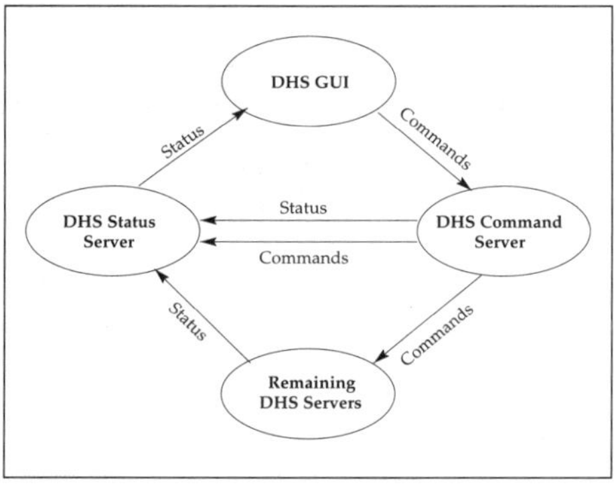

Figure 1. DHS GUI/DHS server communication model.

The primary functions of the DHS graphical user interface (GUI) are to display the current status of the DHS and to provide a means of controlling the DHS. Status information is displayed textually and graphically on the various components of the GUI. The DHS is controlled through various commands, which are typically requested from the GUI by selecting buttons.

The DHS GUI behaves like an application which is external to DHS, despite being the controller and monitor of the DHS. The GUI communicates directly to two of the eight servers described above: the Status and Command servers. Figure 1 shows the channels of communication between the DHS and the DHS graphical user interface.

The entire DHS GUI was written in Tcl/Tk (Ousterhout 1994) and uses the OCSWish (Walker & Gilles 1998) interpreter. The OCSWish Tcl/Tk interpreter was provided by the Gemini International Project Office, Software and Controls group. One facility the OCSWish interpreter provides are mechanisms for sending commands to DHS servers and for the reception of status information. The status information may be received from an Experimental Physics and Industrial Control System (EPICS) database or directly from the DHS Status server. The GUI also uses the object oriented facilities provided by the [incr Tcl]/[incr Tk] Tcl/Tk extension to encapsulate some of its components. The ESO GUI guidelines (Ballester & Comin 1993) were used to determine the general layout and design of the GUI.

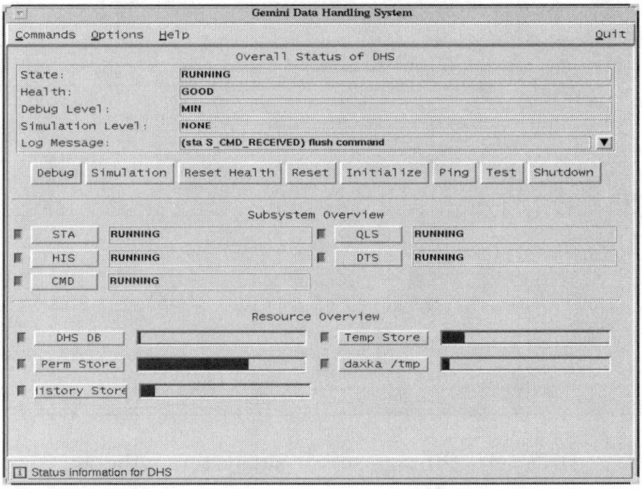

Figure 2. DHS GUI main window.

The main window of the DHS console is shown in Fig. 2. This window is the only window shown when the Console is initially displayed. On the main window there are buttons that send commands to the entire DHS, such as a *shutdown* command. The main window also displays the current DHS status and the resource usage.

Each DHS subsystem server can be controlled from its subsystem window. A subsystem's window is displayed when the appropriate button is selected in the "Subsystem Overview" panel on the main window. The subsystem window also displays status information which pertains only to the particular subsystem. An example of a subsystem window is shown in Fig. 3, the Storage Server's subsystem window. The Storage Server is one of the more complex subsystem windows. The top half of the window is the same for all subsystems - standard status information plus buttons for executing the standard commands. The bottom half is a series of buttons which control the creation of media, see Dunn (1999a) for more details.

3. Design Critique

As with all design models there are advantages and disadvantages. One of the most obvious advantages is the minimization of connections. There are only two connections the user interface must maintain; one to the Status Server and one to the Command Server.

A second advantage is the ease with which the command interface and status interface have each been encapsulated into a single module/object. This was easy to implement since each has a single contact point. External software, such as the Console, should not be forced to have full knowledge of the DHS design in

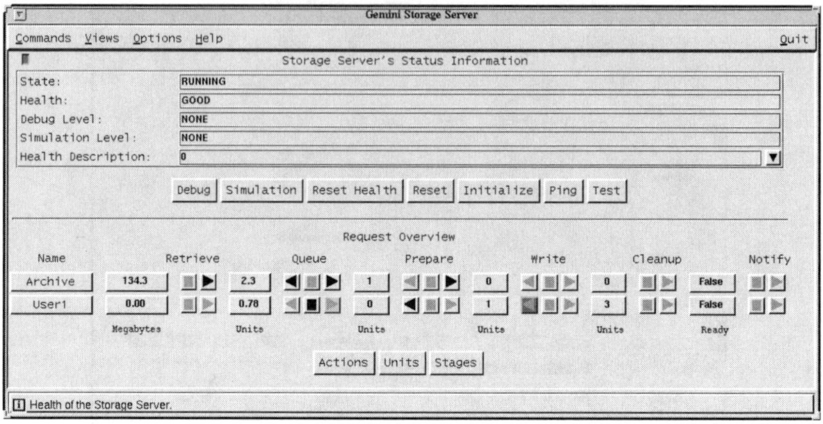

Figure 3. DHS GUI Storage server subsystem window.

order to interface with it, and by having only two servers to be concerned with this is achieved. Also, the encapsulation makes maintenance and debugging of the DHS Console software easier.

The third advantage to the design is the DHS Console is able to monitor and control multiple server programs instead of using a separate GUI for each server. Screen real-estate is a precious commodity, and by having an single GUI for the entire system the real-estate required is drastically reduced.

The minimization of connections is also a disadvantage. If anything goes wrong with the status server the GUI will stop reporting status information for the entire DHS. Similarly, if the command server stops processing commands the GUI can no longer control the DHS. Having a single point of failure is an undesirable situation but the advantages of this design outweigh this disadvantage.

References

Ballester, P. & Comin, M. 1993, ESO Graphical User Interface Common Conventions (GEN-SPE-ESO-00000-0266) (Garching: ESO)

Dunn, J., Cockayne, S., Gaudet, S., Jaeger, S., & Albrecht, M. 1999a, this volume, 265

Dunn, J., Jaeger, S., Hill, N., Gaudet, S., & Cockayne, S. 1999b, this volume, 167

Hill, N., Gaudet, S., Dunn, J., Jaeger, S., & Cockayne, S. 1999a, this volume, 163

———, 1999b, this volume, 155

Ousterhout, J. K. 1994, Tcl and the Tk Toolkit (Reading: Addison-Wesley)

Walker, S. & Gillies, K 1998, OcsWish Technical Manual (ocs.ocs.015-OcsWishTechnical/02, Gemini Project)

Astronomical Data Analysis Software and Systems VIII
ASP Conference Series, Vol. 172, 1999
D. M. Mehringer, R. L. Plante, and D. A. Roberts, eds.

The Gemini Quick Look System

Norman Hill, Séverin Gaudet, Jennifer Dunn, Shannon Jaeger, Steve Cockayne

National Research Council Canada/Herzberg Institute of Astrophysics, 5071 West Saanich Rd. Victoria B.C. Canada, V8X 4M6

Abstract. The Gemini Data Handling System (DHS) developed by the Canadian Astronomy Data Centre (CADC) is required to provide a "Quick Look" display system that allows near real-time display of image data collected by the instruments of the Gemini Telescopes. The Gemini Quick Look System allows data to be displayed on multiple display tools, possibly on different systems, allows the display of data from multiple data streams, and allows images to be sent incrementally (in chunks). The Quick Look Tool is based on the European Southern Observatory (ESO) Skycat display tool, and expands on the functionality provided by Skycat. This paper describes the design and function of the Gemini Quick Look System.

1. Introduction

The Quick Look System of the DHS is intended to provide a near real time display of data collected by the Gemini instruments, and to display data produced by the DHS Data Processing system. The initial version of the DHS, including a working version of the Quick Look System was accepted by the International Gemini Project Office in August, 1998. A second version of the DHS is tentatively scheduled to be delivered in May 1999. This will include some of the enhancements to the Quick Look System described section 3. Other aspects of the Gemini DHS are described in Dunn et al. (1999), Gaudet (1996), Gaudet et al. (1999), Hill et al. (1999), and Jaeger et al. (1999).

2. Quick Look Data Flow

Figure 1 depicts the data flow through the Quick Look System. The components that manipulate the data are divided into three sections:

- **Data Sources** are the sources for the original data which are all external to the DHS. Data sources may include all Gemini instruments, the Gemini Observatory Control System (OCS), and may include non-science instruments like the Guide Camera. Data for a single dataset may be sent by more than one data source and each source may send the data for a single dataset in more than one chunk. It is also possible several data sources to be simultaneously sending data to the DHS. The data structure is described in detail in Dunn et al. (1999).
- **Data Handling System** are the subsystems of the DHS which run on the DHS computers at the summit. All data received by the DHS Data Server

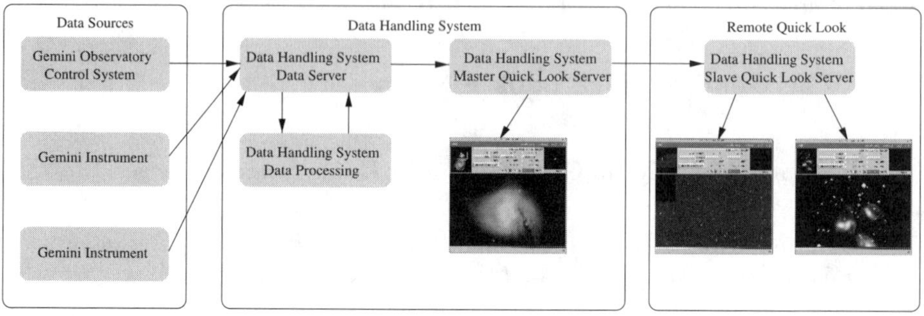

Figure 1. Quick Look Data Flow.

are forwarded to the Quick Look Server for display, and to the DHS Data Processing system for pre-arranged data reduction. Data created by the Data Processing system may be sent back to the Data Server for archival storage or to be forwarded to the Quick Look Server.

- **The Remote Quick Look** components are the parts of the DHS which would run on any remote computer system that is displaying Quick Look data. These systems may be computers dedicated to Quick Look, or they may be computers primarily used for other Gemini functions, which have Quick Look as a secondary function (the systems running the OCS for example).

2.1. The Quick Look Server

The master Quick Look Server receives data from the Data Server and routes the data to appropriate Quick Look Tools and slaved Quick Look Servers. Data routing is based on Quick Look stream names, which are assigned to a dataset by one of its data sources (see section 2.3.). Data is sent to Slave Quick Look Servers using the IMP (Shortridge 1997) message passing system, and to the Quick Look Tools using memory mapped files.

If Quick Look data is to be displayed on computer systems other than the one running the master Quick Look Server, a slave Quick Look Server must be run on the second system. The master Quick Look Server transfers only one copy of each dataset to each slave Quick Look Server, and so even if more than one Quick Look Tool is displaying a dataset, only one copy of the data will be transferred over the network. Having a separate Quick Look Server on each computer also allows memory mapped files to be used to transfer data between Quick Look Server and Quick Look Tools, and avoids having to transfer the entire image over the network each time a chunk is received, which maximizes Quick Look Tool display speed.

2.2. The Quick Look Tool

The DHS Quick Look Tool is a Tcl/Tk program based on the ESO Skycat display tool (Albrecht et al. 1997). It uses the OcsWish Tcl/Tk interpreter provided by the Gemini Project (Walker & Gillies 1998). All existing Skycat functionality is retained.

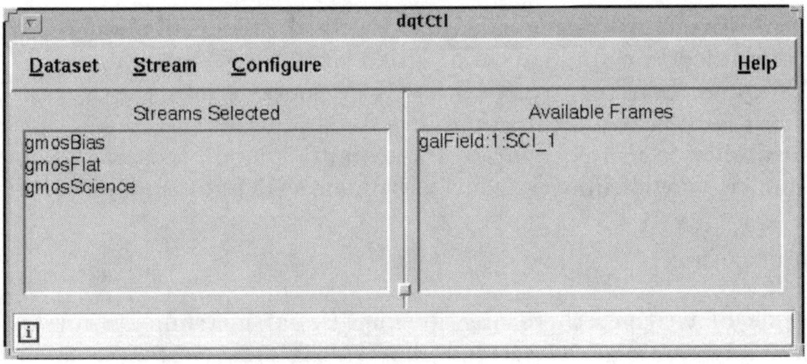

Figure 2. Quick Look Control Panel.

The Quick Look Tool merges data chunks into a single coherent image and displays the most recently received dataset and frame. See Dunn et al. (1999) for more information about chunking. The control panel shown in Fig. 2 was added to Skycat to control the new Quick Look functions. The control panel:

- Shows the list of currently subscribed streams (see section 2.3.).
- Allows the list of subscribed streams to be modified.
- Displays a list of frames available for the current dataset.
- Allows a frame to be selected for display.
- Allows the Quick Look Tool configuration to be modified.
- Allows a specific dataset to be retrieved from the DHS Data server and displayed.
- Allows the currently displayed frame to be saved to the Gemini archive.

2.3. Quick Look Streams

Data are directed to appropriate Quick Look Tools based on Quick Look stream names. A set of Quick Look stream names are assigned to each dataset by one of its data sources. Each Quick Look Tool individually subscribes to a set of Quick Look streams. The initial list of subscribed streams are determined from the Quick Look Tool configuration file, and the user may change the subscribed streams with the control panel (section 2.2.). When the Quick Look Server receives data for a dataset, the data is forwarded only to those Quick Look Tools which have subscribed to at least one of the Quick Look streams associated with the dataset.

Quick Look streams allow simple and flexible control over the routing of data to Quick Look Tools.

3. Future Enhancements

There are several enhancements to the Quick Look System which will be added in the near future:

- The newest version of Skycat allows overlay graphics to be stored in the input files. This feature will be used to allow data displayed by the Quick Look System to contain graphics and graphics overlays.

- Simple image processing as data is received. This will allow removal of flat or bias effect, or application of a bad pixel map.
- The Quick Look control panel will be merged into the Skycat interface using the Skycat plug-in capability.
- The ability to merge separate frames into a rough mosaic. This will allow a unified view of images from instruments with multiple detectors.

4. Conclusions

At the time of writing, there are no operational instruments for the Gemini telescopes, however we have tested the Quick Look System with instrument simulators. We have met our throughput requirements of:

- 10 frames per second for 128 × 128 pixel images.
- 1 frame per second for 512 × 512 pixel images.
- .5 frame per second for 1024 × 1024 images.

The Quick Look System is now being used by instrument developers and by Gemini integration and testing staff for testing and development of Gemini instruments.

A high frame update rate will allow instruments that collect data continuously (e.g., the Gemini Guide Camera) to display data with the appearance of continuous motion. It is expected that the throughput of the Quick Look System will be high enough that the Gemini Guide Camera will be able to use the Quick Look System as a real time display, eliminating the need for a dedicated real time display system.

References

Albrecht, M. A., Brighton, A., Herlin, T., & Biereichel, P. 1997, in ASP Conf. Ser., Vol. 125, Astronomical Data Analysis Software and Systems VI, ed. G. Hunt & H. E. Payne (San Francisco: ASP), 333

Dunn, J., Jaeger, S., Hill, N., Gaudet, S., & Cockayne, S. 1999, this volume, 167

Gaudet, S. 1996, in ASP Conf. Ser., Vol. 101, Astronomical Data Analysis Software and Systems V, ed. G. H. Jacoby & J. Barnes (San Francisco: ASP), 388

Gaudet, S., Hill, N., Dunn, J., Jaeger, S., & Cockayne, S. 1999 this volume, 3

Hill, N., Gaudet, S., Dunn, J., Jaeger, S., & Cockayne, S. 1999, this volume, 155

Jaeger, S., Dunn, J., Cockayne, S., Gaudet, S., & Hill, N. 1999, this volume, 159

Shortridge, K. 1997, Interprocess Message Passing (IMP) System, (AAO/IMP_MANUAL_8, Drama Software Report 8) (Sydney: Anglo-Australian Observatory)

Walker, S. & Gillies, K. 1998, OcsWish Technical Manual, (ocs.ocs.015-OcsWishTechnical/02) (Tucson: NOAO)

Astronomical Data Analysis Software and Systems VIII
ASP Conference Series, Vol. 172, 1999
D. M. Mehringer, R. L. Plante, and D. A. Roberts, eds.

Handling Chunks of Image Data in the Gemini Data Handling System

Jennifer Dunn, Shannon Jaeger, Norm Hill, Séverin Gaudet, Steve Cockayne

National Research Council Canada/Herzberg Institute of Astrophysics, 5071 West Saanich Rd., Victoria, BC, Canada V8X 4M6

Abstract. The Data Handling System (DHS), developed for the Gemini project by the Canadian Astronomy Data Centre, accepts observational data from multiple sources. There may be data descriptions, image data in chunks (to be composed into a single image), and multiple images for one dataset, all being received at different times. All information for a single dataset is written to one Flexible Image Transport System (FITS) file. The methods developed for handling and assembling data chunks are presented.

1. System Overview

The role of the DHS in the Gemini Control System is to manage the flow of data from the instruments on the Gemini Telescopes. DHS is data-driven and is composed of a collection of servers that work together to act on any data it receives from any Gemini principal system. See Hill et al. (1999a) for more information on the DHS clients and servers.

The Data Server is responsible for handling data put and get requests and is responsible for managing the data. This paper focuses on the handling and assembling of datasets by the DHS Data Server.

1.1. Communication

All communication with the Data Server is provided by the DHS library. The library provides mechanisms to clients for establishing connections to the Data Server, creating and decoding dataset structures and sending and receiving dataset structures from the DHS Data Server over the Gemini Data LAN. The library is built on the DRAMA Interprocess Message Passing (IMP) and Self-defining Data System (SDS) packages, developed by the Anglo-Australian Observatory. See Hill et al. (1999b) for more details on DHS communications.

2. Data Contributors & Receivers

The Data Server receives data from multiple sources and sends data to several of the DHS subsystems. Figure 1 shows some of the expected contributors of

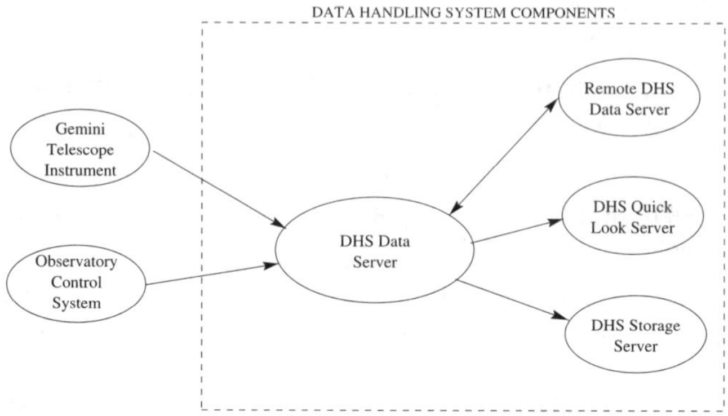

Figure 1. DHS Data Server Context Diagram.

data to the Data Server and the DHS subsystems which will receive data from the Data Server.

2.1. Contributors

The multi-threaded capabilities of the Data Server allow it to handle

1. multiple contributors/sources to a single dataset,
2. different datasets at the same time,
3. and multiple contributors to the same dataset at the same time.

Here, a dataset is considered to be data from a single OBSERVE command.

The flexibility built into the Data Server allows attributes and/or data to be sent to the Data Server all at once or in chunks. When all data have been received by all contributors, the Data Server will assemble the chunks of the jigsaw puzzle to construct a complete dataset.

2.2. Receivers

Each piece of information, whether it is a chunk or complete set, is immediately converted into the Flexible Image Transport System[1] (FITS) format and sent from the Data Server to the Quick Look Server. The Quick Look Server will then send these data to its Quick Look Tool(s), which display the images graphically (Hill et al. 1999b). The DHS Storage Server will receive notification of complete datasets that are destined for permanent storage.

3. Data Formats

Data are sent to the DHS as an SDS structure, which is a hierarchical format. Hierarchical structures are necessary to support the many kinds of data which the Gemini instruments may generate. Data are then stored in a flat FITS format by the Data Server. The SDS format is used because it is part of the message protocol used between DHS and external sources. FITS was chosen

---

[1] `http://heasarc.gsfc.nasa.gov/docs/heasarc/fits.html`

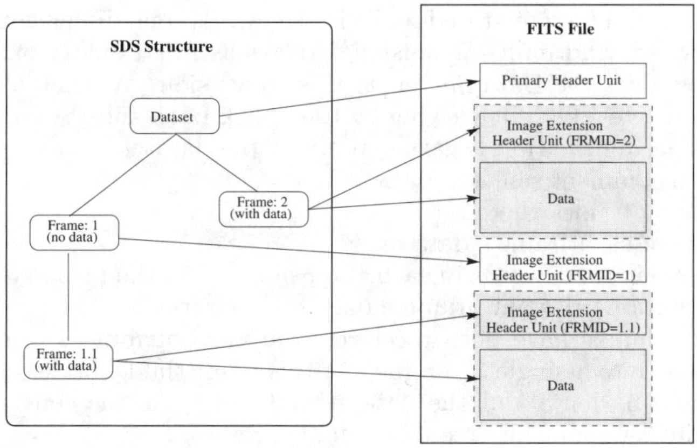

Figure 2. SDS to FITS mapping.

because it is the standard file format used within the astronomical community. All FITS file manipulations use the `cfitsio` library (Pence 1995).

A dataset is the data resulting from an OBSERVE command. A dataset generally consists of header data describing the dataset and one or more frames. A frame is header data and a single array of pixel data. The frame can represent a single exposure or several exposures. It can also be bad pixel masks, quality data, variance data, etc. Frames contain a frame identifier, a frame name, attribute/value data describing the frame, an optional single data array, and an optional set of sub-frames. The hierarchical definition of frames allows related frames to be grouped in a logical way. Header data is a collection of attribute/value pairs which describe a dataset or a frame within a dataset. These data structures are designed to allow for the transmission of arbitrarily complex data.

4. Conversion Process Details

One of the difficulties encountered when designing this system was determining how to preserve the hierarchical information within a flat FITS file format. Figure 2 shows how the SDS hierarchy and data are mapped into FITS files. Note that since data can arrive in any order, Frame 2 actually appears before Frame 1 in the final FITS file.

Each dataset is stored in a single FITS file. A FITS image extension will be created for each frame found in the dataset. The hierarchical information is stored in the FITS keyword `FRMID` which is a value such as 1, 1.1, or 2. Note that if a frame has no associated data then there will be no data following the FITS image extension header.

The Data Dictionary is a configuration file used by the Data Server to map the SDS attributes to FITS keywords. The Data Dictionary is maintained by the instrument support scientists. The properties of the header attributes include the specification of how the data will be stored in the FITS file. For example, single value attributes are stored as FITS keywords, one-dimensional arrays as indexed keywords and multi-dimensional arrays as FITS binary extensions.

Now how does the Data Server do this conversion? As each chunk arrives, it is converted into FITS for viewing by the Quick Look subsystem. During this first conversion information is gathered about the dataset:

1. which instrument sent the data,
2. how many frames there are,
3. the size of each frame's data array,
4. the number of keywords in each extension and in the primary header unit,
5. and the frame IDs and frame names for each frame.

Once all chunks have been received from all contributors, the chunks are then converted into a single FITS file. At this point the Data Server has all the statistical information about the dataset and can make the conversion for the final FITS final without any reallocation of memory.

5. Conclusion

The Data Server was designed long before any of the Gemini instruments were available for testing. As a result, one of the main design criteria was to make it as flexible as possible in order to meet all potential instrumentation needs. With flexibility came complexity. There are multiple configuration files to be filled out prior to running, with the most important being the Data Dictionary. The use of the Data Dictionary has avoided hard-coding attribute/keyword pairs leaving the Gemini staff and instruments to choose them as they wish.

In November 1998, the Gemini Data Handling System had its "first dark" picture using the DHS with a Gemini High Resolution Wavefront Sensor. It is called "first dark" because the calibration lamp was switched off, and it obtained a dark image contaminated with room light. First light for the Gemini system is scheduled to be January 1999. One advantage of using the DHS is that the DHS Data Server will convert the data and send it on to the Quick Look subsystem and display each new frame automatically.

References

Hill, N., Gaudet, S., Dunn, J., Jaeger, S., & Cockayne, S. 1999a, this volume, 155

———, 1999b, this volume, 163

Pence W. 1995, Fitsio User's Guide, Goddard Space Flight Center

Astronomical Data Analysis Software and Systems VIII
ASP Conference Series, Vol. 172, 1999
D. M. Mehringer, R. L. Plante, and D. A. Roberts, eds.

The SCUBA Data Reduction Pipeline: ORAC-DR at the JCMT

Tim Jenness, Frossie Economou

Joint Astronomy Centre, 660 N. A'ohōkū Place, Hilo, HI 96720

Abstract. The ORAC data reduction pipeline, developed for UKIRT, has been designed to be a completely general approach to writing data reduction pipelines. This generality has enabled the JCMT to adapt the system for use with SCUBA with minimal development time using the existing SCUBA data reduction algorithms (SURF).

1. Introduction

ORAC-DR (e.g., Economou et al. 1998; Economou et al. 1999) is a flexible and modular pipeline originally developed by the United Kingdom Infrared Telescope (UKIRT) for the on-line reduction of data from infrared instruments as part of their ORAC project (Bridger, Wright & Economou 1998). This paper describes our experiences with using ORAC-DR at the James Clerk Maxwell Telescope (JCMT) for the reduction of SCUBA (Submillimetre Common User Bolometer Array) data.

The ability of the ORAC-DR pipeline to handle data from instruments as disparate as an infrared grating spectrometer and a submillimetre bolometer array is an indication of the flexibility of the system design.

2. Recipes

Recipes in ORAC-DR consist of a series of data reduction steps (primitives) containing instructions for the reduction of data. Recipes may be written or changed by support scientist with the aid of an editor or, shortly, a drag-and-drop-type GUI.

This is an example of a recipe used for reducing SCUBA photometry data:

```
_PRE_PROCESS_
_FLAT_FIELD_
_EXTINCTION_CORRECT_
_CLIP_BOLOMETERS_ NSIGMA=3.0
_REMOVE_SKY_NOISE_JIGGLE_  BOLOMETERS=all MODE=median
_CALC_PHOT_RESULT_
_CALIBRATE_PHOTOM_DATA_
_PRINT_RESULTS_PHOTOM_FRAME_
_PHOTOM_GROUP_
```

3. Primitives

ORAC-DR primitives contain perl code for the manipulation of data in a given state. They are written in an object-oriented fashion, manipulating objects associated with the individual data frames as well as groups of observations. The steps that involve actual processing of data (as opposed to housekeeping etc tasks) are done via a messaging request to an algorithm engine resident in memory.

Here is an example of the _CLIP_BOLOMETERS_ primitive used by the above recipe. SCUBA has multiple sub-instruments, that simultaneously take data at different wavelengths (e.g., at 450 and 850 microns). This primitive performs a median clipping of the data from all sub-instruments associated with the current frame object. The actual clipping is performed by a task (in this case `scuclip`) in the SURF monolith. SURF (SCUBA user reduction facility, Jenness & Lightfoot 1998a,b) is written in FORTRAN and accepts ADAM messages. The details of the messaging are hidden from the primitive writer, who is typically a software engineer supporting an instrument.

```
# default the NSIGMA parameter if not specified
$nsigma = ($_CLIP_BOLOMETERS_{NSIGMA} || 5.0);

# loop over sub-instruments
for ($i = 0; $i < $Frm->nsubs; $i++) {
  # Generate output filename from the input
  ($in, $out) = $Frm->inout("_clip", $i+1);
  # Execute the action
  $Mon{surf_mon}->obeyw("scuclip",
      "in=$in out=$out nsigma=$nsigma");
  # Update the filename of the current frame object
  $Frm->file($i+1, $out);
}
orac_print "Bolometers clipped at $nsigma sigma\n";
```

Because primitives always manipulate objects associated with the pipeline, they are order-ignorant, i.e., no assumptions are made about the file number, file name, or filename convention. These behaviours are all handled by instrument-specific classes, thus allowing the re-use of primitives elsewhere in different recipes for different instruments.

4. Adding New Instruments

The object-oriented design of ORAC-DR allows new instruments to be added simply by sub-classing existing instrument objects and providing modified methods as needed. The only change required to the core pipeline code is to add the class definitions for the new instrument.

This section describes the modifications necessary to add SCUBA support to ORAC-DR.

4.1. Algorithm engines and messaging system

A single file is provided for each instrument that defines the necessary data reduction packages and the related messaging system. For SCUBA, it is necessary to use the SURF package (this provides the SCUBA specific data processing tasks) and the Starlink KAPPA data reduction system (Currie & Berry 1998) (for general purpose data file manipulation). Both these packages use the NDF data format and the ADAM messaging system.

4.2. Frame definition

A SCUBA observation (or Frame) differs from a UKIRT observation in a number of ways and these differences are handled by the ORAC::Frame class. The SCUBA variant (ORAC::Frame::SCUBA) must do the following:

- Handle the different file naming convention
- Provide translation from SCUBA FITS keywords to ORAC keywords
- Cope with multiple sub-instruments

The first two are fairly simple to implement, but the final issue comes about because SCUBA can observe at up to 4 wavelengths simultaneously with all the data being stored in a single file. Mid-way through the data reduction a file is generated per wavelength and the pipeline must keep track of these files separately. Each primitive is then designed to process each sub-instrument.

4.3. Group handling

The SCUBA observing system does not have an inbuilt notion of an 'observation group'. The major driving force behind the adoption of a pipeline for SCUBA was to provide a way of co-adding separate observations automatically in order to reduce the load on a service observer. The SCUBA implementation of ORAC-DR differs from the UKIRT system in that it must calculate group membership itself from header information (source name, observing mode, position etc) rather than simply querying the header for a group name. To deal with the occasion when an observer does not want an observation to be included in the group (even when ORAC-DR thinks it should be included) a facility is provided for removing certain observations from group operations.

Other differences between SCUBA and UKIRT groups relate to the sub-instrument problem. The group handling primitives must co-add all members of the group at all available wavelengths. Group handling is particularly complex in SCAN/MAP data reduction since there is the added complication of combining observations taken with different chop configurations

4.4. Calibration

Calibration of submillimetre data is completely different to near-infrared/optical calibration and a completely new calibration class was required for SCUBA. Two types of calibration are required:

- Correction for atmospheric extinction.
- Compensation for instrumental gain variations

The atmospheric extinction (sky opacity) can be determined by performing skydip observations (i.e., measuring the sky brightness temperature at different elevations and fitting the data with an atmospheric model) at a single wavelength and then extrapolated to the current wavelength.

Gain calibration is more complicated and has, as yet, only been implemented for photometry observations. In this case a flux calibrator (usually a planet) is observed and a gain derived that is used for further observations.

4.5. Recipes and primitives

New recipes and primitives had to be written since SCUBA is completely different to a near-infrared/optical instrument but this was possible without having to construct a data reduction pipeline from scratch. The major effort in adding SCUBA to ORAC-DR has therefore been entirely focused on the actual data reduction recipes rather than on infrastructure.

5. Conclusions

The JCMT have adopted UKIRT's ORAC-DR data reduction pipeline for processing SCUBA data. This pipeline is now in use at the telescope and provides useful information on the progress of observing in near-real-time. The modular and object-oriented design of the ORAC-DR pipeline has allowed the addition of SCUBA to be made without significant expenditure of effort on pipeline infrastructure. Programming effort has concentrated on writing data reduction recipes and is now benefiting from the decision to share pipeline code with another telescope.

The pipeline has been designed so that it can be used off-line or on-line and it is intended that observers will use the system at their home institution or as a quick-look system for database retrieval.

References

Bridger, A. B., Wright, G. S., & Economou, F. 1998, in ASP Conf. Ser., Vol. 145, Astronomical Data Analysis Software and Systems VII, ed. R. Albrecht, R. N. Hook, & H. A. Bushouse (San Francisco: ASP), 292

Currie, M. J. & Berry, D. S. 1997, KAPPA – Kernel Application Package User's Guide (ver. 0.13) (Starlink User Note 95, CCLRC/Rutherford Appleton Laboratory)

Economou, F., Bridger, A., Wright, G. S., Rees, N. P., & Jenness, T. 1998, in ASP Conf. Ser., Vol. 145, Astronomical Data Analysis Software and Systems VII, ed. R. Albrecht, R. N. Hook, & H. A. Bushouse (San Francisco: ASP), 196

Economou, F., Bridger, A., Wright, G. S., Jenness, T., Currie, M. J., & Adamson, A. 1999, this volume, 11

Jenness, T., & Lightfoot, J. F. 1998a, in ASP Conf. Ser., Vol. 145, Astronomical Data Analysis Software and Systems VII, ed. R. Albrecht, R. N. Hook, & H. A. Bushouse (San Francisco: ASP), 216

———, 1998b, SURF - SCUBA User Reduction Facility User's Manual (ver. 1.3) (Starlink User Note 216, CCLRC/Rutherford Appleton Laboratory)

Astronomical Data Analysis Software and Systems VIII
ASP Conference Series, Vol. 172, 1999
D. M. Mehringer, R. L. Plante, and D. A. Roberts, eds.

Data Reduction of Jittered Infrared Images Using the ORAC Pipeline

Malcolm Currie, Gillian Wright, Alan Bridger

UKATC, Royal Observatory, Blackford Hill, Edinburgh, UK

Frossie Economou

Joint Astronomy Centre, 660, N'ohōkū Place, Hilo, HI 96720

Abstract. We relate our experiences using the ORAC data reduction pipeline for jittered images of stars and galaxies. The reduction recipes currently combine applications from several Starlink packages with intelligent Perl recipes to cater to UKIRT data. We describe the recipes and some of the algorithms used, and compare the quality of the resultant mosaics and photometry with the existing facilities.

1. Introduction

The UKIRT ORAC (Observatory Reduction and Acquisition Control) project (Bridger et al. 1998) aims to improve the efficiency of observing at UKIRT by providing a modern software suite that will be capable of fully automated observing. ORAC will also make the telescope and instruments easier to use for traditional observing, by, for example, reducing the amount of manual intervention that is needed.

One of the key components is the ORAC Data Reduction system (ORAC-DR). Its goal is to offer observers near real-time data reduction, and hence let observers accurately assess the quality of their data while they are still at the telescope. Some of the benefits are more efficient use of the telescope, a higher publication rate, and better quality data. ORAC-DR uses a modular pipeline with the intent of producing close to publication quality images and spectra.

This paper concentrates on the imaging, describing some of the data reduction recipes, and lists some pros and cons of using other people's software.

2. Recipes for Imaging Data Reduction

We have written a number of common data reduction recipes and matching observing sequences which cover a range of photometry and imaging techniques in use at UKIRT. A recipe is a series of high level instructions such as "make a flat" or "generate offsets". The implementation of each of these instructions is through a Perl script, called a primitive, which calls particular data reduction packages, currently from Starlink, to actually do the work.

Many primitives are generic, so new recipes can be created fairly quickly. Most recipes contain a steering primitive to instruct the other primitives what operations should be performed on that image or usually the group of images to which it belongs, and to classify blank-sky from target frames. The recipes run initialisation scripts to enable history recording and specify detector characteristics and subsets, and all apply a bad pixel mask.

The pipeline uses calibration indices for dark, flat, and sky frames. When a calibration frame is required, the recipe accesses the index to find a suitable image, i.e., one which satisfies a set of criteria stored in a rules file. Typical rules are that the filters match and pick the most recent calibration.

In the summary of the jitter-pattern recipes below, the letters in parentheses following the names are codes for processing steps invoked in the recipe. The codes are D=dark subtracted, F=self-flat, M=make mosaic, O=objects masked, R=automatic registration, S=sky subtracted. Section 3 describes these steps.

`BRIGHT_POINT_SOURCE` (DORM) It assumes that a SKY_FLAT flat-field is sufficiently stable for a significant fraction of a night not to affect photometry.

`EXTENDED_`n`x`m (DSORM) This is experimental for extended objects where there may be little sky in close proximity for background subtraction. It alternates between jittered blank sky and the extended source, mapping the source in a n column by m row grid with 50% overlap of adjacent target frames. The mosaic is built in rows. At the completion of a row, the interleaved blank-sky frames are combined to make a flat, which is then applied to the source frames in that row. The order within rows and the order the rows are observed is such that adjacent frames are not observed sequentially. This randomising aims to reduce systematic errors from the sky subtraction. The interpolated mode of the bracketing sky frames is subtracted prior to flat-fielding. This method of sky subtraction has worked in practice achieving uniformity to 0.15% in good sky conditions. This is a major improvement over the formerly used `IRCAMDR`[1] algorithms applied to the same galaxy.

`JITTER_SELF_FLAT` (DFORM) This is a 5- or 9-point jitter for point sources.

`QUADRANT_JITTER` (DFORM) The original was devised by Lance Miller to search for faint extensions around bright objects, without the need to take separate flat observations, or rely upon the flat's long-term stability. The jitter pattern moves the target source to approximately the centre of each quadrant, and the cycle is repeated to the desired integration time. A preliminary flat is derived from a median of the target frames with the quadrant containing the source excluded. The improved version uses this as a trial flat and masks all the prominent objects to make a superior flat (see the Object Masking description in §3.). The refinement reduces the noise around the central target by 40% and removes artifacts. Sky regions are typically flat to better than 0.01% even in crowded fields.

---

[1] `http://www.jach.hawaii.edu/~sryder/UKIRT/mosaic.html`

`REDUCE_DARK` This just applies the bad pixel mask and files the calibration.

`SKY+JITTER`n (SRM) It is similar to `BRIGHT_POINT_SOURCE`, except that it subtracts the sky rather than a dark frame, and is intended for moderately faint sources. $n = 5, 9$. It files the sky calibration.

`SKY_FLAT` (D) This forms and files a flat from a 5 (or more) point jittered mosaic of blank sky. There is a variant which masks sources within the frame.

3. Features of the Primitives

Primitives are the Perl scripts which actually call the application tasks to do most of the work. Some tasks were designed with CCDs in mind, so a degree of tuning is sometimes needed to allow for infrared instruments, pixel scale, seeing, or filter, for example.

Flat creation. Frames are optionally cleaned to remove extreme outliers (± 3 or 6σ about the mean in 15×15-pixel neighbourhood), normalised, combined pixel by pixel using a median, and the combined array normalised.

Object Masking. After the application of an approximate flat field, an inventory is made of objects having at least 12 connected pixels above one sigma above sky. (The optimum thresholds are still being determined.) The locations, shapes, orientations, and sizes are used to make a mask. The mask is applied to the dark-subtracted frames and a new flat created. The improved flat typically shows a uniformity at $\sim$0.02% of the sky. Systematic errors in the sky, which are a major uncertainty in infrared point source photometry, are also reduced significantly by this algorithm.

Automatic Registration. This makes an inventory of the sources above a threshold in each frame. It then performs a pattern recognition (Draper 1998) to identify common features in jittered frames. If the fraction of common objects is under half or the total is fewer than 3, the registration fails, and so the script resorts to reading the offsets stored in the FITS headers or matching a central bright object. Using telescope offsets can lead to trailed sources, as occurred with the `IRCAMDR` package. The improved registration leads to the detection or more accurate measurement of faint sources.

Mosaicking. This is fully automated. Observers do not have to painstakingly measure centroids and manually tile to form mosaics. In addition to Cartesian shifts derived from the automatic registration, the primitive applies a preset rotation matrix to align the image to the cardinal directions with only one resampling. Experiments are underway to use jitter offsets aligned with the chip and to put the rotation information into the World Coordinate System to minimise artifacts in the data and to save CPU cycles.

The mosaicking uses the `CCDPACK` algorithm (Warren-Smith 1993). Only zero-point shifts of intensity are applied to the resampled frames to create the mosaic. Depending on the recipe, the mosaic may be trimmed to the dimensions of a raw frame. Mosaicking removes virtually all the bad pixels.

Automatic Photometry. This is adaptive. It locates the source, measures its point spread function, selects a suitable aperture size scaled in terms of the FWHM and a concentric sky annulus, integrates the flux within the object aperture, and subtracts the modal sky value. The previous software used a fixed aperture and the median sky from a narrow annulus, and hence could generate less accurate photometry. The script also evaluates a correction for light outside the aperture. The results presented include a standard extinction coefficient and zero-point and a formal measurement error.

4. Pros and Cons of Using Other People's Software

Rather than consisting of large packages dedicated to each instrument, ORAC-DR uses other people's general purpose, data reduction software. This approach reduces maintenance costs, provides greater flexibility (such as to reflect quickly a change in the performance or understanding of an instrument), and has the ability to take advantage of new astronomical software and techniques.

It has its downside too. Sometimes the packages either do not provide what we need for a recipe, or we have to invoke several atomic tasks at the cost of efficiency and intermediate files. For example, the point spread function of UKIRT images are markedly non-Gaussian. The KAPPA/PSF (Currie 1997) only fits Gaussian-like functions, whereas a two-component model is needed. Looking at an extreme case[2] reveals a double hexagonal pattern, for which a general psf-fitting task would never be expected to model. Thus, we expect some recipes will require the odd bespoke application. See Economou et al. (1999) for further discussion on this topic.

Acknowledgments. We thank the Starlink programmers, particularly Peter Draper, for implementing enhancements to packages. We also thank the various astronomers whose data have been used to test ORAC-DR.

References

Bridger, A., Economou, F., Wright, G. S., & Currie, M. J. 1998, in SPIE Proc., Vol. 3349, Observatory Operations to Optimize Scientific Return, ed. P. J. Quinn, (Bellingham: SPIE), 184

Currie, M. J. 1997, KAPPA – Kernel Application Package, (Starlink User Note 95, Rutherford Appleton Laboratory)

Draper, P. W. 1998, CCDPACK – CCD Data Reduction Package, (Starlink User Note 139, Rutherford Appleton Laboratory)

Economou, F., Bridger, A., Wright, G. S., Jenness, T., Currie, M. J., & Adamson, A. 1999 this volume, 11

Warren-Smith, R. F. 1993, in Proc. of the 5th ESO/ST-ECF Data Analysis Workshop, ed. P. Grosbøl & R. Ruijsscher (Garching: ESO), 39

---

[2] `http://www.jach.hawaii.edu/UKIRT/telescope/seeing.html`

Astronomical Data Analysis Software and Systems VIII
ASP Conference Series, Vol. 172, 1999
D. M. Mehringer, R. L. Plante, and D. A. Roberts, eds.

A Broadband Wide-Field Survey on the Isaac Newton Telescope

James R. Lewis, P.S. Bunclark

Institute of Astronomy, Madingley Road, Cambridge CB3 0HA, England

N.A. Walton

Isaac Newton Group, Apartado 321, 38780 Santa Cruz de la Palma, The Canary Islands, Spain

Abstract. The Isaac Newton Group's Wide Field Survey Programme[1] is a multi-colour broadband imaging survey being carried out on the 2.5-m Isaac Newton Telescope at the Observatorio Roque de los Muchachos on La Palma. The survey programme is expected to have a duration of five years during which time more than 200 square degrees of sky will be imaged in up to five colours at a spatial resolution of <1.2 arcseconds. In order to maximize its scientific impact, the data obtained through the survey programme are being made available to the community on a near real-time basis – there is no proprietary period.

We describe the techniques employed to meet the challenge of processing and disseminating the substantial data flows associated with this survey. Key areas addressed are: the pipelines developed to pre-process and process the data (e.g., flux and astrometric calibration) and the automation of routines to fulfill user requests for these data products, implemented through WWW interfaces.

1. Introduction

The 2.5m Isaac Newton Telescope (INT) is the oldest of the three telescopes that comprise the Isaac Newton Group of telescopes on La Palma. In an era when several very large telescope projects will be coming on line in the near future (e.g., Gemini North and South, ESO VLT) a shift in emphasis of the role of the INT towards wide field optical image surveys is clearly highly desirable.

In the Spring of 1997 a new Wide Field Camera was commissioned on the INT. This consists of a mosaic of four EEV CCDs, each having a format of 2148 × 4128 pixels 13.5 microns. A single exposure covers a total of about 0.4 sq. degrees. In order to promote the use of the WFC for large scale survey work, the ING Joint Steering Committee decided to make substantial amounts of observing time (three or more weeks per semester) available for high quality

[1] `http://www.ing.iac.es/WFCsur/`

survey programmes. This paper is a description of 'The INT Wide Angle Survey' one of the survey programmes chosen to partake of this time.

2. The Survey

The INT Wide Angle Survey is the first major fully digital, multi-epoch and multi-waveband wide angle survey over a significant area. The survey parameters cover a large range of areas of topical and future interest. Some of the scientific goals of the survey are:

- **Galactic Studies** including the Halo and disk WD luminosity function, relevant to both DM models and to independent calibrations of the Hubble time; stellar density distributions towards the NGP, to improve extant K_z determinations of the local DM; stellar counts towards the anticentre and other widely spaced directions, to determine the stellar warp and refine models of Galactic structure
- **Clusters of Galaxies**. The aim is to determine the space density and cluster-cluster correlation function over the range $0.5<z<1.0$. Galaxy clusters are the largest gravitationally-bound structures in the Universe, and the study of their abundance and evolutionary history with look-back-time places strong constraints on cosmological parameters and the primordial power spectrum that gave rise to the observed large scale structure.
- **Radio Sources & Radio Galaxies**. Deep optical identification of radio sources allows: accurate counts of different types of host along mJy tracks in the P-z plane, studies of radio source luminosity evolution; multi-band investigation of giant-E standard candles; the largest known sample of low-luminosity RGs with good photometry; large-scale structure from photometric redshifts and cell counts in redshift slices; accurate optical positions of FIRST sources for WYFFOS/2DF followup.
- **Optical Variability**. By surveying an intermediate Galactic latitude 3 $\times$ 3 degree field once per year in the u,g,r,i,z passbands and two 'single shot' WFC fields (0.25 deg^2), 12hr apart in RA, more regularly, we will obtain multi-epoch data for a wide variety of projects allowing: several thousand quasars to be monitored yearly, yielding measures of variability vs. luminosity, redshift, quasar-type etc.; rapid monitoring of a subset at 1% accuracy of $\approx$100 QSOs to B$\sim$23, for studies of quasar variability on very short timescales; detection of distant RR-Lyrae variables for studies of the extent of the Halo.
- **Intermediate redshift Type 1a Supernovae**. Whilst dramatic progress has been made in the determination of the fundamental cosmological parameters (Ω, Λ) in the last two years, the analysis is now limited by systematic errors. Identifying $\sim$20 Type 1a Supernovae in the critical range $0.1<z<0.4$ will allow a detailed treatment of these systematic errors.

Although the current survey is targeted at specific scientific goals, a judicious choice of passbands (the Sloan Digital Sky Survey passbands have been chosen for longevity), exposure times and regions will make these data useful for a much wider audience. The field list of field centres and the science programs these data will serve are listed in the Wide Field Survey WWW site[2]

An important element of all of the INT Wide Field Camera surveys is that the raw data are made available on line to the UK and Dutch communities almost as soon as they are observed. At a later stage semi-processed (debiassed and flat fielded) data are available. Finally, fully reduced (i.e., astrometrically and photometrically calibrated) frames will be produced.

Table 1. Survey Fields

Field	Coordinates (J2000)	l,b (°)	size	bands	multi-epoch	other data[a]
0000+00	00 00.0 +00 00	98,−60	3×3	ugriz		N,F
0230+00	02 30.0 +00 00	169,−53	3×3	ugriz	gi	N,F,X
0340+00	03 40.0 +00 00	187,−41	3×3	ugriz		N,F
Pleiades	03 47.0 +24 00	167,−23	3×3	ugriz	iz	N
0801+4018	08 01.7 +40 19	180,+30	3×3	ugriz	gi	N,F,W
1040+00	10 40.0 +00 00	249,+49	3×3	ugriz	gi	N,F,2DF,X
HDF-N	12 36.0 +62 00	126,+55	3×3	ugriz		N,W,X
Virgo	12 39.0 +12 27	294,+75	3×3	ugriz		N
NGP/Coma	12 51.4 +27 08	0,+90	3×3	ugriz		N,F
ELAIS-N1	16 10.0 +54 30	84,+45	3×3	ugriz		N,F,W,
NEP	18 00.0 +66 00	96,+30	3×3	ugriz		N,W,X
2240+00	22 40.0 +00 00	69,−49	3×3	ugriz	gi	N,F

[a]existing or planned survey material; N:NVSS, F:FIRST, W:WENSS X:X-ray data(ROSAT or XMM)

3. Data Reduction

Data reduction will be done with a specially written data reduction pipeline, **wfcred**. The package is written to take advantage of existing software and as such the vast part of the data manipulation is handled by IRAF. The individual data reduction tasks take the form of a number of Perl scripts. (Perl/Tk is used for the user interface.) These scripts are used to gather information about the nominated files, sorting them into subsets by filter, Julian Date, observation type (e.g., Flat field, Bias or Target), etc. Once the files are sorted into their requisite subsets, then they are sent to IRAF tasks such as **ccdproc** for processing.

The formation of good mean bias frames and in particular mean flat field frames can be tricky. Much care needs to be taken to ensure, for example, that frames with large overexposed regions are not included and thus this should be an interactive process. For these reasons the routines **fftool** and **biastool** have been written. As their names suggest these are interactive tools used for the creating and storing mean bias and flat field frames. Included with these are facilities to help the user decide whether or not to include an exposure (e.g., the ability to examine headers, display frames and do simple statistics). Once the user has narrowed down the list of files to be used in the formation of the mean

[2]`http://www.ast.cam.ac.uk/~mike/casu/WFCsur/WFCsur.html`

frames, the processing is then a straight-forward application of standard CCD processing routines in IRAF. One of the most important aspects of **wfcred** is the ability to have libraries of mean bias and flat field frames. This is crucial as even in the best of observing conditions it is nearly impossible to get a full set of flat field exposures each and every night. Once a mean frame has been created the user has the ability to either store it in a library or discard it.

Central to the way in which **wfcred** works is its ability to group files into subsets according to various criteria (e.g., filter name or Julian date). This is done in a way which is invisible to the user who is presented with the data grouped in the most sensible way depending upon the stage of the reduction procedure. For example in **fftool** it is most likely that the user will want to see the files grouped by filter first. Having decided which filter's flats will be processed the user is then presented with the list of files in that filter subset, but regrouped by Julian date. The subsets presented by this final grouping are obviously the four frames (one for each chip) for a given exposure. When it comes to doing the CCD processing and combining it obviously makes far more sense that all data from a single chip are handled simultaneously. Thus once the user has made the final choice of frames the remaining files are regrouped once more by chip name before being sent to the various IRAF routines.

During early stages of development of **wfcred** it was discovered that the analogue-digital converter for one pair of chips was not behaving linearly. To take account of this effect a 'D to I' conversion routine has been written and incorporated into the pipeline. For later dates when the problem has been resolved the converter can be disabled, but will remain as part of the pipeline for the reduction of older data.

In addition to the removal of instrumental signature, target frames also have an astrometrically accurate world coordinate system defined for them. The relevant section of the Guide Star Catalogue is extracted first and the (α, δ) coordinates are converted to (X, Y) with a rough transformation based on the coordinate centre and the known configuration of the chips. The GSC stars are matched up against the (x, y) positions of the brightest detected images found on the chip. Finally a bi-linear polynomial fit is done to (x, y) and (α, δ) using the IRAF routine **ccmap** with the ZPX projection and a known cubic distortion term. There are typically between 15 and 30 stars on a frame and the RMS residuals are generally 0.5 arcseconds or less.

4. Future Plans

Below is a list of future goals for **wfcred**.

- The automatic removal of fringes. This is especially important on the I and Z frames.
- Automated photometric calibration. The standard set of Landolt fields is being observed in each run. Thus it will be possible to automate the location and extraction of image information for these fields and hence automate the photometric calibration of the program fields at the point where the images are being extracted.

Acknowledgments. We gratefully acknowledge the help of Drs. M.J. Irwin and R.G. McMahon for their help in preparing this paper.

Astronomical Data Analysis Software and Systems VIII
ASP Conference Series, Vol. 172, 1999
D. M. Mehringer, R. L. Plante, and D. A. Roberts, eds.

A Pipeline Tool for CCD Image Processing

Jon F. Bell, Peter J. Young, William H. Roberts, Kim M. Sebo

Mount Stromlo and Siding Spring Observatories, Private Bag, Weston Creek, ACT, 2611, Australia

Abstract. MSSSO is part of a collaboration developing a wide field imaging CCD mosaic (WFI). As part of this project, we have developed a GUI based pipeline tool that is an integrated part of MSSSO's CICADA data acquisition environment and processes CCD FITS images as they are acquired. The tool is also designed to run as a stand alone program to process previously acquired data. IRAF tasks are used as the central engine, including the new NOAO mscred package for processing multi-extension FITS files. The STScI OPUS pipeline environment may be used to manage data and process scheduling. The Motif GUI was developed using SUN Visual Workshop. C++ classes were written to facilitate launching of IRAF and OPUS tasks. While this first version implements calibration processing up to and including flat field corrections, there is scope to extend it to other processing.

1. Context

This pipeline tool[1] was originally developed for the Wide Field Imager (WFI)[2] being developed jointly by MSSSO, AAO and the University of Melbourne, but it can be used to process other CCD image data. It has been developed in a general manner so that other processing tasks contained in IRAF[3] can be easily added. One might envisage adding facilities for calibration processing of 1 D spectra for example. While it has been developed to run in conjunction with CICADA[1] (Configurable Instrument Control and Data Acquisition), software developed at MSSSO, it may also be used offline in a stand alone mode. Figure 1 shows how this tool fits into the overall process.

2. Architecture

To build a pipeline we needed 3 main components: A processing engine, a data managing and scheduling tool, and a graphical user interface. All of these components had to mesh neatly with the existing CICADA environment, shown

[1] `http://msowww.anu.edu.au/computing/cicada`

[2] `http://msowww.anu.edu.au/observing/wfi/`

[3] `http://iraf.noao.edu/`

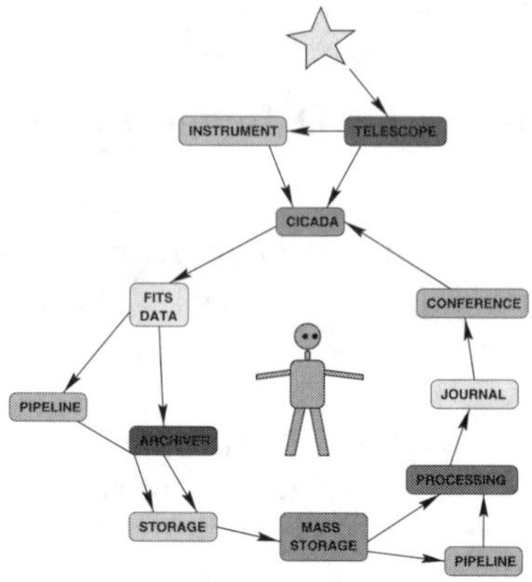

Figure 1. This pipeline tool in context.

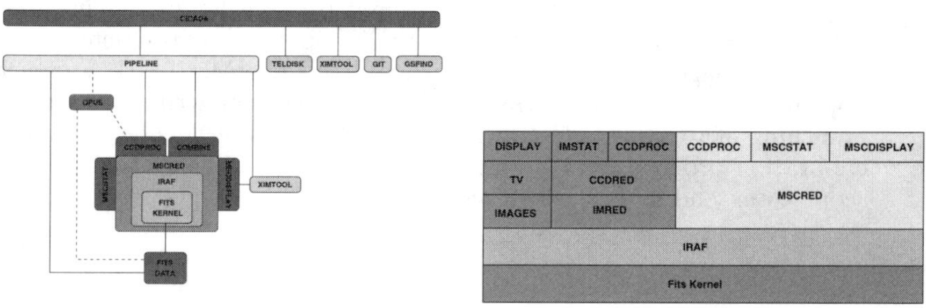

Figure 2. Architecture of Pipeline and Cicada environment (*left*). Inheritance structure of IRAF classes (*right*).

at the top of Fig. 2. They also had to be readily and freely available to all groups involved with the development of WFI. We chose IRAF (see §3.) for the central engine and built a Motif GUI (examples below) with a similar look and feel to CICADA. For a data scheduling and managing tool we experimented with both OPUS and our own code and at present either may be used (as indicated by the dashed lines in Fig. 2). While OPUS performed well, we have only made use of a small subset of its capabilities. Therefore future versions will be based on our own simple scheduler, reducing the complexity of this pipeline tool.

3. Central Engine – IRAF

A key requirement of the processing engine in this pipeline was the ability to process multi-extension FITS (MEF) files. Our investigations revealed that the best means of doing this was to use the recently developed IRAF package mscred. It provides multi-extension analogs of most of the tasks found in the ccdred package of IRAF, as well as a number of mosaic specific tasks. While IRAF tasks can be used in an automated fashion by running the package executables directly, we found it much simpler and more robust to use scripts to run the

IRAF cl directly. The structure of IRAF lends itself very nicely to a multiply inherited OO class structure, in the sense that a task or sub package inherits all the properties of its parent. In order to implement the use of IRAF in our pipeline, we developed a multiply inherited C++ class structure as shown in Fig. 2. This has been done in a very general way, so other tasks and packages can easily be added. As shown in Fig. 2, there is a base class for the FITS kernel, which is inherited by the IRAF class. Packages such as mscred then inherit the IRAF base class, thereby also inheriting the FITS kernel base class. In the present implementation the functions in these classes receive all the necessary information from the GUI or configuration files and write out scripts containing all the necessary details to run each task. Getting information back from IRAF can be another tricky issue when it is run in an automated way. We have approached this issue, by turning the IRAF logging facilities on and using event handlers and timeouts to return information from the log files. When Open IRAF is available it would be possible to use that to directly call IRAF tasks, rather than the method noted above.

4. Typical Processing Steps

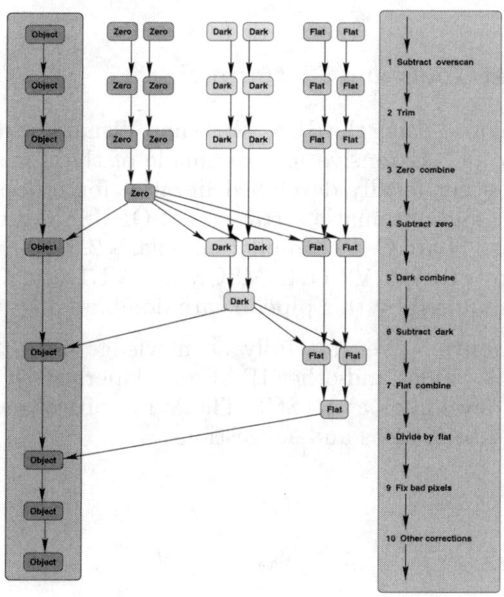

Figure 3. Processing steps presently implemented.

The processing sequence that is presently implemented is shown in Fig. 3, starting at the top and flowing down. The right most box describes the steps involved. Following the successful processing of each image, the image is displayed, along with image statistics, so the user can quickly verify that the processing is proceeding as intended. At present, due to the complicated nature of preparing flat field images for WFI and other wide field CCD mosaics, it is assumed that the calibration images have been pre-prepared (or obtained from a library) and the path on the right is followed for images containing science data. When the formation of flat fields becomes a more reliable process, it will be possible to automate pipelining of the steps shown above as IRAF already supports this.

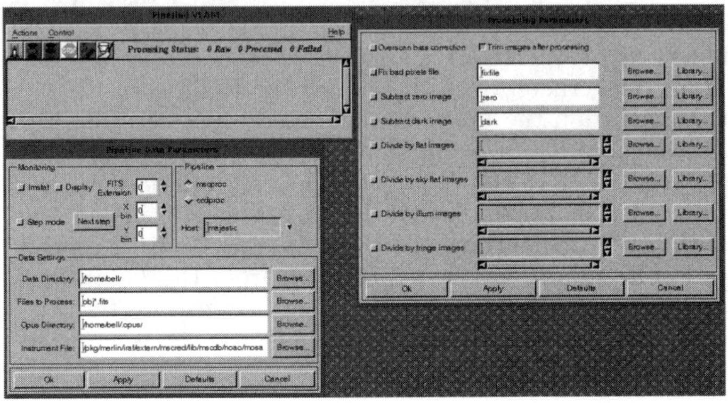

Figure 4. Snapshots of the GUIs.

5. Development Tools and Platform

The GUI was developed using the SUN Workshop Visual Interface Builder V2.0[4], generating C++ code. Extensive use was made of the Tools.h++ 4.0.1[5] class library as well as other locally developed libraries for process communication, exception handling and automatic setting up of OPUS[6]. All modules were built and tested using the Sparc C++ compiler on Solaris 2.6. Versions of third party software used were: IRAF: V2.11.1, MSCRED: V1.1 and OPUS: V1.0. The system resources required by this pipeline are dominated by the data.

Acknowledgments. We gratefully acknowledge very helpful advice and support from Frank Valdes and other IRAF developers at NOAO and Jim Rose and other OPUS developers at STSCI. The WFI software working group also contributed many useful ideas and suggestions.

---

[4]`http://www.sun.com/workshop/visual/`

[5]`http://www.inmark.com/products/`

[6]`http://www.stsci.edu/software/OPUS/`

Astronomical Data Analysis Software and Systems VIII
ASP Conference Series, Vol. 172, 1999
D. M. Mehringer, R. L. Plante, and D. A. Roberts, eds.

On-The-Fly Calibration at STScI

Stephen Lubow, Joseph Pollizzi

Space Telescope Institute, 3700 San Martin Dr, Baltimore MD 21218

Abstract. An on-the-fly calibration system (OTFC) is being developed at STScI. OTFC will allow users to retrieve data that is calibrated at STScI with the latest parameters, reference files, and software. In general, the calibration will be carried out as part of the process of satisfying the user retrieval request. OTFC will allow for a considerable savings in storage space, since calibrated data will (eventually) not be stored in the HST archive. The system allows changes to be made in calibration parameters for individual datasets through the use of a database system. It has features for handling datasets with known problems. The OTFC system must be highly reliable. The calibration processing will be managed by the OPUS system. A prototype system has been developed for WFPC2.

1. Introduction

Currently, Hubble Space Telescope (HST) data are calibrated as they are received at the Space Telescope Science Institute (STScI). Raw and calibrated data are stored in the HST archive (DADS). Frequently, users recalibrate data at their home sites to take advantage of better calibration files or software. To recalibrate, users can obtain the raw data and the recommended reference files from the Starview system. The user must then update the raw data headers with the names of the recommended reference files. The appropriate STSDAS calibration software is then run.

An on-the-fly calibration (OTFC) system at STScI will allow users to obtain data from the DADS archive which is calibrated with the most up-to-date calibration parameters, files, and software. In most cases, this capability requires that data be automatically calibrated at STScI when the data is requested by a user. An OTFC system has been successfully developed for HST data by the CADC/ST-ECF (Crabtree et al. 1995). The requirements for an OTFC system at STScI are somewhat different from the CADC/ST-ECF. More details about the OTFC system can be found on the web at

`http://www.dpt.stsci.edu/otfc/otfc_index.html`

2. Motivation

There are two main motivations for the OTFC system: saving data storage and providing users with improved calibration. These motivations were discussed in the HARP report (Hanisch et al. 1996). HST data are stored in an optical disk jukebox as part of the DADS system. The jukebox system has an online capacity of about 3 TB. The system is essentially full now. We have investigated the possibility of using DVD storage technology, but found that this is not yet ready for our purposes. Furthermore, the Advanced Camera for Surveys is scheduled for installation in HST within the next two years. It will substantially increase the rate at which data are received at STScI and would worsen the storage problem. With OTFC, only the raw data would need to be stored. Since raw data compresses better than calibrated data, there are substantial savings (of about a factor of ten in some cases) by storing only raw data in the archive in compressed form.

The other advantage of OTFC is in providing improved calibration to users. Nearly all (more than 90% of) datasets can benefit from recalibration (although the level of improvement is not always significant). Cameras such as WFPC2 produce dark files that typically should be retroactively used to calibrate observations taken in the prior week. Unfortunately, this means that the original calibration carried out in the STScI pipeline is not the optimal. Furthermore, this less than optimal calibration is stored in the DADS archive. An OTFC system would help solve this problem by providing users with the calibration having the appropriate dark, as soon as it is available.

Currently, instruments that undergo evolution of calibration files or calibration software require users to carry out their own recalibrations at their home sites. It may well be easier for both the users and the STScI support groups to have STScI carry out the recalibration.

Apart from reference files and software, improvements to calibration also can come about through changes to calibration switches and other keywords. Another goal of OTFC is to make such keyword changes possible without rearchiving the data in the optical disk archive.

3. Design Goals

We expect that the OTFC system will calibrate a large number of datasets each day, many times more than are calibrated by the current calibration pipeline for new data. For WFPC2, about 300 datasets are currently requested from DADS each day. Consequently, the system needs to be highly automated, while at the same time be robust. The system should provide a reasonable level of performance. It should not impose a substantial delay beyond the current DADS retrieval times.

Some users have established their own calibration pipelines, while others will sometimes customize the calibrations. To support such users, tools should be provided to determine the latest information about calibration parameters.

Another requirement is that the OTFC system should be compatible with other STScI systems, such as Starview and DADS. In fact, these existing systems will play a role in the OTFC system.

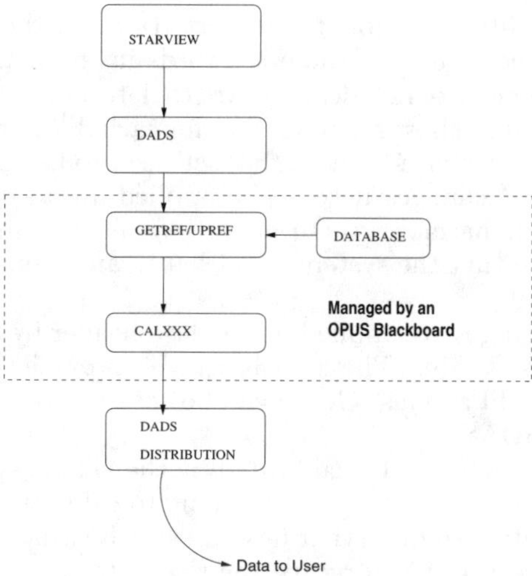

Figure 1. Dataflow for OTFC

4. Components and Dataflow for On-The-Fly Calibration

Figure 1 shows the dataflow for the OTFC system. First, a user requests data through the Starview user interface. This user interface is currently used by the DADS system and will contain extra options for OTFC. The request is passed to the DADS system which retrieves the appropriate data from the optical disk archive and passes that data to the OTFC pipeline. This pipeline is managed by the OPUS blackboard system (Rose 1996), which handles the process control. An enhanced version of the current getref tool queries a database to determine the names of the currently recommended reference files, the recommended calibration switch values, and any changes to other header keywords for the dataset being calibrated. (The header keyword changes are managed by the OTFC Change System which is described in the next section.) It writes its output to a text file called the delta file. The upref tool applies the delta file to update the data headers. The data is then calibrated by the appropriate STSDAS calibration routine. The data is the returned to the user via the DADS distribution service.

5. Change System

To describe changes to calibration keywords relative to the values in the archived raw data, a set of database tables is used. These tables contain entries for each dataset, keyword, and header (in the case of multiple headers). Each entry contains the new and old values of the keyword, the operation (insert, update, or delete) that should be applied to the header, as well as textual comments that describe the nature of the change. In essence, these tables describe how to edit the header. By using a database, we are able to provide a history of changes to keywords that is readily accessible to support staff. Furthermore, these entries are used to change the values for these keywords in the DADS catalog.

The change system also includes software to determine the calibration switch rules (Bestswitch) and the best reference files for each dataset (Bestref). Changes to parameters in the database for each dataset need to be carefully coordinated. For example, keyword changes need to be applied ahead of the determination of reference files. Furthermore, these operations need to be carried out as an atomic transaction so that the system is not left in an inconsistent state if errors occur.

The keyword changes are applied to the data header by an enhanced version of the getref/upref software. These tools will be provided for remote users to carry out their own calibrations. Getref can be accessed through a web interface, Upref is part of STSDAS.

Some datasets should not be calibrated by the OTFC system. For example, the data may be corrupt or it may take too long to calibrate. The OTFC change system contains a database table that describes such situations. The OTFC will not calibrate such data, but will return the raw data to the user.

Currently, for WFPC2, there are 368 keyword changes in the database and 11 datasets have been identified in the database for noncalibration.

6. Plans

We have developed a prototype system for WFPC2. We will release that system to external users in the near future. We will then extend OTFC to handle STIS, NICMOS, and ACS.

Acknowledgments. We thank Dan Durand for informative discussions.

References

Crabtree, D. R., Durand, D., Gaudet, S., & Hill, N. 1995, in ASP Conf. Ser., Vol. 101, Astronomical Data Analysis Software and Systems V, ed. G. H. Jacoby & J. Barnes (San Francisco: ASP), 505

Hanisch, R.J., et al. 1996, HARP: The Hubble Archive Re-Engineering Project Summary Report, STScI report

Rose, J. 1996, in ASP Conf. Ser., Vol. 125, Astronomical Data Analysis Software and Systems VI, ed. G. Hunt & H. E. Payne (San Francisco: ASP), 38

Astronomical Data Analysis Software and Systems VIII
ASP Conference Series, Vol. 172, 1999
D. M. Mehringer, R. L. Plante, and D. A. Roberts, eds.

On The Fly Re-calibration of NICMOS and STIS Data

Alberto Micol[1], Daniel Durand[2], Benoît Pirenne[1], Séverin Gaudet[2], Phil Hodge[3]

Abstract. In 1997 the HST payload was rejuvenated with two new instruments. After an initial period during which the instrument in-flight responses were assessed and understood and the calibration software and data flow stabilised, the CADC and ST-ECF started working together on the extension of the on-the-fly (OTF) calibration pipeline to support NICMOS and STIS. In this article the design and the implementation of such pipeline are shown. The design is based on a usage of database queries, a rule system and an automatic code generation system written in PERL.

1. Introduction

The On The Fly calibration system (OTF), originally implemented by the CADC and the ST-ECF staff is becoming a very important aspect of the HST archive. The main goals behind the implementation of the OTF system were to take advantage of:

- Better (more recent) calibration files
- Better pipeline software
- Much smaller storage area required since only raw files are kept in the archive
- the possibility to offer more calibration steps than originally performed on the data when it was first released

As shown by Lubow & Pollizzi (1999), more than 90% of the datasets located within the HST archive need to be recalibrated! Although sometimes the gain is minimal, the potential of extracting better science from the HST data is clear.

This is even more evident for the data acquired at the beginning of the life of an instrument. In fact, during the first months of NICMOS and STIS operations both the pipeline and the calibration files rapidly evolved as the instrument scientist's understanding of the instrumental responses and environmental conditions improved.

---

[1]ESA/ESO Space Telescope European Coordinating Facility

[2]National Research Council of Canada, Canadian Astronomy Data Centre

[3]Space Telescope Science Institute

2. Recalibration Steps

In order to perform this operation for STIS and NICMOS, major changes to the existing OTF system were required, because of the complexity of these new instruments.

The STSDAS calibration system is "header driven": calibration switches and files, to be used by the calibration pipeline, are listed in the header of the raw file. This is a clever design since it allows to easily cope with changes in the system (e.g., the availability of new calibration steps, or of new reference files) by changing the relevant keywords in the raw file header before running the calibration software.

2.1. RAW Files

The first step in the OTF is to identify the correct list of raw files required for the calibration in order to retrieve them.

2.2. Finding the Best References Files

After retrieving the files, we query the Calibration Database System (CDBS, which is replicated to ECF and CADC from STScI) to get the best references files for a given dataset. This is a complex query since, because of timing differences, sometimes a calibration file is needed but not available yet at our respective centers. We have developed for this case a fall back mechanism allowing us to get the previous best calibration file. Here is an example of the query result:

```
CDBS: shadfile, N/A
CDBS: lamptab,  otab$i9u1550oo_lmp.fits
CDBS: ccdtab,   otab$h1v1158do_ccd.fits
CDBS: darkfile, oref$ibk1356ko_drk.fits
CDBS: wbiafile, oref$iag16560o_bia.fits
CDBS: sptrctab, N/A
CDBS: inangtab, N/A
CDBS: apdestab, otab$hbp11488o_apd.fits
```

The result of this query is stored within the proper ckwxxx pset file which will then be used to edit the raw file header. If a reference file is not available at all, the related calibration switch will be set to omit, independently of the *rule* system (see later).

The list of reference files is built at CADC and checked every day for consistency and completeness. The changes are then replicated to ST-ECF using database and files system synchronisation tools.

2.3. Updating Values of Keywords from the Database

Often, one would find wrong keyword values within the header. Once a HST file has been archived it is impossible to correct its header; instead, STScI updates a database which stores the keyword values. Therefore the next step is to extract the value of each critical keyword from that database.

The mapping of header keywords and the HSTDADS database is kept in a configuration file. SQL queries are built on the fly from it.

This model will be used for the new OTF calibration system that STScI is putting in place for WFPC2.

2.4. Translate the Rules in PERL Code

Table 1. Rule file example for NICMOS.

Instrument	SWITCH	ACTION	RANK	Logical condition
NIC	DARKCORR	OMIT	1	TARGNAME = DARK
NIC	DARKCORR	OMIT	2	FILTER = %BLANK
NIC	DARKCORR	OMIT	3	OPMODE = BRIGHTOBJ
NIC	DARKCORR	PERFORM	4	T = T

The rule file (Table 1) is the key for the STIS and the NICMOS OTF calibration pipeline. It is part of the OPUS system used at STScI to calibrate the observations as soon as the are received from HST. The rule file assembles the knowledge from the instrument scientist about the calibration steps to be performed according to the instrument mode in use.

Although quite cryptic the file is being parsed dynamically when the OTF is started; a PERL script is then built. Here is an example of the perl code created:

```
if ( ( $nicmos_cl{'TARGNAME'} eq "DARK" ) ) {
   $nicmos_cl{'DARKCORR'} = "OMIT";
   $reason = 19;
 } elsif ( ( $nicmos_cl{'FILTER'} =~ /BLANK$/ ) ) {
   $nicmos_cl{'DARKCORR'} = "OMIT";
   $reason = 20;
 } elsif ( ( $nicmos_cl{'OPMODE'} eq "BRIGHTOBJ" ) ) {
   $nicmos_cl{'DARKCORR'} = "OMIT";
   $reason = 21;
 } else {
   $nicmos_cl{'DARKCORR'} = "PERFORM";
   $reason = 22;
 }
```

2.5. Compute Calibration Switches

We have now acquired all the information needed to perform the switch evaluation. For the example case which we have used so far, the decision algorithm is summarised here:

```
For each switch in the rules
   Does the switch need a reference file?
   > no:  switch is set by evaluating the rule
   > yes: did CDBS return a file for this (switch,dataset) key ?
          > no:  missing reference file, switch is set to omit
          > yes: switch is set evaluating the rule, e.g.:
                 Evaluation for DARKCORR is PERFORM
                 Reason being 22
```

The result is put within a cl script which will be executed to alter the raw file header.

2.6. Replace Raw Header Keywords with the New Values and Calibration

Each pset is finally executed via the *putcal* IRAF task, which modifies the header keywords in the specified raw files. The datasets are then ready to be calibrated with the most recent version of the STSDAS calibration S/W.

3. Conclusions

The OTF system contributed (and still contributes) to the reliability of the calibration software: we found and reported problems to the STScI/STSDAS group, which quickly fixed them. As soon as a new version of the calibration software is released we install it in our pipeline. Our archive users, with their archival requests, also contribute to testing the software. The OTF pipeline is used at CADC to produce NICMOS and STIS preview images/spectra of all the available datasets (15 minutes after release date!), further contributing to testing the pipeline.

In other words, the OTF pipeline, being in a never-ending development phase and continuously receiving new reference files, is a dynamic system. An observation calibrated two months ago is different from the one calibrated today. Only at the end of the life of an instrument, when the "final archive" is produced, (i.e., no further development is foreseen), will this process stop and the best(?) calibration pipeline be available to the community.

Since the pipeline is entirely modular, it is quite easy for us to add more to it. One example is the preview pipeline. All the HST preview images and spectra which are used at STScI, ST-ECF and CADC are generated at CADC, as an extension of the on the fly calibration system. Their production, which takes place just after the data become public, contributes to:

- exercise the OTF calibration pipeline
- test the calibration pipeline since it was last used on this dataset (i.e., exactly one year earlier!)
- insure that all the necessary files are present for a potential user.

At the time of writing the HST archive is composed of 338 CDs for the RAW data, and has 18 GBytes of calibration files.

The HST OTF service is available at CADC[4] and ST-ECF[5]. Please contact these respective sites through catalog@eso.org or cadc@hia.nrc.ca.

References

Lubow, S. & Pollizzi, J. 1999, this volume, 187

---

[4]http://cadcwww.hia.nrc.ca/hst/

[5]http://archive.eso.org/archive/hst/

Astronomical Data Analysis Software and Systems VIII
ASP Conference Series, Vol. 172, 1999
D. M. Mehringer, R. L. Plante, and D. A. Roberts, eds.

OAPI: the OPUS Application Programming Interface

Walter Warren Miller III

Space Telescope Science Institute, SESD/DPT, 3700 San Martin Drive, Baltimore, Maryland, 21218

Abstract. The OPUS pipeline processing environment at the Space Telescope Science Institute (STScI) supports the automated conversion of gigabits of raw telemetry from the Hubble Space Telescope (HST) to useful science and engineering products each day. OPUS provides a flexible, fully concurrent, distributed control system in which a set of processes operate. The applications that form the HST data processing pipeline communicate with the blackboard architecture employed by OPUS through a library of C functions. Currently, that library is being ported to C++ as part of an effort to establish and document a standard programming interface to OPUS. The OAPI will facilitate OPUS application development for future HST instruments as well as other spacecraft and ground-based data processing projects.

1. OPUS: An Operational Pipeline System

The OPUS[1] pipeline processing environment at the Space Telescope Science Institute supports the automated conversion of gigabits of telemetry from the Hubble Space Telescope to useful science and engineering products each day. OPUS provides a flexible, fully concurrent, distributed control system in which a set of processes operate (Rose 1998; Rose et al. 1995; Rose, Choo, & Rose 1996; Boyer & Choo 1997; Swade & Rose 1998a). Applications in an OPUS pipeline specify, through simple text files, the criteria under which they are to be triggered by OPUS to perform some action. OPUS takes care of polling the necessary information sources (termed "blackboards") given the application's trigger conditions, and generating the appropriate events when those conditions are met.

The interaction between pipeline applications and OPUS differs depending on whether the application was developed for use with OPUS or is an off-the-shelf software product– OPUS does support both OPUS-savvy and off-the-shelf pipeline applications. In the latter case, a shell process acts as a proxy for communicating with OPUS by executing the "external" application when an event is delivered to it. With the release of the OPUS Sample Pipeline on CD-ROM last year (see Swade & Rose 1998b), it became possible for groups outside of STScI to construct complete OPUS-driven pipelines using external

[1] `http://www.stsci.edu/opus`

applications (numerous groups have done so or are in the process of developing OPUS pipelines on the supported platforms: Solaris, Digital Unix, and Linux).

Without a public application programming interface (API), however, it has not been possible for outside groups to develop software that communicates with OPUS directly (that is, through library function calls). The benefits of designing pipeline applications that communicate with OPUS through library function calls instead of through indirect means may prove critical for certain projects; for example, those where efficiency is a primary concern or where complex interaction with OPUS blackboards is required.

The Data Processing Team[2] recently defined as a high priority goal the full documentation of the programming interface to OPUS. Besides fostering a better environment in which to engineer OPUS-based applications, both in terms of development effort and software support, it provided us an opportunity to standardize the objects which form the core of OPUS in a way that improves the extensibility of the design itself. OPUS already has proven itself to be highly extensible in terms of the variety of pipelines that can be constructed with it (e.g., Rose et al. 1998), but this effort enabled us to abstract the existing object-oriented design in terms of C++ classes that can be used to extend the present capabilities of OPUS while maintaining a common interface to client applications. A well-documented API also makes it possible to offer the library to groups outside of STScI so that they may develop OPUS-savvy pipeline applications and enhanced libraries built upon the base classes included in the library.

2. OAPI Design

The OAPI design aims to satisfy the needs of two groups of software developers. On the one hand, it must serve the programmer who wants to develop OPUS-savvy processes for whom the implementation details of the OPUS library are not important. On the other hand, it must be easily maintainable and offer the flexibility to meet future requirements of OPUS pipelines through expansion or modification of the implementation underneath of the interface. Implicit to both of these needs is the availability of good documentation.

Standard C++ offered us the best means by which to meet the project goals both because of its support for object-oriented design and its ability to coexist with ANSI C at the translation unit level. In particular, since all of the existing software is written in ANSI C, it is trivial to offer the exact functionality for these routines in the OAPI. More importantly, however, is the time gained by not having to translate existing C code to a syntactically different language; this is time better spent enhancing existing algorithms to capitalize on C++'s improvements over C. The choice of C++ as the core programming language for the OAPI is reinforced by the recent adoption of the ANSI/ISO Standard for C++. Along with providing valuable new features to the language, notably the Standard Template Library (STL), developing the OAPI code to this standard

---

[2]`http://www.dpt.stsci.edu/`

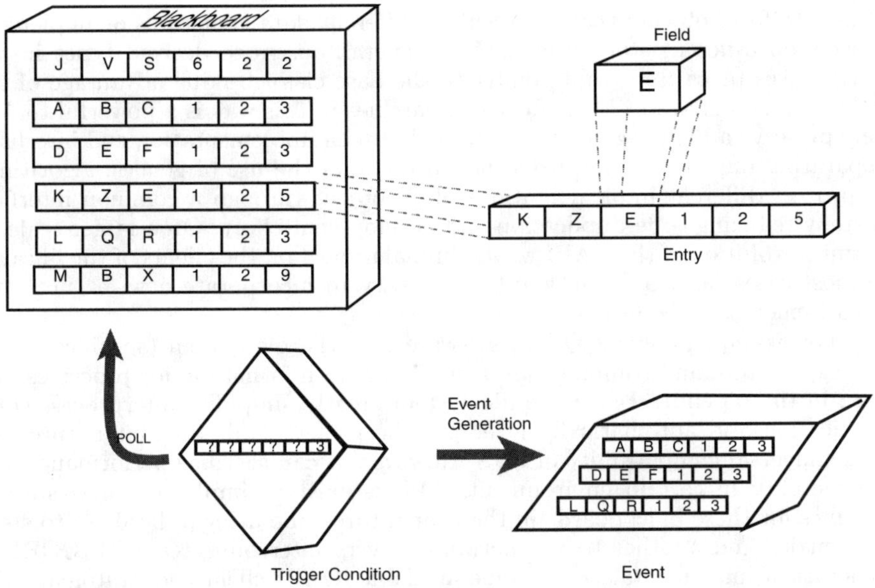

Figure 1. A cartoon representing the relationship between the basic objects in the OAPI. Blackboard objects contain Entry objects, which themselves are composed of Field objects. Polling a blackboard involves constructing a Trigger with an Entry condition, then searching with that condition on the appropriate blackboard. If matches are found, then an Event object is generated.

will ensure the same portability across different platforms as is enjoyed by the current ANSI C implementation.

The existing OPUS software modules are well developed, well tested, and offer a good degree of data hiding. Porting these packages to C++ purely for the added type checking and data hiding that C++ encourages is not a valid reason for undertaking such an effort. What is invaluable to this project is the match between the present software design and the preferred development environment of C++. Object-oriented design (OOD) tends to maximize the separation of interface from implementation. To a large extent, OOD concepts are employed by the original OPUS design and C implementation, but it is left to the developer to initially construct then maintain this philosophy throughout a large number of software packages. C allows the developer to build such a system, but it fails to enforce the rules of OOD and this leads to buggy software and maintenance difficulties. By porting OPUS software to standard C++, we adopt a system that enforces the rules under which we want to develop software. This makes it much easier to expand the capabilities of OPUS while preserving compatibility and to maintain existing software.

The essence of the OAPI design is reflected in a set of base classes that describe the objects that form the core of OPUS (see Fig. 1)– the Blackboard, Entry, and Field classes. More specialized classes are derived from these base classes to form objects that encapsulate different data structures or implement objects on different storage media. In general, access to derived types in the OAPI takes place through a pointer to the base class. Taking advantage of the polymorphic behavior of C++ class hierarchies in this way is a powerful tool to help preserve a high degree of separation between implementation and interface. Separating implementation from interface allows the use of generic algorithms to process different implementations of an object through a common interface thereby reducing code duplication and development effort. On a larger scale, it permits evolution of the OAPI with minimal impact on the clients of the library– in most cases, only a relink will be necessary to incorporate new features and enhancements to the OAPI in client software.

For example, current OPUS software uses the file system (specifically, file names) to store and communicate run-time state information for processes and data in the pipeline. Using the file system greatly simplifies interprocess communication and automatically leverages the robust, fault tolerant nature and near universal compatibility of NFS. However, it can sacrifice performance and extensibility in certain environments. Plans exist to implement new storage schemes for these blackboards in the near future– one using a database to store the entries and another to use network servers interconnected with CORBA. When such implementations become available, OAPI client code automatically will be able to use them without revising any of the algorithms.

Acknowledgments. OAPI design benefited from the insight provided by DPT staff Chris Heller, Jim Rose, Mike Swam, John Schultz, Mary Alice Rose, Daryl Swade, Pete Hyde, Steve Slowinski and Lisa Sherbert. Thanks also go to Allen Farris (STScI) for valuable help with C++ issues, and to Rosa Izela Diaz-Miller for production of the graphics in this article.

References

Boyer, C. & Choo, T. H. 1997, in ASP Conf. Ser., Vol. 125, Astronomical Data Analysis Software and Systems VI, ed. G. Hunt & H. E. Payne (San Francisco: ASP), 42

Rose, J. 1998, in ASP Conf. Ser., Vol. 145, Astronomical Data Analysis Software and Systems VII, ed. R. Albrecht, R. N. Hook, & H. A. Bushouse (San Francisco: ASP), 344

Rose, J., Choo, T. H., & Rose, M. A. 1996, in ASP Conf. Ser., Vol. 101, Astronomical Data Analysis Software and Systems V, ed. G. H. Jacoby & J. Barnes (San Francisco: ASP), 311

Rose, J., et al. 1995, in ASP Conf. Ser., Vol. 77, Astronomical Data Analysis Software and Systems IV, ed. R. A. Shaw, H. E. Payne, & J. J. E. Hayes (San Francisco: ASP), 429

Rose, J. F., Heller-Boyer, C. Rose, M. A., Swam, M., Miller, W., Kriss, G. A., Oegerle, W. R. 1998, in SPIE Proc., Vol. 3349, Observatory Operations to Optimize Scientific Return, ed. P. J. Quinn, (Bellingham: SPIE), 410

Swade, D. A. & Rose, J. 1998a, Proc. of the AIAA/USU Conf. on Small Satellites, (Logan: Utah State Univ.), in press

Swade, D. A. & Rose, J. F. 1998b, this volume, 111

Astronomical Data Analysis Software and Systems VIII
ASP Conference Series, Vol. 172, 1999
D. M. Mehringer, R. L. Plante, and D. A. Roberts, eds.

The STScI HST Astrometry Pipeline

John J. Schultz, Philip Goldstein, Pete Hyde, Mary Alice Rose, Keneth Steuerman, John Baum[1], Rick Perrine[1], Daryl A. Swade[1]

Space Telescope Science Institute, 3700 San Martin Drive, Baltimore, MD 21218

Abstract. An Astrometry Data Processing Pipeline has been developed and deployed at the Space Telescope Science Institute (STScI). This pipeline is responsible for producing Hubble Space Telescope (HST) Astrometry Data Sets used to analyze astrometry observations and the operation of the HST Fine Guidance Sensors (FGS). Prior to the STScI Astrometry Data Processing Pipeline becoming operational, HST Astrometry Data Sets were produced by the Astrometry and Engineering Data Processing (AEDP) System at Goddard Spaceflight Center (GSFC).

This paper describes the processing performed by the STScI Astrometry Pipeline and how the pipeline was designed and developed to reuse existing software components from the STScI OPUS system. OPUS is an STScI developed automated data pipeline system providing a distributed processing environment used to control and monitor applications executing in a sequential order. OPUS is described in detail in other papers[2].

1. Introduction

Prior to early 1998, GEIS format HST Astrometry Data Sets were produced and provided to STScI by the AEDP System at Goddard Spaceflight Center. In early 1998, the HST Astrometry portion of AEDP was discontinued and the STScI Astrometry Pipeline, which produces FITS format datasets, became operational.

The nominal automated Astrometry Pipeline consists of seven sequential processing stages. Each stage, contingent upon successful completion of the prior stage, invokes a specific application to perform a specific function. There are also two stages that can be manually invoked to provide additional functionality. The name and functionality of each stage application is as follows:

1. **ASTPOL** – ingest astrometry observations into the pipeline
2. **SSAST** – produce astrometry observation support schedule files
3. **OMSAST** – produce FITS format Astrometry Data Sets
4. **HTPAST** – save data sets for evaluation by operations personnel
5. **ARCINS** – generate requests to archive incomplete data sets

[1] Computer Sciences Corporation

[2] `http://www.dpt.stsci.edu/dpt_papers/opus_bib.html`

6. **GENREQ** – generate requests to archive complete data sets
7. **INGRSP** – process archive responses from the HST Archive System
8. **ARCTAPE** – archive data sets to tape
9. **ARCCLN** – delete successfully archived data sets from disk

As mentioned earlier, the Astrometry Pipeline was designed and developed to utilize existing components of the STScI OPUS System. In fact, of the above nine Astrometry Pipeline applications, only applications 1, 3, and 4 had to be newly developed. The other six applications are OPUS generic applications that were used as-is with minor modifications to their configuration set-up files. In addition, development of the three new applications incorporated numerous modules from existing OPUS Support libraries containing over 250 modules. This reuse of existing software enabled development of the astrometry pipeline to occur in a minimal amount of time, with a minimal amount of risk, and with considerable confidence that the resultant pipeline would be robust and very reliable.

2. The Astrometry Pipeline

The main responsibility of the STScI Astrometry Pipeline is to produce an Astrometry Data Set for each astrometry observation executed onboard HST. Each Data Set is comprised of four FITS format files; one Science FITS file for each of the three HST FGS, and a Support Schedule FITS file. The Science FITS files contain planned observation information extracted from Science Mission Schedules (SMS) as well as telemetry data accumulated during observation execution time. SMS related information includes data such as observation ids, predicted observation start and stop times, FGS reference positions, and predicted spacecraft orientation. As-executed telemetry data consists of items such as FGS detector photon counts, Star Selector Positions, Star Position Coordinates, and FGS Flag Words. The Support Schedule FITS file contains planned observation information extracted from the STScI Proposal Management Database. This information includes items such as the official target name, the observation proposal id, the name of the observation proposer, information about the instrument performing the observation, and various predicted positional information about the spacecraft.

As described in the Introduction Section, the Astrometry Pipeline is composed of nine applications; seven of which constitute the nominal automated pipeline. Of the seven automated applications, only the first three are involved in producing Astrometry Data Sets; the remaining four deal only with archiving of the data sets once they have been produced. Processing descriptions and additional information pertaining to code reuse for each of the nine applications is provided in following subsections.

2.1. ASTPOL – Ingest Astrometry Observations Into The Pipeline

The Astrometry Pipeline is activated when astrometry data files are received into a specific pipeline input directory. These data files occur in pairs for each astrometry observation executed; an ASCII file containing planned observation information from an SMS, and a binary file containing astrometry telemetry data from when an observation executed. These files are generated and provided to the astrometry pipeline by the STScI HST Observatory Monitoring System (OMS). The ASTPOL application polls for these files and when a pair is found, invokes the next application in the pipeline.

New Modules	Reuse Of OPUS Modules	Percent Module Reuse
1	147	99%

2.2. SSAST – Produce Astrometry Observation Support Schedule Files

This application queries the STScI Proposal Management Database to extract planned astrometry observation information. The extracted information is inserted into ASCII Support Schedule files.

New Modules	Reuse Of OPUS Modules	Percent Module Reuse
0	169	100%

2.3. OMSAST – Produce Astrometry Data Sets

This application actually produces the four FITS Files that comprise each astrometry data set. To produce the Science FITS files, data are extracted from the file pairs provided by the OMS and converted into FITS format. To produce the Support Schedule FITS file, data are extracted from the ASCII Support Schedule file generated by the SSAST application and converted into FITS format. While the FITS files are being produced, data from the OMS provided files are used to compute other observation information not directly provided. This computed information is also included in the FITS files.

New Modules	Reuse Of OPUS Modules	Percent Module Reuse
27	198	88%

2.4. HTPAST – Save Astrometry Data Sets For Evaluation By Operations Personnel

This application makes copies of astrometry data sets and moves them to holding directories where they reside until being evaluated and then explicitly deleted by operations personnel. This application also creates files containing information on the data quality of each processed astrometry observation.

New Modules	Reuse Of OPUS Modules	Percent Module Reuse
1	0	0%

2.5. ARCINS – Generate Requests To Archive Incomplete Astrometry Data Sets

This is the first of the two manually invoked applications not part of the automated astrometry pipeline. It generates requests to have partial astrometry data sets or data set associated intermediate files inserted into the HST archive. This application is used in scenarios where complete astrometry data sets could not be produced.

New Modules	Reuse Of OPUS Modules	Percent Module Reuse
0	85	100%

2.6. GENREQ – Generate Requests To Archive Complete Astrometry Data Sets

This application is the automated pipeline application to generate requests to have complete astrometry data sets inserted into the HST archive.

New Modules	Reuse Of OPUS Modules	Percent Module Reuse
0	131	100%

2.7. INGRSP – Process Archive Insertion Responses From The HST Archive

This application receives responses from the HST archive system indicating the disposition of requests to archive astrometry data sets or portions thereof.

New Modules	Reuse Of OPUS Modules	Percent Module Reuse
0	151	100%

2.8. ARCTAPE – Manually Archive Data Sets To Tape

This is the second of the two manually invoked applications not part of the automated pipeline. It is used to copy astrometry data sets to tape. This application is used in the rare situation where the HST archive is off-line and there is insufficient space to maintain observation data sets on disk.

New Modules	Reuse Of OPUS Modules	Percent Module Reuse
0	124	100%

2.9. ARCCLN – Delete Successfully Archived Data Sets From Disk

This application deletes astrometry data sets from disk after they have been successfully inserted into the HST archive.

New Modules	Reuse Of OPUS Modules	Percent Module Reuse
0	171	100%

3. Summary

The Astrometry Pipeline has been operational at STScI for over 9 months with no software problems whatsoever. There has also been virtually no concerns from the STScI FGS Team about the astrometry data sets being produced. In fact, the new FITS format astrometry data sets contain desired information not contained in the previous GEIS format data sets and are much preferred. In addition, since the astrometry pipeline's underlying OPUS architecture is easily scalable, there are plans to incorporate the FGS Teams' Astrometry Data Set Calibration Software into the astrometry pipeline as an additional application.

From the outset, the STScI Astrometry Pipeline was designed to extensively reuse existing, time-proven components of the STScI OPUS system. This reuse has been the major contributing factor to the timely implementation and success of the STScI Astrometry Pipeline.

Totals for Astrometry Pipeline Software		
New Modules	Reuse Of OPUS Modules	Percent Module Reuse
29	357	92%

Astronomical Data Analysis Software and Systems VIII
ASP Conference Series, Vol. 172, 1999
D. M. Mehringer, R. L. Plante, and D. A. Roberts, eds.

An On-the-fly Calibration Pipeline for the HST WFPC-2 using OPUS

Michael S. Swam, Daryl A. Swade

Space Telescope Science Institute, 3700 San Martin Drive, Baltimore, MD 21218

Abstract. An On-the-Fly Calibration System (OTFC) is being developed for selected instruments of the Hubble Space Telescope (HST) with a primary goal of providing HST Archive users with the most up-to-date calibration in their archive retrievals. The first instrument selected for this project is the Wide-Field/Planetary Camera-2 (WFPC-2), and a calibration pipeline for the WFPC-2 has been developed for OTFC using the OPUS pipeline architecture. This paper will describe the design of the OTFC WFPC-2 pipeline and the many benefits of using OPUS, including distributed multi-processing capabilities, reliable and robust function (the HST pre-archive calibration pipeline has been using OPUS for almost 3 years), high levels of code re-use, ease of integration of third-party products, and short development timescales.

1. Focus

The goals driving the development of an On-the-Fly Calibration system at the Space Telescope Science Institute (STScI) are covered elsewhere (Lubow & Pollizzi 1999). This paper will focus on the technical details of the OTFC pipeline design.

The WFPC-2 was chosen as the first instrument for implementation due to its maturity and its stable calibration software base. Other HST instruments will be added to the system in the future, but WFPC-2 was selected first so that the OTFC support software could be developed with the fewest instrument-driven complications.

2. What is OPUS?

The OPUS architecture, described in detail in other papers[1] was designed and developed by the Data Processing Team at STScI. OPUS provides the framework and tools for building and operating a data reduction system that can spread processing across multiple nodes in a distributed cluster of machines. It supports parallel processing using the blackboard[2] paradigm of interprocess

[1] `http://www.dpt.stsci.edu/dpt_papers/opus_bib.html`

[2] `http://www.stsci.edu/software/OPUS/bb.html`

communication. In addition to processing HST data at STScI, other missions are now using the OPUS software for their pipelines, including FUSE, Integral, and in the future, AXAF and SIRTF. OPUS can address the needs of a variety of missions since it is already written and released (Swade & Rose 1999), is consistently maintained by STScI, and is easily adapted to different requirements.

3. Why OPUS?

The decision to use OPUS for the OTFC pipeline was driven by several main points. The existing, pre-archive calibration pipeline that transforms WFPC-2 telemetry from HST into calibrated datasets for the HST Archive uses OPUS, and has for almost three years. The system has been robust, and provides a graphical user interface for the management and monitoring of the data processing stream, which is very useful in environments with a large data volume.

Another main driver was the concern for system scalability. The parallel, multi-processing capabilities of OPUS provide room for scaling the hardware architecture to meet the demands of the pipeline. Providing a user-friendly service like on-the-fly calibration will virtually guarantee that requests for recalibrated archive data increase. OPUS allows a system designer to add processors and disks with minimal reconfiguration to the system, resulting in a true, scalable architecture. Concerns about the performance of an OTFC system dictate that scalability be addressed, since unacceptable performance could reduce usage.

4. The Pipeline

The OTFC WFPC-2 pipeline was targeted for a relatively short development timescale (less than 1 year). The short timescale accentuated the benefits of code re-use that would be gained by using OPUS, along with existing STSDAS (Space Telescope Science Data Analysis System) tasks. Under OPUS, it is easy to string together a series of programs, shell scripts, and tools, to form a pipeline. Each stage in the OTFC pipeline performs a specific function, and the separation of work allows for greater parallel processing. Each of these pipeline tasks was implemented in a UNIX shell script. Work is accomplished using standard UNIX system commands, OPUS software tools, and STSDAS tools. The pipeline tasks are:

POLWF2 Polls for incoming OTFC requests from DADS (the Data Archive Distribution Service) at STScI.

CPYWF2 Copies the files listed in the OTFC request from a DADS disk to the OTFC system. This may be removed in the future when OTFC and DADS are merged into a single cluster of machines.

TRLWF2 Convert the observation trailer file (a log file of messages from the pre-archive pipeline processing) from FITS format to ASCII, so that OTFC pipeline messages can be appended.

KEYWF2 Update the header keywords of the files in the dataset by running the STSDAS 'getref' and 'upref' tasks. The getref[3] task reads several databases at STScI to obtain a list of the most up-to-date reference files, calibration switches, and keyword fixes. This stage is crucial for producing better calibration products than those currently stored in the HST Archive. The 'upref' tool takes this list of changes (documented in the "delta" file), and applies them to the files in an observation dataset.

F2GWF2 Because the current STSDAS CALWP2 executable requires input in GEIS format, the FITS files retrieved from DADS are converted to GEIS.

CALWF2 The STSDAS calibration executable, CALWP2, is run against the dataset and its updated keyword values, producing new calibration products.

G2FWF2 The GEIS output from CALWP2 is converted back to FITS format, along with the ASCII trailer file and the delta file. DADS only distributes FITS files to the end-user.

RETWF2 Copies the FITS files in the dataset back to the DADS system, via FTP, as dictated in the OTFC request.

RSPWF2 Generates an ASCII OTFC response file for DADS, describing the additions that were made to the fileset by the OTFC pipeline, and indicating the status of the OTFC processing.

The pipeline consists of 9 shell scripts totaling 553 lines of code, with much reuse between modules in the areas of setup and error handling. The existence of generic OPUS software tools and STSDAS tools made this possible. The STSDAS tools perform the bulk of the processing, while the OPUS tools facilitate passing the datasets through the OTFC pipeline. A working prototype of the pipeline was up and running within two weeks and evolved into the production system.

Automated error detection was added to the pipeline in the form of three short UNIX shell scripts. These scripts are reconfigured for each pipeline stage using the resource file feature of OPUS (explained below). They detect errors in processing and send along information to DADS so that the end-user will be informed of a delay in receiving part of their archive retrieval.

5. Benefits of OPUS

The entire OTFC pipeline could be constructed as a series of shell scripts WITHOUT using OPUS, but the benefits of OPUS are numerous. Each pipeline stage can obtain values from environment variables passed in by OPUS from an ASCII setup file, called a resource file. These resource files facilitate changing task parameters without changing code, and allow the same shell script to be used for processing different kinds of data. Parameters including disk directories, file

---

[3] `http://ra.stsci.edu/bps/cdbsdoc/getref.html`

masks, and other instrument-specific information are passed into the script, so that different versions of the script or large IF-THEN-ELSE blocks are avoided.

OPUS provides distributed, parallel, multi-processing capability, so that any stage of a pipeline can be run on any node in the cluster, as long as each node can see the needed data disks. Multiple copies of processes can be running on a single node, or spread across multiple nodes. By processing observations in parallel, the machines in the cluster are more likely to be under a balanced load (CPU and I/O), improving their throughput efficiency. Experiments and experience at STScI have shown that this parallelism increases the throughput of the system when processing a set of observations. Processing of any single observation takes a bit longer, but the staggered pipeline design (much like those used in pipelined CPU chip architectures) provides better throughput, since processing of different observations in different pipeline stages can overlap. For OTFC, the calibration stage (CALWF2) is the bottleneck in the pipeline. Running 3-5 copies of CALWF2 per processor has demonstrated[4] better throughput for the system. Of course, there is a point of diminishing returns, where adding more copies of processes result in more swapping behavior than additional useful work. Experimentation provides the best measure for this threshold where performance starts to degrade.

OPUS multi-processing also allows built-in redundancy, so that if a dataset manages to crash one of the OTFC processes, other copies of the same process are running that can continue to operate on new datasets. This redundancy also applies to a full system crash. If hardware problems bring down one of the processors in a cluster, processes distributed to other machines through OPUS can continue to run and maintain a level of system throughput.

6. Conclusions

At this time, the OTFC WFPC-2 pipeline is in formal testing and will be fielded for in-house retrievals within STScI in January 1999. After a period of satisfactory in-house testing, the WFPC-2 instrument team at STScI will decide when it is appropriate to open up WFPC-2 OTFC retrievals to the full scientific community.

The next phase of the project will be to integrate the recalibration of Space Telescope Imaging Spectrograph (STIS) data into OTFC. The reuse from the WFPC-2 pipeline code achieved in this integration effort is expected to be greater than 90 percent.

References

Lubow, S. & Pollizzi, J. 1999, this volume, 187

Swade, D. A. & Rose, J. F. 1999, this volume, 111

---

[4]`http://www.dpt.stsci.edu/otfc/benchmark_cal.html`

Astronomical Data Analysis Software and Systems VIII
ASP Conference Series, Vol. 172, 1999
D. M. Mehringer, R. L. Plante, and D. A. Roberts, eds.

Automated Software for Slitless Spectroscopy Analysis and Quasar Selection

C. N. Sabbey

Department of Astronomy, Yale University, New Haven, CT 06520-8101

Abstract. The QUasar Equatorial Survey Team (QUEST) is using a 16 CCD drift scan camera on the 1 m Venezuelan Schmidt telescope to collect 30 GB of objective prism data per night. Fully automated software running on a Linux Pentium II reduces and analyses the data faster than it is acquired. Relatively bright spectra in the prism data are automatically detected and used to determine coordinate transformations to the corresponding direct images. The direct image coordinates are used to place spectrum extraction boxes, set the zero point of the wavelength scale, and predict and correct for spectrum overlaps. Multivariate analysis techniques are applied for spectral classification and quasar selection.

1. Introduction

The QUEST collaboration (Snyder 1998) is obtaining broad bandpass (4000 – 9500Å), intermediate dispersion (25 Å/arcsec at Hα) slitless spectroscopy covering 300 deg^2 per night to $m_B \approx 19$. Processing the $\sim 10^6$ spectra contained in the nightly 30 GB of image data, however, places unique demands on the analysis software. The software must be fast and fully automated, reliably extract, deblend, and wavelength calibrate spectra that are often not detectable until later coaddition, and take advantage of the broad bandpass (continuum information) to improve quasar selection.

2. Coordinate transformation and spectrum extraction

By using a direct image catalog of sky coordinates and magnitudes, the locations of the spectra and their contamination by neighboring objects can be predicted. We currently use a catalog generated from the POSS/STScI Digitized Sky Survey (DSS), with the help of the Sextractor (Bertin & Arnouts 1996) and WCSTools (Mink 1997) software packages.

Relatively bright spectra in the prism data are automatically detected and their coordinates are input to a triangle matching program (Valdes et al. 1995) to determine the linear coordinate transformation to the direct catalog. The spectrum position in the dispersion direction is defined by the point of half maximum brightness at the red end sensitivity cutoff. Although this definition will depend on spectral type (by ± 2 pixels), we use it to establish the mean offset and not to set the zero point on an individual basis.

The relatively bright spectra are automatically detected by convolving the image with a matched filter (a 2-D template of the red end cutoff), and then searching for vertical strings of relative maxima that monotonically increase in brightness. Strings of some minimum length (e.g., ten pixels, possibly diagonally connected) reliably locate the distinctive red end cutoff and hence the entire spectrum (which is $\approx$ 250 pixels long). Object detection algorithms used in photometry (i.e., connecting pixels above a brightness threshold) are less useful due to the extreme overlap problem and the segmentation of an object into multiple detections depending on the spectral energy distribution.

The software applies the coordinate transformation to the direct catalog, predicting the spectrum locations to within $\approx 0\farcs3$ (except near photographic plate edges in the DSS data). This accuracy is required to reliably extract faint spectra and set the wavelength zero point to within $\approx$ 0.2%. The accuracy of the transformation is monitored throughout the drift scan by using the $\approx$ 50 brightest spectra per $0\fdg6 \times 0\fdg6$ image frame as "guide stars" (least squares fits to the position residuals refine the transformation). The extraction of 1-D spectra is currently done using a box of width $1.2 \times$ FWHM and uniform weighting.

3. Background determination

The background level is determined by a technique commonly used for crowded fields — image histogram mode estimation with iterative k-sigma clipping of the histogram wings (Da Costa 1992). A fast and straightforward method for implementing this uses the cumulative histogram, `c`, and the weighted cumulative histogram `w` (i.e., the contents of each histogram bin are first multiplied by the bin number). At each iteration, the mean, 25th percentile (`q25`), median (`q50`), mode, and sigma (`sig`) are obtained for the clipped histogram (i.e., between bins `b` and `e`) as follows:

```
nincr = c[e] - c[b-1]
mean  = (w[e] - w[b-1]) / nincr
q25   = ibsearch(c[b-1] + nincr / 4)
q50   = ibsearch(c[b-1] + nincr / 2)
mode  = 3 * q50 - 2 * mean
sig   = 1.482 * (q50 - q25)
```

where `ibsearch` indicates a binary search of the cumulative histogram with an interpolated output index. Convergence (when the mode changes by less than some threshold) occurs rapidly by using `q25` and `q50` to form a robust approximation of the histogram σ. This method is $\sim$ 10 times faster than techniques that require fully sorting the data. First masking out the pixels within the spectrum extraction boxes did not improve the background estimation.

4. Overlap prediction and correction

A spectrum is considered to be overlapped if a significant number of the wavelength bins have more counts due to neighboring spectra in the image than the spectrum itself. Whether a given spectrum is overlapped depends on the relative magnitudes and positions of the neighboring objects, their spectral types and the spatial profile. In the QUEST objective prism data, almost 30% of the 19th magnitude spectra have $\geq$ 25% of their bins dominated by neighbors. However, the software predicts the amount of contamination as a function of wavelength to attempt deblending and allow the unrestricted use of clean wavelength bins.

The software builds a contamination map using the positions and magnitudes of objects in the input sky catalog, and empirical 2-D template spectra produced from bright, uncrowded spectra. Currently a single, representative late type spectrum is used although an improvement would be to create templates as a function of color. The template spectra are added into the map with an intensity scaling of $10^{(m/(-2.5))}$. For each spectrum extracted from the data, the ratio of the contamination map to the 2-D template for that spectrum gives the (multiplicative) contamination correction at each pixel.

This method is fast, easy to implement, and handles arbitrarily complicated overlaps of many objects without excessive approximation. In addition, the overlap calculation can be done independently of the spectrum extraction pipeline. Thus, the overlap calculation does not need to be repeated for each drift scan to be coadded, and the calculation can be improved at any time without repeating the spectrum extraction. Because the input sky catalog contains ID numbers for all objects to be extracted, it is straightforward to match each spectrum to its contamination array (and coadd spectra from different CCDs and nights).

5. Quasar selection

With the broad bandpass of the spectra, spectral classification and quasar selection can be improved by using the continuum information in addition to spectral features. A reasonable approach is to find the smallest RMS deviation between each spectrum and a series of template spectra (or similar technique based on cross-correlation). However, two significantly different spectra can have the same RMS deviation from a given template (e.g., one spectrum has an emission feature while the other has a slightly different continuum slope). In other words, the best matching template is selected using a distance criterion, independent of direction, in flux–wavelength space.

Figure 1 shows an alternative using Principal Component Analysis (PCA) to provide a fast, automated measure of the deviation of a given spectrum from the common energy distributions. This allows the selection of quasars (and other unusual objects) to fainter magnitudes than line-only searches without making assumptions about the variance of quasar spectra. In practice, the representative set of common spectra is defined from the data itself (by clustering analysis on a sample of relatively bright, unblended spectra).

References

Bertin, E. & Arnouts, S. 1996, A&AS, 117, 393

Cristiani, S. & Vio, R. 1990, A&AS, 227, 385

Da Costa, G. 1992, ASP Conf. Ser., Vol. 23, Astronomical CCD Observing and Reduction Techniques, ed. S. Howell (San Francisco: ASP), 90

Gunn, J. E. & Stryker, L. L. 1983, ApJS, 52, 121

Mink, D. 1997, in ASP Conf. Ser., Vol. 125, Astronomical Data Analysis Software and Systems VI, ed. G. Hunt & H. E. Payne (San Francisco: ASP), 249

Snyder, J. A. 1998, in SPIE Proc., Vol. 3355, Optical Astronomical Instrumentation, ed. S. D'Odorico (Bellingham: SPIE), 635

Valdes, F., Campusano, L., Velasquez, J., & Stetson, P. 1995, PASP, 107, 1119

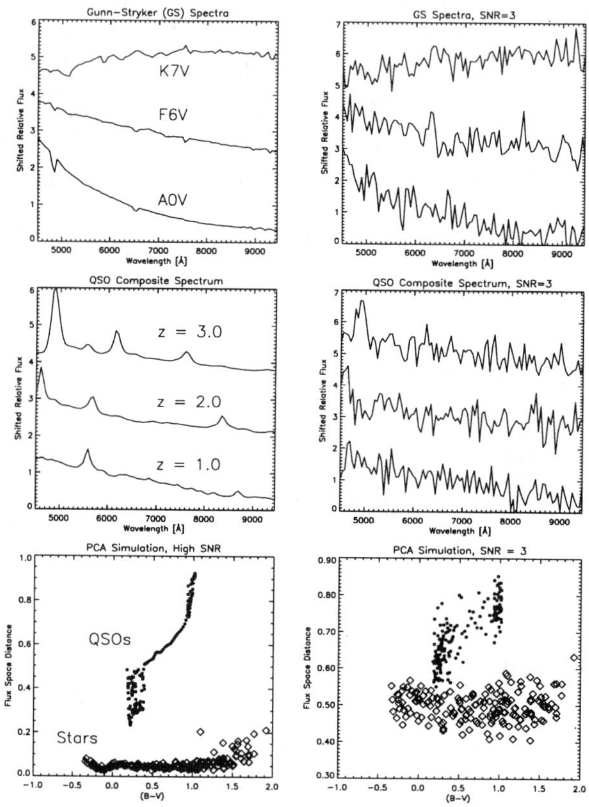

Figure 1. A simulation is shown using PCA to select uncommon (non-stellar) energy distributions. The Gunn & Stryker (1983) stellar spectra and a composite QSO spectrum (Cristiani & Vio 1990) for $0.5 < z < 4.5$ are rebinned to 50 Å bins in the 4500 to 9500 Å bandpass. PCA is applied to the GS spectra to establish a new basis in which the first 5 dimensions account for $> 99\%$ of the stellar spectral variance. A measure of the deviation of a given spectrum from stellar is obtained by projecting it into the new basis and calculating the flux space distance to the volume formed by the N (e.g., 5) low order dimensions.

Part 7. Archiving and Information Services

Astronomical Data Analysis Software and Systems VIII
ASP Conference Series, Vol. 172, 1999
D. M. Mehringer, R. L. Plante, and D. A. Roberts, eds.

A Search and Discovery Tool – AMASE

Cynthia Y. Cheung

NASA Goddard Space Flight Center, Astrophysics Data Facility, Greenbelt, MD 20771

Nick Roussopoulos, Stephen Kelley

University of Maryland, Institute of Advanced Computer Studies

James Blackwell

Raytheon STX Corporation

Abstract. The Astrophysics Multispectral Archive Search Engine *(AMASE)* is a metadatabase prototype built using object-oriented database methodology which allows mission data to be searched easily using scientific parameters. We discuss the implementation issues that arise in integrating heterogeneous data sources, the form and contents of metadata, the creation of cross-system object identifiers, specific tools to handle bulk loading and incremental data loading, interoperability with relational databases, and our future work to extend the capabilities of *AMASE.*

1. Introduction

The importance of multi-spectral research in solving fundamental problems in astrophysics is undeniable. It is generally advantageous, for example, to use multi-wavelength observations to interpret a class of celestial objects. However, NASA archival data are usually not organized to support multi-spectral research. The data are products of space missions and are stored in the archives according to mission-specific parameters, in disparate mass storage systems, and in different formats. This makes it quite challenging to locate data for research that cuts across mission or spectral boundaries. The Astrophysics Multispectral Archive Search Engine *(AMASE)* aims at providing a uniform multi-mission and multi-spectral interface, enabling *global* searches of these heterogeneous, distributed data holdings using astronomical classification and complex spatial queries. Users are able to locate relevant mission data for research without detailed knowledge of the missions beforehand. The URL for the *AMASE* homepage is `http://amase.gsfc.nasa.gov/`.

2. The *AMASE* Design & Implementation

AMASE uses object-oriented database (OODB) techniques to "merge" astronomical data from heterogeneous sources. These techniques *encapsulate* the existing data, metadata, and associated documentation and bibliography into an abstract *Astronomical_Object*. The flexibility and modeling capability of OODB allow for diversity and conplexity within a common framework. The relationship between objects can be captured in the database schema and implemented at the time of data loading. The *Astronomical_Object* can be characterized by very complex and rich data types, and can take on multiple classification to reflect both its intrinsic scientific nature and the instrument parameters connected with scientific data acquisition. The *Astronomical_Object* acts as a "directory" pointing users to, for example, the catalogs in the Astronomical Data Center containing information about the celestial object and the mission data in the distributed archives with different observatory-instrument settings. When a user queries for science data, the search criteria are issued against the scientific parameters of the *Astronomical_Object* while the response is given in terms of mission data descriptors. The mission data descriptors identify the relevant mission data sets and give the archive location and retrieval information. In this way, *AMASE* provides a scientific view into the distributed data archives without changing the existing underlying mission-oriented structure.

AMASE is implemented using the commercial object-relational product *Informix-Illustra DBMS.* It offers the dual advantages of an OODB and a relational database, with complex and user-defined data types that are not normally supported in a relational database, a standard SQL non-procedural query language, the robustness of a storage manager, a query optimizer, and the maturity of the technology provided by relational systems.

3. Building the Knowledge Base

To support queries using scientific parameters, *AMASE* must first build an astronomical "knowledge base". The domain knowledge needed to describe modern astronomical classification structure is captured from experts in the different disciplines. The heritage and relationships of different mission data products are likewise captured from the appropriate mission scientists. Once the basic classification schema is defined, the knowledge base is populated with valid astronomical measurements. These are generally available in published astronomical catalogs and can be captured by automated procedures. But, naturally, the data in the knowledge base will be subject to revision in the course of scientific inquiry: new data will be acquired from future missions, new object cross–identifications will be made, resulting in a better understanding of physical phenomena, and possibly, a new object classification scheme. So a mechanism for incremental buildup of knowledge and subsequent knowledge evolution is built into the *AMASE* architecture, enabling it to become a true search and discovery engine.

Loading an OODB is much more involved than loading a relational database. Correlations between objects are created and maintained by the database *at load time.* Linkages are represented by object identifiers (OIDs) in the database. We choose to use common astronomical names as the cross-catalog and cross-mission OIDs in *AMASE.* The dual use of astronomical names to refer to celestial objects as well as database objects provides a friendly user query mechanism. The OIDs are often not available when the load file is written, because the detected sources of an observation are not yet identified. In such cases, the semi-automated bulk loading procedure will create *incomplete* objects with surrogate identifier when observations are first loaded, then *complete* the object by adding correct pointers to the relevant astronomical characteristics when the target is identified. We are building incremental bulk loading utilities and user utilities to support updates of object characteristics, to "merge" objects upon positive cross-identification, and to separate objects into components when they are resolved by high-resolution observations.

4. Accessing Remote Databases

The major archive sites of the NASA astrophysics missions are geographically distributed. Their data holdings are frequently updated as new observations are taken and new data are released to the public archive. *AMASE* must provide access to data at these archive sites and maintain current information about their data holdings. The schema of each remote archive database is first mapped to the *AMASE* schema. Then a small but sufficient amount of metadata is ingested to create a "partially materialized" view (Roussopoulos, Economou, & Stamenas 1993) in the *AMASE* database. An alternative is to create "virtual" object in *AMASE* that is "pointer-based" and populated with references to the data values and locations in the remote archives (Roussopoulos 1982). Batch updates of these derived views are carried out by a special utility built using the Sybase Gateway of the *Informix-Illustra DBMS.*

5. Query Capabilities

There are two main ways to query the *AMASE* database: by object name and by position. The query by astronomical name illustrates the advantage of an OODB implementation. Since all information that pertains to a requested object is linked at load time, the execution is very efficient. The positional information is returned as part of the search result and is not used at all during the search.

To efficiently process queries based on spatial position, we build spatial indexes on every database object that contains spatial attributes. We use $packedR-trees$ (Roussopoulos & Leifker 1985) to significantly reduce the search selection set and are extending their use to support the spatial ordering of objects and data discovery based on object position.

Since there is no "natural" ordering of multi-attribute data, especially spatial data, results are returned in the order they are retrieved from the index. To obtain the result based on "closeness" to the query object, most query implementation will select the data first, write the results to a temporary object, then sort them for presentation. But the cost in time and space for this approach will be prohibitive for queries with a large number of results. To address this issue, we are developing a "K-Nearest Neighbor" access method that will generate and return the K objects closest to a given spatial position in order (Roussopoulos, Kelley, & Vincent 1995).

6. Future Work – Data Discovery Utilities

A principal goal of *AMASE* is to support astrophysics research by providing a position-based "data mining" facility. The first implementation of this facility will support field-of-view (FOV) correlation. We would use an instrument FOV to scan through the *AMASE* database to extract information for off-target objects, correlate them to other data (from other catalogs/observations) via spatial (primary) and other attributes, then present the results to a scientist for review.

Other data discovery utilities include view-caching and user-defined types. Query results can be cached into the database and used for repetitive queries and/or refinement. An astronomer can build his/her own profile of interested objects from the database. Results cached in views can be accessed much more rapidly as they are compact and require a much narrower search. To support user-defined classes, the system must have a utility to rearrange the class hierarchy and its implied inheritance, and to assign object identifiers to newly derived class object. We plan to use "materialized views" as the mechanism for supporting user-defined types (Roussopoulos 1991; Roussopoulos et al. 1993). Part of this work was funded by an earlier NASA AISR grant.

7. Summary

The *AMASE* prototype is currently populated with selected data products from six NASA astrophysics missions, as well as various astronomical catalogs. Users can search for information by specifying either object name and/or various positional information. The search can be further qualified by selecting the object type, the spectral region and the mission. Sorting and visualization capabilities will soon be available through the ADC Data Viewer (Kargatis et al. 1999). In the coming year, we shall provide the capabilities to interoperate with other data archives and develop utilities for user-defined views, positional data mining, and object cross identification.

References

Kargatis, V. E., Shaya, E. J., Blackwell, J. H., Borne, K. D., White, R. A., & Cheung, C. Y. 1999, this volume, 217

Roussopoulos, N. 1982, ACM Trans. on Database Systems, 7(2), 258

———, 1991, ACM Trans. on Database Systems, 16(3), 535

Roussopoulos, N., Economou, N. and Stamenas, A. 1993, IEEE Trans. on Knowledge and Data Engineering, 5(5), 762

Roussopoulos, N., Kelley, S., & Vincent, S. 1995, Proc. of the 1995 ACM SIGMOD Intl. Conf. on Management of Data, ed. M. J. Carey & D. A. Schneider (San Jose: ACM Press), 71

Roussopoulos, N. & Leifker, D. 1985, Proc. of the 1985 ACM SIGMOD Intl. Conf. on Management of Data, ed. S. B. Navathe (Austin: ACM Press), 17

Astronomical Data Analysis Software and Systems VIII
ASP Conference Series, Vol. 172, 1999
D. M. Mehringer, R. L. Plante, and D. A. Roberts, eds.

Web-Based Tools For Exploration of ADC Data Holdings and NASA Data Archives

V. E. Kargatis, E. J. Shaya, J. H. Blackwell, K. D. Borne

Raytheon ITSS, Code 631, Goddard Space Flight Center, Greenbelt MD 20771

R. A. White, C. Y. Cheung

NASA Astrophysics Data Facility, Goddard Space Flight Center, Greenbelt MD 20771

Abstract. The NASA Astrophysics Data Facility has developed several new tools to facilitate the browsing and visualization of astronomical catalog and journal table data at the Astronomical Data Center (ADC), and to find associated observational data at various data archives. The ADC Viewer allows users to select, cull, and manipulate ADC catalogs and journal tables, and display them in a plotting tool called Catseye. IMPReSS is a tool that displays aperture silhouettes within a specified sky region of a variety of space-based missions, and then links directly to the archived data held at external sites. All three tools are interoperable with other astronomical resources.

1. Introduction

The Astronomical Data Center (ADC) at the Goddard Space Flight Center (GSFC) holds the largest computer readable collection of catalogs, observing logs, and journal tables in the Western Hemisphere. The scientific content of these holdings is very heterogeneous, ranging from astrometric data on stars to a variety of data on quasars and clusters of galaxies. Together with the CDS in Strasbourg, we maintain and expand our collections and make them available to the world community, scientists and laymen alike. A suite of tools available at the ADC facilitates the browsing and visualization of the collections, and allows researchers to refine and enhance their scientific objectives.

2. Tools

Users can access three independent but linked tools for browsing astronomical tables and observing logs. The Viewer access catalogs and journal tables and allows selection and display of table subsets. CatsEye plots 2-D scatterplots of fields from Viewer tables. IMPReSS plots observation "footprints" of space-based missions on a region of the sky and links directly to the archived data.

2.1. Viewer

The Viewer is a Perl-based web interface that allows users to select, cull, sort, save, and download over 2500 catalogs and over 1800 journal tables. Currently, only standardized ASCII tables (the bulk of the ADC holdings) are available through the tool. Through the top page, users can specify a particular catalog, search the ADC, browse the ADC's holdings, or directly access their workspace (described below).

After selecting a table to browse, the user selects the fields to view, and can constrain those fields by numerical range (or regular expression for string fields). The requested table subset is then displayed, and automatically saved in a personal work area ("workspace") identified by a cookie. If the user accepts the cookie (containing only a unique identifier), the workspace is stored on the server for access up to 5 days after last use and can be directly accessed anytime during that period. The user can name, view, download, or delete multiple tables in their workspace. Tables are saved as formatted ASCII, and are downloaded by request in the same format by the user. The Perl code is modularized for easy maintenance and modification.

2.2. CatsEye

CatsEye (shown in Fig. 1) is a Perl- and IDL-based web tool that plots selected fields from Viewer-produced tables in the user's workspace, allowing them to visualize and select catalog data. Users select tables and fields therein to plot as X vs. Y. Overplotting of multiple tables is allowed.

The user's plot selections are processed by Perl CGI scripts and passed to IDL routines, which read in the tables and produce plots, which are translated to GIFs and displayed via HTML. Users can rescale plots as desired and click on points within the plot to access the corresponding full catalog record from the table. These records link to both IMPReSS (discussed below) and NED, provided the record contains spatial coordinates.

2.3. IMPReSS

IMPReSS (shown in Fig. 1) is a graphical interface to astronomical observing logs that presents the user with plane-of-the-sky outlines of pointed observations obtained by space-based telescopes. It searches for NASA missions that have observed within a user-selected region of the sky and time period. Once the user selects the missions they are interested in, IMPReSS displays the color-coded aperture footprints (the fields-of-view, or FOV) on the sky, overlaying multiple observations where necessary. IMPReSS also displays an observing-log-type list of all observations which have met the criteria, and will deliver the user directly to the location of observational data offered by the appropriate archival data center.

IMPReSS uses Perl for the CGI and IDL for observation data searching and display. Observation logs for HST-PC, -WFC, & -WFPC-2, EUVE, IUE, and all available missions from HEASARC are all stored in a simple IDL database that is kept active in an IDL server mode. This enables a search through potentially millions of observations in just seconds. Links to the data are provided through the standard archive access methods offered by each data center.

Taken together, these tools offer a continuous sequence by which users can browse and visualize a catalog and acquire specific data products of interest from the distributed data archives. They can be tremendously useful both in choosing multi-wavelength data for further analysis and in judging what astronomical objects warrant further observation at particular wavelengths.

3. Interoperability

Interoperability with other astronomical data centers and resources figures heavily in the functionality of the suite of tools described here.

The ADC External Query (AEQ) allows a user to access Viewer output through single HTTP GET requests. It uses all the same input and calculation routines of the Viewer, and simply outputs delimited ASCII metadata and data in a predetermined format easily parsed by machine, avoiding all the interactivity of a typical Viewer session. AEQ is intended for institutional access, in which output may be customized or better integrated into other software interfaces. Specific details about the interface and the required metadata formats (which differ for every catalog) are needed to use this interface, and may be acquired through the ADC.

CatsEye, while presently able to plot only Viewer output, uses a simple standardized input table, and we intend to allow external file upload from users in the future, where they can provide a correctly formatted table and produce scatterplots of their own data over the web with ease. CatsEye also provides links to NED for any objects selected when clicking a point on the plot (provided coordinates are available in the table).

IMPReSS offers direct access to the data holdings of various distributed archives centers, and periodically automatically updates any modified observation logs.

IMPReSS and the Viewer are included in the GLU database used by Astrobrowse, allowing automatic acquisition by Astrobrowse agents of the required metadata and query format. IMPReSS, a tool that accepts positional queries, is automatically accessed during any Astrobrowse request.

4. Referenced Web Sites

ADC: `http://adc.gsfc.nasa.gov/`
Viewer: `http://tarantella.gsfc.nasa.gov/viewer/`
CatsEye: `http://tarantella.gsfc.nasa.gov/catseye/cat_frames.pl`
IMPReSS: `http://tarantella.gsfc.nasa.gov/impress/`
AEQ: `http://tarantella.gsfc.nasa.gov/viewer/AEQDoc.html`

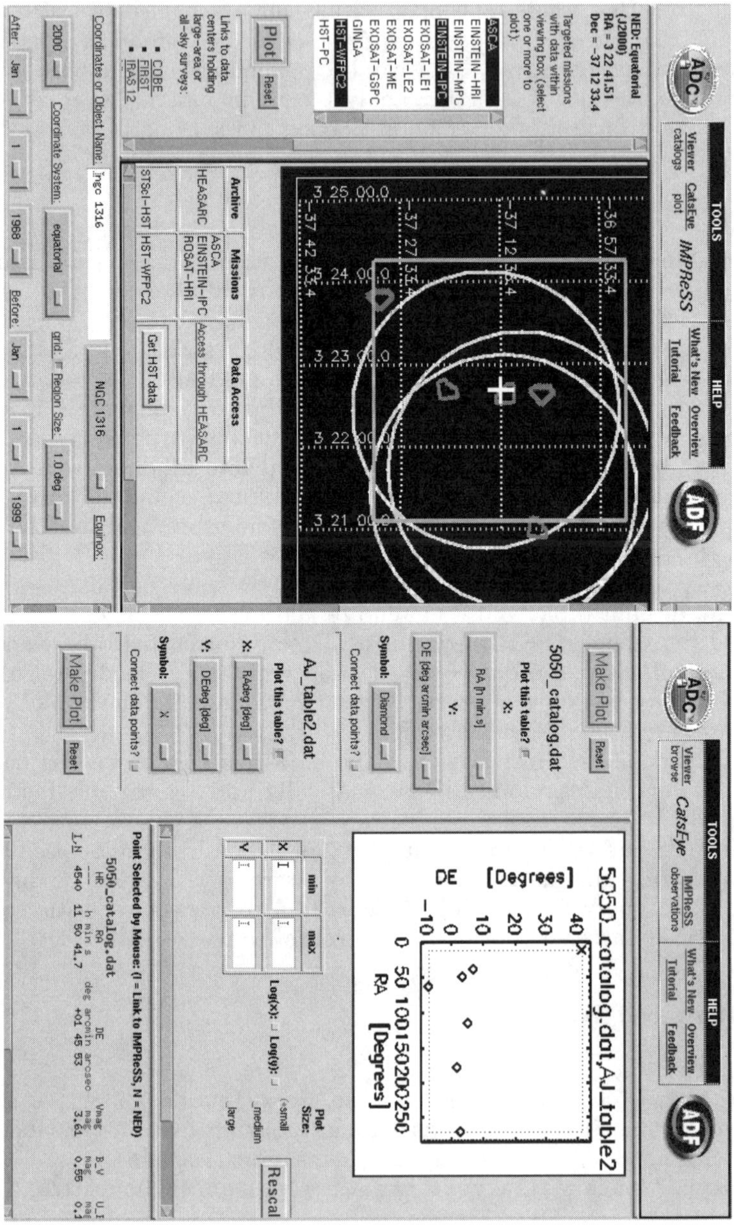

Figure 1. CatsEye (*bottom*) and IMPReSS (*top*) examples. In CatsEye, fields to plot are chosen on the left, the plot is displayed at upper right, and the catalog records of clicked points at bottom right. In IMPReSS, the sky region is chosen at bottom, the available missions at upper left, and the requested sky map and links to archived data at upper right.

Astronomical Data Analysis Software and Systems VIII
ASP Conference Series, Vol. 172, 1999
D. M. Mehringer, R. L. Plante, and D. A. Roberts, eds.

Astrobrowse: a Web Agent for Querying Astronomical Databases

Christina W. Heikkila[1], Thomas A. McGlynn[2], Nicholas E. White

NASA/Goddard Space Flight Center, Greenbelt, MD 20771

Abstract. Astrobrowse is a user agent which significantly streamlines the process of gathering astronomical data on the web. With it, a user can query thousands of resources without having to search through old bookmark lists, deal with out-of-date URLs, or spend time figuring out how to use each resource's unique input form. Given a user's selection of web-based astronomical databases and an object name or coordinates, Astrobrowse will explode queries to all selected databases and collect the results. Since the introduction of the prototype at ADASS VII (McGlynn & White 1998), many changes have been made to the service, from adding JavaScript form validation, to the adoption of GLU (Fernique, Ochsenbein & Wenger 1998) as the mechanism for keeping the resource URLs current.

This software is freely available to anyone wishing to become an Astrobrowse node.

1. Need for Astrobrowse

With the ability to query astronomical databases and catalogs online becoming more and more the norm, it has become daunting to try to keep track of all these queryable services. A bookmark list with all these forms and data centers can become unmanageable. It is also likely to become out of date, since web addresses are frequently changing. In addition, most of these query forms have different formats, requiring the user to learn how to use a different form at each new query site. Often the user must enter positional data in particular formats or epochs: one site may ask for decimal right ascension and declination, while another requires sexagesimal (hh mm ss) coordinates.

To address these problems, the Astrobrowse effort was developed with close cooperation by many institutions. Some of the involved groups are: HEASARC, ADF, STScI, CDS, SAO, IPAC, LEDAS, and CADC. HEASARC[3], STScI[4], and

---

[1]Raytheon ITSS, Email: cwh@lheamail.gsfc.nasa.gov

[2]Universities Space Research Association, Email: tam@silk.gsfc.nasa.gov

[3]`http://heasarc.gsfc.nasa.gov/astrobrowse/`

[4]`http://faxafloi.stsci.edu:4547/starcast/about.html`

CDS[5] each have their own versions of the Astrobrowse software, and SAO and LEDAS are running the HEASARC version.

2. What Astrobrowse Provides

We have already compiled a large list of position-searchable databases on the web. When a new one is suggested, a web form is filled out by the Astrobrowse administrator or by the person suggesting the new site, and the Astrobrowse software converts this to the correct format.

URLs are kept current through the CDS's GLU server software[6]. This software propagates changes in the database descriptions to all Astrobrowse GLU network members, enabling Astrobrowse to use the latest available URLs when querying databases.

Astrobrowse takes care of converting the user's object name or position into the format and epoch required by each remote query service. Considering the amount of time expended to find, load, and fill out a large number of suitable query forms, the time savings for the user become significant, since with Astrobrowse the user only fills out one form to retrieve the same amount of information.

3. How to use HEASARC's Astrobrowse

Using the HEASARC version of Astrobrowse, the user can choose to query from the menus of resources provided, or to create a custom resource list using a search tool. The search tool allows the user to search for only X-ray resources, for example, or only resources serving images, or ones holding data from a particular satelliteions

Once a list of resources is displayed, the user chooses which ones to search via checkboxes, and enters either an object name or coordinates. Astrobrowse then formulates a query to each data service: looking up the object name (if given) via SIMBAD or NED, converting the object coordinates to the format specified by the remote service, and providing additional form input needed by the remote service.

Each query is sent by Astrobrowse in parallel, and as the results are received by the remote services, each is saved on the server machine. The user is shown a framed results page, with an index to the results shown in the left frame, and each individual result shown in the right frame as the user clicks through the index. The user can switch back and forth easily between results, or pop up new browser windows for side-by-side comparisons.

---

[5] `http://simbad.u-strasbg.fr/glu/cgi-bin/astroglu-m2.pl`

[6] `http://simbad.u-strasbg.fr/glu/glu.html`

4. Recent changes to Astrobrowse

4.1. GLU Maintains the Resource Database

The resource description database used by Astrobrowse is maintained by the CDS's GLU network software. This software allows each network member to change the entries owned by them as needed, and GLU distributes the change to all other members. The system is not hierarchical, so if one member institution experiences network problems, no one else is affected. Currently there are eight members in the Astrobrowse GLU network. In addition to the URL and form data needed for the query, the GLU resource description entries carry metadata specifying additional URLs (such as help documents), email addresses of maintainers, server location, etc. The software is freely available from CDS.

4.2. Cookies Store User Choices

HEASARC Astrobrowse sets a cookie which stores the names of the catalogs the user queried last, so if the researcher wants to do a search for data on more than one object, he or she doesn't have to remember which catalogs were selected for the previous source. The cookie expires after one month.

4.3. JavaScript Form Validation

JavaScript functions have been added to the keyword search form and the catalog selection form in HEASARC Astrobrowse. These functions check to make sure the user has specified necessary input in the correct format before submitting a request to the server. Since the form validation is done on the client side, this saves the user time – the user doesn't have to wait for the server to receive the information, decide that there's a problem, and return an error message. This arrangement also reduces the load on the server machine, since only valid requests should be sent to the server. These JavaScript additions have been designed to be cross-browser compatible, and the software will still work with server-side form validation in case users run their browsers with JavaScript disabled.

5. The Future of Astrobrowse

We are investigating allowing non-positional queries in Astrobrowse; already some databases are queryable by object name.

More resources are continually being added to those queryable through Astrobrowse, there are currently more than fourteen hundred.

We have made the HEASARC Astrobrowse software available to download from its web site. We encourage other institutions to install the software and provide the Astrobrowse service.

We have recently been granted funding for a design study on ISAIA (Interoperable Systems for Archival Information Access), which will build on the Astrobrowse concept. ISAIA will have the ability to search on many different metadata parameters, not just position. It will integrate the results received from remote providers into unified tables of data – making interpretation of the data much faster and easier.

References

McGlynn, T. & White, N. 1998, in ASP Conf. Ser., Vol. 145, Astronomical Data Analysis Software and Systems VII, ed. R. Albrecht, R. N. Hook, & H. A. Bushouse (San Francisco: ASP), 481

Fernique, P., Ochsenbein, F., & Wenger, M. 1998, in ASP Conf. Ser., Vol. 145, Astronomical Data Analysis Software and Systems VII, ed. R. Albrecht, R. N. Hook, & H. A. Bushouse (San Francisco: ASP), 466

Astronomical Data Analysis Software and Systems VIII
ASP Conference Series, Vol. 172, 1999
D. M. Mehringer, R. L. Plante, and D. A. Roberts, eds.

The Distributed Astronomical Preprint Service

J. E. Huizinga, Robert J. Hanisch, H. E. Payne, R. L. Williamson

Space Telescope Science Institute, 3700 San Martin Drive, Baltimore, MD 21218

Abstract. We have developed a prototype distributed preprint service. The system is hierarchical, with identical software at each node. Site maintenance tools written in Java allow information providers to enter new preprint metadata and to update them when preprints are published. A web site at each node is built on-the-fly from the metadata, allowing both browsing and searching. Parent servers can gather metadata from child servers, allowing the central server to index everything, and to notice new entries. Queries return preprint identifiers ("prepcodes") which resolve to on-line journals in the case of published preprints.

1. Introduction

Most new peer-reviewed astronomy research literature is now available in electronic form on the World Wide Web. Electronic journals offer features not possible in paper versions: searches, cross-reference links, forward references, and machine readable tables. Electronic submission has streamlined journal production, and dramatically reduced the time between acceptance and publication. Agreements between the publishers of the major astronomy journals allow links from one journal to another, in a collaboration called *Urania*[1].

Against this background, the astronomical preprint has made a much less successful transition to electronic distribution. Preprints are the historical channel for rapid communication of results in astronomy. Preprints have traditionally been produced in an institutional framework to demonstrate an organization's quality and vitality. But isolated institutional preprint web sites make it difficult for a user to discover new and interesting preprints. The centralized Los Alamos National Laboratory astrophysics preprint archive[2] is one solution of the discovery problem. But this service remains to be integrated into the on-line astronomical literature identified with *Urania*.

We are developing a distributed system for integrating preprint collections maintained at participating institutions. Institutional participation provides the best assurance of correctness and currency, distributes the workload and resource requirements, and satisfies an institution's desire to display its own achievements. The system will allow users to locate documents anywhere in

[1] `http://www.aas.org/Urania/`

[2] `http://xxx.lanl.gov/archive/astro-ph/`

the collection by means of a single query, or to search through the holdings at a single institution. The system integrates preprints into the on-line literature by tracking a preprint through to final publication: a hyperlink to a preprint will resolve to the on-line final paper, once it is published. The system could eventually be integrated with journal production by providing manuscripts for referees and production staff. In return, journals could automatically supply the final URL of each preprint. At the moment, however, we are working on a system to facilitate and preserve the preprint tracking being done by hand by librarians.

2. A Prototype System

Our prototype preprint service is a hierarchical system. A central authority tracks participating institutions. Each institution can maintain a number of document collections, and may also grant authority over parts of its "preprint namespace" to subordinate or "child" servers, at the departmental level, say, and so on. Each node in the prototype system has identical software, although "lite" sites are also envisioned.

The practical key to the *Urania* system was a standard naming scheme, known as "bibcodes," for journal articles. Our preprint identifiers, or "prepcodes," reflect the hierarchical nature of the system, and are modeled on the HANDLE[3] system, used by the NCSTRL[4] project, for example. Preprint identifiers consist of four components:

astro.pp.stsci.stsci1281

discipline doctype institution prepid

The first field allows for collections in other disciplines. Doctypes might include other "gray literature" categories, like technical reports and observatory manuals. The institution code is assigned by the central authority, to avoid duplication, while the prepid field is largely arbitrary.

Metadata for each document are keyed by preprint identifier. Metadata include a title, authors and affiliations, an abstract, pointers to author contributed TeX and PostScript files, if present, and the location of the final published paper. The metadata are indexed for searching, and used to dynamically generate most web pages. Metadata are in a format that is nearly XML. Other metadata files describe each collection, and encapsulate parent-child relationships.

At each node of the system, a name resolver provides access to the metadata associated with each prepcode in its assigned name space. Prepcodes at child nodes are passed down the hierarchy to child servers, and unrecognized prepcodes are passed up the hierarchy to the parent server. The central server forwards valid prepcodes to the appropriate institutional server, and catches invalid prepcodes.

---

[3] http://www.handle.net/

[4] http://www.ncstrl.org/

3. Prototype System Components

We envision a system with four components: (1) tools that information providers can use to update and maintain their document collections, (2) tools for creating, maintaining, and searching the indexes, (3) a user query interface, and (4) a notification service. Our prototype system contains versions of the first three:

Site Maintenance and Preprint Entry Tools. Preprint entry tools have been implemented in Java for use as applications or as Web applets. They allow easy entry of new preprints into an institute's collection, and allow entries to be updated, e.g., to reflect publication. Web site pages are generated on-the-fly from the metadata. Page layout is template-driven, allowing easy customization of the appearance at each institution. In the prototype, page generation is handled by Perl CGI scripts, with metadata maintained in DBM databases. Conversion to Java servlets and a database like MySQL are being considered.

Index and Search Tools Preprint metadata are indexed for fielded and full text searches with CNIDR's Isite[5] package, familiar to us from our ASDS (Hanisch, Payne, & Hayes 1994) project. A search returns preprint identifiers of matching documents, which are passed to a name resolver to obtain the document. The name resolver is implemented as a Perl CGI script. Prepcode passing is accomplished by having the name resolver return HTTP redirect instructions to the user's web browser.

A parent server can collect metadata from its children, allowing it to create an index spanning all collections in its name space, and allowing the central authority to index all collections. A user viewing an institutional site and wishing to search a larger collection of documents is directed to the search page at the parent site. This mechanism could be the basis of a notification service, which would take note of new items.

User Query Interface The prototype query interface is similar to our ASDS query interface, and makes use of Isite's "virtual database" concept, which allows a single query to be presented to a set of physically distinct index databases as if they were a single database. At each level of the system hierarchy, an index is constructed for each local collection and for each child site. A dynamically created search page gives users the option of searching all of these databases or any one of them.

The overall architecture is sketched in Fig. 1. The "Top Server" will have many child "Institute Servers," each of which can also be a parent of servers at a lower (departmental, for example) level. That is, the diagram can be expanded both horizontally and vertically, as indicated by dots. Arrows indicate the exchange of metadata and name resolver requests between parent and child, and the user's ability to query the central server or local servers. Site management tools are shown in the upper right of the figure.

---

[5] `http://www.cnidr.org/ir/isite.html`

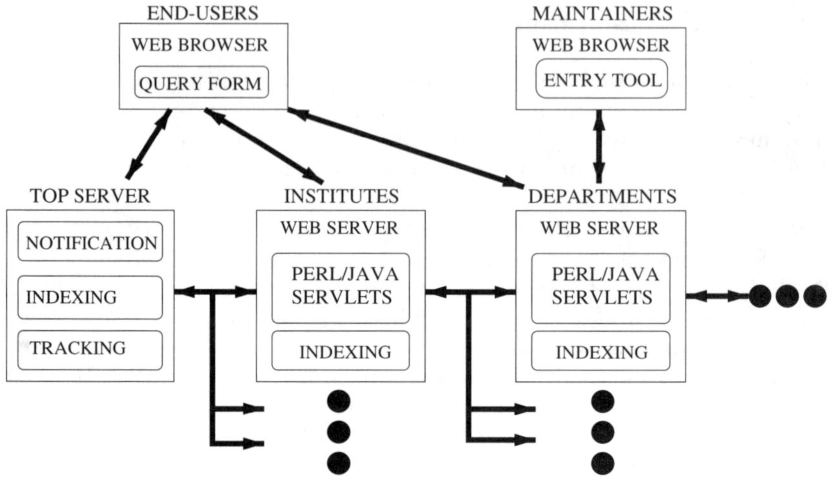

Figure 1. The architecture of a distributed preprint service.

4. Future Work

The prepcode passing between name resolvers is not robust against an intermediate resolver being down. We are considering exchanging metadata about the collections, so that each name resolver can pass prepcodes directly to the responsible server, perhaps using GLU (Fernique, Ochsenbein, & Wenger 1998). Other concerns include ensuring that the system will work in the Windows NT environment, used by many librarians (GLU and our customized Isite are currently unavailable on this platform), database issues, and an all-Java implementation.

Acknowledgments. This project is funded by the NASA Applied Information Systems Research Program (NRA 96-OSS-10) under grant NAG5-3942.

References

Fernique, P., Ochsenbein, F., & Wenger, M. 1998, in ASP Conf. Ser., Vol. 145, Astronomical Data Analysis Software and Systems VII, ed. R. Albrecht, R. N. Hook, & H. A. Bushouse (San Francisco: ASP), 466

Hanisch, R. J., Payne, H. E., & Hayes, J. J. E. 1994, in ASP Conf. Ser., Vol. 61, Astronomical Data Analysis Software and Systems III, ed. D. R. Crabtree, R. J. Hanisch, & J. Barnes (San Francisco: ASP), 41

Astronomical Data Analysis Software and Systems VIII
ASP Conference Series, Vol. 172, 1999
D. M. Mehringer, R. L. Plante, and D. A. Roberts, eds.

ALADIN: A Reference Tool for Identification of Astronomical Sources

François Bonnarel, Pierre Fernique, Françoise Genova, James G. Bartlett, Olivier Bienaymé, Daniel Egret, Joseph Florsch, Houri Ziaeepour

CDS, Observatoire Astronomique de Strasbourg, UMR 7550, 11 rue de l'Université, 67000 Strasbourg, France,
Email: bonnarel@astro.u-strasbg.fr

Mireille Louys

ENSPS, Strasbourg

Abstract. ALADIN is the CDS interactive sky atlas used for source identification and interpretation, multi-wavelength cross-identification, and data quality control. ALADIN includes an image server which contains reference images from the Schmidt survey DSS–I, soon from DSS–II, and images digitized by the MAMA-CAI machine with higher resolution for crowded fields. It allows for the overlaying of information extracted from the SIMBAD database, the VizieR catalogue server, and private catalogues. The X–Window interface, available at CDS and at several French institutes, includes a source extraction facility as well as image compression solutions for image transfer. The new step is to provide access to images of objects from the various CDS services on the WWW. Linked to this quick view facility, a JAVA interface includes interactive graphical processing and source list overlay.

1. Introduction

Cross-correlation of data from different origins is certainly a major challenge for astronomy, particularly for multi–wavelength studies (e.g., Ortiz et al. 1999). One aspect of the problem is to implement tools allowing users to browse through distributed on–line resources. Astrobrowse (Heikkila, McGlynn, & White 1999) and AstroGLU (Egret, Fernique, & Genova 1998) are examples of discovery tools, allowing one to explore a large set of selected databases. In parallel, links permitting the users to navigate easily between on–line information of different kinds are developing rapidly to and from observational data, databases, bibliographic databases, published journals, etc.

The Centre de Données astronomiques de Strasbourg (CDS, see e.g., Genova et al. 1998) manages several reference databases, including SIMBAD, the catalogue service with the VizieR Catalogue Browser, the *Dictionary of Nomenclature of Astronomical objects outside the solar system*, and bibliographic services and tools, with mirror copies of the ADS bibliographic database and of

The Astrophysical Journal, *The Astronomical Journal*, and *The Publications of the Astronomical Society of the Pacific.* The links between all these services and with external databases are provided through the Générateur de Liens Uniformes (GLU), developed by CDS (Fernique, Ochsenbein, & Wenger 1998), which is also used in the Astrobrowse project.

The ALADIN project aims at producing a comprehensive software tool for combining astronomical data by providing simultaneous access to digitized sky photographs, catalogues and databases. The visual overlay of these kinds of information facilitates direct comparison of data from different origins.

2. The ALADIN Image Database

The Digital Sky Survey (DSS-I) (Lasker 1994) forms the core of the image database. The low resolution and light data compression (factor of 10) permit storage covering the full sky on a reasonable number of CD ROMs. These CD ROMS are stored at CDS in a juke–box connected to the image server. The second epoch survey (DSS-II) (Lasker et al. 1998) will be progressively integrated into the system in the near future.

For usage which requires higher resolution images, for example, when working in crowded regions of the sky or in areas with deep observations, the CDS has also built an archive of images digitized with 0.66 arcsec pixels (10 microns), in partnership with the MAMA facility (Guibert 1992) at the *Centre d'Analyse des Images* at the Observatoire de Paris (mainly ESO-R and SERC plates). These full resolution images are available for the southern Galactic Plane, the Magellanic Clouds, and a few smaller regions of specific interest. The archive is implemented on an optical disk juke–box with a storage capacity of 500 GB.

3. The ALADIN User Interface

An *X-Window interface* was developed, which overlays images from the image database with objects from SIMBAD, the VizieR service (including large catalogues and surveys such as GSC, TRC, ACT, or USNO A–2.0), and user provided lists of objects. A list of objects on the image can also be extracted and saved. The user may select a subset of catalogues from VizieR, save the information of interest in local files, and access SIMBAD and VizieR information for the plotted objects. The X–Window ALADIN interface has been made available at CDS and in several French institutes.

More recently, Web access to the image database has also been developed. The first objective was to provide access to reference images of the sky, for other CDS services, such as SIMBAD, the catalogue service, or the bibliography. This is done through the *ALADIN Previewer* which is implemented as a specific service[1] in the CDS Web hub. The images are compressed in JPEG–8 bit format in order to maintain a good response time (the volume transmitted is less than 30 kB). At a later stage, we plan to provide images compressed with the MR–1 package (PMT method, e.g., Louys et al. 1999).

---

[1] `http://aladin.u-strasbg.fr/aladin/alapre.pl`

The first link to other services is from the CDS bibliographic service to give access to an image of all the individual objects which are cited in the keywords of a published papers. For each object, two links are built, one to the information in SIMBAD, the other to the ALADIN image (using the coordinates from SIMBAD). The accuracy of the nomenclature found in the keyword list is checked beforehand. Similar links are also built and checked for object names tagged by authors in the *Astronomy and Astrophysics* abstracts. Links to the ALADIN Previewer will soon be installed in SIMBAD and then in the catalogue service.

In parallel, the development of a Java tool allowing one to overlay SIMBAD and catalogue information on the image and to perform interactive graphical processing is near completion. *ALADIN Java* extensively uses the ability to stack planes containing information from different origins (the image, SIMBAD objects, catalogue data, text and marks added by the user, etc.) and to overlay some or all of them. Interactive links to the complete information from SIMBAD and VizieR are also available. The ALADIN Java interface is implemented in two versions: a Java applet directly processed by the user Web browser and a stand–alone Java application requiring the installation of a Java virtual machine. Some limitations in the functionalities of the Java applet arise from the security restrictions of Java, namely from the fact that local resources cannot be accessed from Java applets. The stand–alone application (i.e., installed on a local machine) allows the user to also plot his or her own object list, provided that it is properly formatted. The user can also save ALADIN images locally in FITS format. For performance reasons, the image object extraction functions are not available in the Java interface at the present time (they are present in the X–interface). They may be added later, or provided by the image server.

4. Conclusion

Recent development of Web access to the ALADIN image database opens new powerful links from bibliographic information or databases to reference images of the sky. The imminent release of the Java interface for interactive sky atlas functionalities will soon provide easy access to a powerful identification tool, also allowing easy implementation of future improvements, which will accompany the evolution of the Java language and environment.

References

Egret, D., Fernique, P., & Genova, F. 1998, in ASP Conf. Ser., Vol. 145, Astronomical Data Analysis Software and Systems VII, ed. R. Albrecht, R. N. Hook, & H. A. Bushouse (San Francisco: ASP), 416

Fernique, P., Ochsenbein, F., & Wenger, M. 1998, in ASP Conf. Ser., Vol. 145, Astronomical Data Analysis Software and Systems VII, ed. R. Albrecht, R. N. Hook, & H. A. Bushouse (San Francisco: ASP), 466

Genova, F., et al. 1998, in ASP Conf. Ser., Vol. 145, Astronomical Data Analysis Software and Systems VII, ed. R. Albrecht, R. N. Hook, & H. A. Bushouse (San Francisco: ASP), 470

Guibert, J. 1992, in Proceedings of the Conference on Digitised Optical Sky Surveys, ed. H. T. MacGillivray & E. B. Thomson (Dordrecht: Kluwer), 103

Heikkila, C. W., McGlynn, T. A., & White, N. E. 1999, this volume, 221

Lasker, B. M. 1994, in Proc. of IAU Symp. 161, Astronomy from Wide–Field Imaging, ed. H. T. MacGillivray & E. B. Thomson (Dordrecht: Kluwer), 167

Lasker, B. M., & STSci Sky-Survey team 1998, BAAS, 192, 6403

Louys, M., Starck, J. L., Mei, S., Bonnarel, F., & Murtagh, F. 1999, A&A, submitted

Ortiz, P. F., Ochsenbein, F., Wicenec, A., & Albrecht, M. 1999, this volume, 379

Astronomical Data Analysis Software and Systems VIII
ASP Conference Series, Vol. 172, 1999
D. M. Mehringer, R. L. Plante, and D. A. Roberts, eds.

The Multimission Archive at Space Telescope

Damian Christian, Faith Abney, Thomas Comeau, Robert Hanisch, Jill Harrison, Catherine Imhoff, Richard Kidwell, Timothy Kimball, Karen Levay, Paolo Padovani, Marc Postman, Joel Richon, Myron Smith, Randall Thompson

Space Telescope Science Institute, Baltimore MD, 21218

Abstract. We present an overview of the recently created Multimission Archive at Space Telescope (MAST). The Hubble Data Archive has expanded to provide easy access to non-HST data sets. MAST includes the following: the International Ultraviolet Explorer (IUE), the Extreme Ultraviolet Explorer (EUVE), Copernicus (OAO-3), ASTRO HUT, WUPPE and UIT data sets, and VLA FIRST data. The Digitized Sky survey is also accessible through MAST. Data from the Far Ultraviolet Spectroscopic Explorer (FUSE) currently scheduled for launch in early 1999, ORFEUS 1 & 2 EUV spectrometer data, and Voyager-1,2 UVS data will be added in the near future. Two 480-platter Plasmon CD-ROM jukeboxes provide data storage and easy access to the MAST data. Data hierarchy, ingest and retrievals, plans for expansion, and features of the web interface are also presented. Users can access data via a WWW interface at `http://archive.stsci.edu/mast.html`.

1. Introduction to MAST

The Space Telescope Science Institute (STScI) holds the Hubble Data Archive (HDA), which includes all Hubble Space Telescope (HST) observations. As of November 1998, the HDA contains over 5.8 Terabytes of science and engineering data representing over 250,000 exposures of more than 20,000 targets. In the past few months the volume of archived data has reached average rates of about $4 - 5$ GB/day, with an average of 20 GB/day retrieved by archive users.

The STScI archive recently expanded by providing access to non-HST data sets. STScI was chosen as one of three major science archive research centers created to provide users with easy access to archival data. The Multimission Archive at Space Telescope (MAST) focuses on the optical, UV and near-IR wavelength ranges, and complements the High Energy Astrophysics Science Archive Research Center (HEASARC) at GSFC and the Infrared Science Archive (ISRA) at IPAC. MAST data sets include extreme ultraviolet, ultraviolet, and selected radio data sets listed below. These data sets may be accessed at the MAST homepage[1] and include:

[1] `http://archive.stsci.edu/mast.html`

International Ultraviolet Explorer (IUE) Contains over 104,000 spectral images of 10,000 individual astronomical sources in the wavelength range 1200–3350 Å.

Copernicus (OAO-3) Far- (900–1,560 Å) and near- (1,650–3,150 Å) ultraviolet spectra of 551 objects.

Extreme Ultraviolet Explorer (EUVE) Extreme UV (70–760 Å) spectroscopic observations 400 sources, mostly galactic.

ASTRO Missions Various UV data sets from the ASTRO 1 and 2 shuttle missions, including

- the Hopkins Ultraviolet Telescope (HUT)
- the Wisconsin Ultraviolet Photo-Polarimeter Experiment (WUPPE)
- the Ultraviolet Imaging Telescope (UIT)

Faint Images of the Radio Sky at Twenty-centimeters (FIRST) A 20 cm (1.4 GHz) radio survey covering over 10,000 square degrees to a flux limit of 1 mJy. Access is provided to the radio images and the source catalog, currently having approximately 437,000 entries.

Digitized Sky Survey (DSS) Digitized photographic Sky Survey plates from the Palomar and UK Schmidt telescopes.

1.1. MAST Interface

The MAST resources are available via a simple WWW interface at the URL given above. A user may choose to search a region of the electromagentic spectrum for images or spectra, or may go directly to the search page for a specific mission. Archival holdings may be searched by:

- Name (with SIMBAD or NED as a name resolver)
- Position (α, δ, and radius)
- Object category
- Observation specifics: date, instrument, filters, etc.

1.2. MAST Cross-Correlations with Astronomical Catalogs

The potential use of the MAST archive is greatly improved by allowing users to search more than one data set at a time and to also allow the archive holdings to be compared with astronomical catalogs. Cross correlations can be performed using:

- The Hipparcos stellar catalog (1997, ESA SP-1200)
- The Active galactic nuclei catalog (Veron-Cetty & Veron 1996; Padovani, Giommi, & Fiore 1997)
- The Abell galaxy cluster catalog (Abell, Corwin, & Odowin 1989)
- User supplied catalogs

2. MAST Holdings

2.1. Data Storage and Access

Currently, most of the MAST data is available over the web and selected data sets can be quickly down-loaded. The IUE and EUVE data sets currently reside at NSSDC and HEASARC, respectively. The IUE data sets are currently being ingested at STScI and will be stored on 2 Plasmon CD-ROM jukeboxes. WORM media was chosen for its cost effectiveness and ready user access. The CD-ROM option is about 50% cheaper than a DLT library, but provides enough space for the current non-HST MAST archive and anticipated expansion. The MAST data network consists of a SUN Enterprise 450 server, a large RAID disk array, and 2 Plasmon 480 slot CD-ROM jukeboxes.

2.2. Data Organization

MAST has made every attempt to include sufficient documentation and background information to make data products as useful as possible. The MAST holdings for specific missions generally include:

Project Description General descriptions of the mission and instrumentation.

Data Processing How the data were reduced and calibrated.

Data Description Documentation on data characteristics: wavelength coverage, resolution, field of view, and anomalies. Data uncertainties and database field descriptions.

Data Format All data are in FITS (Flexible Image Transport System) format. Documentation includes a general description of the file contents and header information.

Mission-specific analysis software IUEDAC IDL, UIT BDR software written in C and Fortran-77, HUT IRAF software, and EUVE EUV1.8 IRAF software.

3. Future Plans

Additional ultraviolet and optical data sets will be incorporated into MAST in the future, including data from the Far Ultraviolet Spectroscopic Explorer (FUSE) currently scheduled for launch in early 1999. FUSE will cover the wavelength range of 905–1195 Å with a resolution of 30,000. MAST will archive Voyager-1 and Voyager-2 UVS data, which covers the 950 to 1350 Å range. MAST also anticipates expanding its support of ground-based archiving with the addition of data from the National Optical Astronomy Observatories Mosaic Imager. As demonstrated by Valdes et al. in their poster at this conference, the Mosaic Imager will perform sub-arcsec imaging of large fields of view and perform surveys in standard photometric bands.

Acknowledgments. STScI is operated by the Association of Universities for Research in Astronomy, Inc., under NASA contract NAS5-26555. Support for MAST for non-HST data is provided by the NASA Office of Space Science via grant NAG5-7584 and by other grants and contracts.

References

Abell, G. O., Corwin, H. G., & Olowin, R. P. 1989, ApJS, 70, 1

Padovani, P., Giommi, P., & Fiore, F. 1997, Mem. Soc. Astron. It., 68, 147

Veron-Cetty, M.-P. & Veron, P. 1996, A Catalogue of Quasars and Active Nuclei (7th ed.) (ESO Scientific Report No. 17)

Astronomical Data Analysis Software and Systems VIII
ASP Conference Series, Vol. 172, 1999
D. M. Mehringer, R. L. Plante, and D. A. Roberts, eds.

A System for On-line Access to the GSC II

Andrea Baruffolo, Leopoldo Benacchio, Lucio Benfante

Osservatorio Astronomico, vicolo Osservatorio 5, 35122 Padova, Italy

Abstract. The second Guide Star Catalog (GSC II) will be a catalog of positions, proper motions, magnitudes and colors for stars and galaxies complete down to $V = 18.5$ and is expected to be completed by the end of 1999. Besides its operational nature, the GSC II will also be a powerful tool for galactic and extragalactic research. However, with an expected number of records of the order of two billion, this catalog will pose problems for efficiently accessing the data. In this paper, we describe our on-going effort aimed at building a management system for the GSC II, based on a new technology Object Relational Database Management System, that will allow researchers to fully exploit the scientific content of this catalog. We thus give an overview of the architecture of the system, the data model being developed, the extensions we are making to the database engine to support astronomical data, and the access interfaces currently under development.

1. Introduction

The second Guide Star Catalog (GSC II) will be the successor of the first astrometric and photometric catalog built to support the operations of the Hubble Space Telescope. It will be a catalog of positions, proper motions, magnitudes, and colors for stars and galaxies complete down to V=18.5 whose first release is expected to be completed by the end of 1999.

Besides its operational nature and thanks to its performances and all-sky coverage, the GSC II will be a powerful tool for galactic and extragalactic research (Lasker et al. 1995). Because of its collaboration with the GSC II project, the Italian astronomical community will be provided with a copy of the catalog as soon as it is completed.

With an expected number of records of the order of two billion, this catalog will pose problems for efficient access to the data. In this paper we describe our on going effort aimed at building a management system for the GSC II, based on a new technology Object Relational Database Management System, that will allow researchers to fully exploit the scientific content of this catalog. This project is a logical continuation of our previous research in the field of astronomical database management systems (Baruffolo & Benacchio 1998a, 1998b).

The final system is expected to allow users to efficiently perform searches in the GSC II and some other related catalogs by the means of a Web-based user interface, display results graphically, and save them locally for further anal-

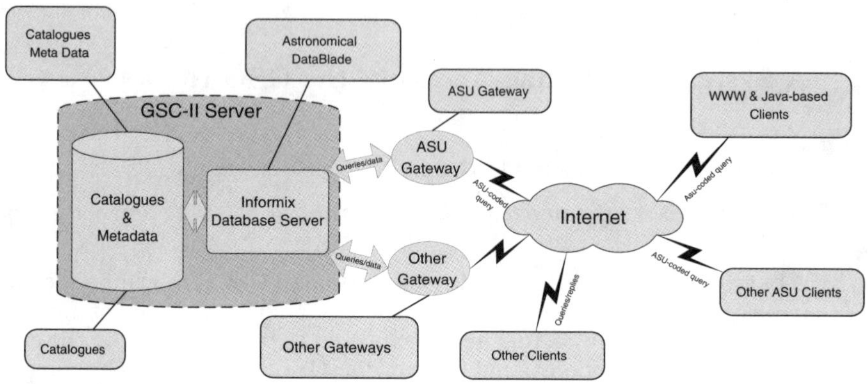

Figure 1. Overall architecture of the GSC II on-line system.

ysis. Gateways will also be provided to allow a variety of clients to access this database.

In Fig. 1 we give an overview of the architecture of the system, where the following elements are highlighted:

1. content of the database;
2. the database structure being developed;
3. the extensions we are developing to the database engine to support astronomical data and queries;
4. access gateways under development or planned;
5. access interfaces currently under development.

In the following sections we briefly describe the various components that make up the GSC II on-line system.

2. Catalogues

At present, only the catalogues which are currently being used or built in the GSC II project are foreseen to be included in the database. They are: HST Guide Star Catalog I & II, Hipparcos and Tycho Catalogues, Guide Star Photometric Catalog 1 & 2, Positions and Proper Motions (North and South), Astrographic Catalogue (AC 2000), ACT Reference Catalog. Other catalogues may be added in the future depending on users' requirements.

3. Catalogues' Metadata

All catalogues in the collection are described in a metadata structure. All applications will access the database through the metadata, so that they will work independently of any catalogue-specific structure. A collection of Java classes for metadata access has been developed and is used in applications written in this language. Using metadata will also simplify the addition of new catalogues to the collection.

4. Astronomical DataBlade

All the catalogues in the collection are managed by means of the Informix Object-Relational DBMS. The database engine can be extended by means of DataBlade modules. These are software modules that plug in to Informix Dynamic Server with Universal Data Option and extend its capabilities in order to store, retrieve and manipulate new data types, besides the *primitive* ones provided with the server.

We are currently developing an astronomical datablade for Informix DS/UD that supports astronomical data (e.g., coordinates), query predicates (e.g., search-by-cone), and R-tree based indices that speed up the execution of queries containing astronomical predicates.

5. ASU Gateway

An Astronomical Server URL (ASU, Albrecht et al. 1996) based gateway is being developed. The ASU is a proposed specification of a syntax for the encoding in a URL of queries to astronomical databases. Our gateway is able to translate queries coded following the ASU specification into queries to the DB and to return results in various formats. It is being developed as a Java servlet. This will allow platform independence and maintain a persistent connection to the DB server for performance reasons. Any client able to submit queries coded following the ASU specification (e.g., ESO's SkyCat) will thus be able to submit queries and retrieve results.

6. WWW & Java-based Clients

We are currently prototyping simple WWW-based clients able to submit ASU encoded queries to the on-line system. These clients are implemented as Java applets. In this way, users are not required to install special client software and more interactivity is possible at the user workstation level (e.g., input validation, name resolving, etc). After this prototyping phase, we plan to design and build more sophisticated interfaces to allow users to submit complex queries to the database possibly in a graphical way, as opposed to the now common form-based interfaces.

7. Other Clients/Gateways

We are following the recent efforts directed at giving users integrated access to world-wide distributed astronomical data collections. We foresee implementing other gateways (e.g., Gamiel, McGrath, & Plante 1998) to allow a variety of clients to query the GSC II on-line system. Our efforts will be directed to provide open access to the system by means of standard conforming interfaces. In our view, this is the most effective way to maximize our system's utilization.

References

Albrecht, M., et al. 1996, "Astronomical Server URL", `http://vizier.u-strasbg.fr/doc/asu.html`

Baruffolo, A., & Benacchio, L. 1998a, in ASP Conf. Ser., Vol. 145, Astronomical Data Analysis Software and Systems VII, ed. R. Albrecht, R. N. Hook, & H. A. Bushouse (San Francisco: ASP), 382

———, 1998b, in SPIE Proc., Vol. 3349, Observatory Operations to Optimize Scientific Return, ed. P. J. Quinn (Bellingham: SPIE), 274

Gamiel, K., McGrath, R., & Plante, R. 1998, in ASP Conf. Ser., Vol. 145, Astronomical Data Analysis Software and Systems VII, ed. R. Albrecht, R. N. Hook, & H. A. Bushouse (San Francisco: ASP), 375

Lasker, B. M., et al. 1995, in Future Possibilities for Astrometry in Space (ESA Publ. SP-379), 173

Astronomical Data Analysis Software and Systems VIII
ASP Conference Series, Vol. 172, 1999
D. M. Mehringer, R. L. Plante, and D. A. Roberts, eds.

Numerical Observatory: a Web Site for the Retrieval, Analysis, and Visualization of Simulated Cluster Data

Greg Daues, Christopher Currie, Peter Anninos,
Larissa Kohler, John Shalf, Michael Norman[1]

National Center for Supercomputing Applications, University of Illinois Urbana-Champaign, Urbana, IL 61801

Abstract. We have developed a user-friendly, highly interactive, "Cluster Archive" web site (`http://sca.ncsa.uiuc.edu/`) that permits users to retrieve, analyze, and visualize archived data from extremely high-resolution numerical simulations of more than 100 X-ray clusters, spanning several orders of magnitude in luminosity and mass. The Cluster Archive site provides fast and efficient client software in the form of Java applets and CGI scripts to connect with computationally intensive server software. These scripts allow fast data selection, transfer, extraction, and the execution of various analysis routines. The site also supports server side interactive 2D Java and VRML-based 3D visualizations.

1. The Digital Archive and Web Interface

The NCSA Laboratory for Computational Astrophysics, together with members of the Cosmology groups at the University of Missouri, the Massachusetts Institute of Technology, and the New Mexico State University, are performing roughly 100 extremely high resolution simulations of X-ray galaxy clusters (with several decades variance in mass). The clusters form in a self-consistent numerical treatment of dark matter and baryonic fluid with a Cold Dark Matter (CDM) initial spectrum and a flat Friedmann background cosmology (Norman & Bryan 1998). These simulations use Adaptive Mesh Refinement (AMR) to achieve unprecedented resolution in the important central cores of the clusters where the densities are highest and where X-ray emissions are the greatest. AMR works by dynamically spawning finer-grained subgrids in the regions of the domain where additional resolution is required (Berger & Oliger 1984). The dynamic and automated generation of subdomain grid hierarchies of varying sizes, resolutions, and locations complicates the data structures in terms of size and information content.

Our goal is to make available to the scientific community the results of the simulations. However, making this kind of data available presents significant technological challenges because of the size and complexity of the data structures involved. Typically the results of these simulations are thousands of multiply-embedded and cross linked AMR grid structures distributed across hundreds of

[1] Astronomy Department, UIUC

thousands of files totaling hundreds of Gbytes. Most users cannot store this amount of data on their local disk, even for a single cluster. Furthermore, they would also not have the processing speed and memory to do the analysis. In order to use the data effectively, users must have access to basic analysis and processing software on an appropriate computing server.

To meet this need, we have developed a "numerical observatory" in a workbench style system that allows users to interact with archived simulation data over the Web. This environment supports the data retrieval across several machines at the NCSA, and between those machines and the local client systems. The observatory further supports dynamic and interactive multidimensional visualization and basic analysis tools to extract specified physical attributes of the cluster systems.

When considering what type of visualization tools to provide to the client's local machine, platform independence and "thin" client visualization software (i.e., software that is small and quick to download) are critical requirements for this project. This is especially crucial for the 3D visualizations. We successfully addressed these concerns in two ways. First, we render 3D objects on the server side. Client applets downloaded from the site allow the user to specify which cluster to visualize, which field quantities to download, and the type of visualization, such as an isosurface. The applet then connects back to a CGI script via the HTTP protocol. This protocol allows the user's applet input to be POSTed to the CGI script (Kaplan 1997).

The script then takes the user's preferences and launches a server-side VTK-based rendering program. (VTK[2] is a freely available visualization toolkit for 3D computer graphics, image and volume processing, and visualization). By rendering on the server, this achieves the goal of having a "thin" client (i.e., small, fast applet) on the user's local machine.

Second, although visualizations created by VTK are not platform independent, VTK does provide classes to convert the visualizations into a VRML 2.0 format and then write them to a file. However, instead of writing to a file, our CGI scripts write the contents to the standard output and the server sends them back to the applet. The applet then simply opens a new browser window, specifies the content type (i.e., vrml), and the browser then loads the appropriate VRML-viewer plug-in. Since most browsers running under many different operating systems have the ability to display VRML-based graphical representations, this allows us to achieve a platform independent way of displaying complex 3D visualizations.

In short, the site leverages widely and freely available technology to deliver high-quality 3D visualizations to a variety of platforms without overwhelming the processing power and storage capacity of a client's local machine. In Fig. 1, we display a screenshot of a working session of the current Cluster Archive site illustrating the multiple isosurface generating features of the VRML 3D visualizations.

The requirement of lightweight client visualization software is also met by Cluster Archive site tools that implement lower-dimensional visualizations through Java graphics. Specifically, the client applet allows the user to launch

[2] `http://www.kitware.com/vtk.html`

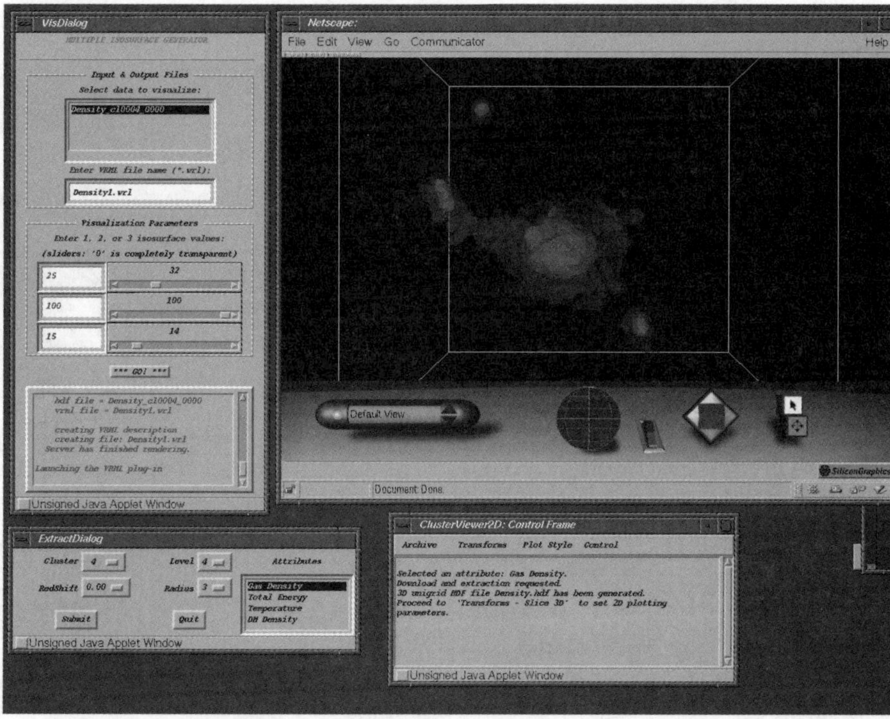

Figure 1. Screenshot of a working session of the Cluster Archive site illustrating VRML-based 3D visualization. Components of a Java applet allow user control of field quantities for data retrieval and 3D multiple isosurface visualization. Three different isosurfaces of constant gas density for a typical cluster are shown.

CGI scripts that sample subsets of large 3D Hierarchical Data Format (HDF) files that are retrieved from the archive storage and reside on the server side. Only these smaller subsets of the data are transferred over the network for processing by the client software. One capability of the client applet is the construction of 2D color contour plots of the sampled data. A screenshot displaying a working session of the web site implementing these plots is shown in Fig. 2. In keeping with the consideration of processing speed and memory on the client side, the client applet allows users to control the size and number of images generated. Archive tools for other forms of lower-dimensional visualizations such as 2D projections of 3D data and 1D line plots are also under development.

2. Future Enhancements

Future developments of the Cluster Archive site will focus on implementing Remote Method Invocation (RMI) to establish a more persistent and unified compute environment, allowing for finer controlled access of objects on the server.

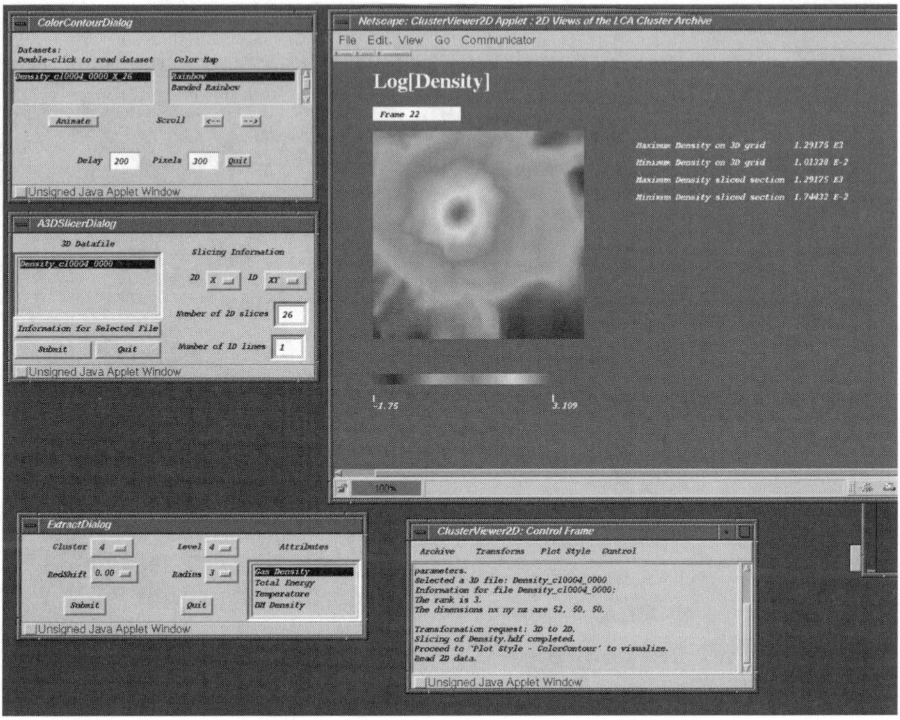

Figure 2. Screenshot of a working session of the Cluster Archive site illustrating Java-based 2D visualization. The components of the Java applet allow the user to specify parameters for data retrieval, sampling of the data into lower-dimensional subsets, and 2D color contour visualization with multiple color maps and zooming features. A plot of the gas density for a slice through the core of a typical cluster is shown.

We will also add links to other electronic literature and develop software to interact with existing digital libraries of observed data. This will necessitate adding support for more data types and formats, as well as the construction of additional analysis tools to allow direct comparison with observed data.

Acknowledgments. This work was supported in part by NASA ATP grant NAG 5-7404.

References

Berger, M. J. & Oliger, J. 1984, J. Comp. Phys. 53, 484

Kaplan, L. 1997, Java Report, 5, 51

Norman, M. & Bryan, G. 1998, Phys. Rev. Lett., 81, 3815

Astronomical Data Analysis Software and Systems VIII
ASP Conference Series, Vol. 172, 1999
D. M. Mehringer, R. L. Plante, and D. A. Roberts, eds.

Spiders and Camels and Sybase! Oh, My!

Irene Barg, Anthony J. Ferro, Elizabeth Stobie

University of Arizona, Steward Observatory, NICMOS, Tucson, AZ 85721

Abstract. The Hubble Space Telescope NICMOS Guaranteed Time Observers (GTOs) requested a means of sharing point spread function (PSF) observations. Because of the specifics of the instrument, these PSFs are very useful in the analysis of observations and can vary with the conditions on the telescope. The GTOs are geographically diverse, so a centralized processing solution would not work. The individual PSF observations were reduced by different people, at different institutions, using different reduction software. These varied observations had to be combined into a single database and linked to other information as well.

The NICMOS software group at the University of Arizona developed a solution based on a World Wide Web (WWW) interface, using Perl/CGI forms to query the submitter about the PSF data to be entered. After some semiautomated sanity checks, using the FTOOLS package, the metadata are then entered into a Sybase relational database system. A user of the system can then query the database, again through a WWW interface, to locate and retrieve PSFs which may match their observations, as well as determine other information regarding the telescope conditions at the time of the observations (e.g., the breathing parameter).

This presentation discusses some of the driving forces in the design, problems encountered, and the choices made. The tools used, including Sybase, Perl, FTOOLS, and WWW elements are also discussed.

1. Introduction

During a meeting involving many of the NICMOS GTOs, considerable interest was expressed in establishing a database containing the PSF star observations obtained as part of GTO programs. A centralized database was seen as a good way of making the existing PSF observations available to everyone on the team. This paper discusses how this PSF database and the user interface evolved.

2. Design Issues

Submissions to the database of processed PSF images are the responsibility of the individual team members. These team members are geographically dispersed, may have used different data reduction software, and may have named their images differently.

A network interface using CGI forms addressed all these issues. Submissions could be made remotely. Users would only need a Web browser to access the submission form. To enforce commonality among submissions, the submission form was structured in a way to ask specific questions regarding reduction procedures and details about the PSF star. The submission form needed to be straight forward and easy to use, it had to impose standards (file naming conventions, minimum descriptive information), and it had to be as short as possible without omitting key information about the PSF star. Our overall design goal was to automate the maintenance of the database as much as possible. Figure 1 illustrates the submission and retrieval process in which a PSF star observation travels from the submitter, is checked and ingested at University of Arizona (UofA), and is later retrieved and sent from the UofA to the requester.

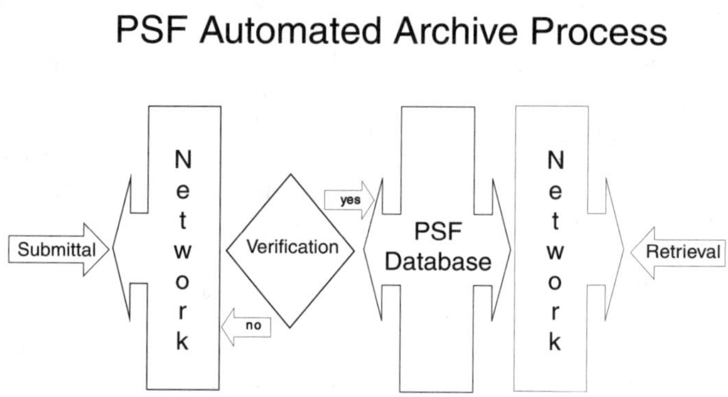

Figure 1. PSF Automated Archive Process

3. Submission Process

3.1. PSF Submittal Form

The PSF Star Submittal form is fairly detailed and requires some effort on the part of the submitter. They are asked to enter information about the PSF star and provide details of their data reductions. Some of the data entered are used as search parameters on the PSF Data Retrieval form, so it is important that the submission form be completed as completely as possible. A small text window in which to enter comments is included. Comments can be used to add any information that is important, but was not included in the form. Information provided on this form is transferred to the FITS headers of the submitted data file as HISTORY keywords. File naming standards were adopted and other standards were enforced through the form itself by requiring entries on selected fields.

3.2. PSF Submittal Checking and Database Population

Upon completion of the submission form, the data are automatically FTPed from the submitter's site to the database server at the UofA. An automated message is sent via e-mail to the user indicating the status of the transfer (received, failed, etc.). A copy of this notification goes to the NICMOS Database Administrator who then runs a set of Perl scripts that perform the following functions.

- Verification
  1. Verify the PSF submission (authorize user, files complete, etc.).
  2. Verify that submitted PSF files conform to the FITS standard.
  3. Check for minimum keywords.
- Build Metadata
  1. Extract selected keywords from the submission form and FITS headers to build the metadata that will populate the PSF database.
  2. Compress and move the PSF data file to the local UofA archive.

4. Retrieval Process

4.1. PSF Data Retrieval Form

This Web interface allows users to search the PSF archive for data that match criteria such as camera, filter, spectral type of the PSF star, breathing ephemeris (relative focus), and x- and y-position on the array. A successful search will present the user with a table of PSF star images from which they may select the PSF's they wish to retrieve. Figure 2 shows the results of such a search.

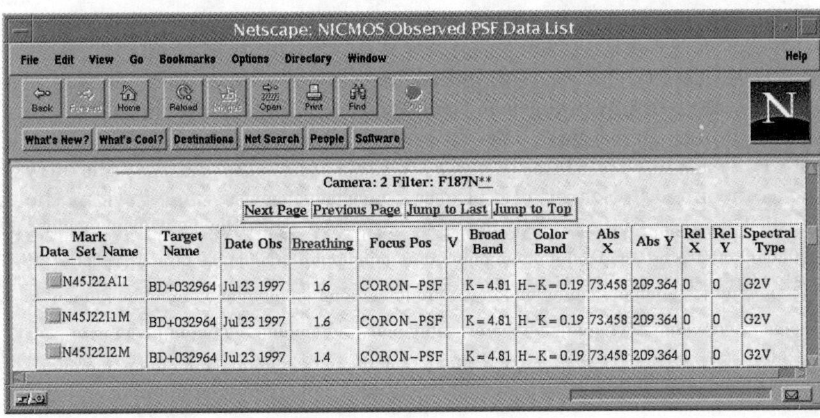

Figure 2. Results from a PSF Retrieval search.

With the exception of the *breathing* parameter, all of the data displayed from the results of the search represent data from the PSF database. The *breathing* parameter is a measurement of the change in the focus due to the temperature changes (breathing) of the telescope. The amount of this change varies and is highly time dependent. However, this value is not known at the time of the observation and therefore is not found in the FITS headers. There are two steps for determining relative focus (breathing) at a given time:

1. Look up the nominal value for the observation date, either in MJD, day of year, or days since 1 Jan 1994.
2. Look up the modeled variation from this nominal value for the exact time of the observation and add this value to the nominal.

The resulting value is in secondary mirror microns and is relative to best focus at WFPC2 PC (zero microns). Calculating this value presents a problem because the nominal value and modeled variation are embedded in two separate Space Telescope Science Institute (STScI) HTML documents that are updated occasionally as the model is improved. A means to extract the necessary information from the most recent versions of these documents became a programming challenge that was met by using a Perl module called `libwww-perl` (LWP). LWP allows us to locate the required two documents by extracting the links from the parent HTML page, fetch the required documents, and parse the necessary information on the fly. A description of how LWP accomplishes this would make a paper of its own.

5. System Software

This entire project was written using Perl V5.004 utilizing the following modules: `CGI.pm`, `Sybperl`, `libnet`, and `libwww-perl` (LWP). The verification scripts, also written in Perl, utilized a set of FITS utility programs from the FTOOLS package created by the High Energy Astrophysics Science Archive Research Center (HEASARC). The SQL server is Sybase[1] SQL Server V11.02.

6. Conclusions

The goal to automate the entire PSF submittal process drove our decision to use Perl as our programming language. As shown in Fig. 1, the submittal process involved communicating with the WWW, transferring data across a network using FTP, verifying the data using FTOOLS, and then adding the new PSF entries to our our SQL Server. The Perl programming language gave us the tools to provide the glue between all of these separate systems to make automation possible. To view some of the PSF forms, go to the (NICMOS[2]) Web page and follow the links to the Database Page.

---

[1] SYBASE is a registered trademark of Sybase, Inc.

[2] `http://nicmos.as.arizona.edu/`

Astronomical Data Analysis Software and Systems VIII
ASP Conference Series, Vol. 172, 1999
D. M. Mehringer, R. L. Plante, and D. A. Roberts, eds.

User Defined Functions for the ISO Post Mission Archive

Ekkehard Wieprecht

Max Planck Institut für extraterrestrische Physik, Giessenbachstrasse 1, D-85748 Garching, Germany

Bart Vandenbussche

Katholieke Universiteit Leuven, Instituut voor Sterrenkunde,Celestijnenlaan 200 B, B-3001 Heverlee, Belgium

Erich Wiezorrek, Helmut Steinle, Otto H. Bauer

Max Planck Institut für extraterrestrische Physik, Giessenbachstrasse 1, D-85748 Garching, Germany

Christophe Arviset, Jose Hernandez, Arno M. Plug, Richard Saxton

ISO Data Centre, ESA SSD, Villafranca, 28080 Madrid, Spain

Abstract. A major problem with designing archives in astronomy is defining the data resolution. In most cases, the smallest element in an archive is the object "observation". Attributes of this object contain meta-information about the observation that should be quickly queriable and are thus stored in the database engine. The data from the observation are stored external to the database engine as one big object. During the design of the archive, it is often difficult or even impossible to anticipate what meta-information of an observation should be queriable.

We present a solution for this problem as adopted for the Post Mission Archive of the Infrared Space Observatory (ISO) mission. Expert users can create User Defined Functions (UDFs) that extract virtual attributes from files external to the database engine. These virtual attributes can then further be used in queries inside the database engine.

We discuss the concept of User Defined Functions, system design, and implementation.

1. The ISO Post Mission Archive

During ISO's (Kessler et al. 1996) operational period (November 1995 to May 1998), its four instruments obtained a wide variety of astronomical data at infrared wavelengths from 2.4-240 microns. About 25,000 observations are stored in the ISO Post Mission Archive (Saxton et al. 1998). This archive provides online access to the data, support documentation, and software through a WWW interface.

The user interface is HTML and JAVA based and issues SQL queries to the database via jConnect[1] for JDBC. The basic interface allows for selection of observations defined by standard criteria such as coordinates, source name, observing mode, exposure time, wavelength, etc. Users can also select data by engineering, trend, and housekeeping parameters.

2. Virtual Data Resolution with User Defined Functions

As with all astronomical databases, data resolution is a critical issue in the ISO Post Mission Archive. The smallest object in the database is often an "observation" which is stored as a file external to the Database Management System (DBMS) or as a Binary Large Object (BLOB) in the DBMS. Only meta-information such as coordinates, average flux, pointing, etc., and relations to other objects like observer, instrument, etc. are stored and queriable (SQL, OQL, etc.) inside the DBMS.

For many instrument analysis and scientific purposes, queries and correlations on a much smaller scale are needed, for example, finding trends in detector noise for a specific instrument status, measuring line fluxes of one particular line of many sources in a vast set of broadband spectra, or comparing dark current with glitch activities. Anticipating all queries that will be needed during the design phase of the database is impossible. A scheme where all relevant data are implemented as attributes in the database is therefore impossible. This need could be accommodated in a database system where not only meta-data is stored in a queriable form, but also the data proper. In the case of spectra, a table containing wavelengths and fluxes would then be defined in the database.

Present "off the shelf" RDBMSs (Relational Database Management Systems) seem not to be robust enough to handle typical sizes of present astronomical databases. Therefore, for the ISO Post Mission Archive, the concept of virtual data resolution with user defined functions has been designed and implemented as a novel database paradigm.

The traditional division is kept where only meta-information about observations and related files is stored in the relational database; the data files themselves are stored external to the database. But the user is allowed to define functions which extract attributes from any data file in the archive. This includes the possibility of running Unix commands (sed, awk, grep, etc.), TCL, or Perl scripts on the ASCII files. Also, Interactive Data Language (IDL[2]) functions can be run on the FITS files. Therefore, it is possible to use features provided by the instrument interactive analysis systems which are coded in IDL.

---

[1]jConnect is a trademark of Sybase Inc.

[2]IDL is a trademark of Research Systems Inc.

3. System Design

The definition of a User Defined Function is entered in the system via a web interface. A UDF takes one file as input, optional additional input parameters, and outputs the virtual attributes for that file.

A defined UDF can be executed on a number of data files. The list of data files on which the UDF has to be run and the additional input parameters are prepared by the user in a temporary database table. The UDF is executed on every filename in the table and the extracted virtual attributes are stored in a second table. The second table can be either inserted again in the database where it can be used for further queries or it can be e-mailed to the user.

4. Implementation - How the ISO PMA User Uses UDFs

If an expert user needs additional attributes of files (e.g., observations) that are not provided by the standard RDBMS scheme he might use UDFs. The interface to define and execute user defined functions is web-based. Figure 1 gives an overview of the interface structure. In order to take advantage of UDF processing, the user needs a UDF user identification and password.

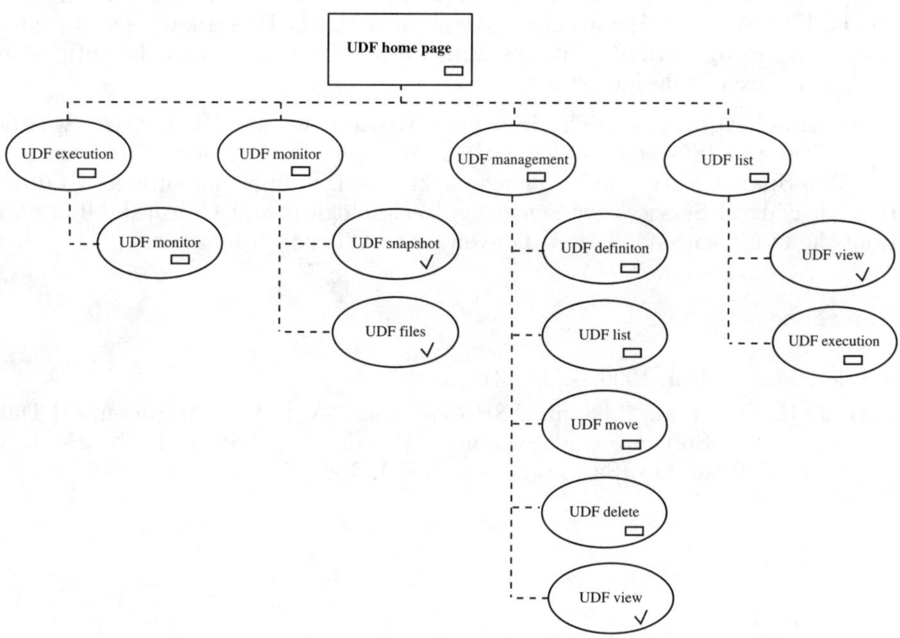

Figure 1. UDF web pages.

The UDF list page gives an overview of existing UDFs in the public and private areas. But the user can also write his/her own new UDF and install it in his/her private area on the server side. Further specifications of the UDF (e.g., input parameters) have to be given using the UDF definition page. This specifications can be displayed by the UDF view page. If a UDF is stable and useful to the community, it can be moved from the private UDF area to the public area by the UDF manager using a web page interface, `UDF move`.

The UDF execution page is used to start the UDF process. The user must specify the fixed input parameters. He also must specify if he wants to select required files via an SQL statement creating a temporary RDBMS table or use an existing database table. The required files are ordered to guarantee most efficient file access and are copied to the work directory, taking into account internal database limitations. A UDF works on one file at a time. The result (table with filenames and virtual attributes) is available as a temporary RDBMS table, which can be used for further SQL processing or sent via e-mail. The UDF process can be monitored using the UDF monitor page where the user might get additional information such as a snapshot of the result or which files have been copied from the archive at a given time. The UDF manager can monitor all the processes and start, stop, or even kill them.

5. Conclusion

We presented concept, design, and implementation of User Defined Functions on the ISO Post-Mission archive. We showed that UDFs allow users to query the archive using virtual attributes, some of which cannot always be anticipated during the archive design phase.

Acknowledgments. The ISO Post Mission Archive (ISO PMA) was developed at the ISO Data Centre, ESA Astrophysics Division, Villafranca del Castillo, Spain. Bart Vandenbussche acknowledges financial support from the Belgian Federal Services for Scientific, Technological and Cultural Affairs and from the Onderzoeksfonds K.U.Leuven grant OT/94/10

References

Kessler, M. F., et al. 1996, A&A, 315, L27

Saxton, R. D., et al. 1998, in ASP Conf. Ser., Vol. 145, Astronomical Data Analysis Software and Systems VII, ed. R. Albrecht, R. N. Hook, & H. A. Bushouse (San Francisco: ASP), 438

Astronomical Data Analysis Software and Systems VIII
ASP Conference Series, Vol. 172, 1999
D. M. Mehringer, R. L. Plante, and D. A. Roberts, eds.

Using a Java Web-based Graphical User Interface to access the SOHO Data Archive

I. Scholl, Y. Girard, A. Bykowski

Institut d'Astrophysique Spatiale, CNRS/Université Paris XI, Bâtiment 121, F-91405 ORSAY, France, Email: scholl@medoc-ias.u-psud.fr

Abstract. This paper presents the architecture of a Java web-based graphical interface dedicated to the access of the SOHO Data archive. This application allows local and remote users to search in the SOHO data catalog and retrieve the SOHO data files from the archive. It has been developed at MEDOC (Multi-Experiment Data and Operations Centre), located at the Institut d'Astrophysique Spatiale (Orsay, France), which is one of the European Archives for the SOHO data. This development is part of a joint effort between ESA, NASA and IAS in order to implement long term archive systems for the SOHO data.

The software architecture is built as a client-server application using Java language and SQL above a set of components such as an HTTP server, a JDBC gateway, a RDBMS server, a data server and a Web browser. Since HTML pages and CGI scripts are not powerful enough to allow user interaction during a multi-instrument catalog search, this type of requirement enforces the choice of Java as the main language. We also discuss performance issues, security problems and portability on different Web browsers and operating systems.

1. Introduction

SOHO is a satellite dedicated to the study of the Sun, built as part of an ESA and NASA cooperation. It operates with 12 instruments on-board that produce about 1.5 TB for 2 years of mission. These are operating in the frame of joint observing programs but their data and catalogs are not homogeneous. SOHO scientific and engineering data are currently stored at NASA/GSFC and at MEDOC/IAS (Scholl 1996). The goal of the current joint effort between ESA/NASA and MEDOC teams is to define a unique catalog for all data as well as a unique user interface to access them independently of their physical location. After a common catalog definition (Scholl & Sanchez 1998), the software development was split between the two teams. ESA/NASA team was in charge of the catalog data ingestion programs while, at MEDOC, we designed and developed the software that is responsible for catalog and file access. Because of the international character of this mission, our basic requirement was to develop a software that is independent of the archive architecture and which provides users with web access capabilities. Moreover, the amount of data and their complexity can lead to heavy queries producing a large volume of results.

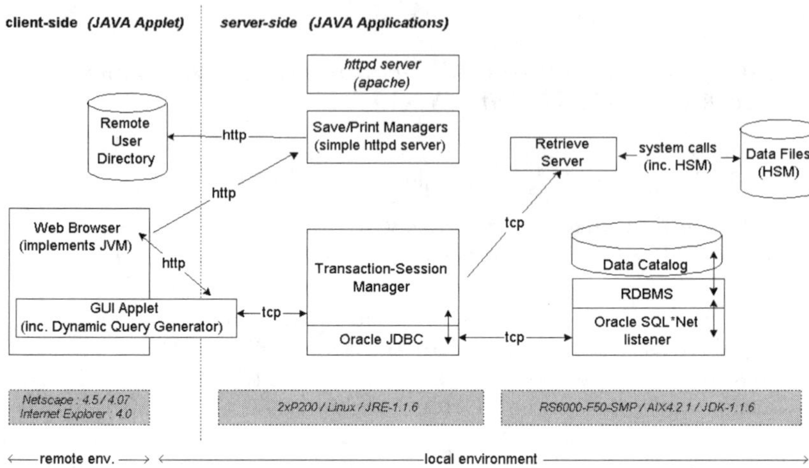

Figure 1. System Architecture.

Then, ergonomy, good response time and easy browsing functions were among our main goals.

2. System Architecture

The catalog access system is pure Java software[1]. It is organized as a 3-tier architecture (see Fig. 1). The client side is a Java applet while the server side is composed of several stand-alone Java applications.

The applet is stored on the HTTP server directory tree to be loadable by the user's Web browser, that must be JDK-1.1 compliant. The client implements the graphical user interface (GUI) functions such as:

1. Query creation and submit,
2. Result display,
3. File retrieval submit.

The main application, the transaction and session manager, is a multithreaded Java server located between the client and the RDBMS. Its roles are:

1. Transaction and session management: the server keeps track of each connected client.
2. Security management: it operates with a dedicated, security enhanced protocol in order to protect the content of the RDBMS from abusive and malicious use.
3. Query management: the server receives the content of the user's query and forwards it to the RDBMS.
4. Result management: in order to speed up result displays, the query result is stored on the server that only sends to the client the number of rows it needs for one display.

[1] `http://www.medoc-ias.u-psud.fr/new_archive/`

The client, the server and the other applications (listed above) communicate using TCP/IP sockets. Due to Java security restriction, the GUI code must be located on the host where the transaction server runs, and therefore both must be located on the HTTP server. Additional components may be run either on this machine or on a separate one. They are the following:

1. The RDBMS: accessed using an ORACLE JDBC driver,
2. The 'retrieve server': dedicated to file retrieval and packaging procedures,
3. The 'save server' is a simple HTTP server used to save retrieved files on the client host,
4. The 'print server' is another simple HTTP server used to allow the user to save/print the result of a query.

The transaction server provides other tiers with a Java API that implements its protocol.

3. Catalog Access

This part of the software involves only the GUI applet, the transaction manager and the RDBMS. The GUI allows the user to compose a query to be sent to the catalog. The first time the user loads the home page of this program, the client establishes a connection with the server. A timeout mechanism keeps this session alive during a pre-defined time of inactivity. Among others, that makes possible to get back the resources the user has been allocated.

The query is built dynamically using a dictionary which is the only part of the software that would be changed if the database structure changes. Each selection enriches the query in order to restrict the set of data (rows) returned.

Once the user has committed his choices, the resulting query is sent to the server which completes it in order to figure out if the data may be accessed by the user. The query is forwarded to the RDBMS via a JDBC driver; the results, received via the same channel, are temporarily stored on the server. Then, the content of the first page can be sent to the client before the end of the transfer. These buffers have several goals: lighten the client, decrease response time, allow the user to browse data by using next/previous (for the same view) and back (for different views) functions without accessing the database each time. Several parameters such as buffer size, number of maximum back operations, etc. can be tuned if necessary regarding the host computer capabilities.

4. File Retrieval

The retrieve operation involves several components at different levels. Figure 2 shows them as well as the sequence of operations necessary for the file retrieval process. The main problem encountered for this part of the software is related to Java security restrictions that don't allow an applet to write on a file system. In order to make the browser to allow this operation, the saving must be done outside the JVM. For user convenience and general security reasons we use an HTTP download instead of collecting login and password from users in order to send files using FTP. The request to the corresponding HTTP server, the save server, is an URL hiding a certificate coming from the transaction server. This

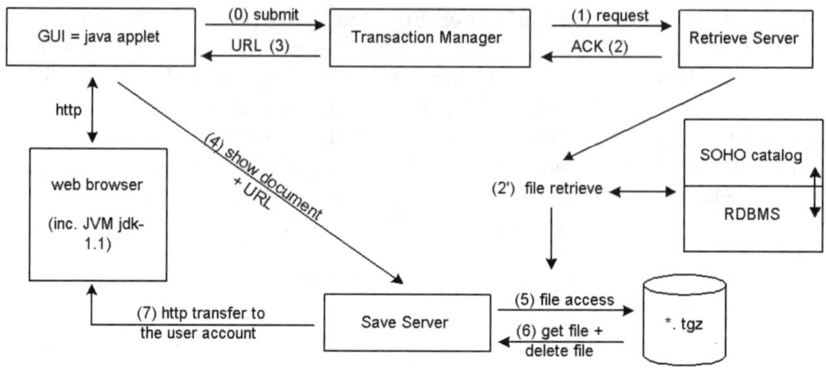

Figure 2. File Retrieval Data Flow.

ensures that the save server will not deliver the precious data to inadequate persons.

The retrieve process is independent of any kind of storage system: for example, at MEDOC we have implemented an HSM architecture with disks and cartridges while the GSFC has on-line disks.

5. Conclusion

In this paper we have presented the first version of a catalog and file web-based access tool. It has been a 10 man-month development. The main problems encountered were due to the Java language instability and the lack of portability of applets on different web browsers and platforms.

Our current work aims to get better performances in tuning the database access and result management. Our future works will be oriented to new facilities enhancing current search criteria as well as adding different parameters (like other ground-based or space solar observatories keywords). We will also follow Java evolutions such as JDK-1.2 with JFC which comes with more powerful components that improve performances and should simplify our code (i.e. RMI communications, direct print capabilities, light weight graphical components).

Acknowledgments. We are grateful to all future users who have tested this software. Special thanks go to Luis Sanchez and George Dimitoglou from ESA-NASA team and Jean-Luc Orcesi from MEDOC team for their technical contributions.

References

Scholl, I. 1996, in SpaceOps'96 (ESA Publ. SP-394) (Munich: ESA), 384

Scholl, I. & Sanchez L. 1998, "User Requirement Document for the SOHO Archive", `ftp://ftp.medoc-ias.u-psud.fr/pub/archive/urd.ps`

Astronomical Data Analysis Software and Systems VIII
ASP Conference Series, Vol. 172, 1999
D. M. Mehringer, R. L. Plante, and D. A. Roberts, eds.

Webmail: An Automated Web Publishing System

David Bell

NOAO[1], *P.O. Box 26732, Tucson, AZ 85726*

Abstract. A system for publishing frequently updated information to the World Wide Web will be described. Many documents now hosted by the NOAO Web server require timely posting and frequent updates, but need only minor changes in markup or are in a standard format requiring only conversion to HTML. These include information from outside the organization, such as electronic bulletins, and a number of internal reports, both human and machine generated. Webmail uses procmail and Perl scripts to process incoming email messages in a variety of ways. This processing may include wrapping or conversion to HTML, posting to the Web or internal newsgroups, updating search indices or links on related pages, and sending email notification of the new pages to interested parties. The Webmail system has been in use at NOAO since early 1997 and has steadily grown to include fourteen recipes that together handle about fifty messages per week.

1. Introduction

Publishing a document on the World Wide Web often involves several steps, which may be performed by multiple parties:

- Authoring – The document's content is created.
- Editing – Textual markup is added, and document may be split up.
- Posting – The document is copied to a Web server.
- Referencing – Links to the document are added or updated on other pages.
- Advertising – The document's presence is announced.

In many cases most or all of these steps can be completed in an automated manner. By reducing the need for direct human involvement, such documents can be posted and updated quickly and accurately, even by individuals without any experience with HTML.

2. Description

The Webmail software is installed in a dedicated account on the Web server which has been given write access to any directories or files it must update. On

[1]National Optical Astronomy Observatories, operated by the Association of Universities for Research in Astronomy, Inc. (AURA) under cooperative agreement with the National Science Foundation

the author's end, posting consists merely of sending an electronic mail message to this account. In some cases, a pre-agreed subject line is used to identify the document. For messages originating from interactive programs, this step is performed with a simple "Submit" button.

Procmail first filters the message by running a MIME-decoder, if needed. This has been particularly useful for handling messages sent from some PC-based mailers which unnecessarily encode plain text attachments, often without the sender's knowledge. Procmail then checks against a series of recipes to identify the message type. These include a list of addresses permitted to post or update the document, and usually information about the expected subject line or contents of the message. Procmail logs and archives all submissions and passes the message contents and type to a Perl script for further processing. Unidentified messages are forwarded for human review.

The Perl script includes recipes for each message type. Standard methods include Web posting, posting to newsgroups, archiving previous versions, and sending email announcements to distribution lists.

Specialized recipes are used for the more sophisticated documents. There is not always a one-to-one correlation between email messages and Web documents – a single message might be split into many pages or used to update just small sections of existing pages. Links may be added or updated on several related pages. For documents that are part of a searchable Web database, index files are also updated.

3. Examples

Webmail was launched in early 1997 for NOAO's internal posting of IAU Circulars, but has since grown to include fourteen recipes that together process about fifty messages per week. Recipes range from simple HTML-wrappers to those performing sophisticated parsing and reformatting. Some examples are described below.

- IAU Circulars – Message are indexed for searching, posted to an internal newsgroup, and saved as plain text. A CGI script on the NOAO Intranet is used to search circulars and add HTML wrappers and links to related circulars. A useful feature of this approach is that the links may point to circulars that arrived at a later time with updated information.

- Telescope Schedules – CTIO and KPNO schedules are sent from a *MS Access* database in a column-formatted table encompassing six months of dates for all telescopes. Individual months are split out of the message, formatted into HTML tables and cross-linked. For each scheduled project, links are provided to a page with detailed information including a full list of investigators and the project's title and scientific abstract. Minor schedule changes may occur several times each semester, so any changed months are announced by email to a distribution list. Example schedules may be found at `http://www.noao.edu/kpno/forms/tel_sched/` or `http://www.noao.edu/ctio/forms/tel_sched/`.

- Conference Room Schedules – These are maintained on a PC running *Schedule+* and are output as HTML and mailed. Webmail fixes some of the markup and splits and recombines weekly schedules into more useful formats for posting on the NOAO Intranet. A single message updates over twenty Web pages.

- ACM Colloquium page – This is one of the simplest recipes Webmail processes, but may be of interest since the author is from outside NOAO. A pre-formatted HTML document is simply posted to the NOAO server at `http://www.noao.edu/noao/acm.html`.

4. Conclusion

We have found the Webmail approach to be quite useful and powerful, yet perhaps so simple as to be easily overlooked. For pre-formatted documents needing little or no markup, new recipes can be set up with just a few lines of procmail and Perl code. The resulting process is more efficient than passing each document through a human Webmaster, and more secure than giving all authors direct access to the Web server. For documents requiring extensive parsing and markup, the benefits of an automated system are even greater.

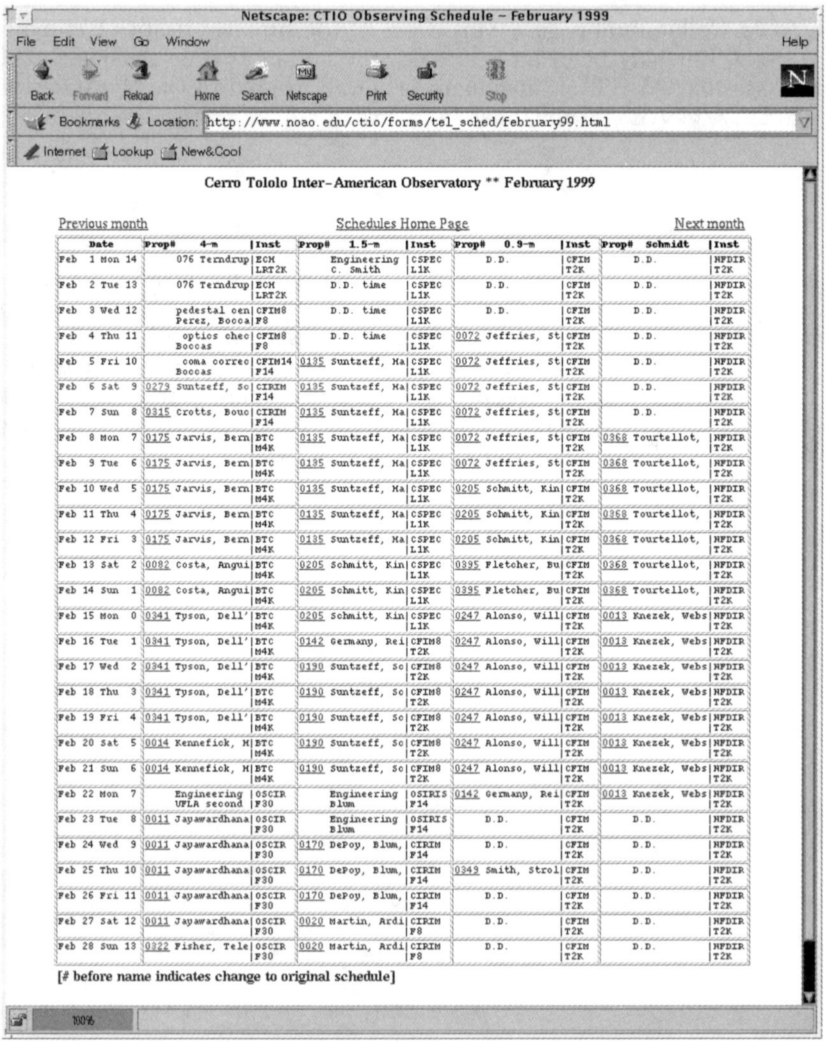

Cerro Tololo Inter-American Observatory ** February 1999

Previous month | Schedules Home Page | Next month

Date	Prop# 4-m	Inst	Prop# 1.5-m	Inst	Prop# 0.9-m	Inst	Prop# Schmidt	Inst
Feb 1 Mon 14	076 Terndrup	ECH LRT2K	Engineering C. Smith	CSPEC L1K	D.D.	CFIM T2K	D.D.	NFDIR T2K
Feb 2 Tue 13	076 Terndrup	ECH LRT2K	D.D. time	CSPEC L1K	D.D.	CFIM T2K	D.D.	NFDIR T2K
Feb 3 Wed 12	pedestal cen Perez, Bocca	CFIM8 F8	D.D. time	CSPEC L1K	D.D.	CFIM T2K	D.D.	NFDIR T2K
Feb 4 Thu 11	optics chec Boccas	CFIM8 F8	D.D. time	CSPEC L1K	0072 Jeffries, St	CFIM T2K	D.D.	NFDIR T2K
Feb 5 Fri 10	coma correc Boccas	CFIM14 F14	0135 Suntzeff, Ma	CSPEC L1K	0072 Jeffries, St	CFIM T2K	D.D.	NFDIR T2K
Feb 6 Sat 9	0279 Suntzeff, Sc	CIRIM F14	0135 Suntzeff, Ma	CSPEC L1K	0072 Jeffries, St	CFIM T2K	D.D.	NFDIR T2K
Feb 7 Sun 8	0315 Crotts, Bouc	CIRIM F14	0135 Suntzeff, Ma	CSPEC L1K	0072 Jeffries, St	CFIM T2K	D.D.	NFDIR T2K
Feb 8 Mon 7	0175 Jarvis, Bern	BTC M4K	0135 Suntzeff, Ma	CSPEC L1K	0072 Jeffries, St	CFIM T2K	0368 Tourtellot,	NFDIR T2K
Feb 9 Tue 6	0175 Jarvis, Bern	BTC M4K	0135 Suntzeff, Ma	CSPEC L1K	0072 Jeffries, St	CFIM T2K	0368 Tourtellot,	NFDIR T2K
Feb 10 Wed 5	0175 Jarvis, Bern	BTC M4K	0135 Suntzeff, Ma	CSPEC L1K	0205 Schmitt, Kin	CFIM T2K	0368 Tourtellot,	NFDIR T2K
Feb 11 Thu 4	0175 Jarvis, Bern	BTC M4K	0135 Suntzeff, Ma	CSPEC L1K	0205 Schmitt, Kin	CFIM T2K	0368 Tourtellot,	NFDIR T2K
Feb 12 Fri 3	0175 Jarvis, Bern	BTC M4K	0135 Suntzeff, Ma	CSPEC L1K	0205 Schmitt, Kin	CFIM T2K	0368 Tourtellot,	NFDIR T2K
Feb 13 Sat 2	0082 Costa, Angui	BTC M4K	0205 Schmitt, Kin	CSPEC L1K	0395 Fletcher, Bu	CFIM T2K	0368 Tourtellot,	NFDIR T2K
Feb 14 Sun 1	0082 Costa, Angui	BTC M4K	0205 Schmitt, Kin	CSPEC L1K	0395 Fletcher, Bu	CFIM T2K	0368 Tourtellot,	NFDIR T2K
Feb 15 Mon 0	0341 Tyson, Dell'	BTC M4K	0205 Schmitt, Kin	CSPEC L1K	0247 Alonso, Will	CFIM T2K	0013 Knezek, Webs	NFDIR T2K
Feb 16 Tue 1	0341 Tyson, Dell'	BTC M4K	0142 Germany, Rei	CFIM8 T2K	0247 Alonso, Will	CFIM T2K	0013 Knezek, Webs	NFDIR T2K
Feb 17 Wed 2	0341 Tyson, Dell'	BTC M4K	0190 Suntzeff, Sc	CFIM8 T2K	0247 Alonso, Will	CFIM T2K	0013 Knezek, Webs	NFDIR T2K
Feb 18 Thu 3	0341 Tyson, Dell'	BTC M4K	0190 Suntzeff, Sc	CFIM8 T2K	0247 Alonso, Will	CFIM T2K	0013 Knezek, Webs	NFDIR T2K
Feb 19 Fri 4	0341 Tyson, Dell'	BTC M4K	0190 Suntzeff, Sc	CFIM8 T2K	0247 Alonso, Will	CFIM T2K	0013 Knezek, Webs	NFDIR T2K
Feb 20 Sat 5	0014 Kennefick, H	BTC M4K	0190 Suntzeff, Sc	CFIM8 T2K	0247 Alonso, Will	CFIM T2K	0013 Knezek, Webs	NFDIR T2K
Feb 21 Sun 6	0014 Kennefick, H	BTC M4K	0190 Suntzeff, Sc	CFIM8 T2K	0247 Alonso, Will	CFIM T2K	0013 Knezek, Webs	NFDIR T2K
Feb 22 Mon 7	Engineering UFLA second	OSCIR F30	Engineering Blum	OSIRIS F14	0142 Germany, Rei	CFIM T2K	0013 Knezek, Webs	NFDIR T2K
Feb 23 Tue 8	0011 Jayawardhana	OSCIR F30	Engineering Blum	OSIRIS F14	D.D.	CFIM T2K	D.D.	NFDIR T2K
Feb 24 Wed 9	0011 Jayawardhana	OSCIR F30	0170 DePoy, Blum,	CIRIM F14	D.D.	CFIM T2K	D.D.	NFDIR T2K
Feb 25 Thu 10	0011 Jayawardhana	OSCIR F30	0170 DePoy, Blum,	CIRIM F14	0349 Smith, Strol	CFIM T2K	D.D.	NFDIR T2K
Feb 26 Fri 11	0011 Jayawardhana	OSCIR F30	0170 DePoy, Blum,	CIRIM F14	D.D.	CFIM T2K	D.D.	NFDIR T2K
Feb 27 Sat 12	0011 Jayawardhana	OSCIR F30	0020 Martin, Ardi	CIRIM F8	D.D.	CFIM T2K	D.D.	NFDIR T2K
Feb 28 Sun 13	0322 Fisher, Tele	OSCIR F30	0020 Martin, Ardi	CIRIM F8	D.D.	CFIM T2K	D.D.	NFDIR T2K

[# before name indicates change to original schedule]

Figure 1. A telescope schedule page produced by Webmail.

Astronomical Data Analysis Software and Systems VIII
ASP Conference Series, Vol. 172, 1999
D. M. Mehringer, R. L. Plante, and D. A. Roberts, eds.

Study to Install an Information System at the University of the Istanbul Astronomy and Space Sciences Department

S. Gülseçen, H. Gülseçen, H. Çalışkan, T. Ak

University of Istanbul, Faculty of Science, Astronomy and Space Sciences Department, 34452 University, Istanbul, Turkey

Abstract. University of Istanbul Astronomy and Space Sciences Department's current computing system which no longer adequately meets the needs of its users—astronomers—has to be re-designed to operate like an Information System. Other major factors that have convinced us to design a new system were the establishment of the National Observatory of Scientific and Technical Research Council of Turkey and an urgent need for on-site scientific data analysis systems. Installation of an Information System started with Initial Investigation and Feasibility Study phases. The needs of users, the scope of the problem, the priority assigned to the project and the resources needed to modify the existing system were accurately estimated through a number of questionnaires. The major modification made was increasing the computing power of the system with two new workstations: DEC Alpha and HP PW. The General and Detailed Design phases determined the technical design of the system. The system is now in its implementation phase.

1. Introduction

The age we live in and the technology we use force us to process, collect, use, distribute and share data, information and knowledge as quickly as possible. To do this, we have Data Processing Systems (DPS), Management Information Systems (MIS) and even Knowledge Based Systems (KBS) that may show intelligent behavior, thanks to progressive work in Artificial Intelligence (AI) (Gülseçen & Gülseçen 1995).

On the other hand, astronomical data is coming from different sources in great amounts. Computer hardware is increasing its capability and computer software is becoming more powerful and more user-friendly so that less experienced people can use it. Technological advances such as widespread access to the Internet are always within reach. Effective Knowledge Management is fast becoming a very important strategic issue. (Raitt et al. 1997). The only thing we must do is to integrate all these properly and carefully. One way to do this is to install an Information System (IS) in an Astronomy Department to act like a "manager" between the department staff and their computer system, their intranet, Internet, data reduction and analysis systems and various astronomical data sources such as observatories and large astronomical databases.

The aim of this paper is to give a general view of work done to improve computing facilities at University of Istanbul Astronomy and Space Sciences Department (hereafter UI-ASSD) and to move toward a new IS, based on improved computer capabilities.

2. The Need for an IS at UI-ASSD and its Features

The Astronomy and Space Sciences Department is now located on the University of Istanbul Campus at Beyazit. The academic staff of department are actively engaged in a wide range of observational and theoretical research on the dynamics and morphology of stellar atmospheres, peculiar stars, binary stars, cataclysmic variables, interstellar column densities, three-colour photometry (RGU), solar prominences and galaxies and cosmology. Data is collected from all over the world and daily sunspot drawings and chromospheric observations are still carried out. Astronomers are beginning to obtain time on the telescopes at the National Observatory of Scientific and Technical Research Council of Turkey (TUG)[1] located approximately 50 kilometers north of Antalya, Turkey, where there are two joint facilities: 1.5 and 0.40 meter telescopes.

The computational facilities at UI-ASSD based on two different kinds of systems (VMS and DOS/Windows) were not enough to support the theoretical and observational research. Scientists in the department obtained large amount of data but there was no possibility to reduce and analyze it on-site. The absence of at least one Unix-based workstation with sophisticated graphics capability was a barrier to utilizing important software for astronomical data processing and analysis, general scientific data analysis applications, document preparation, WWW and Internet facilities and user support. The establishment of the TUG in 1997 has forced astronomers in UI-ASSD to be persistent in requesting a new system that will link them to this observatory for remote observations and to get data on-line.

The study to install the IS of UI-ASSD started with the updating of existing hardware. Later, an IS Team was established. The main goal of IS Team is to improve the computing facilities of department in the same way some observatories and astronomy departments did, in other words using them as models. For this purpose, a visit to the Asiago Astrophysical Observatory in the Italian Pre-Alps, operating jointly with the Astronomical Observatory of Padova, was organized to gain specific knowledge about a highly successful computing system. The fifteen days spent in Asiago resulted in a better understanding of the software needed by professional astronomers and about the operating systems required to make them run. Hints and suggestions about the new system were later adopted by the UI-ASSD were the natural evolution of this fifteen day stage as part of a major Padova-Istanbul collaboration. The features of the new system were determined as follows (Gülseçen et al. 1998):

- work with faster processing power
- employ a multi-user and multitasking OS (Unix and its flavors)
- store large amounts of data
- run most diffused astronomy software (IRAF, MIDAS, Super Mongo,...)
- process all types of data
- have a good access to Internet
- have a good access to all astronomy resources
- have high resolution display capability.

---

[1] `http://astroa.physics.metu.edu.tr/tug/home.html`

3. Development of the IS of UI-ASSD

3.1. What is IS?

An IS is an arrangement of interdependent human and machine components and procedures that interact to support the information or business needs of an organization and the system's users. System Development Life Cycle (SDLC) is the formal process by which organizations build computer-based information systems. An SDLC is used as a guideline in directing and administering the activities involved in establishing system requirements, developing the system, acquiring hardware and software, and controlling development costs. System development can be divided into six phases: Analyze Current System, Define New System Requirements, Design New System, Develop New System and Have Users Test It, Implement New System and Evaluate Performance of New System and Maintain System (Hutchinson & Sawyer 1994; Leeson, Rowe, & Poindexter 1985).

3.2. Development of IS of UI-ASSD: Phase 1 to Phase 4

Because the process of developing a new system costs a great deal of time, energy and money, a steering committee has been formed to help decide how to start. At the end of January 1997, the IS team initiated a project to develop the new system step by step using SDLC. In order to get financial support the project has been proposed to the Research Fund of Istanbul University and at the same time we requested financial support from some computer companies.

Phase 1: In order to gain clear understanding of the existing system and to be able to study user requirements, many interviews and questionnaires have been utilized under the study named "Study to Define Current and Future Information Processing Profile of UI-ASSD". All scientists who had visited European and American Observatories and made their reduction and analysis work there using IRAF, MIDAS and other software were asked to participate in the study of user requirements. They were asked to describe everything they used: manuals, time allocated, features of the hardware used, outputs etc. After information was gathered was analyzed to identify problems and opportunities which will improve the existing system discussed in Section 2.3. Phase 1 concluded with a report which summarized the current system analysis and gave recommendations for the modular structure of new system.

Phase 2: In Phase 2 new system requirements have been defined in greater details. In order to determine what type of computer modifications are needed, a number of astronomy professionals from other Turkish universities have been interviewed and a visit to AsAO has been planned (see Section 2.3). Phase 2 concluded with a Detailed Study and System Requirements Report (Gülseçen et al. 1998). This report provided the basis for the final determination of the completeness and accuracy of the system requirements as well as the economic and practical feasibility of the new system.

Phase 3: New systems must be designed to operate within a framework of controls which constitute a system of safeguards that protect a computer system and data from accidental or intentional damage, input and output inaccuracies, and access by unauthorized persons (Hutchinson & Sawyer 1994). In this phase, design of the new system started and all system design documentation (system flowcharts, block diagrams, forms etc.) has been completed, organized and assembled. Controls involving the physical environment of the additional hardware—limiting access to building, computer room, doors and hardware itself—have been described.

Phase 4: the acquisition of hardware (DEC Alpha pw433 and HP-PW) and installation of system software (Linux, DEC Unix 4.0 and HP-UX 10.20) and freely available and distributable data analysis software (HP-UX/IRAF, Linux/MIDAS) were the main activities of this phase. After the initial training of some users, the early test was done and the results used in two proposals presented at a NATO-ASI meeting held in Çeşme-Turkey in September 1998.

4. Conclusions and Future Work

As many similar projects in astronomical community, the goal of this one is to build a system with which users can effectively combine resources in order to do scientific computing. First we studied user requirements, methods and procedures, communications needs, existing hardware and software and their interaction. User participation in this study assisted the IS team members by expanding their thinking about the components being studied and was of great importance. It has been determined that using questionnaires rather than interviews is more reliable.

The study has not yet been finished. The last two phases are under way. With the completion of these phases, a transition from less powerful computing system to efficient Unix based computing system will be made.

Acknowledgments. We wish to thank Prof. R. Barbon for his kind invitation S. Gülseçen to AsAO, HP-TÜRKIYE for their financial support and Entegre Bilgisayar Sistemleri Ltd. Şti. for their technical assistance. This work was supported by the Research Fund of the University of Istanbul. Project Number(s): 1048/031297, 1114/010598, 1110/010598, 952/090597

References

Gülseçen, S. & Gülseçen, H. 1995, in Proc. on Conf. for Systems Engineering and Defense Applications (Ankara: Kara Harp Okulu), 727

Gülseçen, S., Saygaç, T., Passuello, R., & Rigoni, A. 1998, Reorganization and Reconfiguration of Information System of Istanbul University Observatory Taking the Padova-Asiago Observatory Information System as a Model, (Tech. Rep. No. 1, Dept. of Astronomy & Space Sciences, Instanbul Univ.)

Hutchinson, S. E., & Sawyer, S. (ed.) 1994, Computers and Information Systems, (Chicago: Irwin)

Leeson, M., Rowe, S. H., Poindexter, S. E. 1985, Systems Analysis and Design (2nd ed.), (Chicago: Science Research Associates)

Raitt, D., Loekken, S., Scholz, J., Steiner, H., & Secchi, P. 1997, ESA Bull., 92, 112

Astronomical Data Analysis Software and Systems VIII
ASP Conference Series, Vol. 172, 1999
D. M. Mehringer, R. L. Plante, and D. A. Roberts, eds.

Archive Storage Subsystem for the ESO VLT

Jennifer Dunn, Steve Cockayne, Séverin Gaudet, Shannon Jaeger

National Research Council Canada/Herzberg Institute of Astrophysics, 5071 West Saanich Rd., Victoria, BC, Canada V8X 4M6

Miguel Albrecht

European Southern Observatory, D-85748 Garching bei Munchen, Germany

Abstract. The Archive Storage Subsystem (ASTO), developed by the Canadian Astronomy Data Center, is part of the On-Line Archive Facility in use by the European Southern Observatory (ESO) Very Large Telescopes (VLT) and New Technology Telescope (NTT) Archive System. It is composed of several programs that allow for the preparation and writing of media in parallel. Its function is to respond to requests to create archive and user media on magnetic tape, CD-ROM, or DVD. Parts of this system have also been integrated into the Gemini Telescopes Data Handling System. The design and function of this system are presented.

1. System Overview

The Archive Storage Subsystem (ASTO), developed by the Canadian Astronomy Data Center, is part of the On-Line Archive Facility in use by the European Southern Observatory (ESO) Very Large Telescopes (VLT) and New Technology Telescope (NTT) Archive System. It is composed of several programs that allow for the preparation and writing of media in parallel. Its function is to respond to requests to create archive and user media on magnetic tape, CD-ROM, or DVD. Parts of this system have also been integrated into the Gemini Telescopes Data Handling System. The design and function of this system are presented. The ASTO system supports writing of two types of requests:

Archive data are for long term use and will generally be available for on-line retrieval. Archive data can be grouped together based on attributes such as data type (science data, calibration, etc.). A directory structure is supported for archive requests. Typically, archive data are constantly received and they are stored when there are enough data to fill a media volume (e.g., CD-ROM). The media volume is then ingested into an on-line data system where multiple users can view and/or retrieve the data.

A user, perhaps an astronomer or observer, may have a collection of files that they wish to have written to media. This system provides the user with a means of producing a unit of media (e.g., tape or CD-ROM) containing the data which they can keep. The user request can specify the type of media, the files to write, and if the files are to be compressed. The specification of a directory structure is also supported. A user can request the media to be made at any time, even many months after the observation has been carried out. A GUI has been created but is not discussed here.

Initially data are made available to ASTO by placing them into ASTO's configured magnetic disk area called the Data Stage. Information about the files are also put into ASTO's database. The SYBASE database is also used as persistent store for the status of a request and to provide the ability to rollback actions if a problem occurs.

2. Data Stage Design

Three Data Stage areas are required on magnetic disk:

1. The Archive Data Stage is the magnetic disk area which contains files destined for archive media.
2. The User Data Stage is the magnetic disk area which contains files destined for user media.
3. The Media Data Stage is the disk area used to prepare a media volume for writing.

The Data Stage areas are distinct and can be placed on different disk partitions. This will ensure that, for example, archive data do not consume all disk which would prohibit the preparation and writing of a media volume. The number of media devices that can be simultaneously written is typically constrained by the disk space available in the Media Data Stage and number of devices available for writing.

3. Database Design

ASTO maintains a database which it uses and accesses. The table names are configurable and contain information such as

1. detailed information about files available locally (name, file size, status, compression information, and format),
2. and information about the creation of media (type of request, source of the request, date requested, number of copies, type of media to write, etc.).

4. Media Creation

The various components of the ASTO system manage the writing of media. Each request to write media must proceed through a series of steps called actions. Figure 1 illustrates the actions and rollbacks that are performed when writing media system. It is possible to rollback at any stage prior to the cleanup stage.

The actions are defined as follows:

1. Ingest. This involves the placement of files into the ASTO Data Stage and ingestion of these files into the ASTO storage database. This action is performed either by ASTO or by the VLT On-Line Archive System

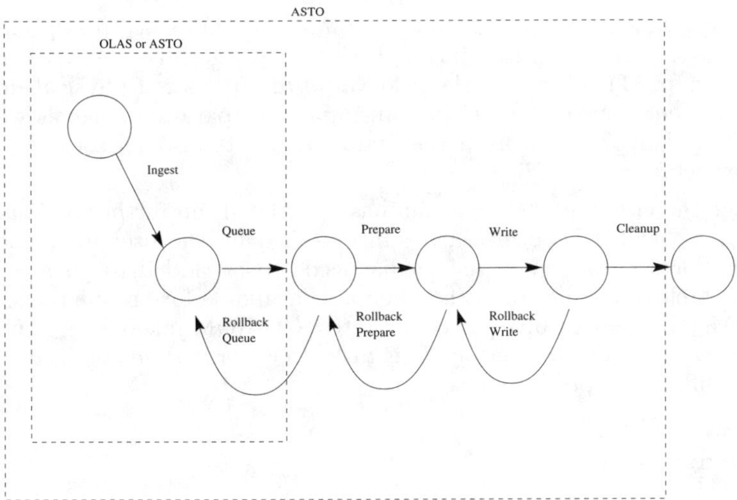

Figure 1. ASTO actions.

(OLAS). At this point, a request will have a name and be assigned a unique identifier.

2. Queue. Files are grouped together which are to be written onto the same physical unit of media, whether the media be CD-ROM, DVD, or tape. For example, a single request may contain enough data to fill three CD-ROMs. Each of the CD-ROM's worth of data is considered to be a unit, and each unit is assigned a unit name that is a combination of the media name and a unit counter. This grouping allows the preparing and writing of the CD-ROMs to be run in parallel, rather than sequentially. An algorithm is used to ensure that, if possible, directories are not split between CD-ROMs. At the completion of each of the following steps, the database is updated to reflect the current state of a request. This action is performed either by ASTO or by OLAS.
3. Prepare. The preparation stage involves whatever is necessary to write the media. For example, the burning of a CD-ROM involves the creation of the physical image before writing. This can be a time consuming process; thus, being able to run this step in parallel with other preparations streamlines this stage. The software which performs the preparation is configurable. For example, a CD-ROM physical image could be created using either the commercially available `Gear`, or the public domain `mkisofs` CD-ROM authoring software.
4. Write. This step performs the actual writing of the medium. For a CD-ROM, this would involve the burning of the prepared physical image from the Data Stage to the physical medium. The script that performs the writing is configurable. For example, `Gear` or `cdrecord` can be used to write the previously prepared physical volume to a CD-ROM. This system supports the writing of multiple copies to multiple devices. A logical device

name, which corresponds to one or more actual devices, is specified. The device is reserved at the start and released at the end of the writing process.

5. Cleanup. This involves the removal of any prepared physical image and any other information created during the preparation stage as well as the removal of the files from the Data Stage. It also releases the reserved device(s).

Once the creation of user media has completed, no further actions need to be taken by ASTO. The media is simply taken by the user for personal use. However, for archive media the media needs to be added to an archive. For example, archive media would then be made available for on-line retrievals. For CD-ROMs, this may involve placement into a CD-ROM jukebox. ASTO includes several executables which perform this task. In general, different executables are used for different types of media.

5. Configuration

ASTO is a highly configurable system. Examples of the information that can be configured are:

1. paths to the different Data Stage areas,
2. the database server name, database name, and database tables names,
3. names of scripts to use for the preparation and writing of a given media type,
4. media unit names, which are a combination of the media name plus a unit counter, which is a number padded with zeros. The default number of zeros can be configured,
5. recognized types of compression,
6. and recognized file formats and their associated extensions.

6. Conclusion

ASTO's ability to perform archive or user media requests has been proven. Parts of this system have been integrated into the Gemini Telescopes Data Handling System. The flexibility of the system allows for many different configurations. Because the actions to produce the media are sufficiently distinct, recovery due to a file error is easily handled. In addition, the ability to write and prepare the media in parallel produces a more time efficient system. A prototype version has been delivered and is in use at the NTT and VLT telescopes as well as ESO headquarters in Garching.

Acknowledgments. To my family, Rodger, Jade and Marina Dunn, for keeping me sane and to the intelligent people I work with.

Astronomical Data Analysis Software and Systems VIII
ASP Conference Series, Vol. 172, 1999
D. M. Mehringer, R. L. Plante, and D. A. Roberts, eds.

The Prospects of DVD-R for Storing Astronomical Archive Data

Benoît Pirenne[1], Miguel Albrecht[2]

European Southern Observatory, Karl-Schwarzschild-Str. 2, D-85748 Garching bei München, Germany

Jörg Schilling

Research Institute for Open Communication Systems, Berlin

Abstract. The Digital Versatile Disk (DVD), heralded as the successor of the CD-R as a direct-access, cheap archive media has been long to materialize. This paper reviews the current situation regarding DVDs and assesses the suitability and maturity of the technology for astronomical data archives. Considerations such as cost per unit of capacity, durability, format issues and compatibility are addressed. The current projects involving DVD technology at ESO are presented.

1. Introduction

In the past 10 years the photon collecting efficiency – and hence the data throughput – of astronomical equipment has been multiplied by almost two order of magnitude: larger mirrors (×4), vastly superior efficiency of CCDs (×3); better use of scheduling tools to make better use of the time available (×2). Moreover, the quest for better and better resolution has also spawned CCD mosaics that can produce extremely large images (×5). New infrared detectors with their rapid sequences of exposures are also challenging data acquisition systems.

On top of the multiplication factor described above, the cumulative effect caused by systematic archival of all observations complicates the problem faced by archivists. Also, using new observing facilities is now expensive enough to make data worth keeping (almost) forever.

2. Towards Higher Density

To address the challenges posed by this dramatic increase, technologists need to re-assess the means at their disposal in the area of data storage. Up until recently, archives were a mere "bit bucket": data was kept on some off-line medium, but hardly ever retrieved. Pioneered by IUE (Wamstecker 1991) and

[1]Space Telescope – European Coordinating Facility

[2]Data Management Division

later by the Hubble Space Telescope (Pirenne et al. 1993), a new era came were archives were a more lively system: data was stored with the idea that any particular exposure would be retrieved many times over, during and after the lifetime of the instrument which produced it (Crabtree et al. 1996).

A third generation archive is probably necessary for the early years of the next decade where we will see requirements for rapid access to huge quantities of digital data, coming from different instruments, different observatories and different epoch. These data will have to be rapidly re-calibrated, associated, analyzed and cross-correlated prior to its distribution to the final users. This is triggering the need for fast access to thousands of science and calibration exposures, each of which occupying perhaps dozens of mega bytes.

3. From CDs to DVDs

For the past 3 to 5 years, the needs of our archives have been satisfied with the existing: low cost, vendor independant, medium density, long-lifetime media materialized by the "CD-Recordable" (CD-R). This medium is indeed extremely cheap ($1/volume), the recording system is also very affordable and is available from multiple vendors. Moreover, large jukeboxes capable of making hundreds of such CD-Rs available quasi on-line are also rather affordable.

When the DVD technology was announced, it really looked like the logical step forward for growing archives. Unfortunately, the technology has only recently been made available and real testing could only be started this Summer.

Another complication, explaining perhaps some of the delay, was due to the competition between potential vendors of the technology: everyone was coming up with its own version of the technology. As a result, the confusion was great amongst potential buyers faced with unclear choices: DVD-R, DVD-ROM, DVD-RAM, DVD-R/W, DVD+RW, ...

So far, those who had to make a choice for a large part turned away from the technology and started to explore other avenues:

- huge disk farms: Reliability and hence backup is a big issue, just next to cost
- large tape robots (cost is very interesting, but access time is a concern, together with lifetime)
- CD-R juke boxes. Mastering and maintaining thousand of individual CD-R is becoming non-trivial and involves a significant management overhead.
- Magneto-optical disks in a juke box: Probably a reasonable choice in the eyes of many, thanks to the fact that the technology is there and well-know, with reasonable capacity growth prospects.

So far a reasonable choice in this area has been the CD-R. Comparisons are shown in Table 1.

In the ESO/ST-ECF archive, the idea of non-erasable, passive media, located in jukeboxes has always seemed very attractive and has been successfully implemented and used for a number of years. Therefore, a "jump" on the first available similar technology from the DVD world was very tempting.

In early 1998, the only such system available was the DVD-R system from Pioneer, with the following positive and negative characteristics:

- (+) same form-factor as the well-known CD ($5\frac{1}{4}''$);
- (+) higher density (3.95GB/volume on one side – 3.7GB real);
- (−) the current 3.95GB density is a preliminary version: the nominal 4.7GB density is to be available in the first half of 1999;
- (−) one vendor only for the recording equipment (Pioneer);
- (+) several vendors for reading equipment (Matshita, Pioneer, Panasonic);
- (+) reading devices also support CD-Rs;
- (+) jukeboxes are being developed from various vendors and can accomodate a mixture of CD-R and DVD-R in the same unit;
- (+/−) the media are available (only) from two manufacturers (Pioneer and TDK)

Table 1. Currently available random access, high-density media and their respective costs.

medium type	description	USD/GB	avail?
0.6GB/vol $5\frac{1}{4}''$ CD-R	Jukebox allows for online, no operator-required access.	$85/GB	Yes
5.2GB/vol $5\frac{1}{4}''$ MO	Double-sided medium, need juke box with media flipping capability	$40/GB	Yes
5.2GB/vol $5\frac{1}{4}''$ DVD-RAM	Double-sided medium, need juke box with media flipping capability	$40/GB	Yes (?)
3.7GB/vol $5\frac{1}{4}''$ DVD-R	Single-sided medium, simple juke box	$46/GB	Yes (reliability?)
4.5GB/vol $5\frac{1}{4}''$ DVD-R	Single-sided medium, simple juke box	$38/GB	Q2 '99
9.0GB/vol $5\frac{1}{4}''$ DVD-R	Double-sided medium, need juke box with media flipping capability	$20/GB	Q4 '99?
disk farm	lots of large disks, backup on DLT	$64/GB	yes
RAID farm	lots of (level-5) RAID arrays, no backup	$77/GB	yes
tape robot	DLT juke box	$26/GB	yes

Magnetic disk systems are RAID (Redundant Array of Inexpensive Disks) systems: their cost is increased to account for a necessary backup on say tapes or for the setup of a given data redundancy level.

4. Current experiments at ESO

In early 1998, our major problem was the lack of Unix-based software to drive the DVD-R recorder. This problem was solved by Jörg Schilling, co-author of this article, who adapted his existing popular public-domain code `cdrecord` to deal with the DVD-R device as if it was simply a larger CD-R. As a matter of fact, the file system written on the new DVD-R medium is still an ISO9660-compliant format, as we did not have the time –or the real need– to compose a UDF file system, which would have been the necessary format to call our medium "DVD". Again, our goal was to record data for our own archive, not for redistribution.

With this system in place, we could start preparing media. Simultaneously, of course, it became clear that we could not start a large-scale mass storage on the new medium without having juke box system to make them available online. This was the sore point as only one company (Pioneer) had only a 100-slot model available in the third quarter of the year. With the promise of other vendors working on larger and more promising models we decided to wait a bit more before purchasing such a device.

In the mean time, we recorded a number of media and tried to fully check them. Roughly 30 media with various type of astronomical datafiles were recorded, mostly as GNU-compressed FITS files. This brought us the advantage that we could fully check not only the file system consistency, but also any possible read-write errors in the files themselves, using the "test" option of the decompression command. In unix, this gave a the following simple medium verification command:

```
find . -type f -print -exec gunzip -t {} \;
```

Using this combination, we could uncover occasional incomplete file reads and "CRC" errors in the files, giving us clues for the errors. With this method, applied to the same disks read with various reader devices, we concluded that all reader devices were not born equal: we could only get error-free reads with the `Panasonic LF-D101` and with the `Pioneer DVD S101` recorder. We still think the recording is only marginal and a better recording device might alleviate our problems.

Fortunately, the DVD jukebox we are currently in the process of acquiring will be equipped with the devices that have proven capable of reading the media we have already produced.

5. What if it fails?

The DVD(-R) technology is brand new and has not yet been planted in firm grounds. In particular, the multiplicity of its variants, the commercial battle amongst vendors proposing different incompatible recordable formats, are only favoring one thing right now: consumer confusion and defiance.

If it would turn out that DVD-R technology does not have a future, we will have to find an alternative route store our data. We want therefore to keep all options open and have decided to choose a juke box system that will accomodate simultaneously most kinds of CDs and DVDs. This will protect the investments we might have done in any technology so far while keeping the possibilities open for such improvements as double-sided media etc.

If all else fails, we will still be able to re-use this particular juke box system and retrofit it to support any kind of $5\frac{1}{4}''$ recording medium.

6. Conclusion

In looking early on at the DVD technology, we have tried to simultaneously satisfy our need to support the data volume growth that current and future astronomical instruments will impose on us, while continuing with our existing data storage paradigm: cheap, non-erasable, passive, $5\frac{1}{4}''$ media.

The early experiments have been plagued with reliability problems and have both slowed the adoption of the technology and cast some doubts about its future.

As we will have to select a method for storing the data we will receive anyway, we have decided to make the kind of careful technological investment that will enable us to change our minds in time, without having to re-consider the entire investment and set of decisions.

References

Crabtree, D., Durand, D., Gaudet, S., Hill N., & Pirenne, B. 1996, in ASP Conf. Ser., Vol. 101, Astronomical Data Analysis Software and Systems V, ed. G. H. Jacoby & J. Barnes (San Francisco: ASP), 505

Pirenne, B., Benvenuti, P., Albrecht, R., & Rasmussen, B. F. 1993, in SPIE Proc., Vol. 1945, Space Astronomical Telescopes and Instruments II, ed. P. Y. Bely & J. B. Breckinridge, (Bellingham: SPIE), 83

Wamsteker, W. 1991, in Databases & On-Line Data in Astronomy, ed. M. Albrecht & D. Egret (Dordrecht: Kluwer), 35

Astronomical Data Analysis Software and Systems VIII
ASP Conference Series, Vol. 172, 1999
D. M. Mehringer, R. L. Plante, and D. A. Roberts, eds.

Formatting Journal Tables in XML at the ADC

Edward Shaya[1], Jim Blackwell, Jim Gass, Vincent Kargatis, Gail Schneider, Kirk Borne

Raytheon ITSS, 4400 Forbes Boulevard, Lanham, MD, 20706

Cynthia Cheung, Richard White

Code 631, GSFC, Greenbelt, MD, 20771

Abstract. The eXtensible Markup Language (XML) is a powerful, internet-ready storage format for data. It provides a precise and clear data description language that is both machine- and human-readable. Documents marked up with XML can be easily parsed for efficient and focused searching. XML's expanded hyperlinking power can be used to enhance connectivity among archive documents. It is fast becoming the standard format for business and information transactions on the Internet and it is an ideal common metadata exchange format.

The Astronomical Data Center (ADC) plans to develop scripts to explore and display table data and metadata. The ADC is also developing XML authoring tools for scientists to convert their data and metadata into XML and to create conversion software for the ingest of journal tables into the ADC XML archive.

1. Introduction

The Astronomical Data Center (ADC)[2] at NASA's Goddard Space Flight Center has a 20-year history of preparing, documenting, validating, and distributing computer-readable versions of published astronomical data tables and catalogs. It maintains an extensive archive of such data and its services are heavily used by the community. There are two major problems facing the Astronomical Data Center (ADC).

- The ADC needs to prepare to manage the impending delivery of about 1,000 published scientific tables per year from the major astronomical journals. Documentation needs to be extracted from the articles to describe the tables and the tables must be converted to machine-readable format.
- At present, it is difficult for computer programs to parse the ADC's plain text documentation of catalogs and journal tables by information content

[1]U. of Maryland, Department of Physics, College Park, MD 20742

[2]`http://adc.gsfc.nasa.gov/`

(for example, the author or wavelength band). Precision searching and smooth navigation require intelligent information tagging.

The eXtensible Markup Language (XML) is an ideal solution to the major challenges above. XML is a powerful Internet-ready structured format for documentation (metadata) and data. It allows high-precision searching, enhanced web navigation, and seamless file exchange between users.

2. The problem of archiving and navigating thousands of tables

The ADC is very concerned with the need to improve the process of making published journal data tables computer-readable. However, the current journal submission and publication process is strongly oriented to the production of human-readable tables. The formats and organization of journal article submission do not fully support the automatic creation of data tables that can be easily read by data analysis software.

Although journal tables are available on the Web right now, their display formats are incompatible for loading the data into analysis or database environments. Some of these common problems are: 1) the inclusion of footnotes into the data 2) words, like 'glitch', embedded in columns of numbers 3) columns with multiple usage and 4) flags embedded with the numbers, like '10:'.

Probably the most difficult task is to extract sufficient information from the journal articles so that tables in the archive can be used correctly and intelligently. A single paragraph description of each table needs to be written (hopefully by the author) specifically for this use and complete information on the units, formats, blanks handling, etc. needs to be included in the metadata.

Given the expected growth rate, in just a few years, finding a particular datum from a table or even finding a particular table will require an unacceptable amount of effort. A way must be found to smoothly navigate between related tables, where the relationship can be defined by the user. A new way is needed to search very large archives, and it needs to be implemented in the broadest possible sense.

3. The wonders of XML

The essential concept of XML is to label, or 'mark up', the information content of a document such that computers can easily parse the document structure and retrieve the data and metadata it contains. A simple example of this is the following: `<author>` Ed Shaya `</author>`. Such tagging of content allows for focused and efficient automated scanning for information. This exceedingly simple concept, having been standardized in the metalanguage XML and accepted by the World Wide Web Consortium (W3C)[3] as a Recommendation in February 1998[4], is causing a revolution on the Internet, especially at inventory and data centers.

---

[3]`http://www.w3c.org/`

[4]`http://www.w3.org/TR/1998/REC-xml-19980210`

• **Defining the structure** – The rules for a set of XML documents are set down through the specification of a Document Type Definition (DTD). Since the structure can be fully prescribed, automated procedures can be used to validate the structure of the documents.
• **Hyperlinks** – XML has a companion schema called eXtensible Linking Language, XLink, (now a W3C working draft) for advanced hyperlinking between documents that allows for a much richer hypertext linkage. For example, XLink links can be type-cast by area of interest and users can choose to activate only specific link types. Another example is having a menu appear with multiple links to chose from when the mouse cursor goes over a certain set of words.
• **Navigation aids and vocabularies** – The Resource Description Framework (RDF) is an XML compatible schema that provides a robust and flexible architecture for supporting website descriptions and vocabulary definitions. Its primary purpose is to aid in locating data on the Web in a way that is understandable to the browsers. It could allow data centers to query each other about the nature, structure, and format of their holdings and services. The ADC needs tools with these features to properly manage the large volume of diverse tables and catalogs that will come with the new pipeline from the AAS Journals.

3.1. Some Details

XML is a version of SGML that will make the ingestion of the publisher's SGML marked up tables a straightforward process.
• **Defining the structure** – The rules governing how information is organized between the start and end tags are defined in a file called a Document Type Definition (DTD). The DTD rule for `<field>` might be:

```
<!Element field (name, description?, unit, format, blank?, chars?, min?, max?,
validMin?, validMax?)>
```

This declares that nested between the tags `<field>` and `</field>` there will necessarily be a name, unit and a format declaration. The other metadata attributes (description, unit, etc.) are optionally contained within field. This is signified by the question marks. The significance of the DTD is that it defines the order and organization of an acceptable archive document and XML provides checks that all documents conform to the rules.
• **Parse and display** – To translate an XML document into something that can be meaningfully displayed on an output device requires an intermediary. The, eXtensible Style Sheet (XSL)[5] is a full programming language for the display of XML documents. An XSL processor executes a series of rules which match tags or patterns and assigns actions to take place when a match is made. The assigned actions can be different depending on the particular device type or resolution that is used for display. One can choose various ways to display XML documents. Some examples are tree diagrams, directed acyclic graph (arrows), ASCII stripped of tags. One can direct the output to RTF (for Microsoft Word, WordPerfect, etc.) or direct it into database format (Excel, Access, Borland 123, etc.). In addition XSL can be used to translate XML into HTML which

---

[5] `http://www.w3.org/TR/WD-XSLReq`

can then be viewed with a web browser and be augmented with a Cascade Style Sheet (CSS).

• **Authoring tools** – We are examining automated methods for scientists to convert their data and metadata into ADC XML format for archive delivery. And, we are creating an SGML-to-XML application that will enable an automatic translation of tables from the major astronomical journal presses into the ADC XML archive. Since XML is derived from SGML, this should be straightforward, but certainly not trivial.

• **Image formats** – We need to investigate the pros and cons of translating standard image data to XML. It may be the case that image and observational raw data are best described by previously defined standards such as FITS, CDF, netCDF, and HDF. We may need to define XML-MIME types for each of these to make suitable use of these older standards as well.

• **Lists and arrays** – We must find how to best represent lists or tables of objects. Information like keywords, authors, and in fact the data tables themselves, can be tagged in several different ways: by individual item, list of items, or multi-dimensional array. We will find which way is optimal for each element, and will determine how to unambiguously delimit items within lists and arrays. Tables are a special issue because at this time there is not an XML standard table element. In our example DTD[6], we have accepted a simplified version of the `<array>` element from TecML (an application independent technical Markup Language written by the Open Molecular Foundation[7] to describe 2-D data.

4. Search tools

We can make use of off-the-shelf search tools as back ends to our website browsing tools. Search tools could be modified to be aware of our tag names and standard field names, and searches for these could be optimized by keeping up-to-date indices of occurrences. In fact, with all of the information well organized and clearly tagged, one can make great use of simple C language programs (like sgrep) that search for tags nested within tags and output their contents. It will become possible for users to bypass the ADC web tools and do searches based on their own software or intelligent agents. This is a major advantage because it allows users to analyze the data in new and innovative ways.

5. Conclusion

XML will probably be used in many areas within astronomy beyond catalogs and tables. It will take a united effort to make the most use of XML's power to organize and interoperate. We will be working with the AAS, the U. of Chicago Press, the AstroBrowse community, the Astrophysics Data Centers Coordinating Council (ADCCC), NED (NASA/IPAC Extragalactic Database), and our sister organization CDS to create a community standard for XML documents.

---

[6] `http://messier.gsfc.nasa.gov/xml/acatalog_dtd.html`

[7] `http://www.ch.ic.ac.uk/omf/`

Astronomical Data Analysis Software and Systems VIII
ASP Conference Series, Vol. 172, 1999
D. M. Mehringer, R. L. Plante, and D. A. Roberts, eds.

An Application of XML and XLink Using a Graph-Partitioning Method and a Density Map for Information Retrieval and Knowledge Discovery

Damien Guillaume, Fionn Murtagh

Université Louis-Pasteur, University of Ulster

Abstract. We have defined an XML language for astronomy, called AML (Astronomical Markup Language), able to represent meta-information for astronomical objects, tables, articles and authors. The various AML documents created have links between them, and an innovative tool can cluster the documents with a graph-partitioning algorithm using the links. The result is displayed on a density map similar to Kohonen Self-Organising Maps. AML and its advantages will be briefly described, as well as the clustering program, which is one of the many possible applications of AML.

1. Introduction

XML (Extensible Markup Language; Cover 1998; Bray, Paoli, & Sperberg-McQueen 1998) is a new markup language similar to SGML (Standard Generalised Markup Language), but simpler and having other advantages over SGML. It is becoming a new standard for publishing information on Internet and will in many cases replace HTML. XLink (a part of XLL, Extensible Linking Language; Maler & DeRose 1998) is the XML language used in XML documents to create links between the documents, and is also better than the simple link used in HTML.

We have defined an XML language for astronomy, called AML (Astronomical Markup Language), able to represent meta-information for astronomical objects, tables, articles, authors and images. The different AML documents created have links between them, and we have created an innovative tool to cluster the documents with the use of the links, and display the result with a density map similar to Kohonen Self-Organising Maps.

After briefly introducing XLink, we present here the specific AML features, the issue of the display of clustered documents, the problem of partitioning the graph of the links, and we finally give an example of the maps we can now create.

2. XLink

XLink, a standard mechanism to link XML documents, is still under development. However, its objectives and features are clearly defined, and a W3C working draft proposes a syntax for it. The main feature of XLink is that it allows the definition of links between more than two documents. The links are also labeled so they have semantics, and minimum information can be specified about the desired browser behaviour when a link is used. While in HTML different tags are used as links (simple **<A href="...">** links, or links to images, for instance), XLink will provide a unifying linking method for XML documents.

3. AML Features

AML is defined by a single file describing the XML tags and their hierarchy, called a "DTD". But, as AML describes different kinds of documents, it is composed of different parts: the first part is a meta-data block, which is used for all AML documents. This gives information about the AML file itself: the AML version, the date of creation, the author and so on. The other parts describe the different "AML objects": astronomical object, table, set of tables, article, person and image.

We have developed a Java applet to be used as an AML browser, able to display AML documents with a nice interactive user interface, and allowing the use of the links between the documents as easily as hypertext links.

There are currently two types of links between AML documents: links by URL, and links by identifier. A link by URL can link together two AML documents such as an article and an author, and are usually bidirectional and included in the two documents. Links by identifier use a resource name instead of a URL, and specify the AML object of the linked document. A list of servers serving AML documents for the given AML object can then be used to let the user choose a server. A new URL is created and from that the link can be used as a link by URL. This is like using URNs (Uniform Resource Names), but we are using a system called GLU[1] (Fernique, Ochsenbein, & Wenger 1998) to resolve the URNs, and this system is currently only used in astronomy and has some specific features.

The AML DTD, together with some examples, interfaces to get AML documents from different servers, and the AML browser, can be found at **http://strule.cs.qub.ac.uk/~damieng/these/**.

4. Discovering Knowledge by Using the Links: a Problem of Visualisation

Different existing servers place astronomical documents at the disposal of Internet users, e.g., ADS, NED or Simbad, and one can use the links between the astronomical objects and the articles to surf from one server to another. However, the documents have more and more links, and when a document has a hundred links, it becomes impossible for the user, unless (s)he is very patient, to look at all the links and to summarise the information.

The most common displays for displaying linked documents, trees and graph representations, are not ideal here: a tree would be too big and wouldn't show well the links that are not in the tree structure, and usual graph representations become unreadable with more than 30 documents.

---

[1] http://simbad.u-strasbg.fr/glu/glu.html

5. Self-Organising Maps

Self-Organising Maps (SOM) displayed as density maps are more and more used as a convenient display solution for information retrieval. In the astronomy field in particular, Kohonen SOM have been used to display summarised information on the content of astronomical bibliographic databases and allow the user to search for documents by subject and get similar documents at the same time (Poinçot, Lesteven, & Murtaugh 1998).

The density map shows clusters of similar documents, with a colour corresponding to the density of documents in each square of the map. For each square, a title (difficult to compute automatically) indicates the subject of the documents in the space of the map.

In the case of the Kohonen SOM built with astronomical articles, the clustering is done with the keywords associated with the articles. In the case we are working on, the relationship between the documents is expressed with links between them, which is very different, but the same user interface could be used to display the clusters of documents.

6. An Efficient Graph-Partitioning Algorithm

To cluster the documents using their links, the problem is not an agglomerative question, because we already have a tree (plus some links between the nodes that cannot be displayed on a classic tree). It is instead a graph-partitioning problem: we have a graph of connected nodes (the documents), with weighted edges (links) between them, and we want to partition this graph so that it can be displayed on a density map. This problem of clustering documents with their links is more interesting with XML and XLink than with HTML, because we have weighted links, the weights corresponding to the semantics of the link: the semantics are specified with XLink but not HTML.

Many proposals have been made related to this NP-hard (Garey & Johnson 1979) problem, such as simulated annealing algorithms (solving the local minima problem by accepting non-improving solutions with a probability based on the *temperature*; Sheild 1987, Johnson et al. 1989), and noising algorithms (solving the local minima problem by adding noise to the data, decreasing with the number of iterations; Sharon & Hudry 1993). A recent proposal for a modified noising algorithm by Sudhakar & Murthy (1997) seems to be one of the best solutions known to us. We used this algorithm, which is rather fast, and gives good results. The two important problems this algorithm is solving, are to obtain balanced clusters, and to minimise the number of cut links between the clusters. A constant (α) is used to provide a trade-off between these two constraints, so that the cost function can be expressed as: $T = I + \alpha * C$, I being the total imbalance between the clusters, and C the weighted number of cut links.

7. Examples

Any AML objects could be used for clustering (currently: astronomical objects, articles, people, images, tables and set of tables), but we need to have access to a database giving AML documents as a result of a query. Currently, the

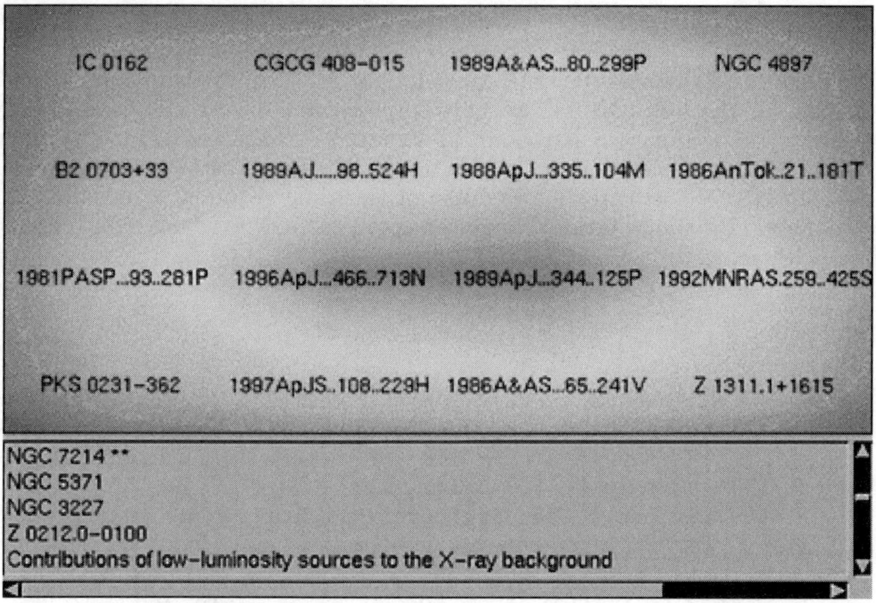

Figure 1. Map for MRK334.

only servers supplying AML documents are: Simbad, NED and ADS. We can also use the database of astronomers created by Chris Benn (La Palma) and Ralph Martin (Royal Greenwich Observatory, Cambridge) to get people-related information. We use 3 different AML objects: astronomical object, article and person, and we have a 3-dimensional space of relations between these objects.

Here is a map example, for MRK334 (an astronomical object). The map is composed of 16 sets of documents, and each set is labeled with the identifier of the document having the most links with other documents in the same set (this may not be the most representative). In the case of an article, it's a 19-characters code called a "bibcode", which is a common identifier for the astronomical bibliography. If the user clicks on a space of the map, the corresponding list of documents appears below the map, with the full title of the article in case of an article. When the user clicks on a document in the list, the content is shown in the AML browser.

Another interesting feature of this program is the selection of two documents, for the whole set of the documents retrieved, being most similar to the first document. The comparison is done with the content of the files, using DTD-specific functions thanks to the XML formatting. For instance, the most similar document for MRK334 is the one for NGC7214, because they are both Seyfert galaxies and many articles have been written about the two objects together. This is a typical "knowledge discovery" application, and it goes much further than what was previously possible with HTML: instead of comparing single words using their frequencies in the texts, it becomes possible to do complex numerical comparisons.

8. Conclusion

Using the ideal features of XML for information retrieval, and its associated language for the links, XLink, we have implemented a new tool for knowledge discovery and used it with astronomical documents. This tool seems very useful, and the only drawback is that, as it is using distributed resources dynamically, it cannot be used in real-time because of the time required to download the documents. Its computational requirements are, however, not far from real-time, with access time dominating processing time.

References

Bray, T., Paoli, J., & Sperberg-McQueen, C. M. 1998, "Extensible Markup Language (XML) 1.0: W3C Recommendation 10-February-1998", `http://www.w3.org/TR/REC-xml`

Cover, R. 1998, "Extensible Markup Language (XML)", `http://www.oasis-open.org/cover/xml.html`

Fernique, P., Ochsenbein, F., Wenger, M., 1998, in ASP Conf. Ser., Vol. 145, Astronomical Data Analysis Software and Systems VII, ed. R. Albrecht, R. N. Hook, & H. A. Bushouse (San Francisco: ASP), 466

Garey, M. R. & Johnson, D. S. 1979, Computers and Intractability, a Guide to the Theory of NP-Completeness, (New York: Freeman)

Johnson, D. S., Aragon, C. R., McGeoch, L. A., & Sheron, C. 1989, Oper. Res., 37(6), 865

Maler, E. & DeRose, S. 1998, "XML Linking Language (XLink): W3C Working Draft 3-March-1998", `http://www.w3.org/TR/WD-xlink`

Poinçot, P., Lesteven, S., & Murtagh, F. 1998, A&AS, 130, 183

Sharon, I. & Hudry, O. 1993, Oper. Res. Lett., 14, 133

Sudhakar, V. & Siva Ram Murthy, C. 1997, Integration, the VLSI Journal, 22, 101

Astronomical Data Analysis Software and Systems VIII
ASP Conference Series, Vol. 172, 1999
D. M. Mehringer, R. L. Plante, and D. A. Roberts, eds.

Extracting Information from Text Data Bases

Rudolf Albrecht[1]
Space Telescope European Coordinating Facility, European Southern Observatory, Karl Schwarzschild Str. 2, D-85748 Garching, Germany, Email: ralbrech@eso.org

Dieter Merkl
Institut für Softwaretechnik, Technische Universität Wien, Resselgasse 3/188, A-1040 Vienna, Austria, Email: dieter@ifs.tuwien.ac.at

Abstract. The problem of computer assisted literature search is often misstated as the attempt to find, in a repository of electronic literature, a paper which, according to title and keywords, contains the desired information. This assumes that the original author was correct and complete in assigning keywords, and that the search criteria are free of preconceived expectations on the context in which the information might be found. Crossing the boundaries of disciplinary subfields is very difficult in this manner.

To avoid these problems it is necessary to categorize and classify the documents in the text data base using the full text. This will identify papers of interest, and of unexpected cross-relevance. We show the results of experiments using unsupervised classification based on neural networks.

1. Electronic Publishing and Literature Data Bases

In astronomy, several years of major journals are now accessible through the Internet. Within a very short time we will have the current body of astronomical knowledge available for computer processing.

Using appropriate tools and networks we can consider the electronically available astronomical literature to be one huge text data base, consisting typically of technical/scientific articles, written in scientific English and using well defined terminology. In addition to improved access and timely availability electronic publications have the advantage of being searchable. Organizations like the NASA Astrophysics Data System (ADS; `http://adswww.harvard.edu/`) specialize in such services, freeing the user from having to read an increasingly enormous amount of material in order to find the desired information.

[1] Astrophysics Division, Space Science Department, European Space Agency

2. Text Data Mining Using Neural Networks

Artificial neural networks are well suited for tasks that are characterized by noise, poorly understood intrinsic structure, and changing characteristics, each of which are present when dealing with text. Learning techniques are favorable in such an environment compared to algorithmic and knowledge-based approaches.

Neural networks consist of one or more layers of processing elements (neurons). The "program" is stored in a distributed fashion within the "weights" of the connections between processing units. Input patterns presented to the neural network will propagate selectively through it, depending on the different weights of the different connections. Neural networks are inherently parallel computational models. It is possible to "train" the network by changing the weight of the connections such that certain inputs generate certain outputs.

The self-organizing map (Kohonen 1995) is an unsupervised neural network for ordering high-dimensionality statistical data in such a way that similar input items will be grouped close to each other. The utilization of self-organizing maps for text data mining already has found appreciation in information retrieval research (cf. Kohonen et al. 1996; Lagus et al. 1996; Merkl 1997a, 1997b, 1998).

Each of the units i of the self-organizing map is assigned an $n-$dimensional weight vector m_i, $m_i \in \mathcal{R}^n$. The weight vectors have the same dimension as the input patterns (the document representation in our application). Each training iteration t starts with the random selection of one input pattern $x(t)$, which is presented to the self-organizing map and each unit determines its activation. The unit with the lowest activation is referred to as the *winner*, c, of the training iteration, i.e., $m_c(t) = min_i \|x(t) - m_i(t)\|$. Finally, the weight vector of the *winner* as well as the weight vectors of selected units in the vicinity of the *winner* are adapted. This adaptation is implemented as a gradual reduction of the difference between input pattern and weight vector, i.e., $m_i(t+1) = m_i(t) \cdot \alpha(t) \cdot h_{ci}(t) \cdot [x(t) - m_i(t)]$. Geometrically speaking, the weight vectors of the adapted units are moved a bit towards the input pattern. The amount of weight vector movement is guided by a so-called learning rate, α, decreasing in time. The number of units that are affected by adaptation is determined by a so-called neighborhood function, h_{ci}. This number of units also decreases in time.

As the Euclidean distance between those vectors decreases and the weight vectors become more similar to the input pattern, the respective unit is more likely to win at future presentations of this input pattern. The consequence of adapting not only the *winner* alone but also a number of units in the neighborhood of the *winner* leads to a spatial clustering of similar input patters in neighboring parts of the self-organizing map. Thus, similarities between input patterns that are present in the n-dimensional input space are mirrored within the two-dimensional output space of the self-organizing map.

Note that there is no attempt to "understand" the contents of the documents. The goal, at least initially, is "document routing" or "information filtering": identifying those documents which are, or should be, of interest to the user given a particular information profile.

3. Classification of Documents

The general problem of classification is to place N data points into M bins, where $1 < M << N$. The width of the bins is not necessarily M/N and the difference between bins should be meaningful.
There are two different basic classification strategies:

1. Try to reproduce a classification which is already being used, has been shown to be useful, and can be improved by adding new data.
2. Try to discover classes which are intrinsic to the data.

In the case of text mining the desired output pattern is generally not known. The conventional classification system as used in libraries is not applicable, because it is too coarse, and because even documents classified in dissimilar categories (e.g., cosmology and asteroids) can be of relevance to each other because they contain common topics (e.g., spectral analysis)

A possible case with a predefined output pattern is the "user profile": based on a user's previous document retrieval behavior the network can be trained to identify documents which fit the pattern of the user.

4. Results and Plans

ApJ Letters articles from November and December 1997 were merged into an input data set of 128 documents. Full-text indexing of all documents was performed. Words that appear in less than 13 documents or more than 116 documents were excluded. 1451 words (terms) remained. The terms were weighed according to $tf * idf$ (term frequency times inverse document frequency) (Salton & Buckley 1988). The scheme ensures that terms that occur frequently within a document but rarely within the whole collection are assigned high weights, providing best discrimination between documents. 10^4 iterations were performed to train the network.

The output consists of an 8 by 8 element map. Documents contained in a bin are considered similar, similarity decreases with distance between bins. The largest bin contains 10 papers, seven of them on spectroscopy of galaxies, which also was the common denominator with the three outliers (solar, brown dwarf, and one paper on a cool degenerate star); another common denominator is "lensing", which occurred as gravitational lensing in seven papers, and in one more as part of the name of an author, an obvious shortcoming.

Other prominently populated bins contain papers on brown dwarfs (4 out of 4) and solar mass ejections (4 out of 4). However, the potentially most interesting bins are the ones which contain papers with very dissimilar topics, for instance a bin populated by six papers with all different topics (e.g., the Sun, Seyffert galaxies, SN187a). Closer inspection shows, however, that all papers deal with the physics of hot gas in a magnetic environment. The "Subject Heading" keywords of the papers would not have helped: the only common keyword is "ISM" (Interstellar Medium), and it occurs only in three papers. Our system is essentially telling the Seyfert galaxy researcher to read, among others, a paper on solar mass ejections. This is something which Seyfert specialists do not ordinarily do, but which, as it turns out, might be of enormous cross-specialty relevance.

At this point the freedom of the system to determine classification criteria is still too unrestricted: there are too many non-domain specific terms. For example, we find bins with dissimilar papers describing similar techniques for observing or data analysis. While this is still acceptable we also find bins with purely linguistic classification criteria, focusing on terms like possible, probable, theoretical, and systematic. Although these are valid terms they should not be taken into account.

The obvious next step is to perform lexical pre-processing on the input data set. All terms will be compared with a set of valid terms, and only these will be retained. The valid terms can be derived from a training set of representative input data (i.e., a private dictionary), or the Astronomy Thesaurus (`http://msowww.anu.edu.au/library/thesaurus/`) can be used. To further increase the efficiency of the classification all synonyms should be eliminated (like magnitude and brightness). Super-terms can be established and all sub-terms can be replaced by them. The effect will be to turn the input papers into sets of keywords which characterize (but not describe) the papers.

It is obvious that these first experiments are only a small step towards the ideal result. It is equally obvious, however, that strategies such as the one we describe represent the only hope of coping with the enormous amount of published literature and the ever increasing fractionality of the field.

References

Kohonen, T. 1995, Self-organizing Maps, (Berlin: Springer-Verlag)

Kohonen, T., Kaski, S., Lagus, K., & Honkela, T. 1996, in Proc. of the Intl. Conf. on Artificial Neural Networks, (Bochum), 269

Lagus, K., Honkela, T., Kaski, S., & Kohonen, T. 1996, in Proc. of the Intl. Conf. on Knowledge Discovery and Data Mining, (Portland), 238

Merkl, D. 1997a, in Proc. of the European Symp. on Principles of Data Mining and Knowledge Discovery, (Trondheim), 101

——, 1997b, in Proc. of the Intl. ACM SIGIR Conference on R&D in Information Retrieval, (Philadelphia), 186

——, 1998, in A Handbook of Natural Language Processing - Techniques and Applications for the Processing of Language as Text, ed. R. Dale, H. Moisl, & H. Somers (New York: Marcel Dekker), in press

Salton, G. & Buckley, C. 1988, Information Processing & Management 24, 513

Astronomical Data Analysis Software and Systems VIII
ASP Conference Series, Vol. 172, 1999
D. M. Mehringer, R. L. Plante, and D. A. Roberts, eds.

Co-occurrence Evidence for Subject Vocabulary Reconciliation in ADS Databases

Jonghoon Lee, David S. Dubin

Graduate School of Library and Information Science
University of Illinois at Urbana-Champaign

Michael J. Kurtz

Harvard-Smithsonian Center for Astrophysics

Abstract. The NASA Astrophysics Data System gives astronomers all over the world access to over a million abstracts in the areas of astronomy and astrophysics, instrumentation, physics and geophysics. A mixture of indexing vocabularies has limited ADS searchers' ability to conduct precise subject searches, but prospects for a single consistent vocabulary of descriptors are promising. We report results of an ongoing project to reconcile the heterogeneous indexing (keywords, index terms, and subject headings) applied to the existing ADS records. Evidence for term mappings are sought in the consistent assignment of different descriptors to the same documents. We describe the apprehension of this evidence through the use a spreading activation network. Design and deployment of an experimental interface to assess these mappings is now under way.

1. Introduction

This paper reports results in an ongoing project to reconcile heterogeneous indexing vocabularies in the NASA Astrophysics Data System (ADS; Eichhorn et al. 1998). Over 15,000 astronomers use ADS each month to retrieve abstracts and full text articles in the areas of astronomy and astrophysics, instrumentation, physics and geophysics. The ADS databases support a number of different search methods, including access by title, author, and astronomical object name. Users can search on a combined index of the abstract and keyword fields, but cannot (through the main web interface) limit their searches to a single controlled vocabulary of subject descriptors. The merging of the abstract and keyword indexes was a deliberate decision by the ADS administrators, because documents in ADS are indexed with several different descriptor vocabularies. Some documents have been indexed by professional indexers using terms from the NASA Thesaurus. Others have had keywords suggested by authors and approved by journal editors.

Controlled indexing vocabularies overcome the variation in authors' natural language by providing standardized labels for concepts. They also permit searchers to limit their queries to the most important ideas or topics in a document. The mixture of different descriptor vocabularies in ADS defeats the standardization goal, and the merging of the abstract and keyword indexes limits the search precision function of the subject indexing. Descriptors representing identical concepts (or closest counterpart) can stand in several different relationships to each other. For example, counterpart terms may be synonyms (e.g., **Andromeda Galaxy** and **Galaxies: Individual Messier Number: M31**). A descriptor in one vocabulary may be a pre-coordinated combination of terms in the other vocabulary (e.g., **Microwave Background Radiation** vs. **Background Radiation** and **Microwaves**).

An ongoing project at the University of Illinois investigates sources of evidence to support the automatic and/or computer-assisted reconciliation of the heterogeneous indexing in ADS. Two sources of evidence have been investigated: lexical resemblance between descriptors and consistent assignment of descriptors from different vocabularies to the same documents. Use of lexical resemblance evidence is discussed in an earlier paper (Dubin 1998).

2. A Spreading Activation Model for Co-Assignment Evidence

Our source of evidence is a subset of the ADS database, in which each document has been indexed with two or more different vocabularies. The consistent assignment of two or more terms from different vocabularies to the same documents suggests some kind of semantic relationship or connection among the terms. Not every connection represents a context-free pairing of synonymous terms: sometimes a pair of coordinate index terms assigned from one vocabulary imply the assignment of a pre-coordinated descriptor from another vocabulary. We therefore developed a model that accommodates such context-sensitivity.

We developed a spreading activation model, similar to those employed for modeling human associative memory (Collins & Loftus 1975). In our model, the network is composed of three layers: an input term layer, document layer, and output term layer. The activation of terms in the input layer is spread through the network to the connected documents and from there to the output terms. As a result, this model produces a list of terms with their activation levels representing the degree of relatedness to the input term(s) (Lee 1998).

Calculation of the activation level employs a simple weighting rule. The weight assigned to the link between a document and a term is determined by the number of connections and the direction of activation. We use the conservation of activation principle which guarantees that the sum of input activation equals to that of output activation. The activation received by a document from input terms is computed as follows:

$$D_j = \sum_{i=1}^{l} S_i w_{ij} \quad T_k = \sum_{j=1}^{m} D_j w_{jk}$$

In these formulas, S_i is the activation of an input term i; D_j is the activation of a document j; T_k is the activation of an output term k; w_{ij} is a weight between an input term i and a document j; and w_{jk} is a weight between a document j and an output term k.

3. Evaluation

We applied the spreading activation model to investigate connections between two different subject vocabularies employed in ADS: Astrophysical Journal keywords (ApJ) and terms assigned to documents from the NASA Thesaurus (STI)[1]. These two vocabularies, one used by professional indexers and the other by authors, differ in degree of specificity and level of pre-coordination (the ApJ terms include pre-coordinated descriptors).

Experimental materials included two sets of documents, one indexed by 10,200 ApJ terms, and the other by 3,335 STI terms. The ApJ set contained 39,366 documents, and the STI set 22,139 documents. Among these, 14,956 documents were identified as co-indexed. The merged network representation included 4,120 STI term nodes in one term layer, 14,956 document nodes in the middle layer, and 2,305 ApJ term nodes in the other term layer. Two spreading activation models were constructed according to the direction of activation (from source term to target term): ApJ → STI and STI → ApJ.

4. Results

Each term from the source vocabulary was used as an input to the network. The output of the spreading activation process was the list of activated terms from the target vocabulary along with their activation level. For example, when the STI term (**Andromeda Galaxy**) is given as an input node, 171 ApJ terms were identified as activated terms in the output layer. One of the problems here is that the output usually includes a large number of terms most of which seem to be barely related (very low activation level like 0.01). Therefore, a cutoff criterion is needed to differentiate more related terms from less related ones.

We adapted the so-called "Mexican-Hat" function for our cutoff criterion. The Mexican-Hat is the second derivative of the Gaussian curve, and has been successfully used for vision processing, especially edge detection (Charniak & McDermott 1987). The set of output activation levels was convolved with the second derivative of the Gaussian in order to find the point where the slope of the activation value distribution drops most dramatically. Only terms above this cutoff point were selected as mapping terms. The average number of terms above the cutoff was 1.9 out of 135 activated terms for STI → ApJ, and 1.6 out of 354 activated terms for ApJ → STI.

Term mappings identified by the spreading activation model include a variety of term-to-term relationships. They ranges from simple spelling variants to complicated semantic factoring. We were encouraged to observe that the model identified many of the same connections uncovered in our lexical resemblance study, although the current model employs no lexical evidence. Examples of relationships apprehended with the network included:

1. Exact match (**Astrometry → Astrometry**)
2. Spelling variants (**Galactic Clusters → Galaxies: Clustering**)
3. Ordering (**Clusters: Globular → Globular Clusters**)

---

[1]Indexers at NASA's Scientific and Technical Information (STI) Group assigned the descriptors.

4. Pre-coordination (**Cosmic Rays: Abundances → Abundance, Cosmic Rays**)
5. Term omission (**Catalogs → Astronomical Catalogs**)
6. Class relation (**Galaxies: Individual Messier Number M33/M81 → Spiral Galaxies**)

5. Discussions and Future Studies

These preliminary results suggest our model can successfully reveal the complex nature of term relationships between two astronomical subject vocabularies. For example, we note the asymmetric characteristics of term mapping. The difference between STI → ApJ mapping and ApJ → STI mapping is observed for many pair of terms. This bi-directional relationship is well revealed since the spreading activation model is sensitive to the direction of term mapping. Current experiments include a generalization of the term-term mapping approach described in this paper: a many-to-many mapping (in which all the terms for a document are activated simultaneously) affords apprehension of the context-sensitive relationships described earlier.

We are evaluating the usefulness of the spreading activation output as evidence for vocabulary merging. We plan a user study in which expert astronomers evaluate the connections suggested by the model. We will follow up with a user evaluation test, assessing the impact of mergings on search precision. We're also evaluating the scalability of the model with larger databases.

We are developing a visualization tool in order to help users understand the complex nature of term relationship. It will show term mapping structure among participating vocabularies using a graphical interface. A directed graph method will be used to represent the directional information of term relationship.

References

Charniak, E. & McDermott, D. V. 1987 Introduction to Artificial Intelligence, (Reading: Addison-Wesley), 101

Collins, A. M. & Loftus, E. F. 1975, Psych. Rev., 82, 407

Dubin, D. S. 1998, in ASP Conf. Ser., Vol. 153, Library and Information Services in Astronomy III, ed. U. Grothkopf, H. Andernach, S. Stevens-Rayburn, & M. Gomez, (San Francisco: ASP), 77

Eichhorn, G., Accomazzi, A., Grant, C. S., Kurtz, M. J., & Murray, S. S. 1998, in ASP Conf. Ser., Vol. 145, Astronomical Data Analysis Software and Systems VII, ed. R. Albrecht, R. N. Hook, & H. A. Bushouse (San Francisco: ASP), 378

Lee, J. 1998, A Theory of Spreading Activation for Vocabulary Merging, (Grad. Sch. of Library and Information Sci., Univ. of Illinois, unpublished report), (Champaign: Univ. of Illinois)

Astronomical Data Analysis Software and Systems VIII
ASP Conference Series, Vol. 172, 1999
D. M. Mehringer, R. L. Plante, and D. A. Roberts, eds.

The ADS Bibliographic Reference Resolver

Alberto Accomazzi, Guenther Eichhorn, Michael J. Kurtz, Carolyn S. Grant, Stephen S. Murray

Smithsonian Astrophysical Observatory, Cambridge, MA 02138, USA

Abstract. An increasingly larger number of publishers and institutions are using the NASA Astrophysics Data System (ADS) to verify the existence and availability of references published in the astronomical literature. In this paper we discuss the tools and utilities that the ADS is developing to provide the capability to automatically parse, identify, and verify the existence and correctness of citations appearing in astronomical publications. Details of the implementation are presented and some preliminary results from the use of the resolver are shown.

1. Purpose of the Reference Resolver

The capability to interactively access publications cited in scientific papers is of undisputed importance to researchers and librarians, greatly increasing their efficiency and quality of work. Thus it is increasingly important for publishers and data providers to be able to automatically verify the syntactic and semantic correctness of the references listed in their online documents as a way to improve the overall quality and accuracy of the information they are providing. Verifying the existence of a reference through a well-established online database such as the ADS also allows them to publish documents containing hyper-links pointing to stable, unique URIs. Additionally, when verifying references via a "smart" resolver (such as the one described in this paper), they contribute new entries to the citation (or "forward reference") database maintained by the provider of the resolver, allowing the creation of citation links pointing back to their site.

2. Implementation

Generally stated, the problem of resolving a reference to a paper published in the literature can be broken down into the following steps: reference parsing, i.e., identifying the different tokens listed in the reference entry; reference identification, i.e., using the parsed tokens to generate a unique identifier for the reference; reference verification, i.e., querying a database to verify the existence of the reference. The degree of success with which each of these steps is performed is dependent upon the outcome of the preceding steps.

In order to provide a flexible and portable way to represent the dataset being processed, the input data is transformed from a sequence of strings into data structures which hold all the information accumulated during the resolving process. Each of the resolving steps listed above simply takes the data structure generated by the previous step as input, modifies it (typically by adding records to it), and then feeds it to the following processing step. To increase the overall resolving success rate, any ancillary information (or "hints") collected by each of the resolving steps is also passed on to the subsequent task. A priori knowledge about the format or origin of the dataset being resolved can be specified via a configuration file, which sets default values for some of the parameters controlling the processing performed by the reference resolver.

Once all the resolving steps have been carried out on a reference, the generated data structure is serialized as an instance of an Extensible Mark-up Language (XML) document. XML was chosen as the output format because of its flexibility in representing complex data structures and for its increasingly significant role as a standard data exchange format.

2.1. Reference Parsing

The first task performed by the resolver is parsing of the input reference in order to identify the fields that compose the bibliographic reference string. The reference listing conventions adopted by all the major astronomical journals since 1991 instruct authors to format entries in slightly different ways depending on the nature of the publication (see example, §2.2.). Other journals or conference proceedings series may adopt different conventions and syntaxes for listing references, often making it impossible to correctly break down the items composing a reference without a priori knowledge. In addition, to accommodate the needs of editors and publishers who intend to integrate the reference resolver in their electronic publication process, we had to take into account the fact that the input reference strings to be parsed could possibly be formatted and encoded in a few different ways (the most typical ones being LaTeX or SGML), and could make use of publisher-specific macros or entity definitions. Therefore, our implementation of the parser is driven by parameters specified in configuration files which are style- and format-specific.

2.2. Reference Identification

Identifying a reference is the activity of mapping the parsed fields extracted from the reference entry into a bibliographic code (Schmitz et al. 1995), the standard format used by the ADS to uniquely identify bibliographic entities. The set of fields resulting from the successful parsing of a typical reference may include: Authors, Publication Year, Journal or Conference Series name, Publication Volume, Page Number, Conference or Book Title, Editor, and Publisher. In order to be able to successfully resolve references containing journal abbreviations and macro names, the resolver uses external tables to map such names into the journal abbreviations used in bibliographic codes. In order to minimize errors in interpreting these strings, the resolver makes use of information supplied by the user via the input configuration file and any other hints collected from the parsing stage. For instance, by noting that the input string is in LaTeX format, that the journal name is specified as a macro, and that the macro style has been declared to be "AAS" in the configuration file, the resolver will match the journal strings against the ones listed in the AAS LaTeX macro package.

Example Examples of output records from the ADS reference resolver. The input reference strings appear in each record within the **<input>** tags.

```
<record>
<authors>
<item>Drinkwater, M. J.</item>
<item>Wood, P. R.</item>
</authors>
<bibcode>1985mlrg.proc..257D</bibcode>
<hints isconf="1" status="VERIFIED" istex="1" parsed="2" />
<input>Drinkwater, M.J., \&amp; Wood, P.R. 1985, in Mass Loss from Red
Giants, ed. M. Morris \&amp; B. Zuckerman, (Dordrecht: Reidel), 257</input>
<journal>in Mass Loss from Red Giants, ed. M. Morris &amp; B. Zuckerman,
(Dordrecht: Reidel)</journal>
<matches>
<item bibcode="1985mlrg.proc..257D" score="1" />
</matches>
<page>257</page>
<year>1985</year>
</record>

<record>
<authors>
<item>Pijpers, F. P.</item>
<item>Hearn, A. G.</item>
</authors>
<bibcode>1989A&amp;A...209..298P</bibcode>
<hints errmsg="error in input reference, does not exist"
status="IDENTIFIED" errcode="321" istex="1" ismacro="1" parsed="3" />
<input>Pijpers, F.P., \&amp; Hearn, A.G. 1989, \aap, 209, 298</input>
<journal>\aap</journal>
<matches>
<item bibcode="1989A&amp;A...209..198P" score="0.9" />
</matches>
<page>298</page>
<volume>209</volume>
<year>1989</year>
</record>

<record>
<authors>
<item>Allard, F.</item>
</authors>
<bibcode>1997ARA&amp;A..35..137A</bibcode>
<hints errmsg="partial input reference, exact match not possible"
inpress="1" status="VERIFIED" errcode="311" istex="1" authtrim="1"
ismacro="1" parsed="2" />
<input>Allard, F., et al.  1997, \araa, in press</input>
<journal>\araa</journal>
<matches>
<item bibcode="1997ARA&amp;A..35..137A" score="0.55" />
</matches>
<year>1997</year>
</record>
```

2.3. Reference Verification

Once a bibliographic code has been generated by the reference identification step, the following verification steps are attempted:

- Verify that the ADS contains an entry that matches the input bibliographic code and list of authors. If an exact match is found, return it to the user.
- If the input reference is incomplete because not all the necessary information required to compute an exact bibliographic code has been specified (e.g., a paper which is in press), attempt partial matches constraining the results to match the partial information supplied, and return a list of plausible matches if any were found.
- If the input reference appears in a publication source and date range for which the ADS has complete coverage, then something in the input reference is incorrect. Under these circumstances the resolver attempts several

fuzzy matches between the input reference and the ADS databases, returning any records with high correlation scores.

- If none of the previous steps has been successful in identifying a reason for the failed match, the resolver can be asked to attempt to find the best match of the input reference against the ADS databases, by iteratively relaxing the search constraints, and then return any possible matches.

3. Conclusions

We have presented the design and implementation of a reference resolving service which is currently being built by the ADS group. The resolver, currently to be considered a beta release of what is to become the standard server, already allows sophisticated matching of a variety of references against the contents of the ADS database, offering valuable feedback whenever a particular reference cannot be identified and matched. Recent use of the service has reported a success rate of 80% when matching references listed in an ASP Conference Proceedings volume.

Planned enhancements to the capabilities currently implemented include: support for a "strict" mode which would flag any error in the formatting of the input reference, even if its resolution can be completed, and a "loose" mode which would be more lenient in the verification process; providing a stable client library which can be used to seamlessly integrate reference resolution into different electronic publishing environments; and support for additional output formats, in particular the Astronomical Mark-up Language or AML (Guillaume & Murtagh 1999) and all the custom formats currently generated by the ADS search interface.

For more information on the status of the ADS reference resolver, please visit the URL `http://adswww.harvard.edu/pubs/resolver`.

References

Guillaume, D. & Murtagh, F. 1999, this volume, 278

Schmitz, M., Helou, G., Dubois, P., LaGue, C., Madore, B., Corwin, H. G., & Lesteven, S. 1995, in Information & On-line Data in Astronomy, ed. D. Egret & M. A. Albrecht (Dordrecht: Kluwer), 271

Part 8. Data Analysis and Processing Techniques

Astronomical Data Analysis Software and Systems VIII
ASP Conference Series, Vol. 172, 1999
D. M. Mehringer, R. L. Plante, and D. A. Roberts, eds.

Computational Technology for Bayesian Inference

Thomas J. Loredo

Department of Astronomy, Cornell University

Abstract. One of the most important practical differences between the Bayesian approach to statistical inference and the more traditional frequentist approach is the nature of the sums and integrals required to implement each approach. In both approaches, the sampling distribution for the data (i.e., the likelihood function) plays a key role; but frequentist calculations integrate the sampling distribution over the sample space, whereas Bayesian calculations integrate it over hypothesis (parameter) space. The numerous advantages to working in parameter space come at a cost: the required integrals are difficult to calculate. I survey recent developments in computational technology for performing such integrals.

1. Introduction

Statistical calculations consist largely of weighted sums or integrals of probabilities. One of the key practical differences between Bayesian inference and frequentist statistics is that drastically different types of integrals appear in calculations taking these competing approaches to the same problem (Loredo 1992). Consider, for example, estimating the parameters of some model, M, given some observed data, D; denote the parameters jointly by θ. A key quantity appearing in both Bayesian and frequentist integrals is the probability for the data assuming the model to be true and presuming the parameters to be known, $p(D|\theta, M)$. Considered as a function of the data, this is called the *sampling distribution*; considered as a function of the parameters, it is called the *likelihood function*, which we will abbreviate as $\mathcal{L}(\theta)$. A basic practical difference between the methods is that frequentist calculations require integrals of this quantity over the data dimensions (sample space), whereas Bayesian calculations require integrals over the parameter space.

Basing inferences on probabilities calculated by summing or integrating over parameter space brings with it a host of advantages over trying to make inferences using probabilities calculated in sample spaces. In the brief space allotted here, significant discussion of these advantages is not possible. But two positive advantages that are of great practical utility must be mentioned. First, in the vast majority of real applications the parameter space may be divided into two parts, $\theta = (\psi, \phi)$, where interest focuses on ψ, and ϕ consists of "nuisance" parameters necessary for modeling the data but are otherwise uninteresting (e.g., background intensities). In Bayesian inference one can straightforwardly eliminate the nuisance parameters while accounting for their uncertainty simply by integrating the joint distribution for (ψ, ϕ) over ϕ. There is no completely satisfactory method for dealing with nuisance parameters in frequentist statistics. Second, comparison of competing models in Bayesian inference is accomplished by comparing the *average* likelihoods of the parameters for the models, rather than the *maximum* likelihoods used in frequentist statistics. The averaging of the likelihood implements an automatic "Occam's razor" that penalizes a model for the size of its parameter space. There is no counterpart to this in frequentist statistics. Interested readers can consult Loredo (1990, 1992) for more discussion of these advantages, as well as discussion of some of the "negative advantages," that is, weaknesses and problems of frequentist procedures that are avoided by calculating in hypothesis space rather than sample space.

These advantages come with a significant cost: parameter space integrals are difficult to evaluate. This paper briefly introduces modern computational technology that has made such calculations feasible in problems of realistic complexity. I presume of the reader some basic familiarity with common statistical terminology. Readers needing a basic introduction to Bayesian inference will find brief, self-contained coverage in two review articles (Loredo 1990, 1992) or in § 2 of Gregory & Loredo (1992), and a thorough pedagogical account in Devinder Sivia's introductory text (Sivia 1996). Those with access to the World Wide Web can find links to some of these works, and to other Bayesian resources of particular interest to physical scientists, at the BIPS[1] (Bayesian Inference for the Physical Sciences) web site. Evans & Swartz (1995) provide a longer survey for a statistical audience covering topics in § 3 to 5 of this article.

2. Comparing Bayesian and Frequentist Integrals

Consider the common case of parameter estimation with data, D, consisting of N statistically independent samples, d_i. Suppose the model, M, has m parameters (denoted collectively by θ), with $m \ll N$. Independence implies that the likelihood can be written as a product of N terms, $\mathcal{L}(\theta) = \prod_i p(d_i|\theta, M)$. The integrals (functionals) we would need to compute in a frequentist analysis resemble the following:

$$I[f;\theta] \equiv \int d^N D\, f(D) p(d_1|\theta, M) \cdots p(d_N|\theta, M). \tag{1}$$

For example, we can calculate the sampling distribution for some statistic $S(D)$ by taking $f(D) = \delta[S - S(D)]$, or we could evaluate the bias of an estimator $\hat{\theta}(D)$ by taking $f(D) = \hat{\theta}(D)$. A rigorous frequentist procedure usually requires that the value of such an integral not depend on θ (since we do not know what the true value of θ is to condition on). In contrast, the integrals required in a Bayesian analysis resemble the following:

$$I[g] \equiv \int d^{m'}\theta\, g(\theta) p(\theta|M) \mathcal{L}(\theta), \tag{2}$$

where $p(\theta|M)$ is a prior distribution for the parameters, and the integral of interest may be over a subset of the parameter space, so $m' \leq m$. For example,

[1] `http://astrosun.tn.cornell.edu/staff/loredo/bayes/`

$I[1]$ gives us the normalization constant for Bayes's theorem; $I[\theta]/I[1]$ gives us the posterior mean estimate for θ, and taking $g(\theta)$ to be an appropriately located "hat" function allows calculation of the probability in a credible region.

Consider first low-dimensional models, with $m \lesssim 3$. For such models, evaluation of the required integrals is usually much simpler for a Bayesian calculation because the dimension of the integral is so much smaller. Analytical integration may be possible; at worst, the integral can be calculated to any required precision with straightforward numerical quadrature. In contrast, even for the simplest problems, the frequentist calculation is not trivial because of the large dimension of the sample space; one must typically use characteristic functions or integral transforms to reduce the dimensionality. As an instructive example, the reader is invited to compare the Bayesian and frequentist treatments of estimating a Gaussian mean when the noise variance is not known a priori (the solution uses Student's t statistic; cf. Sivia 1996 and Meyer 1975 for Bayesian and frequentist derivations). The final procedures are essentially the same, but the calculations leading to them differ dramatically in complexity.

But now consider higher dimension models, with $m \gtrsim 4$. Rigorous frequentist methods usually do not exist for such models, but approximate procedures are easy to develop using Monte Carlo methods.[2] Despite the large dimension of the integrals, the independence of the d_i makes implementation of such methods almost trivial: one simply samples each d_i value independently and evaluates $f(D)$; repeating this simple procedure a large number of times and averaging the results evaluates the integral. In contrast, unless a lucky (or clever) model choice allows some or all of the dimensions to succumb to analytical integration, numerical calculation of a Bayesian integral can be extremely difficult, even though it may have orders of magnitude fewer dimensions than the frequentist one. The reason is that the likelihood is almost never a product of independent factors for each parameter—inferences for the parameters are correlated, often in complicated ways. This prevents simple Monte Carlo integration as described above. Also, the probability factors comprising the likelihood are typically simple and unimodal as a function of the d_i, whereas their product may have a complicated shape as a function of θ, perhaps with multiple modes. Straightforward numerical quadrature can solve the problem in principle, but in practice the "curse of dimensionality" makes quadrature unfeasible if there are more than 3 or 4 dimensions. The developer of a Bayesian code is in the ironic position of being able to write down the exact answer to a problem that cannot be treated exactly in the frequentist approach, but not being able to actually calculate the numerical value of the answer!

Fortunately, the last ten to fifteen years have seen remarkable developments in practical algorithms for performing Bayesian calculations. We can usefully group these algorithms into three families: asymptotic approximations; methods for moderate dimensional models; and methods for high dimensional models. We discuss these in turn. All of these methods can benefit significantly from

---

[2] One might legitimately complain that we are comparing approximate frequentist calculations with exact Bayesian calculations, and are thus being unfair to the Bayesian approach. Unfortunately, rigorous frequentist methods simply do not exist for many realistic problems (see, e.g., § 20.35 of Kendall & Stuart 1977); we can only compare the complexities of the calculations actually done in practice.

reparameterization of a problem to simplify the structure of the posterior; several of the references below discuss this issue.

3. Asymptotic Approximations

It is often the case that the posterior distribution has a single dominant interior mode (i.e., the mode is not on the boundary of the allowed parameter space). Call this mode $\hat{\theta}$. In the vicinity of the mode, the product of the prior and likelihood can be approximated by a multivariate Gaussian, so we have,

$$p(\theta|M)\mathcal{L}(\theta) \approx p(\hat{\theta}|M)\mathcal{L}(\hat{\theta}) \exp\left[-\frac{1}{2}(\theta - \hat{\theta}) \cdot \tilde{I} \cdot (\theta - \hat{\theta})\right], \tag{3}$$

where $\tilde{I}$ is the (observed) Fisher information matrix, a matrix of second derivatives evaluated at the mode: $\tilde{I} = \partial^2 \ln\left[p(\theta|M)\mathcal{L}(\theta)\right]/\partial^2\theta$ for $\theta = \hat{\theta}$. We can find approximate Bayes factors for model comparison by using this approximation to calculation average likelihoods (normalization constants for parameter estimation):

$$\int d\theta\, p(\theta|M)\mathcal{L}(\theta) \approx p(\hat{\theta}|M)\mathcal{L}(\hat{\theta})(2\pi)^{m/2}|\tilde{I}|^{-1/2}. \tag{4}$$

We can also use the approximation to do the integrals needed to eliminate nuisance parameters. If there are uninteresting parameters, ϕ, and parameters of interest, ψ, first construct a "profile" function for ψ, found by maximizing the prior $\times$ likelihood over ϕ (for each ψ): $f(\psi) = \max_\phi p(\psi, \phi|M)\mathcal{L}(\psi, \phi)$. We can construct an approximate marginal distribution for ψ by normalizing the product of $f(\psi)$ and a factor that accounts for the volume of the ϕ space:

$$p(\theta|D, M) \propto f(\psi)|\tilde{I}(\psi)|^{-1/2}, \tag{5}$$

where $\tilde{I}(\psi)$ is the information matrix for the nuisance parameters, with ψ held fixed. This approximation improves on the profile likelihood, $\max_\phi \mathcal{L}(\psi, \phi)$, the frequentist attempt to handle nuisance parameters best-known to astronomers (see, e.g., § 15.6 of Press et al. 1992).

The use of this kind of approximation originates with Laplace, so these approximations are called *Laplace approximations.* They can perform remarkably well in practice even for modest amounts of data, despite the fact that one might expect the underlying Gaussian approximation to be good only to order $1/\sqrt{N}$, the usual rate of asymptotic convergence to a Gaussian for frequentist approximations. The reason is that the final quantities reported in a Bayesian calculation are always *ratios* of integrals. The leading order errors in the numerator and the denominator typically cancel, so Laplace approximations are usually good asymptotically to order $1/N$ or even higher. Good entry points to the literature on this approximation can be found in Tierney & Kadane (1986), Kass, Tierney, & Kadane (1991), and O'Hagan (1994).

A particular appeal of the Laplace approximation to developers building on existing frequentist codes is that all of the ingredients, apart from the prior factor, are likely to be at hand. Many frequentist codes extremize a likelihood or log-likelihood (e.g., χ^2), and provide an approximate covariance matrix that is usually just the inverse of the information matrix required for the Laplace approximation. The Laplace approximation thus provides a quick entry into Bayesian computation.

Finally, similar reasoning underlies a somewhat cruder approximation that provides a Bayesian counterpart to the familiar frequentist procedure of comparing nested models using likelihood ratios (e.g., differences in minimum χ^2) and counting degrees of freedom. Asymptotically, the Type I error ("false alarm") probability associated with the test is given by the tail area of the χ^2_ν distribution, with χ^2_ν equal to -2 times the log likelihood ratio, and ν equal to the number of new parameters in the more complicated model. In a Bayesian setting, one would perform such model comparison by calculating a Bayes factor. Although the Bayes factor depends on the sizes of the parameter spaces of the models via the priors for the models' parameters, an approximate and automatic Bayes factor has been found to be useful in practice, at least in the early stages of an analysis. Known both as the Schwarz Criterion and as the Bayesian Information Criterion (BIC), it uses a Gaussian approximation for calculating average likelihoods much as was done above, but additionally uses an "automatic" prior with a width roughly corresponding to the width of the individual data factors in the likelihood. The result is that the log Bayes factor can be approximated as

$$\ln B \approx \ln\left[\mathcal{L}_2(\hat{\theta},\hat{\phi})/\mathcal{L}_1(\hat{\theta})\right] - \frac{1}{2}m_\phi \ln N, \tag{6}$$

where model 2 is the more complicated model, with additional parameters ϕ, and m_ϕ is the dimension of ϕ. The log likelihood ratio is adjusted for the differing number of parameters, but in a way that depends (weakly) on the number of data, a dependence absent from the asymptotic frequentist likelihood ratio test (fixing the inconsistency of that test). Kass & Wasserman (1995) discuss this approximation and the interpretation of the resulting approximate Bayes factors.

4. Methods for Low Dimensional Models

The Laplace approximation is an asymptotic approximation. The remainder of the Bayesian integration methods we will discuss are approximate only in the sense of having accuracies limited by computational resources; they are "exact" methods in the sense of not requiring any approximation of the integrand. We first discuss methods for low dimensional models, where "low" in practice means $m \lesssim 10$ or 20. The successful methods are based on two familiar classes of numerical integration techniques that by themselves are of rather limited utility in Bayesian calculations. It is modification and combination of these basic methods that has led to successful algorithms.

The most familiar approach to numerical integration is the use of quadrature rules—approximating an integral as a weighted sum of values of the integrand. The various rules differ in the choices of weights and abscissas. The impressive thing about these rules in one dimension is that one can take advantage of known "smoothness" properties of the integrand (such as membership in a polynomial family, or approximate similarity to a polynomial times a known nonlinear function such as a Gaussian) in order to construct rules with fast rates of convergence. Errors falling as $1/n^2$ or $1/n^4$, with n the number of abscissas, are easily achieved. Unfortunately, this desirable behavior is lost when one extends such rules to many dimensions. The simplest approach is to use a product rule—a combination of 1-d rules for each dimension. This suffers from the infamous "curse of dimensionality;" the number of points needed grows exponentially with dimension, so the errors now only fall like $1/n^{2/m}$ or $1/n^{4/m}$. This makes these rules impractical for all but the lowest dimensions ($m \lesssim 4$). Rules exist that extend the quadrature idea to higher dimensions in a more complicated way; they appear under various names in the literature, including cubature, lattice, and monomial rules. In these rules, the abscissas do not lie on a cartesian grid; instead, they are spread over a more complicated multidimensional lattice. But the curse of dimensionality persists for these methods, though it is weakened.

Another familiar approach to numerical integration is Monte Carlo integration—approximating an integral with an average of values of its integrand chosen randomly according to some rule. The great virtue of this approach is that the error falls like $1/\sqrt{n}$, regardless of dimension (or possibly as quickly as $1/n$ if one uses quasirandom sampling rules). This convergence rate is quite poor for one or two dimensions; Monte Carlo integration takes no account of the smoothness information built into quadrature rules and is thus no competition for them in low dimensions. But as soon as $m \gtrsim 3$, the Monte Carlo convergence rate starts to look attractive compared to that of quadrature. Unfortunately, although the *rate* of convergence is attractive, the actual *size* of the errors can be so large as to make implementation impractical. The size of the error depends on the variance of the integrand. This can be reduced in some cases by reweighting the integrand and adjusting the sampling rule to compensate ("importance sampling"). But such reweighting is extremely hard to design in practice unless $m \lesssim 6$.

Although neither of these approaches is useful for Bayesian integration in more than a few dimensions, the successful methods in current use build directly on them. Two families of methods are particularly useful.

First are *randomized quadrature* methods. These methods resemble quadrature rules, but the abscissas are "dithered" randomly. They can combine the virtues of both quadrature and Monte Carlo methods, while diminishing the drawbacks of those methods. The most useful such methods resemble Gaussian quadrature in that they are best used if you can consider the posterior to be reasonably well approximated by, say, a multivariate Gaussian times multinomials (see, e.g., Monahan & Genz 1997).

Second are *subregion-adaptive quadrature* methods. These methods typically use two low-order lattice rules (say, exact for multinomials of order 5 and 7) to estimate an integral *and its error* (via the difference between the estimates from the two rules) in various subregions of the parameter space. The method is applied recursively in regions with the largest error until the entire integral is evaluated to the desired accuracy. These methods automatically put more quadrature points where most of the posterior probability lies, settling for less accuracy in the unimportant tails of the distribution (which often account for a large amount of the parameter space volume).

Randomized and subregion-adaptive methods exist in the general computational literature, but in recent years there has been research effort devoted to tailoring versions of them specifically for Bayesian integration (e.g., Genz

& Kass 1997). The resulting algorithms are not trivial to code from scratch; but fortunately, free and easy-to-use subroutines are readily available. Particularly noteworthy is the work of Alan Genz. The ADAPT subregion-adaptive quadrature algorithm he developed with Malik is available on the World Wide Web; I have used this subroutine in several Bayesian calculations (see, e.g., Loredo, Flanagan, & Wasserman 1997). In addition, his BAYESPACK subroutine package provides a well-documented unified interface to several FORTRAN subroutines for performing Monte Carlo, randomized quadrature, and subregion-adaptive quadrature for Bayesian problems of moderate dimension. Links to this software are available at the BIPS web site previously mentioned.

5. Methods for High Dimensional Models

The majority of straightforward parametric models have a small enough number of parameters that the methods above are likely to prove adequate for implementation of the Bayesian recipe. But sometimes large parametric models must be analyzed (e.g., in analyses of the cosmic background radiation). Also, nonparametric models, once coded, are for all intents and purposes "mega-parametric" models, with each of the discrete sample points or pixels of the estimated function or image playing the role of a parameter. In such problems, the number of parameters can easily reach 10^6 or more. Given the previous discussion, it may seem that a truly Bayesian analysis in such a context, requiring integrals in a highly correlated mega-dimensional space, is simply out of the question. Surprisingly, such analyses have been done routinely for some years now. The basic concept underlying the successful methods is *posterior sampling*, and the most powerful algorithms for implementing them are *Markov Chain Monte Carlo* (MCMC) algorithms.

The notion of posterior sampling should appeal to those used to coding Monte Carlo algorithms in a frequentist context. To calculate a frequentist integral, one can simulate data, as described in § 2, evaluating the integrand for each full data sample. Posterior sampling proceeds similarly, evaluating Bayesian integrals by "simulating hypotheses" rather than data; that is, by drawing samples of θ from the posterior distribution. The great virtue of this approach is that once one has generated a large number of samples θ_i ($i = 1$ to n) from the posterior, that single set of samples can be used to evaluate a whole host of desired integrals for summarizing the implications of the posterior. The marginal distribution for any parameter or subset of parameters can be found simply by ignoring the values of the uninteresting parameters, and plotting or calculating properties of the set of interesting parameter values. Moments of any function of the parameters (including an indicator function whose "moment" gives the probability in a credible region) can be found by averaging the function over the samples. In the case of nonparametric inference, a movie can be made where each frame shows one m-dimensional θ_i (a particular function or image consistent with the data). Viewing this movie will reveal the level of confidence one should have in certain features of the inferred function; if the feature is consistently present in the samples, it has a high probability of being real. Such probabilities can be quantified by calculating appropriate functionals of the samples. This aspect of the approach is appealing enough that posterior sampling is also worth considering for problems with moderate or even low dimensionality.

The problem is, how can we get the samples in the first place? As already noted, Bayesian integration is complicated precisely because the correlations in parameter space do not make Bayesian integrals amenable to the independent sampling methods underlying frequentist simulation-based calculations.

The simplest method for drawing samples from a complicated probability density function is the well-known *rejection method* (Press et al. 1992, § 7.3). This method depends on the ability to construct a good rejection function that resembles the posterior, but from which we know how to sample efficiently. For problems with dimensions $\lesssim 6$, this is often possible, though it can take a lot of work. But for larger numbers of dimensions, all easily constructed rejection functions waste too much volume, and the overwhelming majority of candidate samples end up being rejected. Thus the rejection method cannot take us into regimes not already accessible by other methods (but it can compete with those methods in the low dimensional regime).

To an audience of physical scientists, a possible solution to the problem appears when it is recast in more suggestive notation. Define a function $\Lambda(\theta)$ according to $\Lambda(\theta) = -\ln[p(\theta|M)\mathcal{L}(\theta)]$. Then the posterior distribution can be written as $p(\theta|D,M) = e^{-\Lambda(\theta)}/Z$, where $Z \equiv \int d\theta\, e^{-\Lambda(\theta)}$. Evaluation of the posterior now resembles two classes of problems familiar to physicists: evaluating Boltzmann factors and partition functions in statistical mechanics, and evaluating Feynman path weights and path integrals in Euclidean quantum field theory. Accordingly, beginning in the mid 1980s, statisticians have mined the computational physics literature for methodology from computational statistical mechanics and quantum field theory on the lattice that could be adapted to evaluate Bayesian integrals. It seems about time this methodology found its way back to the physical sciences.

All of these methods work by constructing a kind of random walk (Markov chain) in the parameter space such that the probability for being in a region of the space is proportional to the posterior density for that region. One starts somewhere (more or less anywhere, though it helps to start at a place of high probability), calculates the posterior density there, takes a small step, and then recalculates the density. The step is accepted or rejected according to some rule, and the process is repeated. The resulting output is a series of points in the sample space, usually discussed and analyzed as if it were a time series. The various methods differ according to the rules used to make the moves in the parameter space, and the rules determining whether or not a step is accepted. It turns out to be extremely simple to invent such rules that are guaranteed to produce accepted steps that correctly sample the posterior. Metropolis et al. (1953) gave the basic algorithm; a discussion of it and some useful modern extensions that is particularly accessible to physical scientists is available in the first section of Toussaint's very readable introduction to methods for lattice QCD calculations (Toussaint 1989), a readable tutorial for statistics students is available in Chib & Greenberg (1995), and a more thorough (but now somewhat dated) review from the point of view of a computer scientist is provided by Neal (1993).

Although the resulting MCMC methods are simple in principle, considerable art is required to implement them in practice. Convergence of MCMC methods can be guaranteed, but it takes some time for the series of samples to reach equilibrium; and once equilibrated, the samples comprising the output time series are obviously correlated. Part of the art of these MCMC methods is in choosing rules that lead to quick convergence and short correlation lengths. Aspects of the structure of a particular problem, such as availability of conditional distributions for single parameters, or availability of derivatives, can be used to produce effective algorithms for that problem. Once the algorithm is settled upon, one must also take care in determining when the resulting Markov chain has reached equilibrium, and in taking into account the correlations in the resulting samples when calculating inferences. Good output analysis is crucial to successful implementation of MCMC methods and must be done with some sophistication; one has essentially created a new inference problem (albeit a simpler one) in the course of solving the original one.

The development and application of MCMC methods has been the most active area of research in applied Bayesian statistics in recent years. As a result, there is a considerable body of knowledge to draw from when constructing an MCMC algorithm and analyzing its output. A recent volume of review and application papers collects some of this wisdom (Gilks, Richardson, & Spiegelhalter 1996); a roundtable discussion collects further informal advice for novice practitioners (Kass et al. 1998). A monograph discusses engineering applications (Ó Ruanaidh & Fitzgerald 1996), and there is an MCMC web site[3] devoted to publicizing developments in MCMC methods. Although there is not yet a "black box" MCMC algorithm with universal applicability (and there is not likely to be one), good and simple algorithms exist with considerable generality (e.g., for image analysis and density estimation with common priors). In any case, the algorithms themselves are so easy to code that experimentation is relatively straightforward.

Two complications are worth mentioning. First, development of MCMC methods for model comparison lags behind that for parameter estimation. Such methods so far tend to be highly specialized to particular problems (though see Neal 1993 for descriptions of general methods based on algorithms for calculating free energies that have not yet been thoroughly explored in the statistics community). Second, the virtues of posterior sampling may tempt one to apply MCMC methods to parametric models with small numbers of parameters; but some caution is appropriate. MCMC methods so far appear to be most successful when the parameter space is in some sense "homogeneous," with each parameter having a similar role in the likelihood and the same physical dimensions (as with image pixels); otherwise, a "metric" is needed in the parameter space, complicating the analysis.

Asymptotic approximations make the transition to (approximate) Bayesian inference relatively painless for those with existing frequentist codes. Subregion-adaptive quadrature, randomized quadrature, and posterior sampling make more exact Bayesian calculations straightforward for models with significant, realistic complexity. These methods have played a significant role in fueling the explosion

---

[3] `http://www.stats.bris.ac.uk/MCMC/`

of interest in the Bayesian approach in applied statistics. In astrostatistics, the last five years have seen a strong growth of interest in Bayesian inference; these modern computational tools will certainly play a key role in expanding this interest even further.

References

Chib, S. & Greenberg, E. 1995, Amer. Statistician, 49, 327

Evans, M. & Swartz, T. 1995, Stat. Sci., 10, 254

Genz, A. & Kass, R. E. 1997, J. Comp. Graph. Stat., 6, 92

Gilks, W. R., Richardson, S., & Spiegelhalter, D. J. 1996, Markov Chain Monte Carlo in Practice (London: Chapman & Hall)

Gregory, P. C. & Loredo, T. J. 1992, ApJ, 398, 146

Kass, R. E., Carlin, B. P., Gelman, A., & Neal, R. M. 1998, Amer. Statistician, in press

Kass, R. E. & Wasserman, L. 1995, J. Amer. Stat. Assoc., 90, 928

Kass, R. E., Tierney, L., & Kadane, J. B. 1991, in Statistical Multiple Integration, ed. N. Flournoy & R. K. Tsutukawa (Providence: AMS), 89

Kendall, M., & Stuart, A. 1977, The Advanced Theory of Statistics (London: Griffin & Co.)

Loredo, T. J. 1990, in Maximum Entropy and Bayesian Methods, ed. P. F. Fougere (Dordrecht: Kluwer), 81

———, 1992, in Statistical Challenges in Modern Astronomy, ed. E. Feigelson & G. Babu (New York: Springer-Verlag), 275

Loredo, T. J., Flanagan, E. E., & Wasserman, I. M. 1997, Phys. Rev. D, 56, 7507

Metropolis, N., Rosenbluth, A. W., Rosenbluth, M. N., Teller, A. H., & Teller, E. 1953, J. Chem. Phys., 21, 1087

Meyer, S. L. 1975, Data Analysis for Scientists and Engineers (New York: Wiley)

Monahan, J., & Genz, A. 1997, J. Amer. Stat. Assoc., 92, 664

Neal, R. 1993, Probabilistic Inference using Markov Chain Monte Carlo Methods, (Tech. Rep. CRG-TR-93-1, Dept. Comp. Sci. Univ. Toronto), `ftp://ftp.cs.toronto.edu/pub/radford/review.ps.Z`

O'Hagan, A. 1994, Kendall's Advanced Theory of Statistics, v. 2B, Bayesian Inference (New York: Halsted Press)

Ó Ruanaidh, J. J. K. & Fitzgerald, W. J. 1996, Numerical Bayesian Methods Applied to Signal Processing (New York: Springer-Verlag)

Press, W. H., Teukolsky, S. A., Flannery, B. P., & Vetterling, W. T. 1992, Numerical Recipes in C (Cambridge: Cambridge Univ. Press)

Sivia, D. S. 1996, Data Analysis: A Bayesian Tutorial (Oxford: Oxford Univ. Press)

Tierney, L. & Kadane, J. B. 1986, J. Amer. Stat. Assoc., 81, 82

Toussaint, D. 1989, Comp. Phys. Comm., 56, 69

Astronomical Data Analysis Software and Systems VIII
ASP Conference Series, Vol. 172, 1999
D. M. Mehringer, R. L. Plante, and D. A. Roberts, eds.

The Pixon Method of Image Reconstruction

Richard C. Puetter

Center for Astrophysics and Space Sciences, University of California, San Diego, La Jolla, CA 92093-0424, USA

Amos Yahil

Department of Physics and Astronomy, State University of New York, Stony Brook, NY 11794-3800, USA

Abstract. The Pixon method is a high-performance, nonlinear image reconstruction method that provides statistically unbiased photometry and robust rejection of spurious sources. Relative to other methods, it can increase linear spatial resolution by a factor of a few and sensitivity by an order of magnitude or more. All of these benefits are achieved in computation times that can be orders of magnitude faster than its best competitors.

1. Introduction

Optimal extraction of the underlying quantities from measured data requires the removal of measurement defects such as noise and limited instrumental resolution. When the underlying quantity is an image, this process is known as image reconstruction (sometimes called image restoration, if the data are also in the form of an image).

The original Pixon method (Piña & Puetter 1993; Puetter & Piña 1993; Puetter 1995, 1997) was developed to eliminate problems with existing image reconstruction techniques, particularly signal-correlated residuals and the production of spurious sources. This was followed by the accelerated Pixon method that vastly increased the computational speed (Yahil & Puetter, in preparation). Recently, a quick Pixon method was developed that is even faster, at the expense of some photometric inaccuracy for low signal-to-noise-ratio features (Puetter & Yahil, in preparation). Nevertheless, the quick Pixon method provides excellent results for a wide range of imagery and, with special-purpose hardware now under design, is capable of real-time video image reconstruction.

Since its inception, the Pixon method in its various forms has been applied to a wide variety of astronomical, surveillance, and medical image reconstructions (spanning all wavelengths from γ-rays to radio), as well as to spectroscopic data. In all cases tested so far, both by the authors and by a variety of other workers, the Pixon method has proved superior in quality to all other methods and computationally much faster than its best competitors.

A patent for the Pixon method is pending, restricting its commercial use. For individual scientific purposes, the original Pixon method is freely available in IDL and C++ from the San Diego Supercomputer Center[1]. The accelerated and quick Pixon methods are sold by our company, Pixon LLC, but special arrangements for their free use in selected scientific projects can be made.

The next sections discuss image reconstruction in general (§2.), describe the Pixon method (§3.), and give some examples (§4.). The HTML and PDF versions of this paper contain numerous additional examples via clickable hyperlinks. The complete set of public image reconstructions, including videos, can be seen on our Web site[2].

2. Image Reconstruction Methods

For data taken with linear detectors, image reconstruction often becomes a matter of inverting an integral relation of the form

$$D(\mathbf{x}) = \int \mathbf{dy} H(\mathbf{x}, \mathbf{y}) I(\mathbf{y}) + N(\mathbf{x}) \quad . \tag{1}$$

Here D are the data, I is the sought underlying image model, H is the point-spread function (PSF) due to instrumental and possibly atmospheric blurring, and N is the noise. To avoid confusion, we use the term image to refer exclusively to the true underlying image or its model. Contrary to common parlance, the data are never called the image. We clearly distinguish between abstract image space and real data space, with the PSF transforming from the former to the latter.

If the PSF is only a function of the displacement, $H(\mathbf{x}, \mathbf{y}) = H(\mathbf{x} - \mathbf{y})$, then the integral in equation (1) becomes a convolution, but in general the PSF can vary across the field. Note that the data need not have the same resolution as the image, or even the same dimensionality. For example, the Infrared Astronomical Satellite (IRAS) provided 1–D scans across the 2–D sky, and tomography data consist of multiple 2–D projections of 3–D images. Another common case is of data consisting of multiple, dithered, exposures of the same field, which all need to be modeled by a single image, possibly with PSFs that vary from one exposure to another.

Image reconstruction differs from standard solutions of integral equations due to the noise term, N, whose nature is only known statistically. Methods for solving such an equation fall under two broad categories. (1) Direct methods apply explicit operators to the data to provide estimates of the image. These methods are often linear, or have very simple nonlinear components. Their advantage is speed, but they typically amplify noise, particularly at high frequencies. (2) By contrast, indirect methods model the noiseless image, transform it only forward to provide a noise-free data model, and fit the parameters of the image to minimize the residuals between the real data and the noise-free data model. The advantage of indirect methods is that noise is supposedly excluded (but see §2.3.). Their disadvantage is the required modeling of the image. If a good parametric form for the image is known a priori, the result can be superb.

---

[1] `http://pixon.sdsc.edu`

[2] `http://www.pixon.com`

If not, either the derived image badly fits the data or, conversely, it overfits the data, interpreting noise as real features. Indirect methods are typically significantly nonlinear and much slower than direct methods.

The Pixon method is a nonparametric, indirect, reconstruction method. It avoids the pitfalls of other indirect methods, while achieving a computational speed on a par with direct methods. To appreciate its advantages we therefore first provide a brief description of some competing direct and indirect methods.

2.1. Direct Methods

In the case of a position-independent PSF, naive image deconvolution is obtained by inverting the integral equation (1) in Fourier space, ignoring the noise:

$$\tilde{I}(\mathbf{k}) = \tilde{D}(\mathbf{k})/\tilde{H}(\mathbf{k}) \quad , \tag{2}$$

where the tilde designates Fourier transform, and $\mathbf{k}$ is the spatial wavenumber. This method amplifies noise at high frequencies, since $H(\mathbf{k})$ is a rapidly declining function of $|\mathbf{k}|$, while $D(\mathbf{k})$, which includes high-frequency noise, is not.

To minimize noise amplification, nonlinear filtering is often applied to the data prior to Fourier inversion. Common filters are those due to Wiener (1949) (see also Press et al. 1992) and thresholding of wavelet transforms (e.g., Donoho & Johnstone 1994; Donoho 1995). The Wiener filter has the advantage of being applied in Fourier space. Wavelet thresholding requires a wavelet transform, thresholding, and a back wavelet transform, all followed by Fourier inversion. Demonstrations of the superior performance of the Pixon method relative to Wiener-filtered Fourier inversions can be seen on our Web site in the reconstruction of the Lena image and an aerial view of New York City.

2.2. Parametric Least-Squares Fit

The simplest indirect image reconstruction is a least-squares fit of a parametric model with a small number of parameters compared to the number of data points. Originally due to Gauss, this method is always superior to other methods, provided that the image can be so modeled. It is equivalent to a maximum-likelihood optimization in which the residuals are assumed to be Gaussian distributed. All models within the restricted parameterization are considered equally likely, and the likelihood function maximized is $P(D|I)$, the conditional probability of the data, given the model image. For this reason, the method is also known as maximum a posteriori probability (MAP) image reconstruction.

2.3. Nonnegative Least-Squares Fit

When no parametric model of the image is known, the number of image model parameters can quickly become comparable to, or exceed, the number of data points. In this case, a MAP solution becomes inappropriate. For example, if the number of points in the image model equals the number of data points, then the nonsingular nature of the linear integral transform in equation (1) assures that there is a solution for which the data, including the noise, are exactly modeled with zero residuals. This is clearly the same poor solution, with all its noise amplification, obtained by the naive Fourier deconvolution.

The above example shows that an unrestricted indirect method is no better at controlling noise than a direct method, and therefore the model image must be restricted in some way. The indirect methods described below all restrict the image model and differ only in the specifics of image restriction.

A simple restriction is to constrain the model image to be positive. Since even a delta-function image is broadened by the PSF, it follows that the exact inverse of any noisy data with fluctuations on scales smaller than the PSF must be both positive and negative. By preventing the image from becoming negative, the noise-free data model cannot fluctuate on scales smaller than the PSF, which is equivalent to smoothing the data on the scale of the PSF.

While smoothing over the scale of the PSF helps to reduce noise fitting, it does not go far enough. The Pixon method (§3.) smoothes further where possible, and is therefore able to eliminate noise fitting on larger scales. Demonstrations of its superior performance relative to nonnegative least squares can be seen on our Web site in the reconstruction of γ-ray imaging in the direction of Virgo, as well as in the Lena and New York City reconstructions.

2.4. Maximum-Entropy Method

Bayesian methods go a step further by assigning explicit a priori probabilities to different models and then maximizing the joint probability of the model and the data:

$$P(I \cap D) = P(D|I)P(I) = P(I|D)P(D) \quad . \tag{3}$$

The model probability function, $P(I)$, is known as the prior. For example, the nonnegative least-squares method (§2.3.) is a Bayesian method with all nonnegative models assigned equal nonzero probability and all other models zero probability.

The rationale behind Bayesian methods is to optimize $P(I|D)$, the conditional probability of the image given the data. This probability is proportional to the joint probability $P(I \cap D)$, since the data are fixed, and $P(D)$ can be viewed as a constant. However, it is important to recognize that, unlike $P(D|I)$ which depends on the instrumental response function and noise statistics, $P(I|D)$ is not known from first principles. In practice, therefore, computing $P(I \cap D)$ requires the specification of a completely arbitrary prior, $P(I)$.

The maximum-entropy method (MEM) is the most popular Bayesian image reconstruction technique. It uses the prior

$$P(I) = \exp\left(-\alpha \sum_i I_i \ln I_i\right) \quad , \tag{4}$$

where the sum is over the image pixels, and α is an adjustable parameter designed to strike a balance between image smoothness and goodness of fit. (See Gull 1989 and Skilling 1989 for a discussion of a "natural" choice for α.)

The sum in equation (4) approximates the information entropy of Shannon (1948), which is maximized for a homogeneous image. By favoring a flat image, MEM therefore eliminates structure not required by the data and suppresses noise fitting. The fundamental difficulty with this approach is that the MEM prior is a global constraint (specifically, the prior is invariant under random scrambling of the pixels). MEM therefore enforces an average smoothness on the entire image and does not recognize that the density of information content in the image varies from location to location. Hence, MEM must necessarily oversmooth the image in some regions and undersmooth it in others. More sophisticated MEM schemes try to remedy this situation by applying separate MEM priors in different parts of the image, or by modifying the logarithmic term to $\ln(I_i/M_i e)$, where M is some preassigned model and e is the base of the natural logarithm (Burch, Gull, & Skilling 1983), but there is additional arbitrariness in the choice of M.

By contrast, the Pixon method (§3.) adapts itself to the distribution of information content in the image. Demonstrations of its superior performance relative to MEM are given below (§4.) and on our Web site in a reconstruction of hard X-ray observations of a solar flare.

3. The Pixon Method

Unlike Bayesian methods, the Pixon method (Piña & Puetter 1993; Puetter & Piña 1993; Puetter 1995, 1997) does not assign explicit prior probabilities to image models. Instead, it restricts them by seeking minimum complexity (Solomonoff 1964; Kolmogorov 1965; Chaitin 1966), which not only enables an efficient representation of the image but is also the best way to separate signal from noise.

In simple terms, the Pixon method implements the principle of *Ockham's Razor* to select the simplest plausible model of the image. Clearly, if the signal in the image can be adequately represented by a minimum of P parameters, adding another parameter only serves to introduce artifacts by fitting the noise. Conversely, the removal of a parameter results in an improper representation of the image, since adequate fits to the image require a minimum of P parameters.

While few would dispute that a model with minimum complexity (also called algorithmic information content) is optimal, in practice it is impossible to find such a model for any but the most trivial problems. For example, one might try to model an image as the smallest number of contiguous patches of homogeneous intensity that still adequately fit the data. While there clearly is such a solution, it is quite another matter to find it among the combinatorially large number of possible patch patterns. And we have not even begun to consider patches that are not completely homogeneous.

The Pixon method overcomes this difficulty in the same practical spirit in which other combinatorial problems have been solved, such as the famous traveling salesman problem (e.g., Press et al. 1992). One finds an intelligent scheme in which complexity is reduced significantly in a manageable number of iterations. After that, the decline in complexity per iteration drops sharply, and the process is halted. The solution found in this manner may not have the ultimate minimum complexity, but it is already so superior to other models that it is worth adopting.

The Pixon method minimizes complexity by smoothing the image model *locally* as much as the data allow, thus reducing the number of independent patches, or Pixon elements, in the image. Formally, the image is written as an integral over a pseudoimage

$$I(\mathbf{y}) = \int \mathbf{dz} K(\mathbf{y}, \mathbf{z}) \phi(\mathbf{z}) \quad , \tag{5}$$

with a positive kernel function, K, designed to provide the smoothing. As in the case of the nonnegative least-squares fit, requiring the pseudoimage, ϕ, to be positive eliminates fluctuations in the image, I, on scales smaller than the width of K. Importantly, this scale is adapted to the data. At each location, it is allowed to increase in size as much as possible without violating the local goodness of fit (GOF). Where the kernel is a delta function, the Pixon element spans one pixel, and the Pixon method reduces to a nonnegative least-squares fit, with the noise-free data model smooth on the scale of the PSF. Where the data allow smoothing on larger scales, however, the Pixon elements become larger. As a result, the overall number of Pixon elements in the image drops, and complexity is reduced.

Complexity can be reduced not only by having kernel functions of different sizes to allow for multi-resolution, but also by a judicious choice of their shapes. For example, circularly symmetric kernels, which might be adequate for the reconstruction of most astronomical images, are not the most efficient smoothing kernels for images with elongated features, e.g., an aerial photograph of a city. Altogether the choice of kernels is the language by which the image model is specified, which should be rich enough to characterize all the independent elements of the image. We have found, in practice, that circular kernels spanning 3–5 octaves in size, with 2–4 kernels per octave, are adequate for most image reconstructions. Additional elliptical kernels are needed only for images with clearly elongated features.

The Pixon reconstruction consists of a simultaneous search for the broadest possible kernel functions and their associated pseudoimage values that together provide an adequate fit to the data. In practice, the details of the search vary depending on the flavor of the Pixon method used. Generally, however, one alternately solves for the pseudoimage given a Pixon map of kernel functions and then attempts to increase the scale sizes of the kernel functions given the current image values. The number of iterations required varies depending on the complexity of the image, but for most problems a couple of iterations suffice.

The essence of the Pixon method is the imposition of the local criteria by which the kernel functions are chosen. For each pixel in pseudoimage space, the combination of the Pixon kernel and the PSF define a footprint in data space. We accept the largest Pixon kernel whose GOF and signal-to-noise ratio (SNR) within that footprint pass predetermined acceptance conditions set by the user. If no kernel has adequate GOF we assign a delta-function kernel, provided that the SNR for its footprint is high enough. If the SNR also fails to meet our condition, no kernel is assigned.

The strength of the Pixon method is in its rejection of features that do not meet strict statistical acceptance criteria. Precisely because of this conservatism, which results in a significant lowering of the noise floor, the Pixon reconstruction is able to find weak but significant features missed by other methods and to resolve all features better. Sensitivity is often improved by an order of magnitude or more relative to competing methods and resolution by a factor of a few.

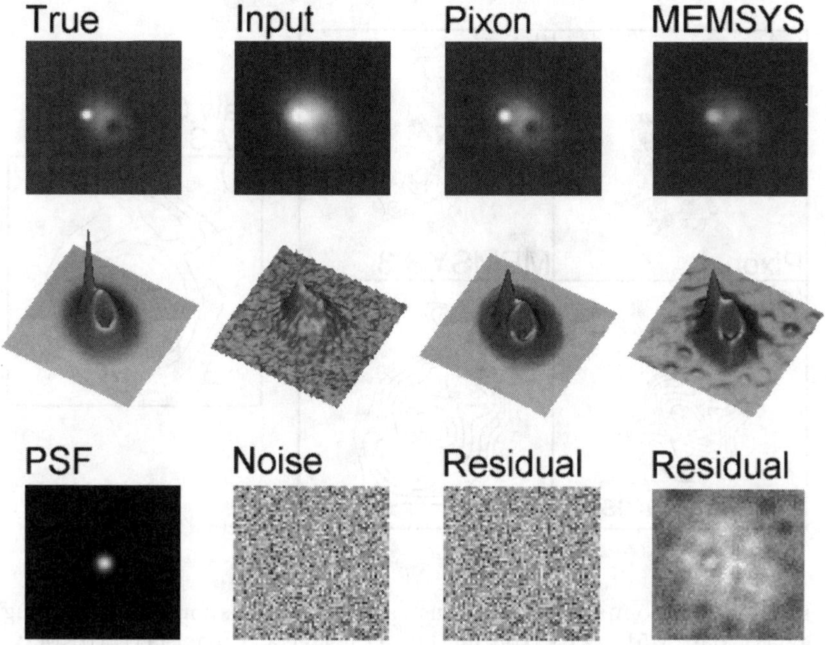

Figure 1. Pixon and MEM restorations, showing PSF (Gaussian, FWHM=6 pixels), true image, Gaussian noise, restored images and residuals. Peak SNR=20; SNR=12 at the bottom of the "hole". (Reproduced from Puetter 1994.)

Note that the SNR required by the Pixon method is not per pixel in the data, but the overall SNR in the data footprint of the Pixon kernel. The Pixon method is just as powerful in detecting large, low-surface-brightness features as small features with higher surface brightness. Acceptance or rejection of the feature is based in both cases on the *statistical significance* demanded by the user. Demonstrations of the ability of the Pixon method to reconstruct images with low SNR can be seen in the mammogram example (Fig. 3 below) and in the reconstruction of γ-ray imaging in the direction of Virgo on our Web site.

Finally, note that with a very terse representation that only has a small number of nonzero pseudoimage values, the image is also represented in a very compressed form. In the field of data compression one normally distinguishes between strong but lossy versus moderate and non-lossy compression. The Pixon method defines a new type of noiseless compression. The signal is preserved and restorable, while the noise is eliminated. This mode of compression is ideal, since compression can easily reach a factor of 100 for a typical image, yet the only loss is that of unwanted noise. A compressed form of the code is now under design; current codes, however, are primarily built for accuracy and speed and do not yet achieve such high compression.

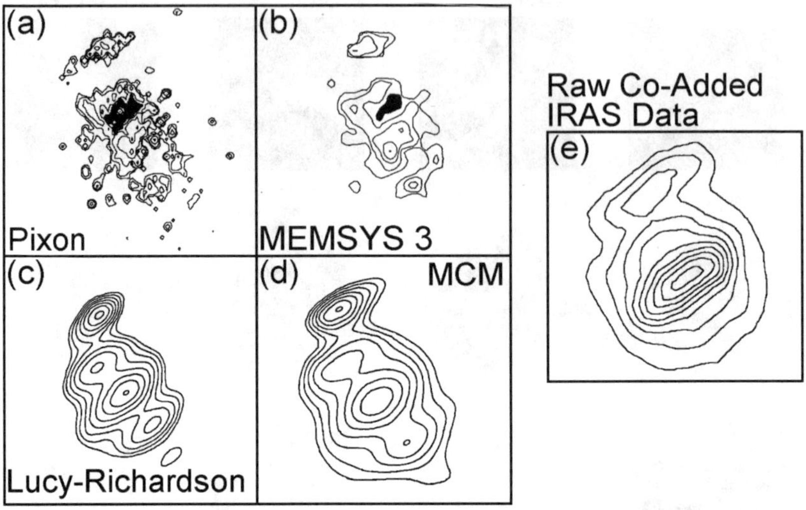

Figure 2. Comparison of reconstruction methods for the interacting galaxy pair M51: (a) Pixon method (Puetter & Piña 1994), (b) Maximum entropy method (Bontekoe et al. 1991), (c) Lucy-Richardson method (Rice 1993), (d) Maximum-Correlation Method, (Rice 1993, used by the HIRES package of NASA's Infrared Processing and Analysis Center), and (e) the raw co-added image.

4. Examples of Pixon Reconstructions

The Pixon method has now been used in a variety of applications by the authors and others (see Puetter 1995, 1997 and references therein). In this section we present a few examples of Pixon reconstructions for a variety of applications. Additional examples can be found in the brochure and figures on our Web site.

4.1. A Synthetic Data Set

The example presented in Fig. 1 (or its color version on our Web site) provides one of the best examples of comparative results of MEM and the Pixon method for a synthetic data set, i.e., a data set with known properties. The MEM algorithm used was MEMSYS 5, the most current release of the MEMSYS algorithms developed by Gull and Skilling (Gull & Skilling 1991). The MEMSYS reconstruction was performed by Nick Weir, and was enhanced by his multichannel correlation method (Weir 1991). As can be seen, the Pixon reconstruction is superior to the multichannel MEMSYS result, and is free of the low-level spurious sources and signal-correlated residuals evident in the MEMSYS reconstruction.

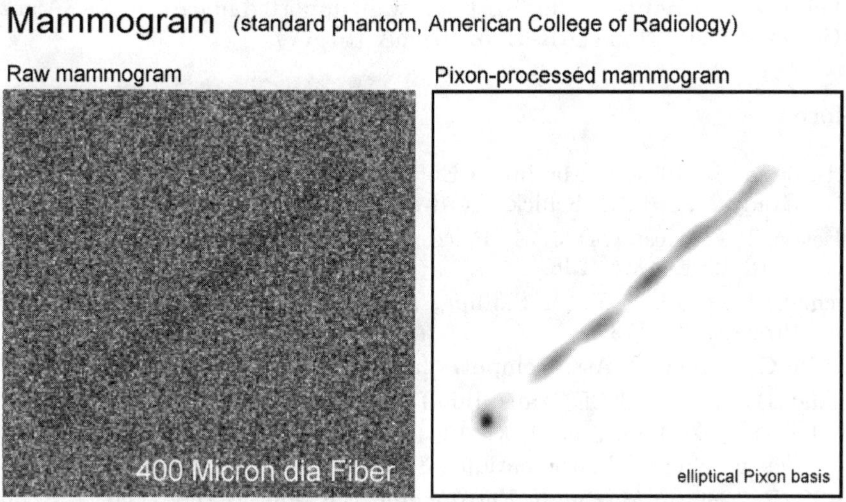

Figure 3. Pixon reconstruction of X-ray mammography data.

4.2. IRAS Imaging of M51

In Fig. 2 we present comparisons of reconstructed images from $60\mu m$ IRAS scans of the interacting galaxy pair M51. This data set was used in an image reconstruction contest at the 1990 MaxEnt Workshop (Bontekoe 1991). As can be seen, the Lucy-Richardson and maximum-correlation (also known as HIRES) reconstructions fail to reduce image spread in the cross-scan direction, i.e., the rectangular signature of the $1'.5 \times 4'.75$ detectors is still clearly evident. They also do not reconstruct even gross features such as the "hole" (black region) in the emission north of the nucleus—this hole is clearly evident in optical images of M51. The MEMSYS 3 reconstruction is significantly better, recovers the hole, and resolves the NE and SW arms of the galaxy into discrete sources. However, the Pixon result is clearly superior. In fact, its sensitivity is a factor of 200 higher than that of MEMSYS 3, and its linear spatial resolution is improved by a factor of 3. The reality of its minute details can be verified by comparing with images at other wavelengths.

4.3. X-Ray Mammography

Presented in Fig. 3 is a Pixon reconstruction of a standard phantom from the American College of Radiology. Here a fiber with a 400 micron diameter was embedded in a piece of material with X-ray absorption properties similar to the human breast. This particular example was selected since it has a low SNR. A family of elliptical Pixon kernel functions was used for the reconstruction. The Pixon method easily detects the presence of the fiber. In some locations, however, the SNR of the fiber is so poor that statistically significant signal is absent. Hence, no detection can be made by the Pixon method.

Acknowledgments. This work was was supported in part by NASA grants NAG53944 (to RCP) and AR-07551.01-96A (to AY).

References

Bontekoe, T. R. 1991, in Maximum Entropy and Bayesian Methods, ed. W. T. Grady Jr. & L. H. Schick (Dordrecht: Kluwer), 319

Bontekoe, T. R., Kester, D. J. M., Price, S. D., de Jonge, A. R. W., & Wesselieus, P. R. 1991, A&A, 248, 328

Burch, S. F., Gull, S. F., & Skilling, J. 1983, Comp. Vis., Graphics, & Im. Process., 23, 113

Chaitin, G. J. 1966, J. Ass. Comput. Mach., 13, 547

Donoho, D. L. 1995, IEEE Trans. Inf. Theory, 41, 613

Donoho, D. L. & Johnstone, I. M. 1994, Comptes Rendus de L'Acadamie Des Sciences Serie I-Mathematique, 319, 1317

Gull, S. F. 1989, in Maximum Entropy and Bayesian Methods, ed. J. Skilling, (Dordrecht: Kluwer), 53

Gull, S. F. & Skilling, J. 1991, MemSys5 Quantified Maximum Entropy User's Manual, (Royston: Maximum Entropy Data Consultants Ltd.)

Kolmogorov, A. N. 1965, Inf. Transmission, 1, 3

Piña, R. K. & Puetter, R. C. 1993, PASP, 105, 630

Press, W. H., Teukolsky, S. A., Vetterling, W. Y., & Flannery, B. P. 1992, Numerical Recipes in Fortran (2nd ed.), (Cambridge: Cambridge Univ. Press)

Puetter, R. C. 1994, in SPIE Proc., Vol. 2302, Image Reconstruction and Restoration, ed. T. J. Schulz & D. L. Snyder, (Bellingham: SPIE), 112

———, 1995, Int. J. Image Sys. & Tech., 6, 314

———, 1997, in Instrumentation for Large Telescopes, ed. J. M. Rodriguez Espinosa, A. Herrero, & F. Sanchez, (Cambridge: Cambridge Univ. Press), 75

Puetter, R. C. & Piña, R. K. 1993, in SPIE Proc., Vol. 1946, Infrared Detectors and Instrumentation, ed. A. M. Fowler, (Bellingham: SPIE), 405

———, 1994, in Science with High Spatial Resolution Far-Infrared Data, (JPL Publ. 95-4), (Pasadena: JPL), 61

Rice, W. 1993, AJ, 105, 67

Shannon, C. E. 1948, in Key Papers in the Development of Information Theory, ed. D. Slepian, (New York: IEEE Press), 1974

Skilling, J. 1989, in Maximum Entropy and Bayesian Methods, ed. J. Skilling, (Dordrecht: Kluwer), 45

Solomonoff, R. 1964, Inf. Control, 7, 1

Weir, N. 1991, in Proc. of the ESO/ST-ECF Data Analysis Workshop, ed. P. Grosbo & R. H. Warmels, (Garching: ESO), 115

Wiener, N. 1949, Extrapolation and Smoothing of Stationary Time Series, (New York: Wiley)

Astronomical Data Analysis Software and Systems VIII
ASP Conference Series, Vol. 172, 1999
D. M. Mehringer, R. L. Plante, and D. A. Roberts, eds.

Algorithms for CCD Stellar Photometry

Kenneth J. Mighell

Kitt Peak National Observatory, National Optical Astronomy Observatories, P.O. Box 26732, Tucson, AZ 85726-67332

Abstract. Basic algorithms for the production of stellar photometry from CCD observations are described and reviewed. Simple algorithms for detecting stars, finding their positions, and estimating the nearby sky background will be presented along with references to the extensive literature devoted to these topics. The basic techniques of aperture and PSF-fitting CCD stellar photometry are discussed in detail. Recent advances in PSF-fitting CCD stellar photometry are described and used to show how accurate V-band stellar photometry with the *Next Generation Space Telescope* could be accomplished with a 8-m primary mirror that is diffraction limited at near-infrared wavelengths.

1. Introduction to CCD Stellar Photometry

The problem of doing accurate stellar photometry with calibrated charge-coupled device (CCD) data amounts to that of condensing the intensity values of hundreds, thousands, or even millions of picture elements (pixels) into a list containing the magnitude and position of each star on the CCD image.

If the stellar field is not crowded, the astronomer can measure the magnitude of each star by doing the digital equivalent of aperture photometry. Stellar photometry becomes much more difficult if one wishes to study crowded stellar fields. The basic assumptions underlying the simple techniques of CCD aperture photometry are frequently not valid in crowded stellar fields and more sophisticated methods like Point-Spread-Function (PSF) model fitting must be used in order to achieve accurate photometry.

2. CCD Aperture Stellar Photometry

The process of determining the apparent magnitude of a star can be surprisingly complex even with a technique as simple as CCD aperture photometry. We start by assuming that our CCD observations have already been flat-fielded and calibrated. The basic process of aperture stellar photometry is, in principle, quite simple:

1. Find the star.
2. Center an aperture of N_A pixels on the star.
3. Add up the electrons within the aperture: S_A (units: electrons e^-).
4. Determine the nearby background flux: B (units: e^-pixel^{-1}).

5. Determine the instrumental magnitude: $m' = -2.5 \log[(S_A - N_A B)/1\,\mathrm{e}^-]$.
6. Determine the aperture correction: $\Delta m'$ (units: mag).
7. Compute the aperture-corrected instrumental magnitude: $m = m' + \Delta m'$.

The important process of transforming instrumental magnitudes to a standard system is extensively described in the literature and will not covered in this contribution due to space limitations.

2.1. Finding Stars

There are many techniques available for the detection of astronomical objects in CCD observations (see, e.g., Fischer & Kochanski 1994, Secker 1995, and references therein). The following is a short introduction to some of the techniques that can be used to detect stars in CCD observations.

The signal of a noisy digital image can frequently be enhanced by suppressing high spatial frequency noise in the image. For modern CCD stellar observations, this generally means suppressing photon and CCD readout noise. This can frequently be accomplished by using small digital low-pass filters like

$$\mathbf{LP}_{3\times3} = \begin{pmatrix} 1/16 & 1/8 & 1/16 \\ 1/8 & 1/4 & 1/8 \\ 1/16 & 1/8 & 1/16 \end{pmatrix}$$

which is commonly used in digital image processing for this purpose. Small digital low-pass filters are useful in finding stars in digital stellar observations that have had the background sky removed (Irwin 1985). The background sky of an image can frequently be approximated by using small median filters. Combining low-pass filters with median filters can be very useful; the LPD (low-pass difference) filter,

$$\mathbf{LPD}_{3\times3}(\mathcal{F}) = \mathbf{LP}_{3\times3}(\mathcal{F}) - \mathbf{MEDIAN}_{5\times5}\left[\mathbf{LP}_{3\times3}(\mathcal{F})\right]$$

works well with *Hubble Space Telescope* WF/PC and WFPC2 images, $\mathcal{F}$, for the purpose of detecting stars and other point sources (Appendix A of Mighell & Rich 1995). The LPD filter is a high-pass frequency filter which is effectively a digital analog of Malin's unsharp photographic masking technique (Malin 1977, 1981, and references therein).

The following listing, for example, is a FORTRAN implementation of a simple peak detector algorithm which identifies any pixel that is greater than any of its eight neighbors:

```
C
C A Simple Peak Detector Algorithm
C
C Copyleft (L) 1998 Kenneth J. Mighell (Kitt Peak National Observatory)
C
      SUBROUTINE PEAKER(IMAGE,NX,NY,PEAKMIN,PEAKMAX)
      INTEGER NX, NY, X, Y, XX, YY
      REAL    IMAGE(NX,NY), PEAKMIN, PEAKMAX, PIXEL, NEIGHBOR
      LOGICAL BINGO
      DO Y = 2,(NY-1)
      DO X = 2,(NX-1)
      PIXEL = IMAGE(X,Y)
      IF ((PIXEL.GE.PEAKMIN).AND.(PIXEL.LT.PEAKMAX)) THEN
        BINGO = .TRUE.
        DO YY = (Y-1),(Y+1)
        DO XX = (X-1),(X+1)
          IF (BINGO) THEN
```

```
            NEIGHBOR = IMAGE(XX,YY)
            IF (NEIGHBOR.GT.PIXEL) THEN
              BINGO = .FALSE.
            ELSE IF (NEIGHBOR.EQ.PIXEL) THEN
              IF ((XX.NE.X).OR.(YY.NE.Y)) THEN
                IF (((XX.LE.X).AND.(YY.LE.Y))
     &              .OR.((XX.GT.X).AND.(YY.LT.Y))) BINGO = .FALSE.
              ENDIF
            ENDIF
          ENDIF
        ENDDO
        ENDDO
        IF (BINGO) WRITE (*,*)
     & 'Found a peak at position (',X,',',Y,') with a value of',PIXEL
      ENDIF
      ENDDO
      ENDDO
      RETURN
      END
C23456
```

This algorithm works particularly well with CCD stellar observations that have had the background sky removed (e.g., LPD-filtered images).

2.2. Determining the Centers of Stars

Many CCD aperture photometry programs require the user to give the position of the star on the CCD frame. Most of these programs offer a choice of centering algorithms to determine the center of a star when provided with only a rough estimate. The review article of Stone (1989) compares the performance of five different digital centering algorithms under a wide range of atmospheric seeing and background-level conditions. It may be useful to create your own centroid algorithm based upon the special requirements of your particular analysis problem. The following listing, for example, is a FORTRAN implementation of a centroid algorithm which produces robust estimates without requiring an estimate of the the nearby "sky" background:

```
C
C A Simple Centroid Algorithm
C
C A modification of the Modified Moment Method (Stone 1989,AJ,97,1227)
C that works well with small apertures.
C
C Copyleft (L) 1998 Kenneth J. Mighell (Kitt Peak National Observatory)
C
      SUBROUTINE CENTROID(X,Z,NN,N,X_IN,X_OUT)
      INTEGER          N, NN ! <-- assumes that 1 <= N <= NN
      REAL             X(NN), Z(NN), X_IN, X_OUT, BIG
      DOUBLE PRECISION DELTA, SUM1, SUM2, DIFF, XX
      DOUBLE PRECISION INTENSITY, POSITION, MINIMUM
      INTEGER          I, J , NITERATIONS
      PARAMETER        (BIG=1E30,NITERATIONS=10)
      MINIMUM = BIG
      DO I = 1,N
        INTENSITY = Z(I) ! <-- Z(I) is the intensity at the position X(I)
        MINIMUM = MIN(INTENSITY,MINIMUM)
      ENDDO
      XX = X_IN
      DELTA = 0D0
      DO J = 1,NITERATIONS
        XX = XX + DELTA
        SUM1 = 0D0
        SUM2 = 0D0
        DO I = 1,N
          POSITION = X(I)
          INTENSITY = Z(I)
          DIFF = MAX((INTENSITY-MINIMUM),0D0)
          SUM1 = SUM1 + (POSITION-XX)*DIFF
          SUM2 = SUM2 + DIFF
        ENDDO
        DELTA = SUM1/SUM2
```

```
      ENDDO
      X_OUT = XX
      RETURN
      END
C23456
```

This algorithm is a simplified version of the one used by my CCDCAP aperture stellar photometry package[1]. If precise absolute or relatives positions from CCD observations are required, then one should investigate the extensive literature dedicated to CCD astrometry.

2.3. Determining the Nearby Background

The background ("sky") flux associated with a star is generally determined by analyzing the distribution of intensities of nearby background pixels. In the case of circular aperture photometry, the background flux is typically determined by analyzing the pixels in an annulus beyond the stellar aperture. The inner radius of the sky annulus is typically several FWHM[2] distant from the center of the aperture in order to avoid the inclusion of contaminating light from the star itself. The width of the annulus is typically large enough so that the annulus contains between ~50 and a few hundred pixels.

Many aperture photometry programs allow the user to set the background flux to be the modal value of the background intensity distribution. The mode of the background distribution is frequently estimated by using the following useful approximation,

$$\text{mode} \approx 3 \times \text{median} - 2 \times \text{mean} \quad (\text{for median} \leq \text{mean}),$$

(Wells 1979; Kendall & Stuart 1958; Haldane 1942; Pearson 1895). This approximation, unfortunately, is known to produce background estimates that are biased towards higher values (e.g., Newberry 1992). Better methods for the estimation of the background are available. Modal estimates of the background distribution *after* the iterative rejection of outlier pixel intensities beyond 2.5–3.0 standard deviations of the mean generally produce reasonable results (Da Costa 1992).

Many algorithms are available and most aperture photometry programs offer the user a choice of several methods to determine the background flux. For example, the popular APPHOT aperture photometry IRAF package offers the user the choice of 11 different ways the background can be estimated (Davis 1987). As always when using analysis software, the astronomer is strongly advised to carefully read the user documentation in order to understand which methodology will produce the best results for a given CCD observation or application.

2.4. Determining the Stellar Signal within an Aperture

The fundamental task of a CCD aperture photometry program is to accurately measure all the electrons, S_A, that fall within an aperture placed on a CCD image. Unless the background flux, B, is exactly zero electrons per pixel, one

[1] `http://www.noao.edu/staff/mighell/ccdcap/`

[2] The Full-Width-at-Half-Maximum of a critically-sampled Gaussian distribution is ~2.35 pixels.

can never directly measure the number of electrons, $S_\star$, from a star within the aperture. One measures, instead, the quantity $S_A = S_\star + N_A B$, where N_A is the aperture of the aperture (in pixels). The instrumental magnitude of an aperture measurement of a CCD observation of a star can thus be defined as

$$m \equiv -2.5 \log \left[\frac{S_\star}{1\,\mathrm{e}^-} \right] = -2.5 \log \left[\frac{S_A - N_A B}{1\,\mathrm{e}^-} \right]$$

where the final form on the right is in terms of observable quantities. By this definition, a stellar intensity of one electron would have an instrumental magnitude of zero. The estimated error of the instrumental magnitude is approximately

$$\Delta m \approx 1.0857 \left[\frac{\Delta S_\star}{S_\star} \right]$$

where $\Delta S_\star$ is the measurement error (in electrons) of the estimated number of electrons from the star within the aperture.

It is clearly important to determine the observable quantities S_A, B, and N_A as accurately as possible. Let us consider the aperture area, N_A, first. If a_j is the area of the jth aperture pixel, then the total area of the aperture, N_A, is simply

$$N_A = \sum_{j=1}^{A} a_j \ ,$$

where A indicates that the summation includes all pixels or partial pixels within the aperture. The area of the jth aperture pixel will be exactly one pixel only when the pixel is completely within the aperture, otherwise only a fraction of the pixel lies within the aperture and the value of a_j is between zero and one ($0 < a_j < 1$ pixel). The area of a *circular* aperture is $A = \pi r^2$ pixels where the radius of the aperture is r pixels. Although it is not difficult to exactly determine the partial pixel areas with square pixels and circular apertures (e.g., Fig. 1), several standard CCD aperture photometry programs only approximate the partial pixel areas. For example, the popular PHOT task of the APPHOT package approximates the circular aperture by an irregular polygon (Davis 1987). While this approximation is generally fine with large apertures, it can sometimes produce large systematic measurement errors for small aperture radii.

Let us now consider the sum of all electrons within the aperture (S_A). If z_j is the intensity (in electrons) of the jth aperture pixel, then one simple approximation of S_A is

$$S_A = \sum_{j=1}^{A} a_j z_j \ . \tag{1}$$

This approximation linearly weights the pixel intensity with the area of the pixel within the aperture. The PHOT task uses a linear pixel weighting algorithm that is very similar to this approximation. The use of Equation (1) implicitly assumes that the Point Spread Function is nearly flat at the edge of the aperture. The PSF is generally nearly flat only at large distances from the center of a star where the encircled-energy function is nearly equal to one (i.e., 100%). The use of Equation (1) is thus generally appropriate for large aperture radii. Wherever

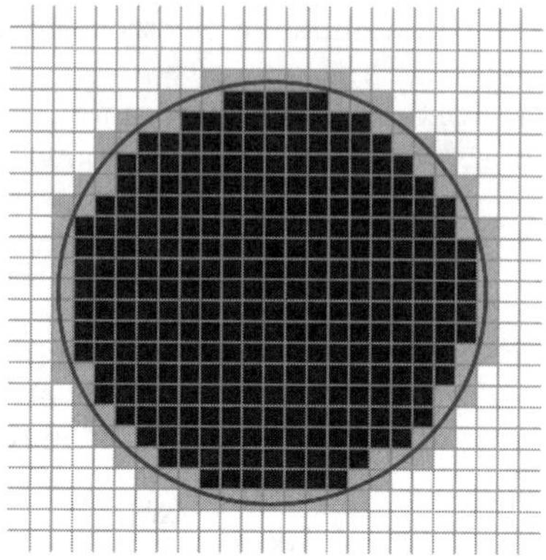

Figure 1. A circular aperture on an array of square CCD pixels. The *black* pixels are entirely within the aperture and the *gray* pixels are only partially within the aperture.

the PSF is a rapidly changing function of radius (e.g., at small radii of seeing-optimized CCD observations) the use of Equation (1) is likely to produce large systematic measurement errors.

One way of reducing this systematic measurement error is to split each pixel into subpixels by using a bilinear pixel interpolation algorithm. One such algorithm is to use the two-dimensional analog of the sinc function: $J_1(\pi r)/2r$. This function, like many others, has the unfortunate effect of degrading the original image by spreading photons (electrons) beyond the original pixel. I created created the QUADPX bilinear pixel interpolation algorithm which splits a pixel into 4 subpixels whose sum is always equal to that of the original pixel (Appendix B of Mighell & Rich 1995). Numerical experiments have shown that using the QUADPX algorithm can reduce the systematic measurement error of critically-sampled Gaussians by a factor of $\sim$6. For example, the photometric error of an aperture radius of 2.0 pixels went from 0.068 mag using Equation (1) to 0.011 mag with the CCDCAP package which implements the QUADPX algorithm.

2.5. Optimal Aperture Size and Aperture Corrections

The best (smallest) stellar photometric errors (i.e., the largest signal-to-noise ratios) are generally obtained with relatively small apertures (see, e.g., Fig. 6 of Howell 1989). Analysis of theoretical CCD signal-to-noise-ratio equations (see, e.g., Newberry 1991, Howell 1992, Merline & Howell 1995, Howell *et al.* 1996, and references therein) shows that large apertures can have large photometric errors when the the total number of stellar photons in the aperture becomes comparable with the total number of background photons in the aperture. Furthermore, a measurement error for the background flux as small as just 1 electron per pixel can by itself produce large photometric uncertainties at large aperture radii. Small apertures, however, can be *too* small when they allow such a small fraction of the star light to be found within the aperture that the photometric error becomes dominated by small-number (a.k.a. counting or Poisson) statistics because little or no signal has been measured.

A Gaussian is a good model for the Point Spread Function of a ground-based CCD observation since the central core of a ground-based stellar profile is approximately Gaussian (King 1971). One can easily show that the optimum signal-to-noise ratio for a Gaussian PSF is obtained for a circular aperture radius of $\sim 1.6\sigma$ (i.e., $r \approx 0.68\,\mathrm{FWHM}$) which contains about 72% of the encircled-energy. Pritchet & Kline (1981) note that the signal-to-noise ratio is fairly insensitive to radius near the "optimal" radius value $\sim 1.6\,\sigma$ for a Gaussian PSF; deviations from the optimal radius by as much as $\pm 50\%$ generally make little difference. Since centering errors will be more critical for smaller apertures than for larger apertures, it is only prudent to err on the larger side by using apertures with radii which are larger than $\sim 0.68\,\mathrm{FWHM}$.

An aperture radius of $r \approx \mathrm{FWHM}$ makes an excellent practical compromise between concerns about systematic centering errors and diminishing signal-to-noise ratios typically obtained with larger aperture radii. By analyzing the CCD signal-to-noise ratio equations, one can show that brighter stars will have larger optimal aperture sizes than do fainter stars. If one must use only one aperture size, then it is clearly advantageous to chose a global aperture size which produces the smallest photometric errors for the faintest stars (i.e., use $r \approx \mathrm{FWHM}$).

Small apertures frequently do not contain all the flux from a star. The amount of the missing star light can found by determining the appropriate aperture correction by measuring nearby bright isolated stars. Howell (1989) and Stetson (1990), among others, describe the process how aperture corrections can be accurately determined using the aperture growth-curve method.

3. CCD PSF-Fitting Stellar Photometry

Consider a ground-based CCD observation of two stars whose stellar images overlap. Assuming we already know the Point Spread Function of the observation, a simple model of the observation will have seven parameters: peak intensities (I_1, I_2), positions (X_1, Y_1, X_2, Y_2), and the background sky level B which is assumed to be the same for both component images. One finds that the parameters are not independent for overlapping stars with the presence of photon and readout noise. The conservation of photon flux will require that if I_1 increases then I_2 must decrease and vice versa for a given value of B. The most accurate photometry possible is obtained when these dependent parameters are fitted simultaneously. Any reasonable model of two overlapping stellar images will be a non-linear function when the positions and peak intensities are to be determined simultaneously. The technique of non-linear least-squares fitting was developed to provide for the simultaneous determination of dependent or independent parameters of non-linear model functions.

Assume that we have a calibrated CCD observation with N pixels and that z_i is the intensity in electrons (e^-) of the ith pixel at (x_i, y_i) with an error of σ_i. Let $model(x, y, ;a_1, \ldots, a_M)$ be a model of the intensity values that has two coordinates (x, y) and M parameters. For notational convenience, let the vector $\boldsymbol{r}_i$ represent the coordinates (x_i, y_i) and the vector $\boldsymbol{a}$ represent all the parameters [i.e., $\boldsymbol{a} \equiv (a_1, \ldots, a_M)$]. Thus the model of intensities will normally be written as $model(\boldsymbol{r}_i; \boldsymbol{a})$.

The measure of the goodness of fit between the data and the model, called chi-square, is defined as

$$\chi^2(\boldsymbol{a}) \equiv \sum_{i=1}^{N} \frac{1}{\sigma_i^2}[z_i - model(\boldsymbol{r}_i;\boldsymbol{a})]^2 \quad .$$

The theory of least squares states that the optimum value of the parameter vector $\boldsymbol{a}$ is obtained when $\chi^2(\boldsymbol{a})$ is minimized with respect to each parameter simultaneously. If the function $\chi^2(\boldsymbol{a})$ is thought of as a surface in the M-dimensional parameter space and if $\boldsymbol{a}_{\mathrm{true}}$ is defined as the optimal parameter vector, then the absolute minimum of that surface is $\chi^2(\boldsymbol{a}_{\mathrm{true}})$. The surface of the function $\chi^2(\boldsymbol{a})$ might be very complicated and the fitting algorithm must be able to converge to believable solutions (answers) even in ill-defined situations.

For some small correction parameter vector $\boldsymbol{\delta}$ we can approximate $\chi^2(\boldsymbol{a}+\boldsymbol{\delta})$ by its Taylor series expansion:

$$\chi^2(\boldsymbol{a}+\boldsymbol{\delta}) = \sum_{n=0}^{\infty} \frac{1}{n!}(\boldsymbol{\delta}\cdot\nabla)^n \chi^2(\boldsymbol{a}) \approx \chi^2(\boldsymbol{a}) + \boldsymbol{\delta}\cdot\nabla\chi^2(\boldsymbol{a}) + \frac{1}{2}\,\boldsymbol{\delta}\cdot\boldsymbol{H}\cdot\boldsymbol{\delta}$$

where

$$[H]_{jk} \equiv \frac{\partial^2\,\chi^2(\boldsymbol{a})}{\partial a_j \partial a_k}$$

the jkth element of the $M\times M$ Hessian matrix $\boldsymbol{H}$ of $\chi^2(\boldsymbol{a})$ [see, e.g., Arfken 1970; Press et al. 1986]. If $\chi^2(\boldsymbol{a}+\boldsymbol{\delta})$ is a local minimum of χ^2, then it can be shown that

$$\boldsymbol{H}\cdot\boldsymbol{\delta} = -\nabla\chi^2(\boldsymbol{a}) \ .$$

By solving this equation for the correction vector $\boldsymbol{\delta}$ it is then possible to determine a better parameter vector $\boldsymbol{a}_{\mathrm{new}} = \boldsymbol{a}_{\mathrm{old}} + \boldsymbol{\delta}$. When the parameter vector $(\boldsymbol{a})$ is redefined to be the better parameter $(\boldsymbol{a}_{\mathrm{new}})$, the Hessian matrix and the gradient of $\chi^2(\boldsymbol{a})$ can then be recalculated to determine a new correction vector $(\boldsymbol{\delta})$. This process repeats until $\boldsymbol{\delta}$ is sufficiently small. The final parameter vector is called the optimal parameter vector, $\boldsymbol{a}_0$, and should be very close to the optimal parameter vector $(\boldsymbol{a}_{\mathrm{true}})$ if the fit is good.

This method will find the absolute minimum of $\chi^2(\boldsymbol{a})$ if the original guess $\boldsymbol{a}$ is near $\boldsymbol{a}_{\mathrm{true}}$. Unfortunately, the original guess of the parameter vector $\boldsymbol{a}$ may not always be very good. For production stellar photometry software it is important that the search for the absolute minimum of χ^2 be both robust and efficient.

3.1. Analytical Models

One can frequently create a realistic intensity model of a ground-based CCD observation of a total of K stars on a non-flat background with a combination of Moffat (1969) functions on a tilted plane:

$$\begin{aligned} model(\boldsymbol{r};\boldsymbol{a}) &\equiv model(x,y;B_0,B_X,B_Y,I_1,X_1,Y_1,\rho_1,\tau_1,\ldots,I_K,X_K,Y_K,\rho_K,\tau_K) \\ &\equiv B_0 + B_X x + B_Y y + \sum_{k=1}^{K} I_k \Psi_k \end{aligned}$$

where the B_0, B_X, B_Y terms describe the tilted-plane model of the background and I_k is the intensity of the kth star which has a Point Spread Function of

$$\Psi_k \equiv \int_{x-0.5}^{x+0.5} \int_{y-0.5}^{y+0.5} \left[1 + \left\{ \frac{(x - X_k)^2 + (y - Y_k)^2}{\rho_k^2} \right\} \right]^{-\tau_k} dy\, dx \,. \qquad (2)$$

The partial derivatives of *model*$(r; a)$ must now be determined:

$$\frac{\partial model}{\partial B_0} = 1, \quad \frac{\partial model}{\partial B_X} = x, \quad \frac{\partial model}{\partial B_Y} = y, \quad \frac{\partial model}{\partial I_k} = \Psi_k,$$

$$\frac{\partial model}{\partial X_k} = I_k \frac{\partial \Psi_k}{\partial X_k}, \quad \frac{\partial model}{\partial Y_k} = I_k \frac{\partial \Psi_k}{\partial Y_k}, \quad \frac{\partial model}{\partial \rho_k} = I_k \frac{\partial \Psi_k}{\partial \rho_k}, \quad \frac{\partial model}{\partial \tau_k} = I_k \frac{\partial \Psi_k}{\partial \tau_k}.$$

If the PSF, Ψ_k, does not have a general analytical solution, it can usually be approximated numerically. For example, if we subdivide a pixel into η^2 subpixels, then Equation (2) can be approximated as $\Psi_k \approx \gamma_k^{-\tau_k}$ where

$$\gamma_k^{-\tau_k} \equiv \frac{1}{\eta^2} \sum_{i=1}^{\eta} \sum_{j=1}^{\eta} \left[1 + \left\{ \frac{([x - \xi + \eta^{-1} i] - X_k)^2 + ([y - \xi + \eta^{-1} j] - Y_k)^2}{\rho_k^2} \right\} \right]^{-\tau_k}$$

and $\xi \equiv (\eta + 1)/(2\eta)$. The new partial derivatives of *model*$(r; a)$ can now be determined as follows

$$\frac{\partial model}{\partial I_k} \approx \gamma_k^{-\tau_k}, \quad \frac{\partial model}{\partial X_k} \approx \frac{2 I_k \tau_k}{\rho_k^2} \frac{(x - X_k)}{\gamma_k^{\tau_k + 1}}, \quad \frac{\partial model}{\partial Y_k} \approx \frac{2 I_k \tau_k}{\rho_k^2} \frac{(y - Y_k)}{\gamma_k^{\tau_k + 1}},$$

$$\frac{\partial model}{\partial \rho_k} \approx \frac{2 I_k \tau_k}{\rho_k^3} \frac{[(x - X_k)^2 + (y - Y_k)^2]}{\gamma_k^{\tau_k + 1}}, \quad \frac{\partial model}{\partial \tau_k} \approx -I_k \gamma_k^{-\tau_k} \ln(\gamma_k) \,.$$

Simulations and practical experience has demonstrated that the $\gamma_k^{-\tau_k}$ approximation with the Levenberg-Marquardt method of non-linear least squares (Levenberg 1944, Marquardt 1963) produces accurate and precise CCD stellar photometry (Mighell 1989, 1990).

3.2. Digital Models

One can create a realistic intensity model of a *Hubble Space Telescope* (*HST*) CCD observation of K stars on a non-flat background with a combination of *digital* Point Spread Functions on a tilted plane:

$$\begin{aligned} model(\boldsymbol{r}; \boldsymbol{a}) &\equiv model(x, y; B_0, B_X, B_Y, I_1, X_1, Y_1, \Psi_1, \ldots, I_K, X_K, Y_K, \Psi_K) \\ &\equiv B_0 + B_X x + B_Y y + \sum_{k=1}^{K} I_k \Psi_k (x - X_k, y - Y_k) \end{aligned}$$

where the B_0, B_X, B_Y terms describe the tilted-plane model of the background and I_k is the intensity of the kth star which has a digital PSF, Ψ_k, that is represented with matrix of numbers with a sum of one. If the kth star is isolated and on a flat background, then the digital PSF, Ψ_k, can be easily derived from the actual image. Alternatively, Ψ_k could be a synthetic PSF computed by PSF modeling software [e.g., the TINY TIM package[3] of Krist (1993, Krist & Hook

[3] `ftp://ftp.stsci.edu/pub/software/tinytim/`

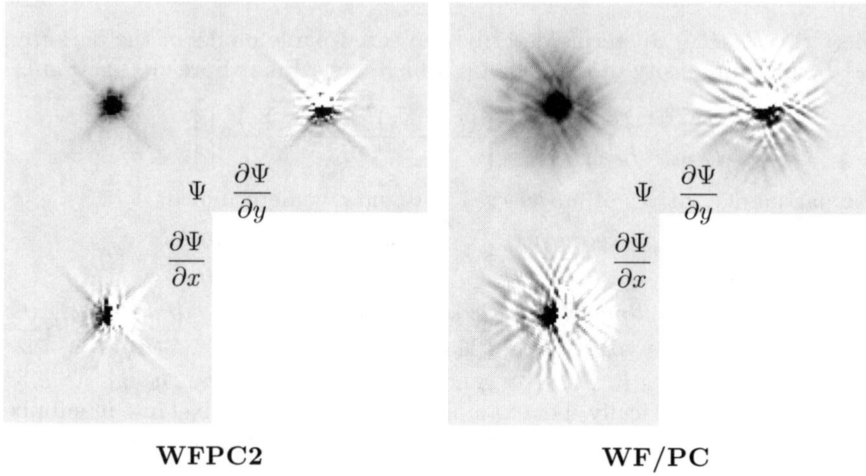

Figure 2. Typical Point Spread Functions (Ψ) of the *Hubble Space Telescope* WFPC2 *(left)* and WF/PC *(right)* instruments (M star, F555W filter) and their respective partial derivatives with respect to the directions x ($\partial\Psi/\partial x$) and y ($\partial\Psi/\partial y$). These PSFs were synthesized using TINY TIM VERSION 4.0B (Krist 1994).

1997)]. With a digital PSF in hand, one can then easily determine the partial derivatives with respect to x and y using standard numerical differentiation techniques (see Fig. 2). The art of CCD photometry with digital Point Spread Functions lies in the implementation details. While the mathematics is the same as for the case of analytical PSFs, the software engineering issues are significantly more challenging.

3.3. Recent Advances

Traditional PSF-fitting CCD photometric reduction packages like DAOPHOT (Stetson 1987) use analytical functions to represent the Point Spread Function. All the major partial derivative computations are computed on the analytical model of the PSF. Any deviations of the real-world PSF from the analytical PSF are then generally stored in a residual matrix which is only used to determine the χ^2 goodness-of-fit.

I have recently demonstrated the feasibility of doing accurate CCD stellar photometry with digital Point Spread Functions. I have developed a new digital PSF-fitting algorithm which does not require a residual matrix because all partial derivative computations are done on the digital PSF itself using standard numerical differentiation techniques. This algorithm has already passed the proof-of-principle stage with the successful reduction of simulated *Next Generation Space Telescope (NGST)* CCD stellar observations (see Fig. 3).

I investigated the performance of CCD stellar photometry with a 1.5-micron diffraction-limited 8-m Next Generation Space Telescope. These simulations used artificial Point Spread Functions for three different 8-m NGST design concepts which were kindly provided by John Krist. Assuming that the 8-m NGST primary mirror has 1/13 wave RMS errors at 1.5 micron, I determined that 90% of the light from a star falls within an aperture radius of 0.1 arcsec — the size of one WF pixel of the Hubble Space Telescope WFPC2 instrument. The three NGST design concepts have nearly identical V-band encircled-energy functions; the degradation caused by the differences between the three *NGST* design concepts is quite negligible when state-of-the-art digital-PSF photometric reduction software is used to analyze uncrowded stellar fields.

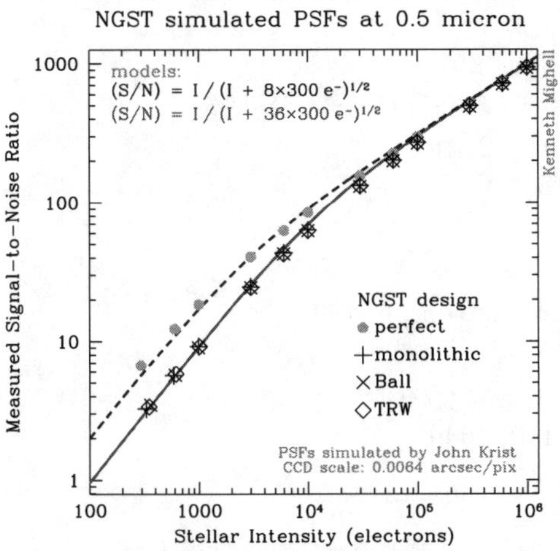

Figure 3. Measured signal-to-noise ratios of simulated *NGST* CCD stellar observations using the Levenberg-Marquardt method of non-linear least squares fitting with *digital* Point Spread Functions. Each mark represents the average signal-to-noise ratio of 100 stars that were simulated using using synthetic V-band PSFs on a flat background of 300 electrons per pixel. The filled circles show the result using a critically sampled V-band PSF for a perfect 0.5-μm diffraction-limited 8-m *NGST* mirror. The V-band results for three different 8-m *NGST* design concepts of a 1.5-μm diffraction-limited mirror.

All the V-band PSFs used in these simulations were sampled at 0.0064 arcsec pixel^{-1} which is the critical V-band sampling rate of a perfect 0.5-micron diffraction-limited 8-m NGST. Better photometric performance for all three NGST design concepts could be obtained by using larger CCD pixels: a pixel size of $\sim$0.013 ($\sim$2 $\times$ 0.0064) arcsec pixel^{-1} should provide an optimal pixel sampling when the 8-m primary mirror is diffraction-limited at 1.5 micron.

References

Arfken, G. 1970, Mathematical Methods for Physicists (2nd ed.), (New York: Academic Press)

Da Costa, G. S. 1992, in ASP Conf. Ser., Vol. 23, Astronomical CCD Observing and Reduction Techniques, ed. S. B. Howell, (San Francisco: ASP), 90

Davis, L. 1987, "Specifications for the Aperture Photometry Package", National Optical Astronomy Observatories, `ftp://iraf.noao.edu/iraf/docs/apspec.ps.Z`

Fischer, P. & Kochanski, G. P. 1994, AJ, 107, 802

Haldane, J. B. S. 1943, Biometrika, 32, 294

Howell, S. B. 1989, PASP, 101, 616

———, 1992, in ASP Conf. Ser., Vol. 23, Astronomical CCD Observing and Reduction Techniques, ed. S. B. Howell, (San Francisco: ASP), 105

Howell, S. B., Koehn, B., Bowell, E., & Hoffman, M. 1996, AJ, 112, 1302

Irwin, M. J. 1985, MNRAS, 214, 575

Kendall, M. G. & Stuart A. 1958, The Advanced Theory of Statistics, Vol. I, (London: Charles Griffin & Co.) pp. 39 and 179

King, I. R. 1971, PASP, 83, 199

Krist, J. 1993, in ASP Conf. Ser., Vol. 52, Astronomical Data Analysis Software and Systems II, ed. R. J. Hanisch, R. J. V. Brissenden, & J. Barnes (San Francisco: ASP), 530

———, 1994, Tiny Tim User's Manual (Ver. 4.0)

Krist, J. & Hook, R. 1997, Tiny Tim User's Guide (Ver. 4.4)

Levenberg, K. 1944, Quart. of Appl. Math., 2, 164

Malin, D. 1977, AAS Photo-Bull., No. 16, 10

———, 1981, J. Phot. Sci., 29(5), 199

Marquardt, D. 1963, J. SIAM, 11, 431

Merline, W. J. & Howell, S. B. 1995, Exp. Astron., 6, 163

Mighell, K. J. 1989, MNRAS, 238, 807

———, 1990, A&AS, 82, 1

Mighell, K. J. & Rich, R. M. 1995, AJ, 110, 1649

Moffat, A. F. J. 1969, A&A, 3, 455

Newberry, M. V. 1991, PASP, 103, 122

———, 1992, in ASP Conf. Ser., Vol. 25, Astronomical Data Analysis Software and Systems I, ed. D. M. Worrall, C. Biemesderfer, & J. Barnes (San Francisco: ASP), 307

Pearson, K. 1895, Phil. Trans., 186, 343

Press, W. H., Flannery, B. P., Teukolsky, S. A., & Vetterling, W. T. 1986, Numerical Recipes, (Cambridge: Cambridge Univ. Press)

Pritchet, C. & Kline, M. I. 1981, AJ, 86, 1859

Secker, J. 1995, PASP, 107, 496

Stetson, P. B. 1987, PASP, 99, 191

———, 1990, PASP, 102, 932

Stone, R. C. 1989, AJ, 97, 1227

Wells, D.C. 1979, in SPIE Proc., Vol. 172, Instrumentation in Astronomy III, ed. D. L. Crawford, (Bellingham: SPIE), 418

Astronomical Data Analysis Software and Systems VIII
ASP Conference Series, Vol. 172, 1999
D. M. Mehringer, R. L. Plante, and D. A. Roberts, eds.

Joint Processing of Radio Data Produced by the SSRT Together With Data of Other Spectral Ranges

V. V. Grechnev, A. T. Altyntsev, S. K. Konovalov, S. V. Lesovoi

Institute of Solar-Terrestrial Physics, Lermontov St. 126, Irkutsk, Russia 664033, Email: grechnev@iszf.irk.ru

Abstract. Some techniques and a set of IDL programs and routines were developed to process data obtained by the Siberian Solar Radio Telescope (SSRT) together with data of other spectral ranges.

1. Introduction

The emergence of abyssal magnetic fields from the Sun induces various physical processes in the solar corona. Radio observations allow one to study the solar corona against the background of the solar disk and augment the measurements of soft and hard X-rays. Radio observations are sensitive to both thermal and non-thermal components of plasma in quiet and perturbed conditions in the corona up to coronal mass ejections. They provide the only method for measuring the coronal magnetic fields because the calculations using photospheric data are not reliable under perturbed conditions. They also provide diagnostics for parameters of high-energy electrons. Radio observations give the only way to study the processes of particle acceleration and energy release in the outer corona.

The variety of emission mechanisms and their dependence on parameters of the emitting electrons make it extremely difficult to study solar processes without information provided by other spectral ranges. Namely, the magnetic configuration given by magnetograms along with the temperature and density distributions given by X-ray observations contribute to the correct interpretation of radio data. Thus, the study of the solar activity based on the radio observations suggests the involvement of multi-spectral data. Thanks to the development of the Internet, astronomers can now, in fact, use data of all ground-based and orbital observatories. On the other hand, the use of the heterogeneous material requires various methods of data processing, coordination (overlay) of solar images in different formats, etc.

Experience shows that data processing is usually done in a few stages. In the first stage, routine procedures are performed (rejecting of records, building and calibrating maps, bringing them to a standard form, etc.). Subsequent steps include previewing, selecting foremost records, correlating them with data obtained by other instruments, measuring the parameters of radio sources, and calculating estimates of the physical conditions and the model.

As a rule, the first level of data reduction is executed by routine instrument-specific software. Subsequent data processing is often carried out under conditions when not only techniques but also the approach to the solution of a particular physical task is not clear. These conditions suggest interactive work when various techniques of data processing can be tested.

An effective tool to solve such tasks is Interactive Data Language (IDL[1]) which has been used by many observatories and institutions for some years. Taking into account its advantages, we have chosen it as the base in solving of our tasks. We present a set of IDL programs and routines which process data obtained by the Siberian Solar Radio Telescope (SSRT) (Smolkov et al. 1986; Altyntsev et al. 1994; Uralov et al. 1998) together with data of other spectral ranges: Nobeyama Radioheliograph (NRH), Yohkoh, BATSE, Kitt Peak, Big Bear, etc. Besides well-known methods, these programs realize some specific techniques developed by us. Some of the programs which perform tasks that are particularly demanding of computer time and resources were written in C++.

2. Conventional Procedures

2.1. Calibrations of the Intensity and the Spatial Position

We usually apply both intensity and spatial self-calibration to the observational data. In complicated cases, solar images obtained by other instruments are also used.

Radio maps are produced by the SSRT using frequency scanning. Hence non-uniformities of frequency characteristics of both the feed and receiver systems result in appearance of "geometric" noise whose spectrum overlaps the signal's spectrum. We cannot employ an external signal for the calibration, so we apply self-calibration for both the feed and receiver characteristics. For this purpose, mean and standard deviation are calculated for each channel within some time interval (usually $\sim$ 1 hour) from which zero values and gains are obtained.

For the intensity calibration, we use the technique developed by Dr. Y. Hanaoka (Hanaoka et al. 1994) which is based on statistical analysis of the image. As two values dominate in solar radio maps (the levels of the sky and the "quiet" Sun), the histogram of any map must contain two prominent peaks. The distance between those peaks is the level of the quiet Sun expressed in corresponding units. Since the SSRT is an equidistant antenna array, there exists a possibility to use not only clean, but also dirty maps because of good coverage of the $(u, v)-$plane. The use of dirty maps for the intensity calibration results in some systematic errors in brightness temperature of compact sources. This error depends on the angular size of radio sources and can reach as high as 20%. In complicated cases, when the SSRT beam is essentially distorted, the calibration can be made in the interactive mode. In this method, two regions are selected in images: the first region corresponding to the sky, and the second to the quiet Sun.

Precise centering of solar images based upon the calculated orientation of the SSRT beam cannot be ensured because of many displacing instrumental factors. We use the well-known cross-correlation technique on the solar map

[1]IDL is the trademark of Research Systems, Inc.

and the model disk. The presence of patterns across images are sometimes due to distortions of the beam. To reduce their influence, the linear decline is subtracted from the image. Here, pixel values in the maps are limited to exclude the influence of bright sources.

The accurate centering of the response of an interferometer operated in the 1-dimensional (1-D) mode ("scans") is hardly possible without use of 2-D images. The centering of SSRT scans can be made, for example, using radio maps of the NRH. In such a case, a 1-D scan is computed from the 2-D image along the proper direction. The contribution of the quiet Sun is suppressed in both images, and then the centering is made referring to compact quasi-stable sources using the cross-correlation technique.

2.2. Overlay of Data Obtained by Different Instruments

To compare images obtained in different spectral ranges (and, of course, by different astronomical instruments), their correlation is needed. It is convenient to do this using an overlaid contour image.

Images to be compared must be projected onto the same grid. In some cases, all needed values are already contained in the data files (e.g., FITS files of NRH, Kitt Peak Vacuum Telescope, etc.). In other cases, both the solar radius and the coordinates of the solar center are found in the interactive mode. Then the images are regridded and brought into coincidence.

It is possible to correlate both SSRT 1-D scans and 2-D images with any 2-D solar images obtained by other instruments.

Of course, if the interval between observations of the solar images to be correlated was ≥ 1 hour, the sources on the solar surface will be displaced. This displacement must be corrected.

2.3. Compensation for the Rotation of the Sun

A rather simple but efficient routine was developed to "rotate" solar images around its polar axis. This transformation is used in solving such tasks:

- "synchronizing" images obtained at different instances,
- creating high-sensitivity images by averaging of frames obtained within a long interval (some hours),
- monitoring the development of active regions. Here, this transformation allows to correct images for the projection distortions.

2.4. Measurements of Radio Source Characteristics and Their Variations

The programs and routines perform measurements of coordinates, sizes, areas, brightness temperatures, and fluxes of radio sources (Konovalov et al. 1997a, 1997b). They also build time profiles of variations for these characteristics. Some programs have widget-based graphical user interfaces which allow one to perform standard measurements easily and efficiently.

3. Special Techniques

Astronomical instruments can create many images during an event which contain large amounts of information. Integral characteristics help to reveal essential features. Along with the total flux measurement, we also use the effective height of sources which can be measured either as a weighted center, or as coordinates where the second derivative has its minima. However, structural variations are masked in this case. Additional features can be detected by means of another integral characteristic: the effective area of an emitting region computed as its flux divided by maximum of the brightness temperature over the image.

4. Conclusion

The programs and routines have been checked under MS Windows 95 and UNIX with IDL versions 3.0.1 through 5.1. They are used extensively in data processing for the SSRT and the NRH. This software was used in a comprehensive investigation of a range of solar events since 1992. We plan to put our IDL programs onto our Web site, `http://ssrt.iszf.irk.ru`.

Acknowledgments. We are indebted to Dr. D. Stern (Research systems, Inc.) for his assistance in updating the programs to IDL 5.0. We are grateful to the ADASS'98 conference organizing committee for the financial aid which has made possible the attendance of the presenter (V. Grechnev) at the conference. We wish to thank the editors of these proceedings for their help with this paper. This work was supported by the Russian projects of RFFI No. 96-02-16648, No. 97-02-16906, and Astronomia.

References

Altyntsev, A. T., et al. 1994, A&A, 287, 256

Hanaoka Y., et al. 1994, in Proc. of the Kofu Symp. (NRO Report No. 357), ed. S. Enome & T. Hirayama, 35

Konovalov, S. K., Altyntsev, A. T., Grechnev, V. V., Lisysian, E. G., Rudenko, G. V., & Magun, A. 1997a, in ASP Conf. Ser., Vol. 125, Astronomical Data Analysis Software and Systems VI, ed. G. Hunt & H. E. Payne (San Francisco: ASP), 100

——, 1997b, in ASP Conf. Ser., Vol. 125, Astronomical Data Analysis Software and Systems VI, ed. G. Hunt & H. E. Payne (San Francisco: ASP), 447

Smolkov, G. Ya., Pistolkors, A. A., Treskov, T. A., Krissinel, B. B., Putilov, V. A. 1986, Ap&SS, 119, 1

Uralov, A. M., Grechnev, V. V., Lesovoi, S. V., Sych, R. A., Kardapolova, N. N., Smolkov, G. Y., & Treskov, T. A. 1998, Solar Physics, 178, 557

Astronomical Data Analysis Software and Systems VIII
ASP Conference Series, Vol. 172, 1999
D. M. Mehringer, R. L. Plante, and D. A. Roberts, eds.

Infrared Jitter Imaging Data Reduction: Algorithms and Implementation

Nicolas Devillard

European Southern Observatory, Munich, Germany

Abstract. Jitter imaging (also known as *microscanning*) is probably one of the most efficient ways to perform astronomical observations in the infrared. It requires very efficient filtering and recentering methods to produce the best possible output from raw data. This paper discusses issues attached to Poisson offset generation, efficient infrared sky filtering, offset recovery between planes through cross-correlation and/or point pattern recognition techniques, plane shifting with subpixel resolution through various kernel-based interpolation schemes, and 3D filtering for plane accumulation. Several algorithms are described for each step, having in mind an automatic data processing in pipeline mode (i.e., without user interaction) as intended for the Very Large Telescope. Implementation of these algorithms in optimized ANSI C (the `eclipse` library) is also described here.

1. Introduction

Jitter imaging is an efficient method used in astronomical infrared observations to take care of sky background issues with a minimum loss of observing time. By observing the same spot on the sky with small offsets around a central position for each exposure, it is hopefully possible to deduce the sky background variations directly by filtering the images, and to separate the astronomical and the sky signals.

A typical acquisition in jitter mode consists in a set of 10 to 100 frames. The first frame is assumed to be centered on the point of interest; the following frames are slightly offset from the first position with offsets no larger than a reasonably small fraction of the detector size (15% for SOFI/ISAAC). The data reduction process consists of sky estimation and subtraction, plane recentering, and finally frame co-addition to a single frame. Each processing step is discussed here in the context of automatic data reduction in pipeline mode (i.e., without user interaction).

2. Poisson Offset Generation

The following procedure is used to generate homogeneous offsets to use for a standard jittered observation. Points are generated with a Poisson sampling law (Wolberg 1990) which specifies that no two points are closer to each other than a minimal distance. This distance is computed by taking into account the size of the (rectangular) domain in which points are generated, and the number of points to generate. The minimal distance between any two points for a set of n points generated in a rectangle of size $S_x \times S_y$ is chosen as:

$$d(S_x, S_y, n) = \sqrt{\frac{S_x S_y}{n\sqrt{2}}} \tag{1}$$

3. Sky Background Filtering

Sky background filtering is done through a method called *sky combination*. The method is an iteration on each pixel position on the detector along a time line. The algorithm can be described for each pixel position as:

- Loop on all planes. For each plane, do the following:
- Loop on all pixel positions on the detector. For each position, do the following:
- Extract an array of $\pm h$ pixel values in time around the current pixel (this produces $2h + 1$ values).
- Scale (divide) each pixel in the array by the (filtered) average value of the plane from which it was extracted.
- Sort this scaled array by increasing pixel value.
- Reject the $rmin$ and $rmax$ pixel values in the sorted array.
- Average the remaining pixel values and assign the result to the same pixel position in the same current plane in the output cube.

Notice that if $rmin = rmax = h$, this method is equivalent to a running median filter; only the central sorted value is kept.

4. Frame Offset Detection

The next step is to measure or acquire the offsets between the frames. It is reasonable to assume that the acquisition orders for telescope pointing are available as ancillary data. If they are not, there should be an automatic way of finding these offsets with a minimal chance of error. Lastly, users may want to use their own method to detect offsets and feed their input into the algorithm for ultimate refining. We discuss here a cross-correlation method used both for automatic offset detection and refining up to subpixel resolution.

4.1. Points of Interest

Cross-correlation techniques need valid points in the image to work properly. We describe here a simple peak detector used to detect such points. In most cases, it should provide a reasonable number of points to use for offset detection.

Create a binary image from the input frame by thresholding with a $K\sigma$ clipping. Instead of using mean and sigma values for this clipping, we estimate these values with a *median* and *average absolute deviation from the median*, to be less sensitive to noise.

The value of K depends on the amount of noise and the number of objects present in the image. For very noisy images, it is best to lower K to a value between 0.1 and 1 (the default is 3).

This thresholding creates a crude map of bright regions in the input image. To clean bad pixels and other such small defects, a binary morphological erosion is then performed. A classical floodfill algorithm is then applied to estimate the centroid of each remaining region, weighted by pixel values of the input image. A list of detected object centers and maximum pixel values in each region is then returned.

4.2. Cross-correlation product

The cross-correlation of two images I_1 and I_2 is defined as the product:

$$\sum_{p_1 \in w_1} \sum_{p_2 \in w_2} p_1 \otimes p_2 \tag{2}$$

where p_1 is the pixel value which index is running over the domain of interest w_1 in the image I_1, and similarly p_2 a pixel value which 2-dimensional index is running over the domain of interest w_2 in the image I_2. The cross-correlation product we use is a sum of squared differences:

$$p_1 \otimes p_2 = \sum_{w_1, w_2} (p_1 - p_2)^2 \tag{3}$$

With this criterion, the point of greatest similarity corresponds to a minimum of difference in the cross-correlation product. A subpixel offset detection is then applied by fitting a parabola to the cross-correlation signal in x and y, and looking for the subpixel minimum. This method is theoretically precise up to a 1/100 pixel precision in the ideal case of noiseless images. The precision obtained with SOFI data is about 1/10 pixel.

In automatic mode when there is no indication of what the offsets between frames are, this cross-correlation technique is used within a wide search zone. When a first estimate of the offsets is available (refining mode), the search area is lowered to a few pixels in all directions to speed up the process.

5. Frame Registration

To shift an image by a subpixel offset, it is necessary to use resampling methods. Given some assumptions about the input signal cut-off frequency, an image is considered as the discretization of a continuous band-limited signal. According to Shannon's theorem, it is possible to completely retrieve this continuous signal given enough samples. It is then easy to *resample* the image, i.e., to construct a new image with samples taken at a new position of interest, namely the required offset. Interested readers are referred to Wolberg (1990).

In the present implementation, we chose to resample images in the image space. The interpolation kernel is based on the 16 closest pixel values, and the kernel definition is based on a hyperbolic tangent function in Fourier space:

$$H_k(f) = \left(\frac{tanh(k(x+0.5))+1}{2}\right)\left(\frac{tanh(k(-x+0.5))+1}{2}\right) \qquad (4)$$

This interpolation kernel is then computed in image space through a Fourier transform. It converges exponentially toward zero, which makes it an ideal candidate for resampling with a small pixel neighborhood.

6. 3D Averaging

The last step of the reduction is to concatenate all registered planes on the third axis into a single plane. The method we use is a filtered 3D average. For each pixel position, we reject outliers along the time axis and average remaining pixels to a single value. This yields the single output frame, which is the result of the data reduction process.

7. Conclusion

Interested readers are referred to a more complete ESO document at `http://www.eso.org/projects/dfs/papers/jitter98/`. The `eclipse` library has been used to implement this algorithm. The home page for this software package can be found at `http://www.eso.org/eclipse`.

References

Wolberg, G. 1990, Digital Image Warping, (Los Alamitos: IEEE Press)

Astronomical Data Analysis Software and Systems VIII
ASP Conference Series, Vol. 172, 1999
D. M. Mehringer, R. L. Plante, and D. A. Roberts, eds.

Combining Undersampled Dithered Data – A Review of the Options

Richard N. Hook, Norbert Pirzkal

Space Telescope – European Coordinating Facility, European Southern Observatory, Karl-Schwarzschild-Str. 2, D-85748 Garching bei München, Germany

Andrew S. Fruchter

Space Telescope Science Institute, 3700 San Martin Drive, Baltimore, MD 21218, USA

Abstract. Many imaging systems in astronomy have pixels which are too large to adequately sample the image falling on them. It is widely known that "dithering" – shifting by small amounts between exposures – can allow some of this lost information to be recovered. This technique is extensively used for HST WFPC2 and NICMOS imaging and is proposed for other systems.

Algorithms have been developed, by the authors and others, to allow the effective combination of such dithered, undersampled data. This paper describes a simple imaging model which allows the limits of dithered imaging to be readily appreciated and briefly reviews available software. The virtues and drawbacks of different methods are presented as a guide to workers with dithered data. An example using HST NICMOS data is given.

1. An Introduction to Undersampled Imaging

Imaging detector systems such as CCDs are often regarded as arrays of little square buckets which count up incoming photons which fall into the appropriate pixel area. An equivalent, more instructive view of the imaging process is to consider the image formation as a convolution of the optical image at the surface of the chip with the "pixel response function" (PRF, typically, but not always a square top-hat function the same size as the pixel) followed by a point-sampling of this new, smoother, continuous distribution. In the undersampled case this second sampling step is not done on a sufficiently fine grid to extract all the detail from the image. Dithered imaging can improve this sampling. However, the initial smoothing from the convolution with the PRF of the detector has already suppressed fine structure in the optical image – this is made worse by the noise – and this lost information cannot be fully recovered. For this reason large pixels with dithered imaging cannot fully replace a fully sampled imaging system.

2. Algorithms

Several methods have been proposed for the reconstruction of the true intensity distribution on the sky from dithered undersampled images. Reviews of the situation a few years back, biased towards the requirements of HST WFPC2 imaging are given in Hook & Adorf (1995) and Adorf & Hook (1995). These describe iterative methods, one (ACOADD) a simple multiple input channel generalization of the standard Richardson-Lucy restoration method and the second a "projection onto convex sets" (POCS) approach. ACOADD was originally developed for combining images with different PSFs but can also be used, with shifted PSFs and sub-sampled output images, for combining dithered under-sampled images. Its results are normally presented as restorations carried to convergence and subsequently convolved with a kernel function which suppresses spurious high spatial frequency information. This allows a choice of the effective output PSF within the constraints of artifacts on one side and destroying too much resolution on the other. The ACOADD method is available in the **stsdas.contrib** package within IRAF.

The requirements of the Hubble Deep Field imaging project late in 1995 led to the development of a direct, non-iterative linear approach to this problem which has become known as "drizzling" and widely used for HST and other data. The method is described in Fruchter & Hook (1997, 1998). It is available within the **dither** package in STSDAS and also from the Web[1]. Drizzling was used for the combination of all the imaging data from the Hubble Deep Field South observations in October 1998.

Recently Lauer (1998) has approached the problem in a different way and has derived a formally correct way of reconstructing an image from multiple dithered undersampled input images which has no loss of resolution at the Nyquist scale. This is achieved by suppressing artifacts caused by aliasing in Fourier space. This paper is also a detailed discussion of the subject in general. Although this approach may well be optimal in some cases, its recent appearance and the lack of a common-user software implementation at present mean that it cannot be included in the comparisons given below.

Both Drizzling and the Lauer method do not attempt to remove the effects of the optical PSF or the PRF of the detector. The images which they produce are hence free of the artifacts which arise when using non-linear methods to attempt to recover higher spatial frequency information which may not exist in the original data. On the other hand they fail to use such information when it does exist – typically in high contrast, high signal to noise regions of well-exposed images. In this regime non-linear restoration-based methods can be valuable.

3. Pros and Cons

The methods described above are in some respects complementary. To guide the prospective user, some of the most important advantages and disadvantages are tabulated here.

---

[1]`http://www.stsci.edu/~fruchter/dither/`

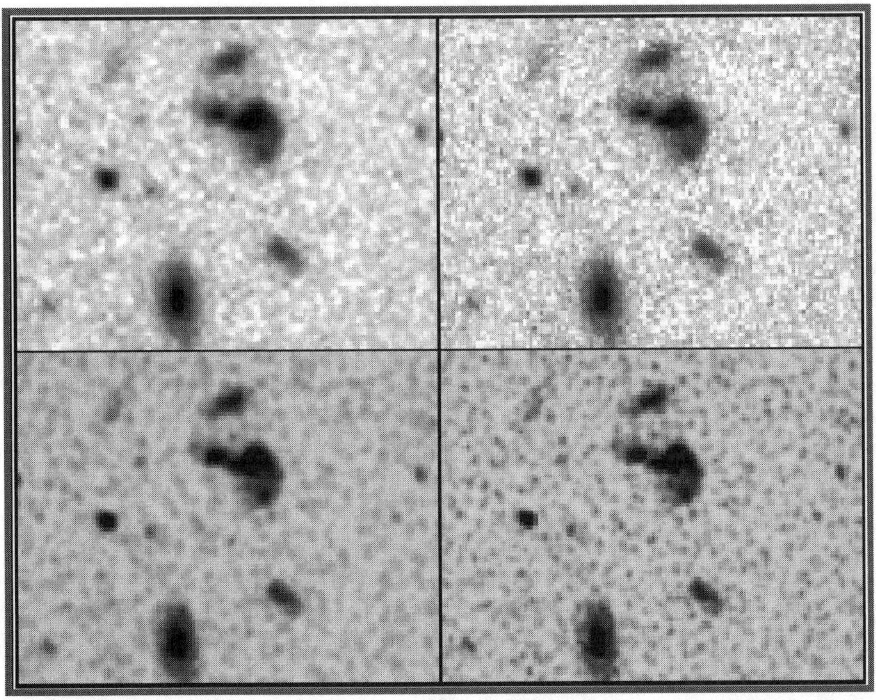

Figure 1. A comparison of combinations of deep HST NICMOS Camera 3 data. There were 9 evenly-spaced dither positions. The upper combinations were done with Drizzle and the lower ones with ACOADD. The upper left had $pixfrac = 1.0$ and the upper right $pixfrac = 0.5$. The lower left was smoothed with a Gaussian of $\sigma = 1.0$ output pixels and that at the lower right with $\sigma = 0.7$. Data courtesy Mark Dickinson (STScI).

3.1. Multichannel Lucy Restoration (ACOADD):

- + can attempt reconstruction of higher spatial frequency information (super-resolution)
- + can create output images with a specified resolution
- + can combine images with different PSFs
- – may introduce photometric bias of point-sources and other artifacts
- – cannot handle geometric distortion
- – creates output images with strongly correlated noise
- – requires good knowledge of the PSF
- – is computationally intensive and iterative

3.2. Drizzling

- + is photometrically faithful and introduces minimal artifacts
- + produces noise correlations only on small scales and in predictable ways
- + can handle arbitrary geometric distortions

- + can handle pixel weighting in a fully flexible way
- + is relatively fast and efficient for large images
- – requires images with matched PSFs
- – cannot achieve super-resolution
- – produces output images have a small amount of space-variant smoothing

4. An Example

To illustrate the relative merits of the two methods compared above, some deep HST NICMOS camera 3 imaging at 9 dither positions were combined using both ACOADD and Drizzle as the combination engine. The results are shown in Fig. 1 for two different parameters in each case. For all images the output image had a pixel spacing which was three times as fine as that of the input. The contrast and slight resolution enhancement offered by ACOADD is clear as is the strongly correlated noise speckling in the background.

References

Adorf, H.-M. & Hook, R. 1995, in Proc. ST-ECF Workshop, Calibrating and Understanding HST and ESO Instruments, (Garching: ESO), 251

Fruchter, A. S. & Hook, R. N. 1997, in SPIE Proc., Vol. 3164, Applications of Digital Image Processing XX, ed. A. Tescher (Bellingham: SPIE), 120

———, 1998, PASP, submitted, astro-ph/9808087

Hook, R. N. & Adorf, H.-M. 1995, in Proc. Calibrating Hubble Space Telescope: Post Servicing Mission, ed. A. Koratkar & C. Leitherer (Baltimore: STScI), 341

Lauer, T. R. 1998, PASP, 111, 227

Astronomical Data Analysis Software and Systems VIII
ASP Conference Series, Vol. 172, 1999
D. M. Mehringer, R. L. Plante, and D. A. Roberts, eds.

Quadrant DC Offset Removal in NICMOS Images Using an Eye-pleasing Criterion

I. Busko

Space Telescope Science Institute, Baltimore, MD 21218, Email: busko@stsci.edu

Abstract. This paper describes a simple technique that cancels the DC offsets often found in Hubble Space Telescope's calibrated NICMOS images. The technique is based on the minimization of features that are very easily detected by the human visual system.

1. The Problem

Hubble Space Telescope NICMOS images often show quadrants with a slight mismatch in DC offset levels after the calibration process. The cause of these residual offsets is still not fully understood, but the images can be empirically fixed by adding arbitrary DC constants to three arbitrarily chosen quadrants in order to bring them to the same DC level of the fourth, "reference" quadrant.

Thus the basic problem in fixing these images is to find the set of three numerical constants that will "fix" the image. But the definition of what is a "fixed" image is arbitrary, and a number of solutions have been proposed. They are all based on some sort of background estimation in each quadrant, using fits to the background histogram, medians, modes, etc. The problem with these procedures is that they are sensitive to image structure and content, and particularly to low–frequency background gradients. And no matter what the procedure is, the quality of the final solution must be evaluated by eye.

2. The Solution

The procedure described here avoids the dependency on human observers, as well as decreases the sensitivity on image content and gradients, typical of other methods. It works by directly building, with an objective criterion, the solution that mostly pleases the human eye. When looking into an image with quadrant DC offsets, the visual system is mostly responsive to the edges between quadrants. The visual system is actually very sensitive to the amount of Fourier power in the highest spatial frequencies associated with the edges. Thus, the eye will pick up as the best among several "fixed" images, the one that has the minimum Fourier power at the quadrant edges.

Based on that idea, a simple algorithm can be implemented that minimizes the Fourier power in the highest frequencies at the quadrant edges, as a function of the three sought DC offsets. Since the algorithm is based only on information of local nature, it can be very robust against large–scale image structure and background gradients. Since only the few pixels along the (internal) quadrant edges must be examined, the algorithm can be potentially very fast as well.

3. The Algorithm

The algorithm basically builds an estimate of the high-frequency Fourier power along the edges between quadrants, and uses a minimization code to get the best set of DC offsets that minimize that power.

The Fourier power is proportional to the squared differences between intensities extracted at each side of an edge. Intensities are extracted from within a band with given width. The exact position where the edge sits (128) is ignored. For the vertical edge, at a single image line, pixels sitting at the left side of the edge, within the band, are averaged into a_1. Pixels sitting at the right side are averaged into a_2. The Fourier power contribution from that image line is just $(a_1 - a_2)^2$. Contributions from all 256 image lines along the vertical edge are added together. The same procedure is applied to the horizontal edge, resulting in a single scalar quantity that estimates the global Fourier power along the edges. The exact Fourier frequency which is measured is a function of the band size. A very narrow band will include power from the highest frequencies only, but the resulting power measure will be more sensitive to noise. Good results were attained with band widths of 3–5 pixels to each side of the edge.

The implemented algorithm also looks for the data quality information from each pixel and can exclude pixels with selected anomalies. It can also exclude pixels above a certain threshold, and can exclude the largest differences when computing the power estimator.

The algorithm then uses the "amoeba" (downhill simplex) minimization method to search for the set of DC offsets that minimize the power estimator. The upper–left quadrant is used as a reference, and the three others are varied until the method converges.

The algorithm, which depends only on a minimal set of parameters, was implemented as a C function that can be easily included into other software. This C function was used as the core of an IRAF task that automatically removes the quadrant DC offsets of any number of NICMOS files.

Figure 1. NICMOS image showing DC offsets between quadrants. Intensity stretch was adjusted to enhance faint background features.

Figure 2. Same image depicted in Fig. 1 after being processed by the DC offset removal algorithm. The algorithm uses the upper–left quadrant as reference and adjusts the DC offsets of the other three. Notice how it is almost totally insensitive to the presence of bright objects. Same intensity stretch as in Fig. 1.

Astronomical Data Analysis Software and Systems VIII
ASP Conference Series, Vol. 172, 1999
D. M. Mehringer, R. L. Plante, and D. A. Roberts, eds.

Data Structure and Software of the UCAC-S Project

Norbert Zacharias[1], Marion I. Zacharias[2]

CTIO, 950 N.Cherry Ave., Tucson, AZ 85719,
Email: nz@ctioj8.ctio.noao.edu

Abstract. The USNO CCD Astrograph Catalog South (UCAC–S) project is a high precision astrometric survey of the entire southern hemisphere for the 7–16 magnitude range, aiming at an accuracy of 20 mas for $8-14^m$ stars. A 4k × 4k CCD is used at the back–end of a 5–element, 0.2 meter aperture astrograph. A 2–fold overlap pattern of 43,839 fields will produce 1.8 terabytes of compressed raw data, stored on tapes and CD-ROMs. The project started in January 1998 at CTIO and is scheduled for 2 years to complete. Customized Fortran routines are used throughout. The data structure, quality control procedures and software layout are described.

1. The Project

The USNO CCD Astrograph Catalog South (UCAC–S) project is a high precision astrometric survey of the entire southern hemisphere for the 7–16 magnitude range in a single bandpass (610/60 nm), (Gauss et al. 1996). The goal is to get positions accurate to 20 mas for stars in the $8-14^m$ magnitude range, and about 70 mas at the limiting magnitude (Zacharias 1997). The ACT (Urban et al. 1998) and the Hipparcos and Tycho Catalogues (ESA 1997) are used as reference (UCAC–R), supplemented by an extragalactic link program carried out in parallel mainly at the Cerro Tololo (CTIO) 0.9-meter telescope. A 4k × 4k CCD is used at the back–end of a 5–element, 0.2 meter aperture astrograph. A 2–fold overlap pattern of 43,839 fields will produce 1.8 terabytes of compressed raw data. Observing started in January 1998 at CTIO and is scheduled for 2 years to complete the southern hemisphere. An extension towards the north is planned. For more details about the project see Zacharias et al. (1997) and the web site `http://aries.usno.navy.mil/ad/ucac/ucac-s.html`.

2. Data Structure

The hardware components and the pixel data flow are shown in Fig. 1. A simple loss–less compression method (I*2 to I*1) is used for the CCD frames throughout

[1]U.S. Naval Observatory, Washington, DC

[2]Universities Space Research Association, Washington, DC

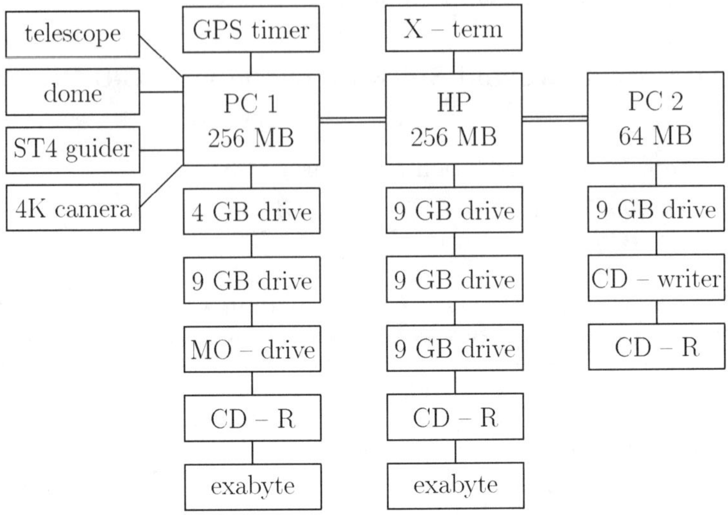

Figure 1. Hardware.

the system including storage on tapes and CD-ROMs. One set of backup tapes is shipped to Washington. The CD-ROM version is convenient for random access (manually) and is believed to be more long–term reliable than the tapes. The detection and object fit parameter data will be about 40 GB, while the final catalog is an estimated 1.5 GB for 40 million stars.

3. Observing Procedures

Observing procedures are outlined in Fig. 2. Abbreviated on–line reductions are performed immediately after a pair of frames (about 125 and 25 sec exposure on each field) have been acquired to obtain quality control statistics, which are displayed about a minute after the frames have been taken, while the automatic observing continues in parallel. The observer has the option to intervene, e.g., for changing the focus or other parameters. Complete reductions are performed the next day in batch mode (Fig. 3).

The schedule for the next night takes into account the entire observing history and adheres to strict quality tolerances for each frame.

4. Software Layout

Steps for the astrometric reductions are shown in Fig. 4. Customized Fortran routines are used throughout. ASCII files and a strict nomenclature are used for auxiliary data handling, while large astrometric data files utilize unformatted direct access structures. The raw data reduction is based on Software for Analyzing Astrometric CCDs (SAAC), (Winter 1999). The Hamburg Block Adjustment Program Package (HBAPP) previously used for photographic astrometry

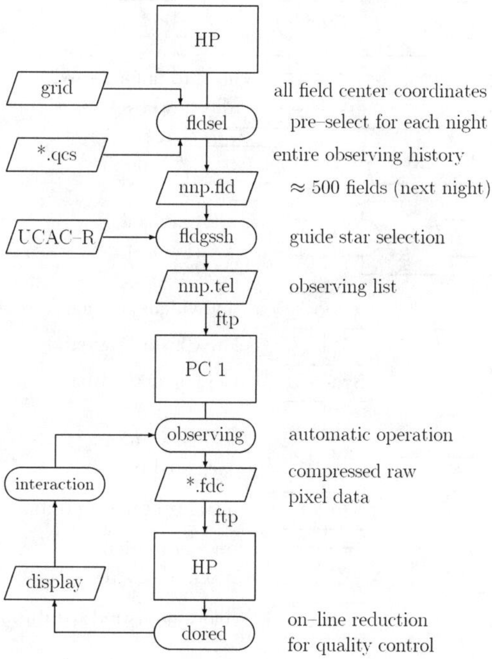

Figure 2. Scheduling and Observing.

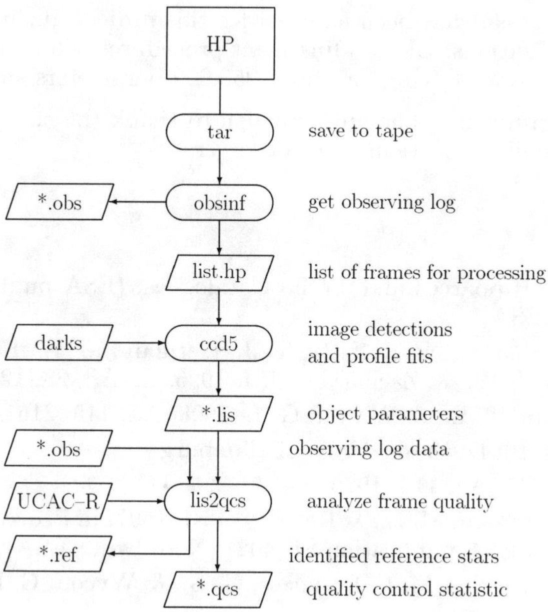

Figure 3. End of night, daily reductions.

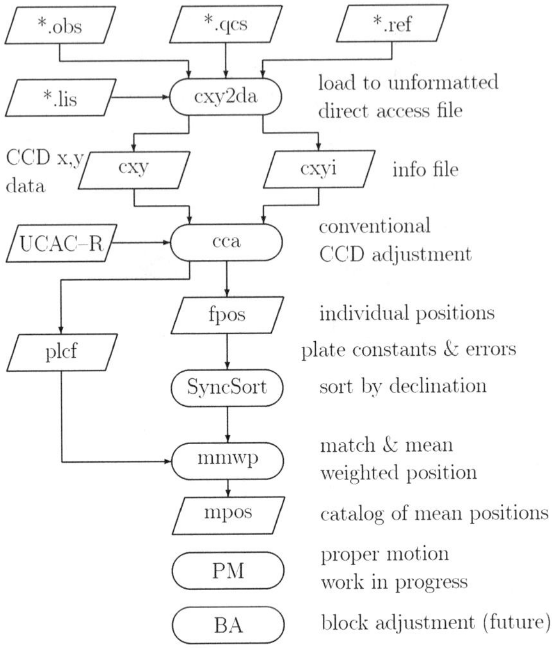

Figure 4. Astrometric reductions.

(Zacharias et al. 1996) has been adopted for this project, including a variety of PGPLOT visualizations. Block adjustment procedures will be used for the final astrometric reduction, solving for about 250,000 parameters simultaneously.

Acknowledgments. The authors wish to thank the entire UCAC-S team and the CTIO staff, in particular the observers.

References

ESA, 1997, The Hipparcos and Tycho Catalogues, (ESA publ. SP-1200) (Noordwijk: ESA)

Gauss, F. S., Zacharias, N., Rafferty, T. J., Germain, M. E., Holdenried, E. R., Pohlman, J. W., & Zacharias, M. I. 1996, BAAS, 28, 1282

Urban, S., Corbin, T. E., & Wycoff, G. L. 1998, AJ, 115, 2161

Winter, L. 1999, Ph.D. thesis, Univ. of Hamburg

Zacharias, N., 1997, AJ, 113, 1925

Zacharias, N., Germain, M. E., & Rafferty, T. J. 1997, in Proceedings Hipparcos Venice 97 (ESA publication SP-402) (Noordwijk: ESA), 177

Zacharias, N., Zacharias, M. I., Douglass, G. G., & Wycoff, G. L. 1996, AJ, 112, 2336

Astronomical Data Analysis Software and Systems VIII
ASP Conference Series, Vol. 172, 1999
D. M. Mehringer, R. L. Plante, and D. A. Roberts, eds.

Automatic Detection of Linear Features in Astronomical Images

Matthew Cheselka

IRAF Group, National Optical Astronomy Observatories, Tucson AZ 85726

Abstract. A new IRAF task has been developed that automatically identifies linear (line-like) features in an image or set of images. Such features could include moving targets taken over a long exposure (asteroids, meteors, satellites, aircraft), bad rows and columns of CCD arrays, featured caused by CCD "bleeding", or diffraction spikes. The linear features can have gaps and still be recognized as a single feature. Identifying a linear feature in the input image requires several steps. First, a list of pixels within a range of data values is found. Second, pixels within this list are examined to determine if they are colinear. Lastly, the task examines the colinear list of pixels and finds pixels which are adjacent within specified parameters. The colinearity detection is done via the Hough Transform (Gonzalez & Wintz 1987). Once features have been identified, information about the features is written to an output file and profile plots are generated. An optional binary mask can also be created in the case where bad CCD rows or columns are being identified, or where particular linear features need to be masked out.

1. Introduction

Computer programs that are able to automatically identify features in images have been shown to be very useful compared with manually attempting the same thing. This is especially true when there are large numbers of images to examine. In addition, identification of linear features in images for the purposes of analysis or masking them out has many applications.

In order to create a function that will do this, we need to understand exactly what a linear feature is. First, a "feature" is an object in an image that has a high enough S/N ratio to be detected. Second, the intensity distribution is such that pixels which make up the feature are located adjacent to one another or have some other kind of distinguishable pattern. A feature is "linear" if these pixels all fall on a line of some width and length.

With these definitions, the Hough Transform is adequate to search for colinear lists of pixel locations and a simple adjacency algorithm is suitable to check for the proper position distribution.

2. Algorithm Description

In order to automatically detect linear features, the first step is to select pixels in the image (Fig. 1, left panel) that fall within a given data value range. The input image is assumed to already be dark subtracted and flattened. The user has the option of selecting one or more ranges if it is determined, for example, that some features will be bright and some will be faint. Each data value range can either be computed from the minimum and maximum data values, in terms of evenly spaced σ ranges above the mean, or in terms of evenly spaced data value ranges. The user must specify the number and the type of ranges desired.

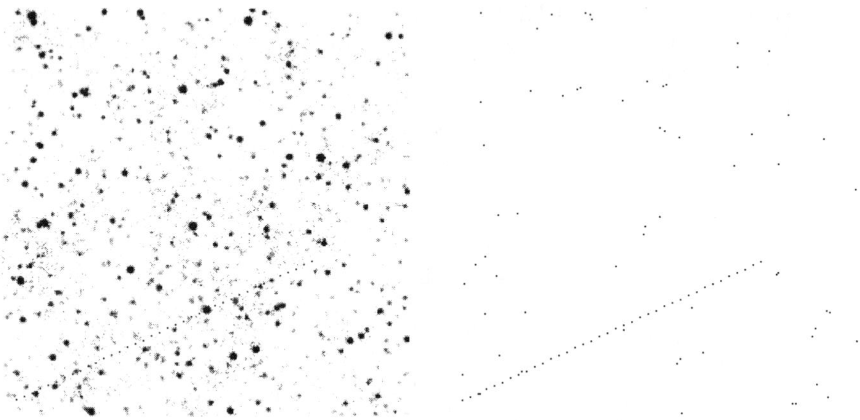

Figure 1. Input image with linear feature (*left*); resulting image after thresholding (*right*).

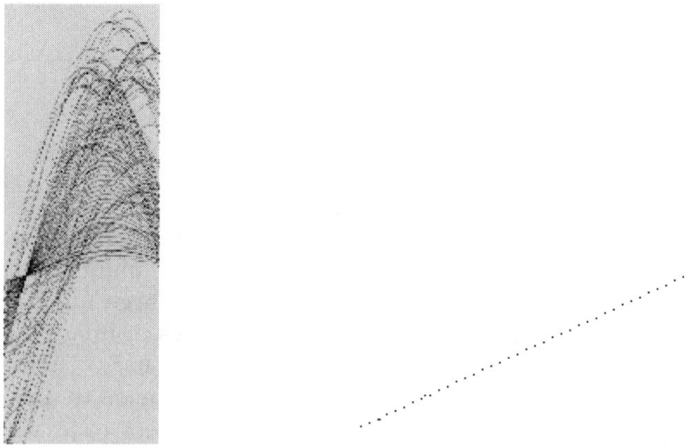

Figure 2. Parameter space "image" of pixel list (*left*); the detected colinear feature (*right*).

For each data range, a list of pixel locations (x,y coordinates) is created and fed into the Hough transform (Fig. 1, right panel). The Hough transform converts these pixel locations into a new parameter space given by the equation:

$$\rho = (x \cos\theta) + (y \sin\theta) \tag{1}$$

Each pixel location (an x,y pair) is therefore represented as a curve (Fig. 2, left panel) in this new parameter space corresponding to the above equation. In this parameter space, the x-axis is θ and the y-axis is ρ.

Pixels in image space that are aligned with one another will have curves that intersect a one particular point in parameter space. The task creates an "image" in parameter space by accumulating these curves and stacking one on top of another. Naturally, points of intersection will have larger values than other points.

Once a parameter space image is created, a simple peak finder will locate the highest peak. The highest value in parameter space corresponds to the maximum number of selected colinear points in image space. The value at that peak corresponds to the number of pixels in image space that are aligned. The location of the peak in parameter space is used to "backtrack" to find the image space pixels that correspond to that peak. Typically, a parameter space image will have several peaks. Each peak corresponds to other locations in image space where pixels have lined up. This fact can be exploited to search for features that are of any user-specified length.

Another aspect of extended linear features is that the image space pixels that describe them are in some way adjacent to one another and have a minimum and maximum length. Undoubtedly, there will be pixels that do not correspond to the feature but are still part of a colinear set. Once such a set is found, the list of pixel locations is sent to an adjacency calculator to determine if a suitable number of them are next to one another. This is where the next set of user-specified parameters comes into play. First, the distance between adjacent pixels must be smaller than a given separation. Second, the length of the feature must fall between a minimum and maximum number of pixels. If all conditions are met, a feature has been positively identified (Fig. 2, right panel). Figures 3 and 4 show other examples of this method.

There was a desire for this task to go beyond just identification by actually doing some simple analysis of the features themselves. First, information about the feature (beginning and ending pixel location, length, position angle, mean intensity and standard deviation) is written to an output file. Second, the user can also optionally create profile plots of each feature and either have them appear inside their `xgterm` window or saved as IRAF graphics metacode files to be examined at a later time. Third, and probably most useful, is that a binary mask (a pixel list file) is created that can then be used by other IRAF tasks as a region identifier for interpolation or other masking purposes.

References

Gonzalez, R. C. & Wintz, P. 1987, Digital Image Processing (2nd ed.), (Reading: Addison-Wesley), 130

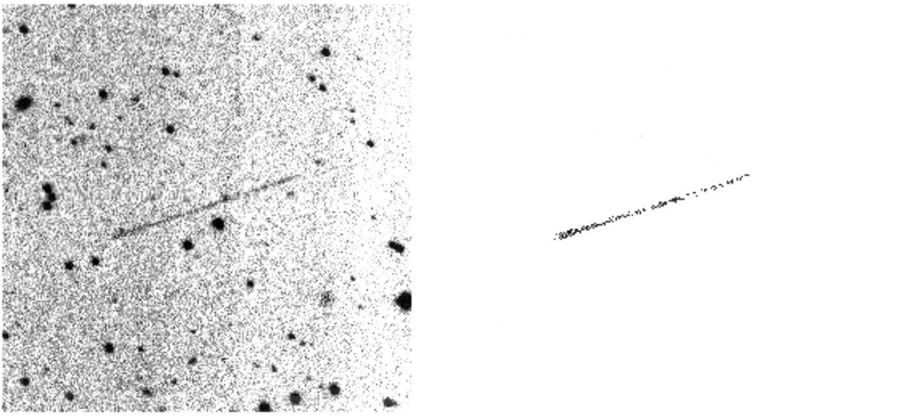

Figure 3. Asteroid streak (*left*); detected linear feature (*right*).

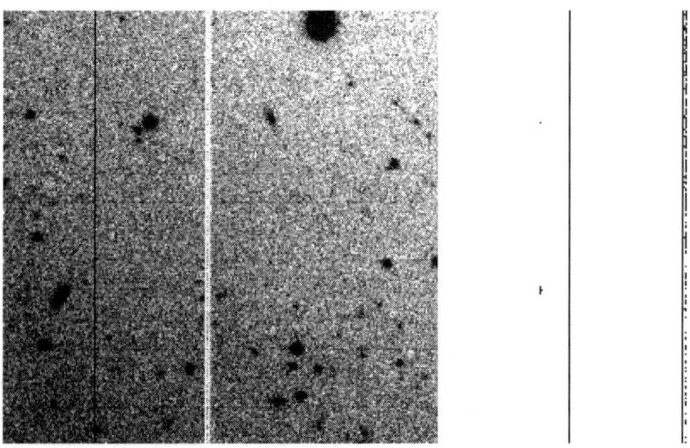

Figure 4. NOAO Mosaic image showing bad columns (*left*); detected linear features (*right*).

Astronomical Data Analysis Software and Systems VIII
ASP Conference Series, Vol. 172, 1999
D. M. Mehringer, R. L. Plante, and D. A. Roberts, eds.

Modern CCD Observations of Moving Celestial Objects: Algorithms and Software for Interactive Processing

Oleg Bykov

Pulkovo Astronomical Observatory, 196140, Pulkovskoe shosse 65, St. Petersburg, Russia

Abstract. The problems of observing asteroids using CCDs are discussed. Their identifications and orbit determinations are considered. Also, a method for quality control of CCD observations is suggested.

1. Introduction

CCD observations of the Small Solar System bodies are rapidly progressing. Every month professional astronomers together with an increasing number of amateurs obtain several hundreds of thousands of asteroid positions and send them to the International Minor Planet Center. Everyone who observes a celestial object using a CCD detects moving celestial objects as a "by product". Usually, tests of such observations and identifications of moving objects are not produced by these observers.

Obviously, it would be advantageous for the MPC to receive observations of asteroids with their identifications. Thus, it is necessary to provide astronomers, especially amateurs, with a package for quick solutions to the control and identification problems immediately after CCD observations have been made.

The original algorithms and software were developed in the Pulkovo Astronomical Observatory to solve the following problems:

- test of any positional observations of celestial bodies, artificial and natural, by analyzing the first and second derivatives of the orbital or approximate coordinate (O-C) residuals,
- calculation of the apparent motion parameters (namely topocentric angular velocity, μ, angular acceleration, $\dot{\mu}$, positional angle, ψ, and curvature, C of the trajectory) by using the spherical coordinate approximation,
- preliminary orbit determination of a celestial body observed on a super short arc of its topocentric trajectory; we use the Classical Laplace Method and the new Apparent Motion Parameters Method created at Pulkovo Observatory for the calculation of the circular and the elliptic orbital elements,
- an identification of the observed celestial body using of known orbital catalogues or our own orbit determinations,
- our own ephemeris service for known and unknown celestial objects from artificial satellites up to the Kuiper Belt asteroids.

We have had much success with the practical applications of our algorithms and software. We use the "CERES Plus" package and "LAPLACE" software in our analysis of moving celestial bodies. The algorithms and major results were published in Russian (Bykov 1989, 1996; Kiselev & Bykov 1973, 1976). Several examples are given here in the Tables.

We hope that our investigations may be useful for near-Earth space control and also for searches of the moving celestial objects through analysis of old and new CCD frames obtained at any amateur or professional astronomical observatory.

2. Examples

Table 1. Accuracy of CCD Observations.

Observatory		Telescope (diameter, focus, field of view)	Detector, catalogue	Number of			Mean error of a single observation for		Int. or Ext. acc.
code	name, country			minor planets	dates	positions	α	δ	
1	2	3	4	5	6	7	8	9	10
801	Oak Ridge Obs., USA	1.5 rfl	CCD GSC	14	27	72	0.34 0.05	0.36 0.06	ext.
704	Lincoln Lab., USA	1.0 rfl	CCD	5	3	16	0.52 0.14	0.50 0.09	int.
711	McDonald Obs., USA	0.76 rfl 2.3 46×46′	CCD GSC	4	4	22	0.32 0.02	0.13 0.09	int.
				51	28	335	0.21 0.03	0.16 0.02	ext.
610	Pianoro, Italy	0.26 rfl 1.0 22×15′	CCD GSC	55	19	204	0.46 0.05	0.51 0.07	int.
				4	8	29	0.61 0.09	1.29 0.80	ext.
689	Flagstaff, USA	0.20 tr.c	CCD GSC	15	9	53	0.29 0.07	0.26 0.04	ext.

A sample of accuracies of CCD observations of the numbered minor planets reported to International Minor Planets Center during 1995-1996.

The CERES software package, which was developed at the former Institute of Theoretical Astronomy (in Russia), was used to calculate (O-C)-differences for nearly 700 minor planets which were observed irregularly and quasi-simultaneously in 1995-96 using CCDs as well as photography at nearly 50 observatories. The accuracy of observations was estimated by means of dispersion of average (O-C) for each type of observations obtained by each telescope. Some results are given in Table 1. Here, the term "Internal accuracy" (int.) indicates that an estimation was derived from analysis of minor planet positions obtained in one night for each of them. "External accuracy" (ext.) indicates that observations made during several successive nights were used only. In the columns 8 and 9, where the main results of the accuracy investigations are presented, the values below accuracy values are the mean errors of our estimates.

3. Example of Accuracies of Modern CCD Observations.

Table 2. (O–C) residuals for the NMP 3035 Chambers.

Obs.	Date (DT) 1998 09	α 22^h49^m	$O-C$ $''$	δ $-6°23'$	$O-C$ $''$	V	*Zen.* °
704	18.25538	60.439	0.74	16.49	−0.92	17.7	40
	18.26829	59.860	0.85	20.89	−0.99	17.7	40
	18.28114	59.211	−0.13	25.30	−1.08	17.7	40
	18.30675	58.062	0.07	33.11	−0.31	17.7	43
Mean	18.27789	59.393	0.38	23.95	−0.82		
Error			±0.24		±0.18		
699	18.27145	59.725	0.98	19.79	1.18	17.4	28
	18.28938	58.936	1.35	25.79	1.18		28
	18.30800	58.068	1.01	32.31	0.91		31
Mean	18.28961	58.910	1.11	25.96	1.09		
Error			±0.12		±0.09		

(O–C) residuals for the NMP 3035 Chambers observed by two telescopes (obs. codes **699,704**) with a CCD during a single night.

Table 3. Error Estimation.

Obs.	Date 1998 09	j	$\dot{\alpha}$ s	O–C $''$	$\dot{\delta}$ $''$	O–C $''$	Err.$\dot{\alpha}$ %	Err. $\dot{\delta}$ %
610	10.89271	3	−45.072	57.40	−404.74	−47.95	7.8	13.4
704	14.28740	5	−48.696	−15.76	−353.99	−4.74	2.2	1.4
699	18.28961	3	−45.345	0.69	−342.64	−7.34	0.1	2.2
704	18.27789	4	−46.472	−17.87	−323.14	11.47	2.6	3.4
704	25.17234	5	−39.320	5.89	−303.09	−3.36	1.0	1.1
704	14 − 25	14						

The first derivatives and estimation of accuracy of data presented in Table 2.

Parameters $\dot{\alpha}$, $\dot{\delta}$, their accuracy characteristics (in percents) and differences (O − C) which may be obtained by linear approximation of the coordinate sets presented in Table 2 and by comparison with the exact values given by the CERES software.

Table 4. Apparent Motion Parameters.

NMP Dates	451 Patientia 1993 04 08.24614	1531 Hartmut 1993 07 23.99637	2302 Florya 1993 07 21.24017
codes	809 (*photo*)	801&587 (*CCD*)	801&587 (*CCD*)
positions	15	7	7
nights	5	3	3
topoc. arc	$0^\circ\!.5$	$0^\circ\!.6$	$1^\circ\!.3$
α calc.	$15^h43^m36^s\!.933$	$21^h15^m23^s\!.125$	$21^h05^m09^s\!.804$
obs.	$15^h43^m36^s\!.985$	$21^h15^m23^s\!.134$	$21^h05^m09^s\!.841$
δ calc.	$-4^\circ\ 27'\ 00''\!.97$	$-4^\circ\ 31'\ 33''\!.41$	$-16^\circ\ 55'\ 30''\!.76$
calc uad obs.	$-4^\circ\ 27'\ 01''\!.75$	$-4^\circ\ 31'\ 33''\!.97$	$-16^\circ\ 55'\ 30''\!.37$
μ calc.	$421''\!.9944$	$774''\!.6957$	$818''\!.9019$
obs.	$421''\!.8901$	$774''\!.9191$	$817''\!.7663$
$\dot{\mu}$ calc.	$16''\!.6739$	$11''\!.7524$	$15''\!.2804$
obs.	$16''\!.7164$	$12''\!.5745$	$15''\!.4006$
ψ.	$289^\circ.3509$	$276^\circ.9167$	$279^\circ.1318$
obs.	$289^\circ.3548$	$276^\circ.8532$	$279^\circ.1641$
c calc.	7.6302	3.4289	1.6596
obs.	7.6576	3.6675	1.6718

Calculation of the geocentric differential parameters for some numbered minor planets and comparison with the actual parameters obtained with a processing of positional observations.

Table 4 shows apparent motion parameters $\alpha, \delta, \mu, \dot{\mu}, \psi, C$ are reliable and can be calculated from the modern position observations. They allow us to obtain the preliminary orbit elements of asteroid.

Acknowledgments. We thank the LOC of the ADASS'98 conference for hospitality and financial support.

References

Bykov, O. P. 1989, Problems of Universe Investigations, 12, 328

———, 1996, in Proc. of IAU Symp. 172, Dynamics, Ephemerides and Astrometry of the Solar System, ed. S. Ferraz-Mello & J.-E. Arlot (Dordrecht: Kluwer), 458

Kiselev, A. A. & Bykov, O. P. 1973, Sov. Aston. J., 50, 1298

———, 1976, Sov. Aston. J., 53, 879

Astronomical Data Analysis Software and Systems VIII
ASP Conference Series, Vol. 172, 1999
D. M. Mehringer, R. L. Plante, and D. A. Roberts, eds.

DASH – Distributed Analysis System Hierarchy

Masafumi Yagi, Yoshihiko Mizumoto, Eiji Nishihara, Michitoshi Yoshida, Ryo Asai, Yoshihiro Chikada, George Kosugi[1], Tadafumi Takata[1], Ryusuke Ogasawara[1]

National Astronomical Observatory, Osawa, Mitaka, Tokyo 181-8588, Japan

Yasuhide Ishihara, Hiroshi Yanaka

Fujitsu Limited, Nakase, Mihama, Chiba 261-8588, Japan

Yasuhiro Morita, Hiroyuki Nakamoto

SEC Co. Ltd., 22-14 Sakuragaoka, Shibuya, Tokyo 150-0031, Japan

Abstract. DASH is a distributed data reduction and analysis system constructed on CORBA. It plays an important role in the SUBARU computer system not only as an offline analysis system but also as a test and development environment of online quick analysis. We abstracted an object from astronomical data reduction/analysis procedures; 'PROCube' (data reduction PROcedure Cube) and implemented it, so that we could share and reuse the reduction procedure. We are also trying to make it possible to retrieve sufficient original data for an analysis semi-automatically in cooperation with SUBARU data archive system (STARS).

1. Introduction

DASH is an object-oriented data reduction and analysis system for 8.2-m SUBARU telescope at Hawaii. In early 1999, SUBARU will obtain scientific first light image and soon it will begin to produce several tens TB data per year. DASH is designed as the SUBARU observatory system to manage, reduce and analyze such a huge amount of astronomical data produced by various instruments. In addition, reproducibility and portability of the reduction, seamless operability on heterogeneous computer system, open architecture, trial-and-error processing, and pipeline processing, and research target oriented analysis are required. To satisfy these requirements, we adopt CORBA as a distributed object environment and construct a distributed data reduction and analysis system (Takata et al. 1998). We have made two prototypes in 1997 (Mizumoto et al. 1998a, 1998b), and we have just tested Ptorotype-3 at SUBARU Observatory in Nov.

[1]Subaru Telescope, National Astronomical Observatory of Japan, Hilo, HI 96720, USA

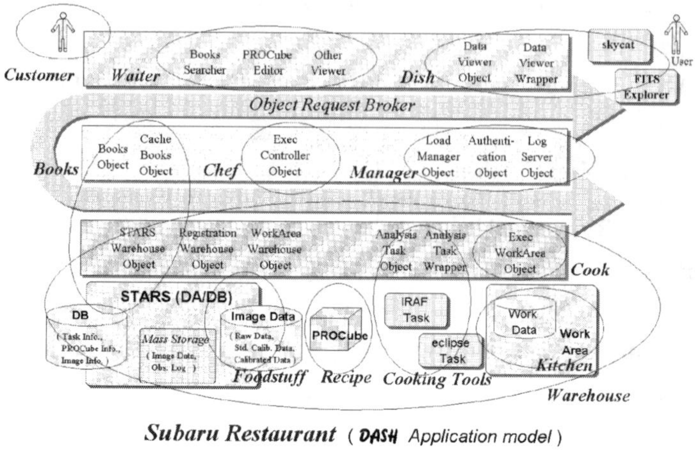

Figure 1. DASH application model.

1998. DASH project would be completed by early 2000, and we have achieved just as we had scheduled so far (Mizumoto et al. 1998b).

2. Components of DASH: SUBARU Restaurant

DASH system was developed with Object Oriented Methodology. When DASH was designed, the system was modeled as a restaurant, with three tiers: UIs, controls, and data. The distributed objects are designed based on the model and the role of each object is understood through this metaphor (Mizumoto et al. 1998b).

3. The trinity of SUBARU software system: the role of DASH

There are three subsystems in SUBARU software system; Subaru Observation Software System (SOSS; Sasaki et al. 1998) at the summit of Mauna Kea, Subaru Telescope data ARchive System (STARS; Takata et al. 1998) with quick-look image system for archive search (Baba et al. 1999) at Hilo base, and DASH.

Before one begins to analyze data in DASH, he or she will search in STARS and retrieve raw data into DASH. With an observation dataset, the retrieval of relevant calibration data can be done automatically. The data are reduced and analyzed in DASH to obtain scientific results. Then, reduced and standard calibration data are registered into archives.

The feedback from DASH to SOSS is two-fold. As we observe with SUBARU, the astronomical results of DASH are directly reflected to the next observation plan. For example, if some find that more DARK frames are needed for precise analysis, the information is reflected to the auto-scheduling system in the SOSS.

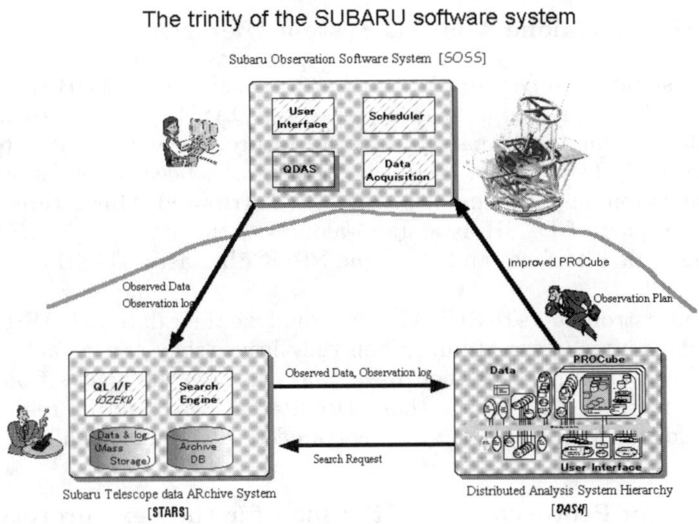

Figure 2. SUBARU software system overview.

The second feedback is improvement of quick data analysis system (QDAS) for quick look. We can use the same PROCube in the quick data analysis system in SOSS. There is a small subset of DASH in the observational system for data processing as a pipeline. An improvement of PROCube in DASH is reflected to SOSS and observer can get quick and precise result of the target at summit during observation. The three subsystems, SOSS, STARS, and DASH thus form a trinity.

4. PROCube: data reduction PROCedure Cube

PROCube is an object that holds all information about data processing (Mizumoto et al. 1998b). It contains data, engines, flow of reduction and analysis, tunable parameters for each engine, a log of reduction, and the current status.

Making a new PROCube is similar to drawing a flowchart of a reduction; each engine has several inputs, outputs and some parameters, which must be provided by the user. In PROCube Editor, this process is visualized; the user can cut & paste the names of data to connect the engines, set parameters for engines, and execute it. Intermediate files are automatically created and deleted when they are no longer needed. As PROCube holds all information about the reduction, the results are always reproducible.

Once a PROCube is created, it could be reused for reduction of another data by simply changing the original data. QDAS in SOSS already uses this PROCube pipeline, using data from DAQ system. In the next prototype, we try to fill the calibration data automatically with observational datasets (Kosugi et al. 1998).

5. SASH: Standalone Analysis System Hierarchy

DASH is designed as a part of SUBARU software system at SUBARU Observatory, Hawaii, however, people may desire to use DASH system at their institute (e.g., in Japan) for their analysis. In the future, we will be able to connect CORBA directly from Japan to Hawaii. Currently, because of rather poor internet connection, objects cannot be directly distributed. Our future plans is to extract some part of DASH as a stand-alone system without CORBA (SASH) so that the same interfaces and the same PROCubes as in DASH could be used in SASH.

Visiting astronomers to SUBARU will analyze their data in DASH at Hawaii for a few days after observation. When they leave, they can detach their environment, bring it along with their data, and continue the analysis on SASH at their home institute. In future, these satellite SASHs could be re-attached on DASH, or form satellite DASH systems connected to the original.

6. Results of Prototype-3 and the plan for the next prototype

With DASH prototype-3, we can reduce and analyze test data of some test instruments semi automatically with PROCubes. User authentication, data and engines sharing in the same proposal group are implemented and tested.

The main targets of next prototype are the stand-alone system (SASH), and the observational dataset. Another important targets are consideration for engines with GUI and development of algorithm for load dispensing. In Prototype-3, only non-interactive engines could be used in PROCubes, though such interactive engines are necessary for some analysis. Only data transfer load is considered in prototype-3, since the data transfer is one of the dominant factors of the system load. In next prototype we will monitor CPU load in some test system, and will change the algorithm accordingly. This DASH prototype-4 will be used and tested with the real data from SUBARU next year.

References

Baba, H., Yasuda, N., Ichikawa, S., Watanabe, M., Horaguchi, T., Hamabe, M., Aoki, W., & Ozawa, T. 1999, this volume, 413

Kosugi, G., et al. 1998, in SPIE Proc., Vol. 3349, Observatory Operations to Optimize Scientific Return, ed. P. J. Quinn (Bellingham: SPIE), 421

Mizumoto, Y., et al. 1998a, in ASP Conf. Ser., Vol. 145, Astronomical Data Analysis Software and Systems VII, ed. R. Albrecht, R. N. Hook, & H. A. Bushouse (San Francisco: ASP), 332

Mizumoto, Y., et al. 1998b, in SPIE Proc., Vol. 3349, Observatory Operations to Optimize Scientific Return, ed. P. J. Quinn (Bellingham: SPIE), 173

Sasaki, T., et al. 1998, in SPIE Proc., Vol. 3349, Observatory Operations to Optimize Scientific Return, ed. P. J. Quinn (Bellingham: SPIE), 427

Takata, T., et al. 1998, in SPIE Proc., Vol. 3349, Observatory Operations to Optimize Scientific Return, ed. P. J. Quinn (Bellingham: SPIE), 247

Astronomical Data Analysis Software and Systems VIII
ASP Conference Series, Vol. 172, 1999
D. M. Mehringer, R. L. Plante, and D. A. Roberts, eds.

The IUE High Dispersion Spectra Processing in IRAF

Petr Škoda

Astronomical Institute of the Academy of Sciences of the Czech Republic, 251 65 Ondřejov, Czech Republic

Abstract. The IUE high dispersion spectra available in NEWSIPS format of IUE final archive can not be directly analyzed in IRAF. A number of different packages like IUEDR (part of Starlink) or IUERDAF (written in IDL) have usually been used for this purpose. However, with the combination of the external IRAF package `iuetools` and several `cl` scripts, batch analysis of a large number of spectra can be performed. We describe the process of obtaining the data, converting it to IRAF format, echelle orders merging, and continuum fitting up to the measurements of radial velocity of selected resonance lines. We also give suggestions for rapid preparation of a convenient plot of line profile series ordered by time and orbital phase. Finally, a brief discussion of advantages and bottlenecks of IRAF approach in regards of other packages is given.

1. Introduction

The spectra obtained by IUE satellite during its long life mission are one of the important sources of information about hot stars. Any astronomer working in the field of Be and WR stars should be able to work with the data in the simple manner without detailed study of instrumental setup, image distortions, camera aging effects etc. For that purpose the best way is taking the NEWSIPS absolutely calibrated data available as MXHI (high dispersion) and MXLO (low dispersion) files of IUE Final Archive. In the following text a brief outline of the necessary reduction steps of IUE data processing is given. An extended version of this contribution (in PostScript) and the necessary `cl` scripts are available at `ftp://ra.stsci.edu/pub/newsips/contrib/`.

2. Obtaining data

The NEWSIPS data may be found at two places – the data processed at Goddard at the NASA server NDADS[1] and at ESA INES server[2]. The ESA server, however, contains only a part of all IUE spectra available, (e.g., for CX Dra there were only 7 of 102 available at NASA). The spectra in ESA archive are re-

[1] `http://archive.stsci.edu/iue/retrievals.html`

[2] `http://ines.vilspa.esa.es/ines/`

processed using new techniques (i.e., improved extraction, echelle order splicing etc.). The data retrieval itself is done by filling in the form for jukebox server including an e-mail address. After some time, the relevant data are copied to the staging disk area and an e-mail notification is sent. Optionally, information necessary to access a private ftp server may be provided and the data from NDADS will be automatically downloaded.

3. Conversion to IRAF formats

As the most important and most complicated data files are the high dispersion spectra, we will describe only MXHI files handling and only SWP camera data. Basic options for IUE data processing are available[3]. For data reduction in IRAF the package `iuetools`[4] must be installed. It consists of two tasks: `mxexpand` and `mxtomultispec`. The `mxexpand` task takes the MXHI file and produces a FITS file with the binary table extension. Information for each order is stored in one row of binary table. The new FITS file may be used for data processing in STSDAS (e.g., plotting with `sgplot` or `igi`), but the majority of users will probably know much better the basic IRAF packages for CCD data analysis in multispec format (`splot, continuum, specplot`). Thus the FITS file must be converted by the `mxtomultispec` task to the `*.imh` multispec file with each order in a separate aperture line. Only the absolutely calibrated fluxes are extracted. Each aperture has the dispersion polynomial fitted and written into WCS header. This file may be already analyzed in `splot` with each echelle order seen as different aperture/beam. A script (`mxconvert.cl`) is provided in order to assist in the bulk image processing.

4. Merging echelle orders

The analysis of each order separately with `splot` may be useful for single line investigation, however the wavelength coverage of one order is very small (about 20 Å) and the increased noise at both ends (caused by ripple correction) will degrade the useful part even more. The order splicing is thus necessary. We decided for the sake of simplicity to splice all orders in the SWP range in one long spectrum. This splicing has the affect that the data in MXHI files are linearized in wavelength but each echelle order is rebinned with different wavelength difference interval. As the IRAF world coordinate system (WCS) requires only one wavelength increment for the whole spectrum, the flux data must be re-sampled — it may introduce some minor error in the position of sharp spectral features. We have, however, come to the conclusion that the influence of re-sampling errors on the radial velocity measurement is negligible in comparison with other effects like the uncertainty of line profile fitting due to the strong noise, aperture pointing errors etc.

---

[3]`http://archive.stsci.edu/iue/da_sw.html`

[4]`http://ra.stsci.edu/newsips/`

For order merging we have adopted the method by Solano[5]. He has shown that in the overlapping region of orders m and $m-1$ of SWP camera the S/N ratio is is higher in one third of it. Using this method we have developed the cl task `mxmerge.cl` which takes the multispec files `*.imh`, copies the selected range of wavelengths (about 0.03 Å smaller at each side to avoid the overlapping uncertainty) for each order into separate files using `scopy` and then merges these separate files into one long spectrum using `scombine`.

5. Fitting continuum and spikes removal

For some purposes (e.g., study of line profile changes) it is convenient to normalize the spectra to the common continuum level. As it is quite difficult to achieve this in some spectral regions due to the line blanketing, we have to estimate some smooth curve along the whole spliced spectra. The IRAF task `continuum` fits well. We have found an optimal setting with spline3 curve of order 5, high_rej=0 and low_rej=2. The continuum sample regions must be manually selected by panning through the zoomed view, however the correct continuum fitting requires adjusting the spline parameters and sampling region selection.

The removal of the bad points is complicated; fortunately the most annoying features like fiducial marks and cosmic ray hits have the shape of a sharp and deep or high spike. So we may use the `continuum` task again, now the order of spline must be very high (about 30) and the sigma rejection boundaries about 4 (both low and high rejection).

6. Examining interesting lines

Standard `splot` fitting routines like Gaussian, Lorentzian or Voigt may be used for fitting line profiles, estimating the position of the line centers etc. Some boxcar smoothing is also helpful.

7. Plotting sets of spectra

The plot of stacked spectra labeled (e.g., with the binary phase) can be achieved easily using the powerful IRAF task `specplot`. Using the graphics redirection operator `>G` the graphics metafiles `*.mc` containing each picture are produced. Now we can fit several pictures to one page using the `gkimosaic` task and print the resulting PostScript file.

8. Transformation to RV scale

The last operation may be the transformation of spectral region to the radial velocity scale with the origin at particular laboratory wavelength. The transformation will be done by the `mkv.cl` script written by Frank Valdes and taken

[5] `http://ines.vilspa.esa.es/ines/docs/esm.pdf`

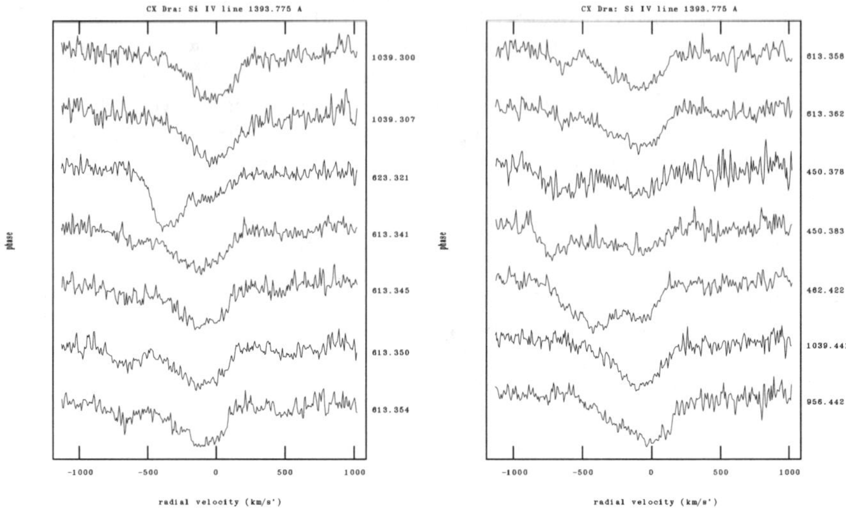

Figure 1. Stacked plot in radial velocity scale

from the *adass.iraf.application* newsgroup. We finally get after processing by `specplot` and `gkimosaic` the stacked plots in radial velocities (see Fig. 1).

9. Conclusions

There are indications, that the NEWSIPS extraction routine introduced some systematic errors[6] into final spectra — mainly for high dispersion SWP data, so for the precise measurements (e.g., changes in the line depth of order of several percent) the re-extraction of the spectra should be done from SIHI files. This is, however, out of scope of this article. We have shown that using the combination of IRAF scripts and its powerful tasks the customized data files and overview plots can be produced for large number of spectra easily. In comparison to the IUERDAF software, there remain some doubts about the precision of linear re-sampling and the most frustrating is the lack of a quality flags handling in present IRAF which would enable the correct removal of all problematic data marked by some quality flag. Such a task can be done in IUERDAF by `patch` routine. On the other side the processing using IRAF can be done quickly with the strong interactive control (e.g., `specplot` task) using only software available free of charge in contrast to the quite expensive IDL[7] system.

Acknowledgments. This work was supported with the subvention by the Ministry of Education, Youth and Sports of the Czech Republic (grant LB98251).

---

[6] `http://hypatia.gsfc.nasa.gov/iue/seville.html`

[7] IDL is the trademark of Research Systems, Inc.

Astronomical Data Analysis Software and Systems VIII
ASP Conference Series, Vol. 172, 1999
D. M. Mehringer, R. L. Plante, and D. A. Roberts, eds.

Optimizing Functions Using ASCFIT

David Grumm

Smithsonian Astrophysical Observatory, 60 Garden Street, MS 81, Cambridge, MA 02138 USA

Abstract. As part of the Smithsonian Astrophysical Observatory AXAF Science Center (ASC) Data Analysis Environment, ASCFIT was designed as a generalized fitting engine to enable users to fit models to the data that will be returned by AXAF, the Advanced X-ray Astrophysical Facility (Doe et al. 1996). I present a description of the use of a development version of ASCFIT as a function optimizer, in which the optimization methods available within ASCFIT are applied to seven benchmark functions of varying complexity. These functions are the 2-dimensional Gaussian function previously investigated by ASC, and six functions commonly used in optimization testing. The optimization methods combining the Powell technique with one of several other techniques successfully converged for all but one of the benchmark functions.

1. Introduction

The library of ASCFIT optimization methods was applied to a set of benchmark functions. The optimization methods are in the OPTIM package written by Mark Birkinshaw, and can be categorized as either "single-shot" routines, or "scatter-shot" routines (Birkinshaw 1998). In single-shot routines (Simplex & Powell), the optimization attempts to converge from the initial state to the global minimum in a continuous fashion. In scatter-shot routines (Grid, Monte Carlo, and Simulated Annealing), the entire parameter space is searched to determine a better minimum than the initial state. The functions were optimized by their minimization and vary from continuous functions with a single minimum to discontinuous functions with many local minima. Some of the methods start with a scatter-shot method, and then use a single-shot method to refine the search, as in Monte-Carlo + Powell. The two Simulated Annealing methods differ in that in method 2 all of the parameters are varied simultaneously, while in method 1 they are varied separately. The methods are:

Simplex	Powell
Grid	Grid + Powell
Monte Carlo	Monte Carlo + Powell
Simulated Annealing 1	Simulated Annealing 1 + Powell
Simulated Annealing 2	Simulated Annealing 2 + Powell

2. Functions Optimized by ASCFIT

The benchmark functions $\mathtt{f}_0$ through $\mathtt{f}_5$ are based on those of Ingber & Rosen (1992), who selected them as being typically used for genetic algorithm benchmarking. The 2-dimensional Gaussian function was chosen as it was previously investigated by ASC. These functions are described below.

Function 2dg(x, y) is a two-dimensional Gaussian function:

$$2dg(x,y) = 1 - Ae^{-\frac{(x-x0)^2+(y-y0)^2}{2\sigma^2}},$$
$$x0 = 30.0, y0 = 30.0, A = 1.0, \sigma = 10.0,\ |x| \le 100.0,\ |y| \le 100.0$$

The function $\mathtt{f}_0$ has coefficients s_i, t_i, d_i, and c defined such that the function is a paraboloid with axes parallel to the coordinate axes, and a set of holes that increase in depth near the origin. The function is very difficult to minimize, having over 10^{20} local minima in which the optimization may become trapped:

$$f_0(x_1,...,x_4) = \textstyle\sum_{i=1}^{4}(t_i sgn(z_i) + z_i)^2 c d_i) \text{ if } |x_i - z_i| < |t_i|$$
$$= \textstyle\sum_{i=1}^{4} d_i x_i^2 \text{ otherwise,}$$
$$z_i = \lfloor |x_i/s_i| + 0.49999 \rfloor sgn(x_i) s_i,$$
$$s_i = 0.2, t_i = 0.05, i = 1, ...4,$$
$$d_i = \{1.0, 1000.0, 10.0, 100.0\},$$
$$c = 0.15,$$
$$-1000.0 \le x_i \le 1000.0, i = 1, ..., 4$$

The function f_1 is a simple sum of squares, with a single minimum:

$$f_1(x_1,x_2,x_3) = \textstyle\sum_{j=1}^{3} x_j^2; \quad -5.12 \le x_i \le 5.12, \quad i = 1,...,3.$$

The function f_2 is the classical Rosenbrock function:

$$f_2(x_1,x_2) = 100(x_1^2 - x_2)^2 + (1 - x_1)^2; \quad -2.048 \le x_i \le 2.048, \quad i = 1,2.$$

The function f_3 is discontinuous and has a single minimum:

$$f_3(x_1,...,x_5) = 30.0 + \textstyle\sum_{j=1}^{5} \lfloor x_j \rfloor; \quad -5.12 \le x_i \le 5.12, \quad i = 1,...,5.$$

The function f_4 is a noisy function, as η is a uniform random variable bounded by $[0,1)$:

$$f_4(x_1,x_2) = \textstyle\sum_{j=1}^{2} x_j^4 + \eta; \quad -1.28 \le x_i \le 1.28, \quad i = 1,2.$$

Function f_5 has 25 local minima, and is optimized with a value of approximately 0.001996 when $x_1 = x_2 = -32$:

$$f_5(x_1,x_2) = \frac{1}{500 + \sum_{j=1}^{25} \frac{1}{j + \sum_{i=1}^{2}(x_i - a_{ji})^6}}$$
$$a_{j1} = \left\{ \begin{array}{c} -32, -16, 0, 16, 32, -32, -16, 0, 16, 32, -32, -16, 0, \\ 16, 32, -32, -16, 0, 16, 32, -32, -16, 0, 16, 32 \end{array} \right\},$$
$$a_{j2} = \left\{ \begin{array}{c} -32, -32, -32, -32, -32, -16, -16, -16, -16, -16, 0, \\ 0, 0, 0, 0, 16, 16, 16, 16, 16, 32, 32, 32, 32, 32 \end{array} \right\},$$
$$-65.536 \le x_i \le 65.536, \quad i = 1,2.$$

3. Results

For each optimization method and function, the initial parameter state was first chosen to be fairly distant from the global minimum, and all ASCFIT default parameters were used (scenario A). The optimization was then run using stricter convergence criteria and the same initial state (scenario B), indicative of the case when one is optimizing a function whose global minimum is not even approximately known (More, Garbow, & Hillstrom 1981). Finally, the optimization was run using the stricter convergence criteria and an initial parameter state chosen to lie much closer to the correct state (scenario C).

Table 1 summarizes the results for scenario B. For each function and technique, the value shown is the normalized final parameter distance, defined as the ratio of the distance between the final parameter state and the global minimum to the maximum distance between any two states within the allowed parameter space. Distances of less than 10^{-6} are shown as 10^{-6}. In general, the results for scenario C (not shown) indicate that, as expected, a more favorable initial state improved the single-shot routines' optimizations of smoothly varying functions. The final column is the execution time for the gaussian, normalized to the time for the simplex method. The result of one technique is plotted in Fig. 1, with the initial parameter distances for scenarios A, B and C.

Table 1. Function optimizations for scenario B.

METHOD	2dg	f_0	f_1	f_2	f_3	f_4	f_5	TIME
Simplex	3E-1	3E-4	2E-5	4E-5	1E-6	3E-1	2E-1	1.0
Powell	4E-6	2E-1	1E-5	2E-5	1E-6	3E-1	3E-1	1.3
Grid	3E-4	1E-6	2E-1	8E-5	6E-2	1E-1	1E-6	76.8
Grid+Powell	4E-6	2E-5	2E-5	1E-4	1E-6	4E-1	1E-4	2435.5
M. C.	6E-4	3E-2	2E-1	1E-3	6E-2	2E-1	1E-6	76.0
M. C.+Powell	1E-6	2E-5	1E-5	7E-5	1E-6	4E-1	1E-6	477.0
S. A. 1	3E-5	2E-4	4E-3	1E-3	1E-6	3E-1	1E-6	35.8
S. A. 1+Powell	1E-6	1E-5	1E-5	6E-5	1E-6	4E-1	1E-6	282.8
S. A. 2	4E-5	2E-4	1E-2	1E-4	3E-2	2E-1	1E-6	1.3
S. A. 2+Powell	2E-5	9E-6	8E-5	1E-6	1E-6	4E-1	1E-6	28.8

4. Discussion

Using optimization scenario C, the global minimum was located for all functions except $\mathtt{f_4}$, in which the routines became trapped in local minima. For this function, the grid and simulated annealing techniques would presumably locate the minimum if strict enough criteria were used; the grid method showed improved optimization for scenario B compared to scenario A. Ingber & Rosen (1992) tested a similar function with 30 variables and obtained comparable results.

The single-shot routines successfully converged for the smoothly varying functions (2dg, f_1, and f_2), particularly when starting from a favorable parameter state. For the discontinuous functions, these routines sometimes became

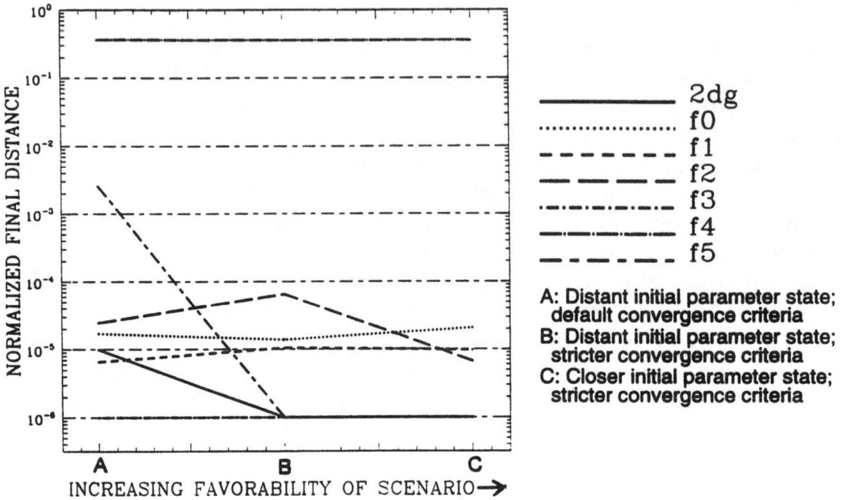

Figure 1. Monte Carlo + Powell.

trapped in local minima. In these cases, the optimizations were not improved substantially by using an initial parameter state closer to the global minimum or by using stricter convergence criteria.

The scatter-shot routines were less successful in optimizing the smooth functions, as they sample the entire parameter space without refining the search near the minimum. However, for both the smooth and the discontinuous functions, the addition of the Powell technique improved the optimization, at the expense of longer execution times (see Table 1). Such methods with refining steps following sampling of the parameter space are to be preferred.

Acknowledgments. This project is supported by NASA contract NAS8-39073. I would like to thank the following individuals for helpful discussions and their review: Mark Birkinshaw, Janet DePonte, Stephen Doe, Diab Jerius, Malin Ljungberg, Dan Nguyen, Aneta Siemiginowska, and Robert Zacher.

References

Birkinshaw, M. 1998, Some Advice on Minimization Routines, (AXAF Science Center Internal Memo)

Doe, S., Ljungberg, M., Siemiginowska, A., & Joye, W. 1996, in ASP Conf. Ser., Vol. 145, Astronomical Data Analysis Software and Systems VII, ed. R. Albrecht, R. N. Hook, & H. A. Bushouse (San Francisco: ASP), 157

Ingber, L. & Rosen, B. 1992, Mathematical and Computer Modelling, 16(11), 87

More, J., Garbow, B. S., Hillstrom, K. E. 1981, ACM Trans on Math. Software, 7, 17

Part 9. Astrostatistics and Databases

Astronomical Data Analysis Software and Systems VIII
ASP Conference Series, Vol. 172, 1999
D. M. Mehringer, R. L. Plante, and D. A. Roberts, eds.

References for Galaxy Clusters Database

M. Kalinkov, I. Valtchanov, I. Kuneva

Institute of Astronomy, Bulgarian Academy of Sciences, 72 Tsarigradsko Chaussee blvd, BG-1784 Sofia, Bulgaria

Abstract. A bibliographic database will be constructed with the purpose to be a general tool for searching references for galaxy clusters. The structure of the database will be completely different from the available now databases as NED, SIMBAD, LEDA. Search based on keyword system will be performed through web interfaces from numerous bibliographic sources – journal articles, preprints, unpublished results and papers, theses, scientific reports. Data from the very beginning of the extragalactic research will be included as well.

References for galaxy clusters (RGC) is continuation of a previous project for collecting all published information for galaxy clusters. There are neither previous attempts nor projects for compiling such database for galaxy clusters and our effort will be to include into the database all of the available bibliography through our system of keywords until the end of 1999. Now over 3,000 entries are included into the preliminary version of the database.

1. Introduction

Clusters of galaxies are the largest gravitationally bound systems in the Universe. They are the most effective tracers of the large scale structure of the Universe, as well as of the formation history and evolution of the largest mass scales.

About 16,000 clusters of galaxies are known at present and there is a large amount of observational data for about 2–3,000. It is true that some information on clusters (catalogs, lists, etc.) can be retrieved from any astronomical data center, but it is impossible more specific information to be obtained. The only exception is NASA/IPAC Extragalactic Database (NED). But despite of the tens of thousands references stored in NED, there is no easy way to find where, e.g., photometry is published for a given cluster.

The main aim of our project is to give the opportunity for queries where data for any observation or theoretical consideration is published or stored. It is reasonable to test the basic concepts which will be incorporated in the database using A/ACO clusters of galaxies (Abell 1958; Abell, Corwin, & Olowin 1989). And this work concerns mainly the data for these clusters.

2. Catalogs of clusters of galaxies. Identifications

The first catalog of clusters of galaxies was published by Abell (1958) for $\delta \geq -27°$ and extended by Abell et al. (1989) for $\delta \leq -17°$. Thus both catalogs are unique in their coverage of the entire sky. We denote the clusters from these catalogs with prefix "A" (which is not the same as in NED). There are A1 ÷ A4076 numbers, about 4050 different clusters. Moreover, ACO contains about 1150 supplementary (S) southern clusters which are either not rich enough or at a great distance to be included in the main catalog.

Many clusters are found by Zwicky et al. (1961, 1963, 1965, 1966, 1968a, 1968b) but for $\delta \geq 0°$. Their total number is 9133 (denoted with prefix "ZC"). Other lists of groups and clusters are presented in the following table.

Table 1. Lists of groups and clusters.

Reference	Prefix	Number of Objects
	Visual procedure	
Klemola (1969)	Kl	44
Sersic (1974)	Ser	several tens
Rose (1976)	Ro	124
Duus & Newell (1977)	DN	710
Snow (1970)	Sn	29
Braid & MacGillivray (1978)	BMc	~ 400
Quintana & White (1990)	QW	267
	Semi-automatic algorithm	
Shectman (1985)	Sh	646
	Abell-like automatic algorithm	
Lumsden et al. (1992)	ED	737
Dalton et al. (1997)	APM	957

Various visual searching procedures for clusters result in a subjective catalogs. While the catalog of Shectman is the first one where a "blind" computer search was performed among the famous Lick counts of galaxies (Shane & Wirtanen 1967; Seldner et al. 1977).

The Abell-like algorithm catalogs are entirely automated and so they are entirely objective. Clusters of galaxies are found also by Jackson (1982), Gunn, Hoessel, & Oke (1986), Willick (1991) and Postman et al. (1996).

Some designations of clusters are proposed and used before A/ACO catalog publication and some of them have clear correspondence to the objects in the A/ACO catalog. For example, A3266 = Ser040/6 = DN0431-616 = APM510 and references for all these objects must point to A3266. So it would be impossible to retrieve the information available for a given A/ACO cluster without making first a cross-identifications of clusters among all available catalogs or lists. Some identifications are presented in the original ACO catalog with some misprints or errors.

New cross-identifications between A/ACO clusters and the other clusters are performed. Abell–Zwicky cross-identifications are taken from Kalinkov, Valtchanov, & Kuneva (1997, 1998).

3. Two ways to retrieve references

The basic idea is to assign two lists to each reference – the first contains the A/ACO clusters and the second contains the keywords describing observational data or results.

Two ways for references retrieval for any cluster are possible – using a glossary of keywords or using a thesaurus which is created for that particular need of clusters of galaxies.

The present version of this project operates with a glossary. Using a thesaurus of the kind compiled by Shobbrook & Shobbrook (1993) would be convenient but it is very hard to create a new thesaurus for such a narrow field like clusters of galaxies. If a usual thesaurus is available then it would be quite simple to find any reference assigned with a narrower term. But it is not possible to use a narrower terms for clusters of galaxies, when there are hierarchical structures (double galaxies, groups, superclusters, substructures and so on).

4. Current status

1. 3000 references are included – journal articles, workshops, symposia, colloquia papers, thesis works.
2. References for all A/ACO clusters of galaxies, together with their cross-identification may be retrieved at present. There are about 5200 clusters.
3. The glossary contains about 300 keywords.

5. Future developments

1. Actualizing the references. At the end of 1999 we expect to entry about 12,000 titles.
2. Including non A/ACO clusters – about 16,000 objects.
3. Extending the glossary.
4. Creation of thesaurus oriented to the field of clusters of galaxies.
5. Free web access to the database

The full version of the paper is available at astro-ph/9811328.

Acknowledgments. This project was initiated ten years ago as a part of a project *"Reference Catalog of Clusters of Galaxies"*. The creation of the database for clusters of galaxies is a complex and time consuming work, having in mind all the recent developments and the pouring observations of clusters of galaxies. Many colleagues have helped us and it is impossible to mention all of them. But we would like to thank particularly H. Andernach, H. Corwin, G. Dalton, G. de Vaucouleurs, J. Huchra, W. Keel, R. Nichol, F. Ochsenbein, A. Omont, G. Paturel, M. Postman, M. Ramella, H. van der Laan, R. Williams.

This research has made use of the NASA/IPAC Extragalactic Database (NED) which is operated by the Jet Propulsion Laboratory, California Institute of Technology, under contract with the National Aeronautics and Space Administration.

This work was supported by the National Research Fund of the Bulgarian Ministry of Education, Science and Technology (contract F721/1997).

References

Abell, G. O. 1958, ApJS, 3, 211

Abell, G. O., Corwin, H. G., & Olowin, R. P. 1989, ApJS, 70, 1

Braid, M. K. & MacGillivray, H. T. 1978, MNRAS, 182, 241

Dalton, G. B., Maddox, S. J., Sutherland, W. J., Efstathio, G. 1997, MNRAS, 289, 263

Duus, A. & Newell, B. 1977, ApJS, 35, 209

Gunn, J. E., Hoessel, J. G., & Oke, J. B. 1986, ApJ, 306, 30

Jackson, R. 1982, Ph.D. thesis, Univ. California, Santa Cruz

Kalinkov, M., Valtchanov, I., & Kuneva, I. 1997, Catalog CDS VII/201, (ADC 7201)

———, 1998, Bull. Inform. CDS, 49, in press

Klemola, A. R. 1969, AJ, 74, 804

Lumsden, S. L., Nichol, R. C., Collins, C. A., & Guzzo, L. 1992, MNRAS, 258, 1

Postman, M., Lubin, L. M., Gunn, J. E., Oke, J. B., Hoessel, J. G., Schneider, D. P., & Christensen, J. A. 1996, AJ, 111, 615

Quintana, H. & White, R. A. 1990, Ap&SS, 173, 265

Rose, J. A. 1976, A&AS, 23, 109

Seldner, M., Siebers, B., Groth, E. J., & Peebles, P. J. E. 1977, AJ, 82, 249

Sersic, J. L. 1974, Ap&SS, 28, 365

Shane, C.D. & Wirtanen, C. 1967, Publ. Lick Obs., 22, 1

Shectman, S. A. 1985, ApJS, 57, 77

Shobbrook, R. M. & Shobbrook, R. R. 1993, The Astronomy Thesaurus, (Epping: IAU)

Snow, T. P. 1970, AJ, 75, 237

Willick, J. A. 1991, Ph.D. thesis, Univ. California, Berkeley

Zwicky, F., et al. 1961, Catalogue of Galaxies and Clusters of Galaxies. Vol. 1. (Pasadena: Caltech)

———, 1963, Catalogue of Galaxies and Clusters of Galaxies. Vol. 2. (Pasadena: Caltech)

———, 1965, Catalogue of Galaxies and Clusters of Galaxies. Vol. 5. (Pasadena: Caltech)

———, 1966, Catalogue of Galaxies and Clusters of Galaxies. Vol. 3. (Pasadena: Caltech)

———, 1968a, Catalogue of Galaxies and Clusters of Galaxies. Vol. 4. (Pasadena: Caltech)

———, 1968b, Catalogue of Galaxies and Clusters of Galaxies. Vol. 6. (Pasadena: Caltech)

Astronomical Data Analysis Software and Systems VIII
ASP Conference Series, Vol. 172, 1999
D. M. Mehringer, R. L. Plante, and D. A. Roberts, eds.

R-Trees for Astronomical Data Indexing

Andrea Baruffolo

Astronomical Observatory, vicolo Osservatorio 5, 35122 Padova, Italy

Abstract. The use of multidimensional spatial access methods (SAMs) has been sometimes advocated in the past for astronomical database applications, in order to speed up the retrieval of data. However, they have had little practical application until now. This can be ascribed partly to the difficulty of integrating new access methods in traditional, i.e., relational, database systems. With the advent of new database management systems, which we call collectively Object Data Management Systems (ODMSs), it is now possible to effectively employ SAMs in building astronomical data management systems. A wide variety of SAMs have been introduced in the past, among these R-trees offer many advantages that make them well suited for astronomical applications. In this paper we describe the features that, in our opinion, make R-trees useful in an astronomical context and how they can be adapted to work with astronomical data.

1. Introduction

The advent of large astronomical databases, particularly catalogues, is posing the problem of giving researchers an efficient and unencumbered access to these large data collections. Finding the answer to this problem has been difficult because traditional database technologies are unable to effectively cope with them (see e.g., Pirenne & Ochsenbein 1991).

The use of multi-dimensional spatial data structures has been advocated to solve this problem, either by building indexes to the main data collection or by physically organizing data on secondary storage media (see e.g., Barrett 1995; Wicenec & Albrecht 1998).

However, it is to be noted that a solution in a database context would be preferable, i.e., we would not like to throw away the advantages brought to us by database technology, in particular query processing and physical and logical data independence.

The recent emergence of new technology databases, that we call collectively ODMSs, now makes it possible to directly support astronomical data and, at least for those system that feature extensible indexing capabilities (Stonebraker 1996), secondary access methods that are *natural* to these datatypes. However, the question remains about what data structure may be appropriate and/or convenient to index astronomical data.

In this paper we describe the features that in our opinion make R-trees an attractive possibility as indexing structures for astronomical data.

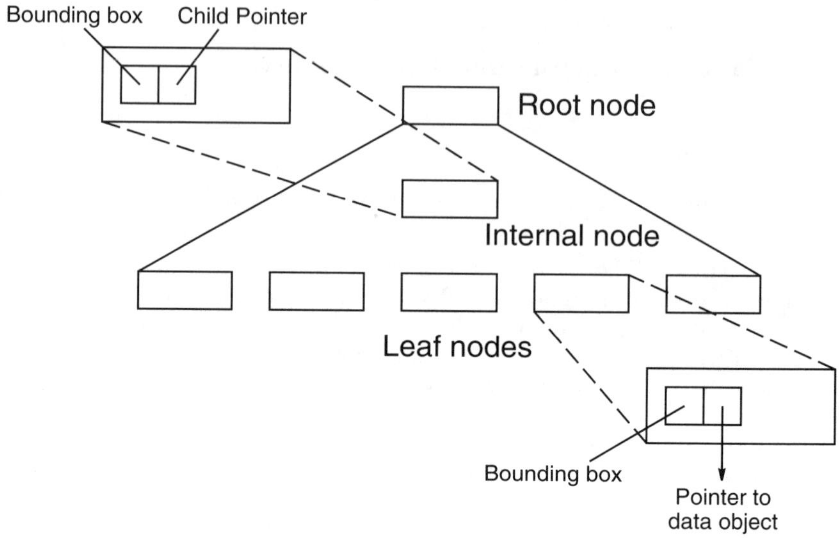

Figure 1. A conceptual view of R-trees.

2. R-Trees

R-trees were originally proposed as an indexing mechanism for non-zero size spatial data in a multi-dimensional Cartesian space (Guttman 1984). Basically, an R-tree is an height-balanced tree in which internal nodes contain couples of pointers to the lower nodes in the tree and the corresponding bounding box, while index records in leaf nodes contain bounding boxes and pointers to the actual data objects. A conceptual representation of an R-tree is shown in Fig. 1. A thorough description of R-trees can be found in the original paper, in this section we concentrate on the issues that, in our opinion, make R-trees interesting for practical implementations of astronomical database management systems.

First of all, although the original definition of R-trees dealt with objects in a Cartesian space and supported only overlap and containment query predicates, it turns out that the basic principle can be generalized to support almost arbitrary data and query predicates (Hellerstein, Naughton, & Pfeffer 1995).

Secondly, besides other spatial search structures that have been proposed in the past and claimed to be more efficient, R-trees are conceptually simple so that they are more easily integrated in a database environment. In fact, they are found both in research (e.g., PostgreSQL, Predator) and commercial DBMSs (e.g., Informix).

Furthermore, there have been many studies aimed at extending R-trees to support parallel (Kamel & Faloutsos 1993a), high concurrency (Kornacker & Banks 1995) and bulk loading (Kamel & Faloutsos 1993b) operations, making this structure particularly attractive for implementation in commercial DBMSs. Although not all these features may be present in the real product, they can in principle be realized.

One problem with R-trees is that it is not possible to guarantee worst-case performance. However, this is a problem they have in common with almost all other multi-dimensional access methods (Freeston 1995). Experience shows that they nevertheless exhibit *good* behavior for most kinds of data (Guttman 1984).

Finally it is worth noting that although R-trees were originally intended to index non-zero size spatial data objects, so that bounding boxes are stored in the leaf nodes of the tree, they can obviously index zero-sized objects (points) and a good implementation of this index structure would adapt to make efficient use of disk space (i.e., store points instead of bounding boxes in the leaf pages of the index).

3. Adapting to Astronomical Data

In astronomical data collections, there are many data that can be though of as points in a multi-dimensional space and are then suitable to be indexed using R-trees (i.e., multi-color photometry). Among these, coordinates have always played a special rôle, so that we shortly describe how R-trees can be used to index an astronomical database on them.

Coordinates on the sky can be (and often are) represented in a database as ordered couples of longitude and latitude. An R-tree based index can be built using, in the internal nodes, bounding boxes made up of two coordinate pairs that represent the minimum and maximum latitude and longitude of all objects contained in the branch that is pointed to. Leaf nodes contain the actual coordinates and pointers to the DB records.

With this tree organization, all *usual* search predicates on coordinates (e.g., point and range query, search by cone) can be applied to the R-trees, provided that proper account is made for the metric (i.e., distance on the sphere), because they are simple variants of the containment and overlap predicates. The same argument applies to nearest neighbor searches for which an algorithm based on R-trees has recently been developed and is general enough to be applied to astronomical coordinates (Roussopoulos, Kelley, & Vincent 1995), in this case also by allowing for the particular metrics on the sphere.

It is worthwhile noting that other representations of coordinates (e.g., by means of triads) and bounding boxes (or, better, bounding predicates, see Wang, Hellerstein, & Lipkind 1998) are possible. In fact, it would be interesting to investigate the various alternatives with respect to disk space requirements and query execution performance.

4. Summary and Conclusions

R-trees, although they may not be the best performing secondary access methods for all data and queries (Kriegel et al. 1990; Wang, Hellerstein, & Lipkind 1998), are convenient to use especially in a database context. This is due to the fact that R-trees based indices, that can be extended to support astronomical data and query predicates, are now found in various research and commercial DBMSs.

Furthermore, advantages over traditional structures (e.g., B-trees) in indexing astronomical data have already been demonstrated in a prototype implementation (Baruffolo & Benacchio 1998a). In fact, we are implementing R-tree based indices in a project aimed at serving a very large on-line Freeston reference catalogue (Baruffolo & Benacchio 1998b; Baruffolo, Benacchio, & Benfante 1999).

References

Baruffolo, A., Benacchio, L., & Benfante, L. 1999, this volume, 237

Baruffolo, A. & Benacchio, L. 1998a, in ASP Conf. Ser., Vol. 145, Astronomical Data Analysis Software and Systems VII, ed. R. Albrecht, R. N. Hook, & H. A. Bushouse (San Francisco: ASP), 382

——, 1998b, in SPIE Proc., Vol. 3349, Observatory Operations to Optimize Scientific Return, ed. P. J. Quinn (Bellingham: SPIE), 274

Barrett, P. 1995, in ASP Conf. Ser., Vol. 77, Astronomical Data Analysis Software and Systems IV, ed. R. A. Shaw, H. E. Payne, & J. J. E. Hayes (San Francisco: ASP), 472

Freeston, M. 1995, SIGMOD Record, 24, 80

Guttman, A. 1984, Proc. of the SIGMOD'84 Ann. Meet., ed. B. Yormark (Boston: ACM Press), 47

Hellerstein. J. M., Naughton, J. F., & Pfeffer, A. 1995, in Proc. of the 21st Conf. on Very Large Data Bases, ed. U. Dayal, P. M. D. Gray, & S. Nishio (San Francisco: Kaufmann), 562

Kamel, I. & Faloutsos, C. 1993a, SIGMOD Record, 21, 195

——, 1993b, in Proc. of the 2nd Intl. Conf. on Information and Knowledge Management, ed. B. Bhargava, T. Finin, & Y. Yesha (New York, ACM Press), 490

Kornacker, M. & Banks, D. 1995, in Proc. of the 21st Conf. on Very Large Data Bases, ed. U. Dayal, P. M. D. Gray, & S. Nishio (San Francisco: Kaufmann), 134

Kriegel, H.-P., Schiwietz, M., Schneider, R., & Seeger, B. 1990, Lecture Notes in CS, 409, 89

Pirenne, B. & Ochsenbein, F. 1991, ST–ECF Newsletter, 15, 17

Roussopoulos, N., Kelley, S., & Vincent, F. 1995 SIGMOD Record, 24, 71

Stonebraker, M. 1996, Object–Relational DBMSs: The Next Great Wave, (San Francisco: Kaufmann)

Wicenec, A., & Albrecht, M. 1998, in ASP Conf. Ser., Vol. 145, Astronomical Data Analysis Software and Systems VII, ed. R. Albrecht, R. N. Hook, & H. A. Bushouse (San Francisco: ASP), 512

Wang, S., Hellerstein, J. M., & Lipkind, I. 1998, Near-Neighbor Query Performance in Search Trees, (UCB Tech. Rep. CSD-98-1012)

Astronomical Data Analysis Software and Systems VIII
ASP Conference Series, Vol. 172, 1999
D. M. Mehringer, R. L. Plante, and D. A. Roberts, eds.

ESO/CDS Data-mining Tool Development Project

Patrcio F. Ortiz, François Ochsenbein
Centre de Données Astronomiques de Strasbourg

Andreas Wicenec, Miguel Albrecht
ESO, Garching

Abstract. ESO and CDS are currently developing the data-mining tools which will allow users to access and combine the information stored in the forms of catalogs at CDS, and catalogs and data at ESO. CDS counts with about 5000 tables handled by the **VizieR** system (the data-mine, based on Sybase), including catalogs which appeared recently in journal articles. ESO has the data collected in the EIS program (deep survey with NTT) and will have in the future data from VLT's observations and other archived instruments in La Silla.

The main goal has been to build a "friendly" user interface so that users located remotely can either submit their own data "tables" (as ASCII files) for comparison or extract information from either ESO or CDS to perform cross correlations in all the parameter space provided by the data catalogs -not restricting the correlations to positional ones. We have decided to use HTTP as the exchange protocol between the Data Mining Facilities (DMF) located at CDS and the ESO servers, as well as between the user and the DMF.

1. Introduction

The objectives of this joint project between ESO and CDS are to build tools which will allow astronomers to access a large volume of information in the form of electronic data with the purpose of cross correlation. Remotely located users can either submit their own data "tables" (as ASCII files) for comparison or extract information from either ESO or CDS to cross correlate by position in the sky or by any of the parameters provided by the data catalogs. An important point has been the development of knowledge structures with the purpose of facilitating the description of the data to provide highly flexible data-mining options.

This paper outlines the main features users will find at the DMF and comments on features which have allowed us to optimize the consulting time. In addition we discuss the mechanisms used to organize the information in order to accept complex queries regarding the selection of the catalogs to use for comparison based on the nature of the objects contained, the "quantities" stored, wavelength coverage, sky region, etc.

2. The Data Mine at CDS

CDS hosts, under the **VizieR** system, around 2000 catalogs with more than 5000 tables, all of them accessible, described and stored using a very uniform system (Ochsenbein 1998; Ochsenbein et al. 1999). VizieR already allows queries by position and other criteria to individual catalogs.

The catalog content consists of: compilation catalogs, surveys, observing logs (from space missions), and tables from journal articles, with both observational and modeled data.

There are other facilities at CDS which could be used as a complement to data-mining: bibliographic references, name resolver (via Simbad), web resource locator (GLU), etc.

Knowledge-detection structures were developed to complement the meta information in the system. These structures were based on **column content** and also on **astronomical object type**.

3. The Knowledge-Detection Structures.

Two knowledge detection structures were developed: one for astronomical object types, and the other for column content. The structure for object type resembles the structure used in SIMBAD, with a four level hierarchy; the source to assign object types to the catalogs and tables is the standardized description file (ReadMe file) developed at CDS and shared now by other data centers and journals.

The structure related to column content was fully developed for this project. It contains 35 main categories and has a four level hierarchy. Categories such as Photometry, Positions, Spectroscopy, Time and Physical Quantities are amongst the most populated.

A **Unified Content Descriptor** (UCD) is then assigned to each of the columns in each of the tables accessed with **Vizier**. We developed an automatic UCD assignation procedure based on column name, column units, and column description.

Because of the importance of the use of UCD's for datamining purposes, we developed tools to assign UCD's to user provided files.

The existence of knowledge structures implies that it is possible, on one hand, to retrieve the names of all catalogs/tables containing any given set of UCD's, and on the other hand, to perform cross correlations with catalogs/tables containing the same UCD's as the ones describing the content of a user file.

As an example, the following table shows the assignment of UCD's to a few "typical" quantities (or columns) one could use for data-mining:

Column:	R.A.	DEC	Name	Vmag	z
Units:	h:m:s	d:m:s		mag	
UCD's:	POS_EQ_RA	POS_EQ_DEC	ID_MAIN	PHOT_JHN_V	REDSHIFT

Figure 1. Cross Correlator options.

4. The Cross Correlator

Figure 1 shows the options provided by the cross-correlation interface. One way to envision datamining is a process in which users want to perform cross correlation operations using the data stored in a Data Mine as the comparison side, against their own data or data extracted from the same or other Data Mine. On the Users's side (reference side) there are several scenarios, such as:

- Consulting about single coordinates (user provided).
- Consulting about a list of coordinates and/or other quantities (user provided in the form of a file).
- Using data extracted from a Data-Mine (CDS, ESO, or somewhere else).

On the Data Mining Facility side (comparison side) CDS offers:

- Find counterparts in any catalog under **VizieR** . By far the most time consuming operation.
- Find counterparts in the Main Catalogs under **VizieR** . The main catalogs is a short list comprising some of the most relevant catalogs or surveys in many fields.
- Find counterparts in your favourite catalogs under **VizieR** . Users may submit a list containing the encoded names of what they consider the most relevant catalogs for their research purposes, and counterparts will be sought in those catalogs only.
- Compare data against ONE catalog under **VizieR** . The user knows well the content of a catalog and wishes to correlate against it. Correlation is open to all numerical quantities.
- Compare against catalogs with similar content under **VizieR** . Catalogs with similar "column content" (based on their UCD's) are open for cross correlation in all dimensions.

Some of the important features developed for this project include the following:

Web based engine which provides connection between the user, CDS, and ESO facilities.

Positional Cross Correlator designed to minimize the time to determine whether a catalog contains information in a certain region of the sky or not.

Quick search mechanism takes advantage of the querying system of **VizieR** to determine the matching elements.

High Flexibility to the user so that a number of alternatives can be selected with ease and speed.

Non positional Cross correlation by all numerical quantities stored in catalogs implies that users will be able to perform N-dimensional space analysis with as many numerical quantities as desired.

5. Practical Applications

During the development stage a number of practical applications were used as tests to measure flexibility and performance of the prototypes, some of the most relevant ones are listed here.

- Search for counterparts in one particular direction in the sky amongst all the catalogs under VizieR or a selected set of them.
- Search for counterparts for a user provided list of objects amongst all the catalogs under VizieR or a selected set of them.
- Self consistency test in any of the catalogs stored in the data-mine, e.g., search for double or multiple quasars in the latest list of AGN's by Veron-Cetty & Veron (1998).
- Positional cross correlation between one catalog in the datamine and one or more catalogs, e.g., galaxy clusters' catalog vs X-ray catalog.
- Search for neighbour objects based on non positional criteria, e.g., based on a user list find all objects with differences in redshift and absolute magnitude less than user defined values.

Other applications have been explored and will be part of the manual for the facilities or be part of the on-line examples available to users.

Acknowledgments. We are grateful to Pascal Dubois and Françoise Genova for valuable comments about the interface and the features one expects to find tools like the one presented in this paper.

References

Ochsenbein, F. 1998, in ASP Conf. Ser., Vol. 145, Astronomical Data Analysis Software and Systems VII, ed. R. Albrecht, R. N. Hook, & H. A. Bushouse (San Francisco: ASP), 387

Ochsenbein, F., Fernique, P., Ortiz, P., Egret, D., & Genova, F. 1999, in Future Generation Computer Systems (Amsterdam: North-Holland)

Veron-Cetty, M. P. & Veron, P. 1998, Quasars and Active Galactic Nuclei (8th ed.) (ESO Scientific Rep. No. 18)

Astronomical Data Analysis Software and Systems VIII
ASP Conference Series, Vol. 172, 1999
D. M. Mehringer, R. L. Plante, and D. A. Roberts, eds.

StarView II: A Customizable Query Generation Tool

Bridget Kennedy, Bruce Mayhew

Space Telescope Science Institute

Abstract. StarView is the user interface to the Hubble Space Telescope (HST) Data Archive. It supports queries to the HST catalog, submission of retrieval requests, data preview, and other archive-related functions. This presentation looks at the StarView II effort to make a flexible, adaptive framework for browsing astronomical catalogs. We discuss user interaction with the customizable form, describe the use of Java Foundation Classes in implementing the Model View Controller design pattern, and present design features that allow simple, robust communication with third-party components added at run-time.

1. Introduction

StarView II addresses two important concerns within the astronomical community: the desire for a flexible catalog browsing tool and the need for an adaptive framework capable of seamless interaction with current and future functional components.

2. Flexible Browsing Environment

Along with standard query forms, StarView II supplies an interactive custom query form building tool – the Customizer. As shown in Fig. 1, the Customizer contains a keyword search panel and hierarchical displays of the data structure. With these catalog data browsing components, the user builds custom query forms through simple point-and-click interaction. All forms maintain their own state information, permitting asynchronous execution of multiple searches. Each query form, along with its specific formatting and state information, can be saved and retrieved using Java object serialization.

StarView II forms implement the Model View Controller (MVC) design pattern through encapsulated data modeling and selectable data views that interact via a controller. Clear division of functionality, and strategic use of interface classes, result in a system that readily accommodates varying data sources and view components.

2.1. Data Models

A simple two-dimensional vector was chosen for the structure of the base data model. Application-specific knowledge of the data structure is contained in a

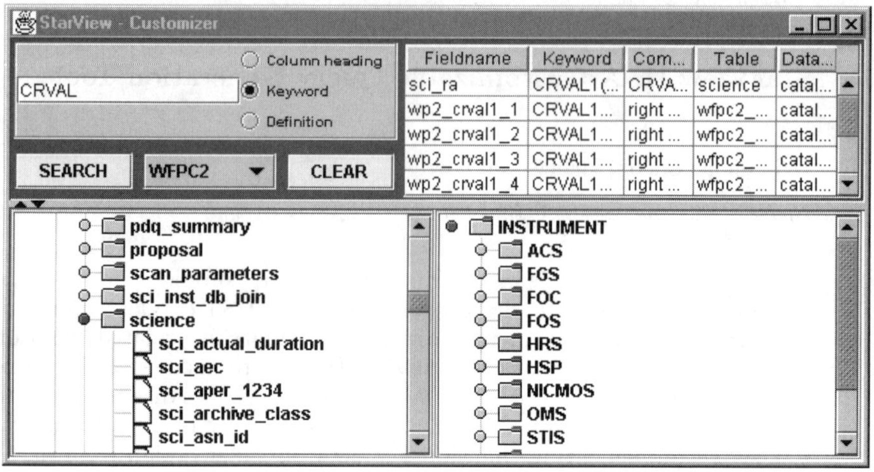

Figure 1. StarView II Customizer

separate loader class that translates the underlying data into the generic two-dimensional vector. This allows the flexibility to replace or subclass the loader class to handle different data sources such as a flat file or internal memory structure. The rest of the StarView II MVC architecture remains unchanged.

2.2. Selectable, Extensible Data Views

The prototype StarView II view components, list, text field, and table, are extensions of Java Foundation Classes (JFC a.k.a. Swing). The MVC architecture built into JFC provides an ideal backdrop for implementing the StarView II design. Using Java interface classes, view components register interest in data model changes. Once registered with the data model, JFC event processing keeps view components up-to-date with data changes.

View components register as listeners with the base data model through a minimally specified interface class. The data model has no additional knowledge of the characteristics of the listener. Another interface expands this strategy by allowing third-party components to assume the role of a StarView II MVC view. The only imposition, i.e., shared code, on third-party code is implementation of the StarView II MVC component interface. The end goal for StarView II is to allow run-time addition of these components.

2.3. Control

Control, or data modification, is effected by a Query Processor object in response to Search and Navigation user interface (UI) components. The Query Processor interacts with a Database object. Each query form can search using a hard coded query, or, as in the case of HST searches, generate the query at run-time. Automated query generation is a complicated system and is isolated to run on a server allowing the option of turning off or bypassing this mechanism on a per form basis.

Data navigation refers to the ability to step and scan through the database using UI controls. In considering navigation, it is important to discuss the StarView II ability to keep various views in synch via registration with a common data model. For example, as a user steps through records in a list box, the data model sends events to all view component listeners. The view components update their displays reflecting data associated with the record number received in the event.

3. Adaptive Functional Framework

Historically, StarView has been used to browse the HST Data Archive. In setting goals for StarView II, we recognized the need for a more flexible, robust browsing tool and the fact that each user has a different set of functional requirements. We accommodate unique functional requirements by providing a plug-and-play framework in which a user can create a version of StarView tailored to specific needs.

3.1. StarView II Interfaces

The concept of Java interface classes provides the basis for communication between StarView II and third-party components.

- Data pull interfaces allow a component to query StarView II for available attributes and data contained within a form. Using this interface, components can be instantiated at run-time and retrieve any portion of the information contained on a screen.
- Data push interfaces allow a component to push generated data back to a StarView II qualification form.
- MVC interfaces allow a component to register for data model changes. The component receives event notification of a change and then uses the data pull interface to obtain a specific attribute location from the data stream.
- The Observer/Observable interface makes use of a user-defined object class for data transfer between applications. StarView II is required to have prior knowledge of the object's structure. This accommodates more complex command interactions (Dolensky, Mayhew, & Kennedy 1999).

4. Proof of Concept: Third-party component interaction

StarView II's adaptive framework was well illustrated while testing the ease of installation of JIPA (Dolensky et al. 1999) and SkyMap.

SkyMap is a component developed independently at STScI that plots HST observations on an AITOFF projection. The association established between SkyMap and StarView II demonstrates a highly interactive relationship achieved through Java interfaces. Our tests exercise bi-directional interfaces that send data to an external MVC view and receive data from a third-party functional component.

As an MVC component, SkyMap registers as a listener to the main data model and receives data change notifications. SkyMap responds to the event

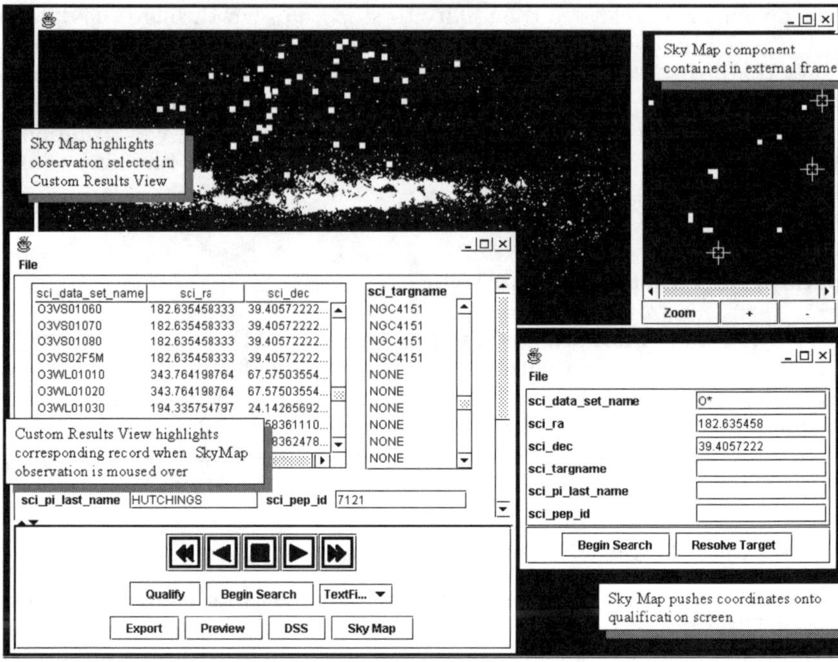

Figure 2. StarView II/SkyMap Interaction

from the data model by highlighting the location for the current record on the AITOFF projection.

As a member of a group of MVC components associated with a particular data model, SkyMap interacts with other view components. When the mouse enters an observation on the SkyMap, it sends a 'set current record' to the main data model. Other view components receive an event from the model and update accordingly, achieving a seamless interactive graphical display.

Finally, the SkyMap and JIPA tests both exercise the ability to push values into StarView II from an external source. As the user points and clicks on the SkyMap display, or the JIPA preview image, the associated coordinates are pushed into Ra and Dec fields on a qualification screen. These pushed coordinate values can then be used for subsequent searches.

References

Dolensky, M., Mayhew, B., & Kennedy, B. 1999, this volume, 454

Astronomical Data Analysis Software and Systems VIII
ASP Conference Series, Vol. 172, 1999
D. M. Mehringer, R. L. Plante, and D. A. Roberts, eds.

Application of the Minimum Determination Algorithm to the Study of the Fine Structure in the Mass-Luminosity Relation and of the Nature of "Overmassive" Stars

Oleg Malkov, Dana Kovaleva

Institute of Astronomy, 48 Pyatnitskaya St., Moscow 109017, Russia

Elena Schilbach

Astrophysikalisches Institut Potsdam, An der Sternwarte 16, D-14482 Potsdam, Germany

Abstract. The available observational data on masses and/or luminosities of binary components and the current models for different chemical compositions of Pop I stars were used to improve the mass-luminosity relation for stars of moderate masses. We have developed a technique of simultaneous minimization of the discrepancies between the predicted and observed masses/radii, assuming equal ages and chemical compositions of binary system components. The method of descending coordinates was used to optimize the observed and theoretical parameters.

This technique was also applied to decompose the low-mass components of several stars suspected to be unresolved binaries. In order to explain the offsets of these stars from the "normal" location in the mass-luminosity plane, we computed the most probable masses and luminosities of their hidden components.

1. Description of the Method

This paper presents a mathematical approach which can be used in investigations of different astrophysical problems. The method is based on the minimum determination technique applied to functions of several variables covering a limited range of values. Generally, the method can be described as follows.

Let p_1, p_2, ..., p_n be n independently observational parameters (IOP) for a given kind of objects. We assume p_1^0, p_2^0, ..., p_n^0 and $\Delta_1, \Delta_2, ..., \Delta_n$ to be their corresponding observed quantities and observational errors, respectively. Further, we define a set of non-directly observable parameters (NDP) q_1, q_2, ..., q_m for the objects under study. In addition, we consider a set of astrophysical conditions relevant for the investigated objects. There are two different kinds of these conditions. First, the strong mathematical relations between parameters $p_i = p_i(q_1, ..., q_m, p_j, ..., p_k)$ (e.g., the mass of a binary is equal to the sum of the component masses); second, the empirical relations between two parameters $Q_i = Q_i(q_j)$ (e.g., mass-luminosity relation).

We define an initial set of parameters (ISP hereafter) selected from NDP and IOP. The remaining parameters from IOP can be computed in accordance with the relations given above. Treating the observational data p_i^0 and their computed values p_i, we can find a new set of ISP which optimally satisfy both the observations and the adopted conditions.

In other words, we have to minimize a function

$$D = \sqrt{\left(\Sigma_{i=1}^{n}\left(\frac{p_i^0 - p_i}{\Delta_i}\right)^2 + \Sigma_{i=1}^{n}\left(\frac{q_i - Q_i}{\Delta_{Qi}}\right)^2\right)}, \quad (1)$$

where Δ_{Qi} is the uncertainty of the relation $Q_i = Q_i(q_j)$.

Minimizing the function D, we can derive optimal sets of IOP and NDP and, consequently, determine more suitable values within their error boxes for observable and non-directly observable parameters. The function D is dimensionless and can be used as a criterion for the accuracy of the solution. To find the minimum of the function D, the method of descending coordinates (or other appropriate numerical algorithms) can be used. An application of the method to a given astrophysical problem assumes a careful construction of the suitable function D and definition of reasonable limits for the included parameters.

Below we present two examples of possible applications.

2. Fine Structure in the Mass-Luminosity Relation

The main objective of the project was to improve the mass-luminosity relation (MLR) for stars of moderate masses by optimal using the available data on binaries components. For this purpose we considered the data (IOP) on masses $(m_1^0 \pm \Delta(m_1^0), m_2^0 \pm \Delta(m_2^0))$ and radii $(r_1^0 \pm \Delta(r_1^0), r_2^0 \pm \Delta(r_2^0))$ for the components of eclipsing binary stars from the Catalogue of Astrophysical Parameters of Binary Stars (Malkov 1993). The corresponding set of NDP includes the age a ($\equiv \log T(yr)$) and abundance z of the components. Assuming a common origin for the components, we may consider their ages and chemical compositions to be equal. As additional conditions, we used theoretical stellar models of Schaller et al. (1992) for different chemical compositions of Pop I stars. The values of $\lg L$ and $\lg T_{eff}$ derived from these models were used to determine the parameters of the relations $r_i = R(m_i,\ a,\ z)$ and the corresponding accuracy Δ_{R_i}.

The function D to be minimized can be written as

$$D = \sqrt{\left(\Sigma_{i=1}^{2}\left(\frac{m_i^0 - m_i}{\Delta(m_i^0)}\right)^2 + \Sigma_{i=1}^{2}\left(\frac{r_i^0 - R(m_i,\ a,\ z)}{\Delta_R i}\right)^2\right)}. \quad (2)$$

As the result, the optimal values of masses and radii as well as their chemical composition (z) and evolutionary stage (a) were obtained for 12 double star systems. For one of the considered binaries (AS Cam), Fig. 1 shows how these parameters correlate with the original observational data.

About 30 other double star systems are included in the project. Extending the investigations to a larger number of binaries will improve the database on optimal masses and luminosities. This will allow us to derive the mass-luminosity relation for stars of moderate masses with higher accuracy.

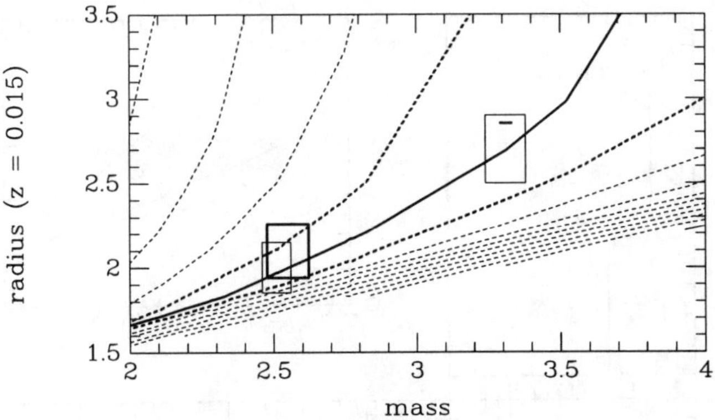

Figure 1. The IOP of the components of the AS Cam system. The error boxes for the observed (*thin line*) and optimal (*thick line*) values of masses and radii are presented. The isochrones for the optimal abundance $z = 0.015$ are also plotted in steps of $\Delta a = 0.15$. The solid line and two heavily dashed lines mark the isochrone of the optimal age of the system and the area of uncertainty ($a = 8.14 \pm 0.15$), respectively.

3. Explanation of the Nature of "Overmassive" Stars

Here we considered stars located in the area of underluminous objects of the MLR. Using the approach described above, we tested whether at least some of them may be considered as photometrically unresolved binaries.

In this case the corresponding IOP are mass $m = m_1 + m_2$ and luminosity $\lg L = \lg(L_1 + L_2)$ of an underluminous star supposed to be a binary. The initial set of observational data $m^0 \pm \Delta m^0$, $\lg L^0 \pm \Delta \lg L^0$ for the low mass stars were taken from Malkov, Piskunov, & Shpil'kina (1997). As additional condition, we used the theoretical mass-luminosity relation for the low mass region represented by a function $m = M(\lg L)$. The width of MLR was assumed to be $\Delta(M) = m * 0.5$. In this case we have to minimize the function

$$D = \sqrt{\left[\frac{m^0 - m}{\Delta m^0}\right]^2 + \left[\frac{\lg L^0 - \lg L}{\Delta \lg L^0}\right]^2 + \Sigma_{i=1}^{2} \left[\frac{m_i - M(\lg L_i)}{\Delta(M(\lg L_i))}\right]^2} \quad (3)$$

The parameters of components were determined for four recently resolved binaries and for two other stars which have been suspected to be multiple (Kovaleva & Malkov 1999). The distribution of probable values of masses and luminosities is plotted in Fig. 2 for one of the resolved stars (GJ 508 A).

Applying our method to the suspected binaries, we derived the parameters of the components which may describe the situation realistically. Nevertheless, we do not claim that these results present the *final proof* for the duplicity or multiplicity of these two stars examined here. The offsets of the stars from the usually adopted MLR may be also explained both by large observational errors and by insufficient knowledge of the actual MLR. The results presented here merely show that the possibility for these stars to be unresolved binaries cannot be rejected.

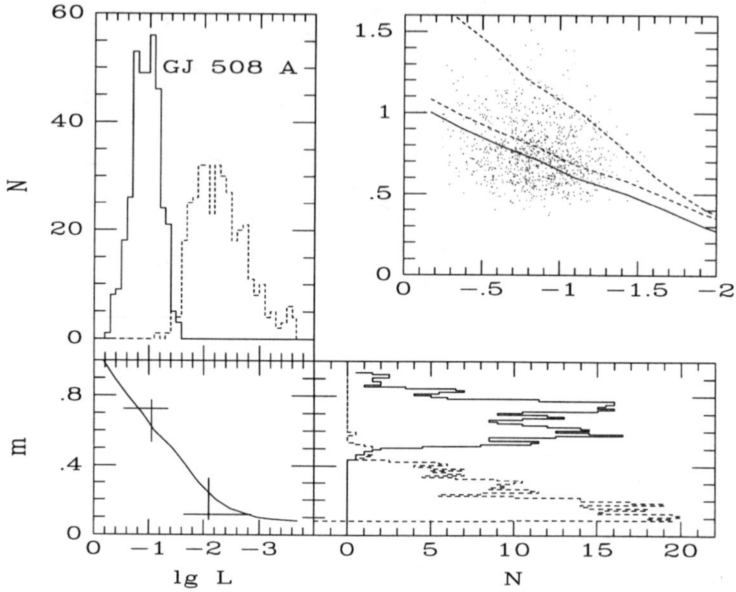

Figure 2. Results for the resolved binary GJ 508 A with unknown parameters of the components. The upper right panel presents the MLR (*solid line*) and the binary area (*between the dashed lines*) where the unresolved binaries should be located. Points show the values m and $\lg L$ which may be accepted for the star due the observational errors. The resulting distributions of masses and luminosities of the main component (*solid lines*) and the secondary component (*dashed lines*) are shown in the lower right and upper left panel, respectively. The predicted masses and luminosities of the components are plotted in the lower left panel together with the theoretical MLR.

Acknowledgments. The authors thank A. Piskunov and A. Tokovinin for valuable comments. OM and DK acknowledge the Russian Academy of Sciences for the support within the "Young scientists support program". OM thanks the Deutsche Forschungsgemeinschaft (project 436 RUS 113/380) and the LOC of the ADASS'98 Conference for the financial support that made this presentation possible.

References

Kovaleva D. A. & Malkov O. Y. 1999, Astronomy Reports, in press

Malkov, O. Y. 1993, Bull. Inf. CDS, 42, 27

Malkov, O. Y., Piskunov, A. E., & Shpil'kina, D. A. 1997, A&A, 320, 79

Schaller G., Schaerer D., Meynet G., Maeder A. 1992 A&AS, 96, 269

Astronomical Data Analysis Software and Systems VIII
ASP Conference Series, Vol. 172, 1999
D. M. Mehringer, R. L. Plante, and D. A. Roberts, eds.

IDL-based Database of Solar Active Regions

D. E. Gary

New Jersey Institute of Technology, Newark NJ, USA,
Email: dgary@njit.edu

V. V. Grechnev, L. V. Shabarova

Institute of Solar-Terrestrial Physics SD RAS, Irkutsk, Russia

A. Vourlidas

G. Mason Univ/Naval Research Lab, Washington DC, USA

M. Nishio

Department of Physics, Kagoshima University, Japan

Abstract. A database on solar active regions has been implemented in IDL (Interactive Data Language). The observational parameters of the regions are stored in a multi-level structure array that is distributed as a standard IDL save file. Convenient access to the data is provided by an application equipped with a graphical user interface (GUI). A variety of search modes are implemented. Full-disk 17 GHz radio maps produced by the *Nobeyama Radioheliograph* provide the basis for the database.

1. Introduction

Researchers of solar activity often need to access solar active region data. Sources of such information include the *Solar Geophysical Data* bulletin and other periodical publications. Searching the printed publications requires substantial time and effort. Although it is possible to obtain these data using the Internet, gathering the data from several different data sources remains difficult.

A convenient alternative would be an "illustrated desk book on active regions", in a database form, placed in the researcher's computer and equipped with a means of searching for desired information as well as printing or saving it in a suitable form. Such a "desk book" should be accessed quickly, allow extraction of the data of interest, and should be easily supplemented and extended.

There is a variety of database file formats from which to choose, but some may be limited to particular platforms or are unfamiliar to the larger research community. We have chosen a widely used, platform independent software environment called Interactive Data Language (IDL[1]) which can save data into a machine-independent format (XDR). The IDL programming language ensures complete compatibility with various operating systems (MS Windows, Macintosh, several UNIX versions, and VMS). So far, we have created an IDL-based database on daily active region data from 1992 July 1 to 1994 December 31 (NOAA 7213 thru NOAA 7821)[2]. This database file has a size of 1 MB.

The database is better utilized when combined with daily full-disk radio maps for the same period. The maps were obtained at 17 GHz in total intensity and circular polarization by the *Nobeyama Radioheliograph* (Nishio et al. 1994) (the radio maps are in standard FITS format and occupy three compact disks[3]). A researcher can easily access the database including the radio maps that provide information on the location of the strong magnetic fields, filaments, prominences, and other solar features.

2. Structure of the database

The database is constructed as a multi-level IDL structure arranged into an array, with each element corresponding to a single day. The structure for each day contains: (1) a field with data relevant to the entire visible disk of the Sun and (2) an array of fields for each active region (space limited to 10 active regions per day). Such a structure ensures that each element of the database can be accessed in a logical and flexible manner.
The common field for each day's entry contains:

1. date and time of observation;
2. solar P and B_0 angles and its optical radius;
3. number of active regions visible on the disk;
4. the peak radio brightness temperature at 17 GHz and its location on the visible disk in both intensity and polarization.

The structure fields for the individual active regions contain:

1. NOAA number;
2. geocentric coordinates (offsets in arc seconds from disk center) and heliographic coordinates (longitude and latitude in degrees relative to the solar equator and meridian);
3. magnetic type (α, β, γ, δ configuration);
4. area (in millionths of disk area);
5. Carrington longitude;
6. characteristics of the leader and follower sunspots: coordinates, 17 GHz radio brightness temperature in intensity and polarization, sunspot magnetic field strength (from *Mount Wilson*).

---

[1]IDL is the trademark of Research Systems, Inc.

[2]The *Yohkoh NOAA Active Region database* was used to locate the nominal position of each region.

[3]Daily radio maps produced by the *Nobeyama Radioheliograph* can be downloaded by anonymous ftp from `solar.nro.nao.ac.jp/pub/map`

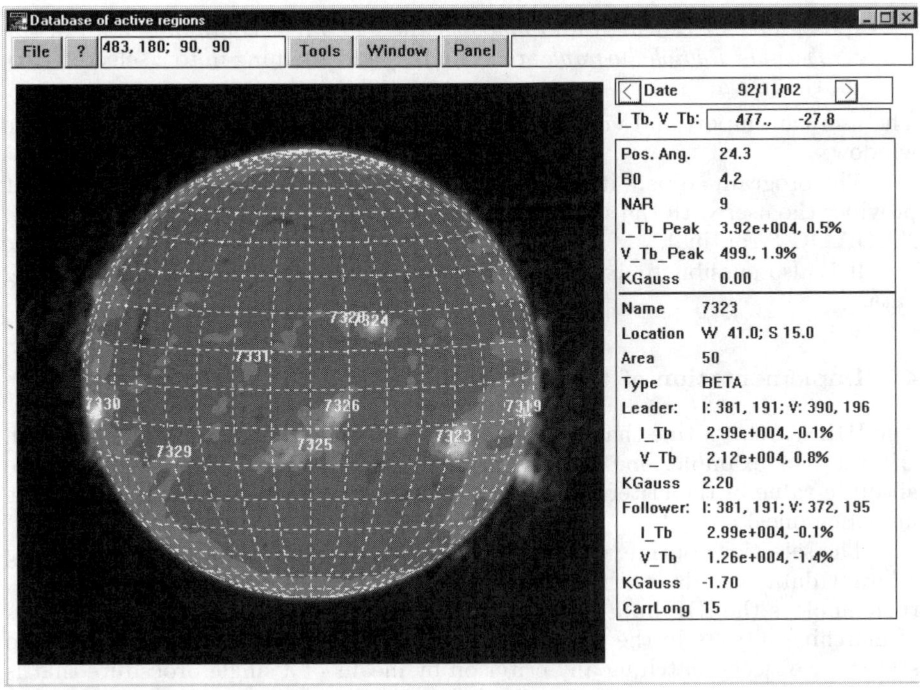

Figure 1. A screen dump of the database control program.

3. The database interface

The database interface is a GUI (Fig. 1), which provides convenient and efficient use of the database. The user can implement various search and selection modes and at the same time can view the Nobeyama 17 GHz radio maps in intensity and circular polarization (if these data are available). The figure shows a screen dump of the database control program. Screen dumps can be made in a graphical file (GIF format) or in a separate graphics window. Data matching a given criterion can be easily selected (e.g., via menu selections and entering values of interest). It is also possible to save data relevant to a selected date or active region into a text file.

In brief, the user has access to the following information:

- position of active regions and their NOAA numbers on the visible disk of the Sun for a selected date (a heliographical grid can be overlaid);
- data for the current date and for a selected active region (the selection is made with the mouse pointer);
- solar coordinates (heliographical and rectangular) under the mouse pointer and (if a radio map is displayed) brightness temperature;
- plots of area, latitude, brightness temperature, magnetic field for the leader and the follower sunspots of a selected active region during its passage across the solar disk, which can be plotted on the screen and saved into a file;

- possibility of viewing radio maps (if the FITS files containing data of the *Nobeyama Radioheliograph* are available) and zooming in to a selected part of the image.

The user can import and view text files and headers of FITS files in a scrollable window.

The program adjusts itself automatically to the resolution of the screen. It provides the user with the standard IDL tools for color table manipulation (e.g., **XLOADCT** routine).

It is also possible to expand the database and to correct or update existing data.

4. Implementation of the search engine and the GUI

The IDL search routine that we have developed is very flexible in setting search criteria. For example, one can search for data matching a given criterion, the absolute value of the criterion, or its inverse (e.g., all active regions except the matching ones).

The selection routine, **db_filter**, first obtains a subset database containing relevant data records, with structure identical to that of the entire database, then employs the IDL built-in function **WHERE** to return the array of indices of matching objects in the database. This implementation makes it possible to search for objects matching any criterion by means of a single procedure that is notable for its transparency and simplicity.

The GUI application is a standard widget-based interface constructed of several functions and procedures that are packaged into a single file (52 kB). This facilitates its installation and use. The program code has been tested successfully under Windows 95 and UNIX with IDL versions from 3.0.1 to 5.0.2. Further development of the application and the database is being planned so that it can accommodate other data, in particular, radio data from the Siberian Solar Radio Telescope (SSRT) (Smolkov et al. 1986).

Acknowledgments. We are indebted to Dr. D. Stern (*Research Systems, Inc.*) for his assistance in updating the program to IDL 5.0. We are grateful to the ADASS'98 conference for the financial aid which has made possible the attendance of the presenter (V. Grechnev) at the conference. We are indebted to staff members of the *Nobeyama Radioheliograph* for 17 GHz solar maps. This work was supported by SCOSTEP-96 grant.

References

Nishio, M., et al. 1994, in Proc. of Kofu Symp., (NRO Report No. 360), ed. S. Enome & T. Hirayama, 19

Smolkov, G. Ya., Pistolkors, A. A., Treskov, T. A., Krissinel, B. B., & Putilov, V. A. 1986, Ap&SS, 119, 1

Astronomical Data Analysis Software and Systems VIII
ASP Conference Series, Vol. 172, 1999
D. M. Mehringer, R. L. Plante, and D. A. Roberts, eds.

Zero-Points of FOS Wavelength Scales

Michael R. Rosa,[1] Florian Kerber

Space Telescope European Coordinating Facility, European Southern Observatory, Garching, Germany

Abstract. We have investigated the internal zero-points of the HST's Faint object spectrograph (FOS) on-orbit wavelength calibration between 1990 (launch) and 1997 (de-commissioning). The analysis is based on cross-correlating about 1200 WAVECAL exposures for the high-resolution dispersers, using as templates those exposures which define the dispersion solutions currently in use by the FOS pipeline. FOS has two channels BLUE/RED using two independent Digicon detectors. For BLUE systematic shifts of the zero-points are present, which amount to a maximum offset of 7 pixels (1.75 diodes) over the entire period. The zero-points for RED modes present an apparently random distribution with a peak-to-peak range of 7 pixels. We discuss the effect of the geomagnetic environment as a possible cause for the observed behaviour and describe the ongoing work to reduce the uncertainty in the wavelength scale.

1. Introduction

The FOS was designed primarily as a faint object spectrophotometer, with less stringent requirements on highly reproducible velocity measurements. On-orbit analysis indicated that the total $1\,\sigma$ error budget of grating wheel (FGWA) non-repeatability, source miscentering and internal to external wavelength scale offsets should not exceed 2 pixels ($\approx$ 0.25 resolution elements) (Keyes 1997). In addition, dispersion relations fitted to calibration lamp exposures regularly obtained for monitoring purposes did not directly show any obvious trends. Therefore, the original set of dispersion solutions was used throughout the entire operational life of FOS as the default in the pipeline calibration.

Lately, indications have been accumulating that FOS wavelength scale zero-points might be much more in error than the figure given above. In conjunction with the ST-ECF FOS Post Operational Archive project, we are reanalyzing all FOS wavelength calibration observations obtained on-orbit.

2. Analysis Method

As the usual polynomial fitting procedures may not be adequate to reliably detect zero-points shift due to masking of the polynomial coefficients, we chose

[1] Affiliated to the Astrophysics Division, Space Science Department, European Space Agency

to cross-correlate the raw, uncalibrated exposures. These cover the entire period between science verification (SV) (August 1990) and the last on-orbit wavelength calibration observation obtained in December 1996 with a single template for each FOS disperser respectively. The template raw calibration lamp exposures are those used for the SV analysis (Kriss, Blair, & Davidsen 1991) which define the default dispersion relations used by the FOS pipeline. Cross-correlation of the data sets yields a very clean peak which can be positioned to better then 0.1 pixel ($3\,\sigma$) using Gaussian profile fitting for all gratings.

3. Results

For the analysis we broke down the samples according to detector, grating and aperture, and plotted the x-shifts versus the Modified Julian Date (MJD) of exposure start. As opposed to expectations (see above), substantial shifts were found for all FOS dispersers, over both short and long time intervals, amounting to a full amplitude variation of 7 pixels peak-to-peak on both FOS channels.

3.1. FOS/BLUE specific

All the FOS blue side dispersers show a very similar and conspicuous pattern with time (cf. Fig. 1). At any given epoch the peak-to-peak scatter of the zero-point offset about the mean trend is approximately $\pm$ 1 pixel (i.e., 10 times the measurement error). The general trend amounts to an average drift of 0.75 pix/year with respect to the zero-point of the pipeline wavelength scale for each disperser. An observation in Cycle 6 therefore may be off by more than 5 pixels.

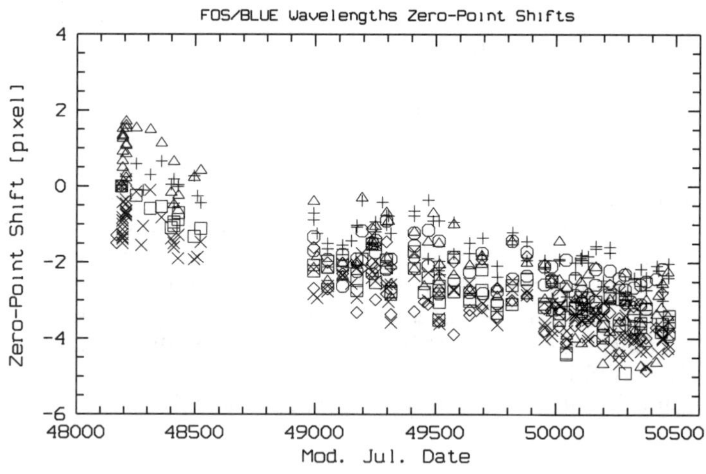

Figure 1. Uncorrected FOS/BLUE wave calibration sets

Due to the apparent similarity in the pattern for all blue dispersers we combined all BLUE grating data into a common dataset, depicted in Fig. 1. This was done iteratively. A first straight combination, was used to define three windows in time, in which the data appear to follow different linear trends.

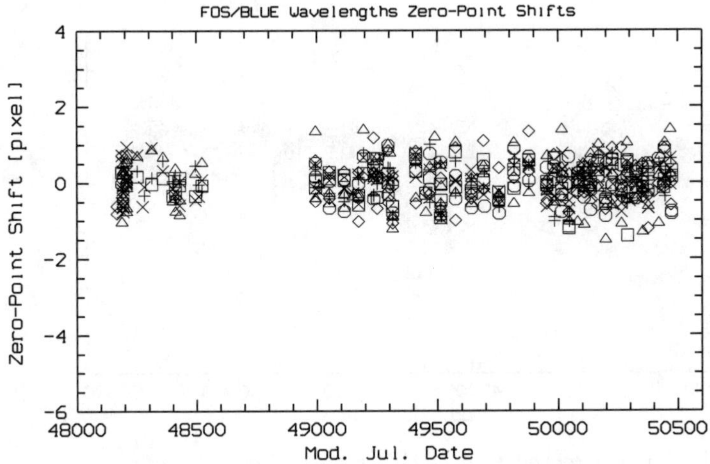

Figure 2. FOS/BLUE wave calibration sets after correction for linear trend and grating specific offset

Interestingly enough, the boundaries of these windows coincide with the start of SV, the activation of on-board GIM (geomagnetically induced image motion) correction, the installation of COSTAR and the de-commissioning of FOS. Linear regressions were fitted to the data in these three windows, and were subsequently used to renormalize the shifts for each grating/aperture combination. The additive renormalization usually was less then 0.2 pix, for all but two gratings. The result is depicted in Fig. 2. There are no apparent slopes in the residuals (average trend – individual disperser), reassuring that this procedure is valid.

3.2. FOS/RED specific

The red side data points appear to present a scatter-diagram with no obvious trends in time and with a full peak-to-peak range of about 7 pixels (Fig. 3). Application of the same procedures used for the blue side slightly reduces the scatter but it remains large ($\approx$ 5.5 pixels) compared to the blue side.

4. Discussion

All results strongly indicate that the main physical source of the problem of shifts in the x-scale (wavelength zero-point shifts) of FOS data is unrelated to either the aperture or the FGWA. It does seem to be a detector-related process instead. We offer the following hypothesis: The long-term drift (see Fig. 1) and short-term, orbital-position related motion of the x-scale on the FOS diode arrays is very likely related to the differing magnetic shielding properties of the FOS Digicons, i.e., another incarnation of the GIM. FOS/BLUE, which had a factor 4-7 better mu-metal shield than FOS/RED, was not very susceptible to x-drifts as HST moved through its orbit. However, the shield could have

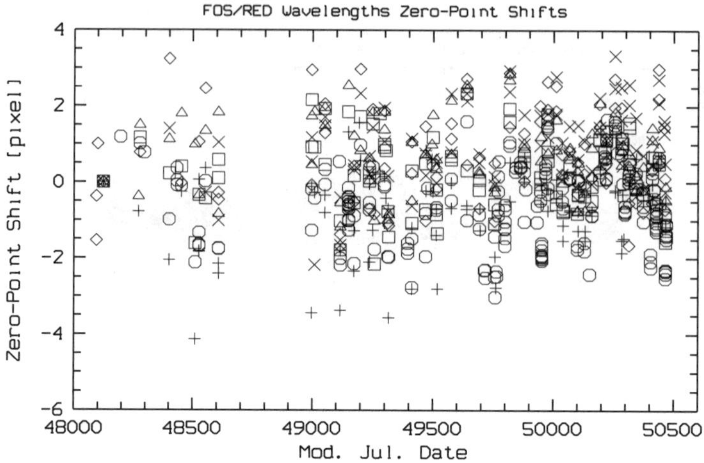

Figure 3. Uncorrected FOS/RED wave calibration sets

acquired its own field over time, producing the long-term linear drifts. A full understanding of the zero-point shifts will require a realistic description of the ambient magnetic environment of the HST and its effect on the FOS detectors. Work to this end is well underway.

5. Quick Fix

A more detailed description of the work presented here is given in the FOS instrument science report (ISR) 149, which is available at the STScI web site:

`http://www.stsci.edu/ftp/instrument_news/FOS/pub/fos_isr149r1`

A script that corrects the zero-point shifts and reduces the scatter to the one pixel peak-to-peak uncertainty level in the Blue is available at:

`http://www.stsci.edu/ftp/instrument_news/FOS/fosbl_wcorr`

The analysis of the cause of the FOS zero-point shifts at the ST-ECF is continuing and procedures providing a full scale correction will be made available once the behaviour of the FOS wavelength scales can properly be described.

References

Keyes, T. ed. 1997, HST Data Handbook, v. II, (Baltimore: STScI)

Kriss, G. A., Blair, W. P., & Davidsen, A. F. 1991, Revised FOS Wavelength Calibration (STScI FOS ISR CAL/FOS 067) (Baltimore: STScI)

Astronomical Data Analysis Software and Systems VIII
ASP Conference Series, Vol. 172, 1999
D. M. Mehringer, R. L. Plante, and D. A. Roberts, eds.

Background Fluctuation Analysis from the Multiscale Entropy

R. Gastaud, J.-L. Starck, M. Pierre

DAPNIA/SEI-SAP, CEA/SACLAY, F-91191 Gif sur Yvette Cedex

Abstract. The entropy measurement has been proposed for image restoration several years ago in astronomy. Different entropies have been used for this purpose. The multiscale entropy has the advantage to introduce a noise modeling, and the concept of information at different level of resolution. It can be used for tracking the presence of signal in a noisy data set. We show in this paper that it can also be used for background fluctuations analysis. A range of examples illustrates the results. We finally apply this method on X-ray simulations (XMM-EPIC) of diffuse large scale structure emission in order to discriminate different cosmological models.

1. Introduction

The concept of entropy has been introduced in astronomy for 40 years especially for deconvolution regularization. Many entropy definitions have been proposed in the past, but they do not take into account the spatial correlation of the image. Some authors tried a pyramidal approach to take a spatial correlation into account (Bontekoe, Koper, & Kester 1994), but this adds complexity. The correct way to do it is to compute the entropy in the wavelet space (Starck, Murtagh, & Gastaud 1998):

$$H(X) = \sum_{j=1}^{l} \sum_{k=1}^{N_j} h(w_{j,k}) \tag{1}$$

with $h(w_{j,k}) = -\ln p(w_{j,k})$. Here h stands for the entropy (or information) relative to a wavelet coefficient, and H for the multiscale entropy (MSE) of a signal or an image. For Gaussian noise, we get

$$H(X) = \sum_{j=1}^{l} \sum_{k=1}^{N_j} \frac{w_{j,k}^2}{2\sigma_j^2} \tag{2}$$

where σ_j is the noise at scale j. We see that the information is proportional to the energy of the wavelet coefficients. This definition of entropy verify the following criteria:

1. *The information in a flat signal is zero.*
2. *The amount of information in a signal is independent of the background.*

3. *The amount of information is dependent on the noise. A given signal Y ($Y = X + \text{Noise}$) does not furnish the same information if the noise is high or small.*
4. *The entropy must work in the same way for a pixel which has a value $B+\epsilon$ (B being the background), and for a pixel which has a value $B - \epsilon$.*
5. *The amount of information is dependent on the spatial correlation in the signal. If a signal S presents large homogeneous features above the noise, it contains less information.*

This definition has been used with success for restoration (Starck & Murtagh 1999).

2. Multiscale Entropy applied to Background Fluctuation Analysis

Previous equation is used to compute an average over the pixels:

$$E(j) = \frac{1}{N} \sum_{k=1}^{N} h(w_j) \tag{3}$$

$E(j)$ gives the mean entropy at the scale j. From the mean entropy vector, we have statistical informations on each scale separately. Having a noise model, we are able to calculate (generally from simulations) the mean entropy vector $E_{\text{noise}}(j)$ resulting from a pure noise. Then we define the normalized mean entropy vector by

$$E_n(j) = \frac{E(j)}{E_{\text{noise}}(j)} \tag{4}$$

Now, this function of the scale can be used to analyse the background fluctuations.

2.1. Detection of sources

Five simulated images were created by adding 400, 200, 100, 50, 0 sources to a noisy image. The background is a Gaussian noise of amplitude 1. The sources are given by:

$$I(x,y) = I_0 \exp\left(\frac{(x-a)^2 + (y-b)^2}{2\sigma^2}\right) \tag{5}$$

with $I_0 = 1$ and $\sigma = 2$. Defining the signal to noise ratio (SNR) as the ratio between the standard deviation in the smallest box which contains at least 90% of the flux of the source, and the noise standard deviation, we have a SNR equal to 0.25: the sources are not detectable. Several tests of this non-detectability have been made: χ^2 test in direct space (i.e., is this pixel above the Gaussian noise?), correlation in direct space (i.e., is this sub-image a psf?), wavelet filter: the routine mrfilter (MR/1 software[1]). The sources are really hard to detect:

---

[1] `http://ourworld.compuserve.com/homepages/multires`

genuine sources are lost in false detections (tens against ten thousands). The mean entropy function of the scale has been computed for these images. The error is estimated by drawing ten times each image. Figure 1 clearly shows that the sources increase the entropy of the image. But it is obvious that the positions of these sources remain unknown.

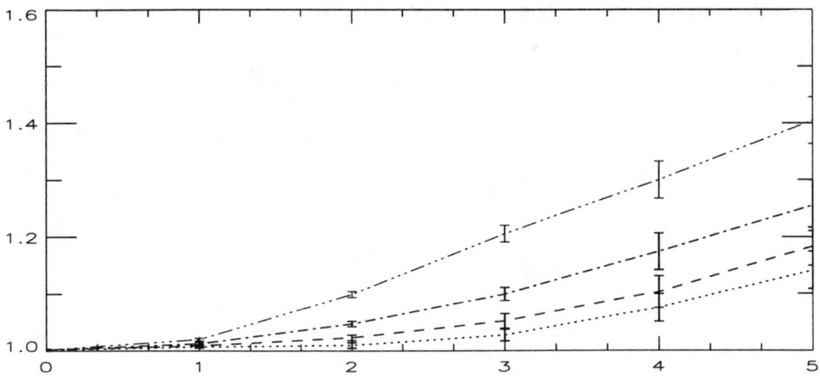

Figure 1. Mean entropy versus the scale of 5 simulated images containing undetect able sources and noise. Each curve corresponds to the multiscale of one image. From top to bottom, the image contains respectively 400, 200, 100, 50, and 0 sources, (which is confused with the x-axis).

2.2. Detection of faint feature

The new satellite XMM (to be launched in year 2000) will open a new era in the study of the sky at the energy from 0.1 to 10 keV. This will enable astronomers to detect large scale structures (hundreds of megaparsecs) and help to discriminate between different cosmological models. The problem is to detect the hot gas in filaments which are expected to be between galaxy clusters. The emission of these filaments, predicted from hydrodynamical simulation models, is very faint and diffuse. On the image of filaments will be superimposed very numerous background and foreground quasars and clusters of galaxies. We try the multiscale entropy method to detect the existence of filaments. Greg Bryan, from MIT, kindly provided simulation of the sky at different cosmological distances (redshift). From the density and the temperature cubes given by the simulations, a photons flux per simulation cell for the bandwidth from 0.4 keV to 4 keV is computed. The image of a filament located at $z = 0.5$ was computed with a pixel field of view of 4 arcsecond. The size of the image is 512×512 pixels. This image contains both clusters and filaments. In addition the point-like images of quasars are simulated using the law $\log(N)$–$\log(S)$ (Hasinger et al. 1998). The two images were summed, then multiplied by the integration time, here 400 kilosecond. Poisson noise, PSF blurring, vignetting are added. Two sets of images have been simulated, with and without the filaments. As previously, the mean entropy function of the scale has been computed. The difference between the two entropies, for the image with filament and without filament, is plotted in Fig. 2. The error bars e were estimated from from five realisations of the images, for both image with(σ_1) and without (σ_2) filaments

$$e = 2.5\sqrt{\sigma_1^2 + \sigma_2^2} \tag{6}$$

This difference is significant at the $2.5\,\sigma$ level (98.76%)

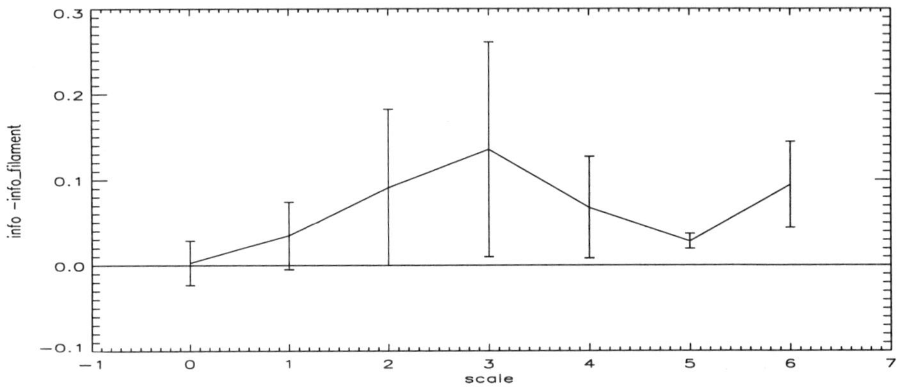

Figure 2. difference of mean entropy for the images with and without filament, error bar $2.5\,\sigma$.

3. Conclusion

We have seen that information must be measured from the transformed data, and not from the data itself. The transformation chosen was a non-orthogonal wavelet transform which is suited to astronomical images. The mean entropy is a good indicator of the presence of undetected sources, or faint extended structures. It is well suited to background fluctuations analysis.

Acknowledgments. We want to thank Greg Bryan who kindly provided us with hydrodynamical simulated data.

References

Bontekoe, T. R., Koper, E., & Kester, D. J. M. 1994, A&A, 294, 1037

Hasinger, G., Burg, R., Giacconi, R., Schmidt, M., Trümper, J., & Zamorani, G. 1998, A&A, 329, 482

Starck, J.-L., Murtagh, F., & Gastaud, R. 1998, IEEE Trans. on Circuits and Systems, 45, 1118

Starck, J.-L. & Murtagh, F. 1999, Signal Processing, in press

Astronomical Data Analysis Software and Systems VIII
ASP Conference Series, Vol. 172, 1999
D. M. Mehringer, R. L. Plante, and D. A. Roberts, eds.

A New Entropy Measure Based on Multiple Resolution and Noise Modeling

Fionn Murtagh

Queen's University Belfast and Strasbourg Observatory

Jean-Luc Starck

CEA, DAPNIA, Saclay

Abstract. A new measure of information is defined which is based on noise modeling and incorporates resolution scale. Properties of this measure are discussed. Use of such an entropy measure for filtering and deconvolution is exemplified. Implications for structuring of, and access to, image archives are noted.

1. Introduction

We show how entropy can be incorporated into a multiscale image processing setting. The latter is a powerful setting for a wide range of image processing operations. Additionally it allows for image noise modeling, and subsequent noise removal.

Immediate applications include image filtering and deconvolution. Other objectives include object finding and definition, and feature characterization. More globally, we also note that such vision modeling will be necessary in future generation image databases.

Background on these methods can be found in Starck, Murtagh, & Bijaoui (1998). Information on a multiresolution image and vision software package, MR/1, can be found at `http://visitweb.com/multires`.

2. Multiscale Entropy Filtering Method

The multiscale entropy method consists of measuring the information h relative to wavelet coefficients, and of separating this into two parts h_s, and h_n. The expression h_s is called the signal information and represents the part of h which is definitely not contaminated by the noise. The expression h_n is called the noise information and represents the part of h which *may* be contaminated by the noise. We have $h = h_s + h_n$. Following this notation, the corrected (i.e., filtered) wavelet or multiscale coefficient $\tilde{w}$ should minimize:

$$J(\tilde{w}_j) = h_s(w_j - \tilde{w}_j) + \alpha h_n(\tilde{w}_j)$$

i.e. there is a minimum of information in the residual $(w - \tilde{w})$ due to the significant signal, and a minimum of information which can be due to the noise in the solution $\tilde{w}_j$.

In order to verify a number of properties, the following functions are proposed for h_s and h_n in the case of Gaussian noise:

$$
\begin{aligned}
h_s(w_j) &= \frac{1}{\sigma_j^2}\int_0^{|w_j|} u \operatorname{erf}\left(\frac{\mid w_j \mid -u}{\sqrt{2}\sigma_j}\right) du \\
h_n(w_j) &= \frac{1}{\sigma_j^2}\int_0^{|w_j|} u \operatorname{erfc}\left(\frac{\mid w_j \mid -u}{\sqrt{2}\sigma_j}\right) du
\end{aligned}
$$

3. Regularization Parameter and Data Model

The regularization parameter, α, can be determined by using the fact that we expect a residual with a given standard deviation at each scale j equal to the noise standard deviation σ_j at that scale. Hence we have an α_j per scale.

If we have a model, D_m, for the data we can use

$$J_m(\tilde{w}_j) = h_s(w_j - \tilde{w}_j) + \alpha h_n(\tilde{w}_j - w_m)$$

4. Regularized Entropy-Based Filtering

The regularized entropy-based filtering algorithm is as follows.

1. Estimate the noise in the data σ (see Olsen 1993; Starck & Murtagh 1998a).
2. Wavelet transform of the data.
3. Calculate from σ the noise standard deviation σ_j at each scale j.
4. Set $\alpha_j^{min} = 0$, $\alpha_j^{max} = 200$.
5. For each scale j do
   1. Set $\alpha_j = \frac{\alpha_j^{min} + \alpha_j^{max}}{2}$
   2. For each wavelet coefficient $w_{j,k}$ of scale j, find $\tilde{w}_{j,k}$ by minimizing $j_m(\tilde{w}_{j,k})$ or $j_{ms}(\tilde{w}_{j,k})$
   3. Calculate the standard deviation of the residual:

   $$\sigma_j^r = \sqrt{\frac{1}{N_j}\sum_{k=1}^{N_j}(w_{j,k} - \tilde{w}_{j,k})^2}$$

   4. If $\sigma_j^r > \sigma_j$ then the regularization is too strong, and we set α_j^{max} to α_j, otherwise we set α_j^{min} to α_j.
5. If $\alpha_j^{max} - \alpha_j^{min} > \epsilon$ then goto 5.
6. Multiply all α_j by the constant α_u (default: 1).
7. For each scale j and for each wavelet coefficient w find $\tilde{w}_{j,k}$ by minimizing $j_m(\tilde{w}_{j,k})$ or $j_{ms}(\tilde{w}_{j,k})$.
8. Reconstruct the filtered image from $\tilde{w}_{j,k}$ by the inverse wavelet transform.

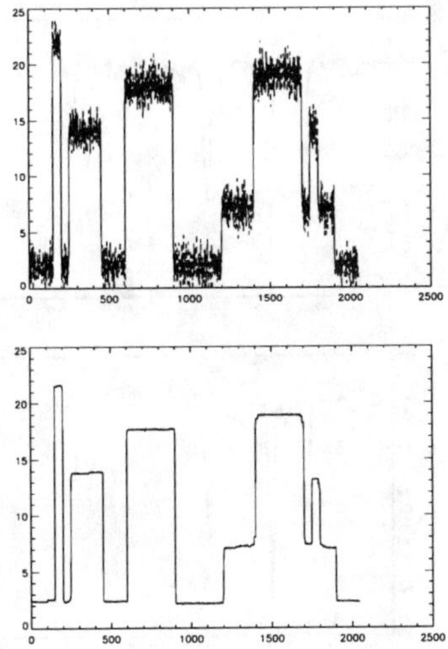

Figure 1. Top, noisy blocks and filtered blocks overplotted. Bottom, filtered blocks.

The minimization of j_m or j_{ms} (step 5.2) can be done by any method. For instance, a simple dichotomy can be used in order to find $\tilde{w}$ such that

$$\frac{\partial h_s(w - \tilde{w})}{\partial \tilde{w}} = -\alpha_j \frac{\partial h_n(\tilde{w})}{\partial \tilde{w}} \tag{1}$$

The idea to treat the wavelet coefficients such that the residual respects some constraint has also been used in Nason (1996) and Amato & Vuza (1998) using cross-validation.

5. Filtering: Examples

Figure 1 shows a difficult case of smooth and sharp transitions. The filtering method described here allows an excellent noise filtering of it to be carried out. Figure 2 shows a spectrum, and an effective noise filtered version.

6. Summary and Conclusion

We have described an integrated approach for specification of information content, handling of noise, and effective implementation. The multiscale entropy is similarly of benefit in deconvolution.

We note that large image repositories *require*

- Noise separation, for compression and quicklook
- Variable resolution over time, for progressive transmission, and variable resolution over space leading to multifoveated images.
- Usable approaches to information characterization.

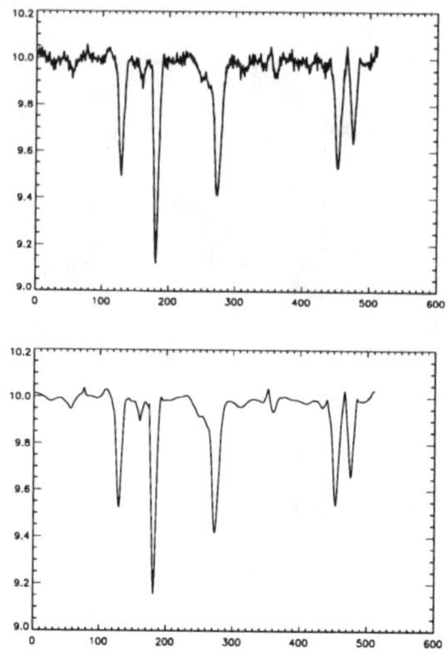

Figure 2. Top, real spectrum and filtered spectrum overplotted. Bottom, filtered spectrum.

Hence information- and entropy-based processing technologies will be necessary for the effective design and implementation of large, future generation, image databases.

Further reading is available in Starck, Murtagh, & Gastaud (1998) and Starck & Murtagh (1998b). The methods described here are available in the multiresolution analysis software package, MR/1, Version 2.0. Details of the MR/1 software package can be found at `http://visitweb.com/multires`.

References

Amato, U. & Vuza, D. T. 1998, Rev. Roumaine Math. Pures Appl., in press

Nason, G. P. 1996, J. Roy. Stat. Soc. B, 58, 463

Olsen, S. I. 1993, Comp. Vis. Graph. Image Proc., 55, 319

Starck, J. L. & Murtagh, F. 1998a, PASP, 110, 193

———, 1998b, Signal Proc., in press

Starck, J. L., Murtagh, F., & Bijaoui, A. 1998, Image and Data Analysis: The Multiscale Approach, (Cambridge: Cambridge Univ. Press)

Starck, J. L., Murtagh, F. & Gastaud, R. 1998, IEEE Trans. CAS II, 45, 1118

Astronomical Data Analysis Software and Systems VIII
ASP Conference Series, Vol. 172, 1999
D. M. Mehringer, R. L. Plante, and D. A. Roberts, eds.

Data Retrieval Software for the USNO-A Catalog: Another Member of the -DARES Family

Oleg Malkov, Oleg Smirnov

Institute of Astronomy, 48 Pyatnitskaya St., Moscow 109017, Russia

Abstract. The recently released USNO-A V1.0 catalog contains astrometric and photometric information for 488,006,860 objects, and is widely used by the astronomical community for all sorts of applications. However, accessing the actual catalog data is not quite straightforward, due to the huge size of the catalog and a somewhat complicated internal format. To facilitate data retrieval, we have created U-DARES, a user-friendly program that lets one look directly at the data in the USNO-A catalog, either as a graphical sky map, a plot, or a simple text table.

Like GUIDARES (Malkov & Smirnov 1994), its predecessor for the Guide Star Catalog, U-DARES can read a sampling of USNO-A data for a given sky region, store this sampling in a text file, and display a graphical map of the sampled region either in projected celestial coordinates (perfect for finder charts) or in the Aitoff projection. It supports rectangular and circular regions defined by coordinates in the equatorial, ecliptic (any equinox), galactic or supergalactic systems. Other features include magnitude cut-offs and specification of the shape of the area.

1. USNO-A V1.0

USNO-A, version 1.0 (Monet 1996) is a catalog of 488,006,860 sources whose positions can be used for astrometric references. These sources were detected by the Precision Measuring Machine (PMM) built and operated by the U. S. Naval Observatory Flagstaff Station during the scanning and processing of the Palomar Observatory Sky Survey I (POSS-I) O and E plates, the UK Science Research Council SRC-J survey plates, and the European Southern Observatory ESO-R survey plates. The PMM detects and processes at and beyond the nominal limiting magnitude of these surveys, but the large number of spurious detections requires that a filter be used to eliminate as many as possible. The inclusion requirement for USNO-A was that there be spatially coincident detections (within a 2 arcsecond radius aperture) on the blue and red survey plate. For field centers of -30 degrees and above, data come from POSS-I plates, while data from field centers of -35 and below come from SRC-J and ESO-R plates.

USNO-A contains right ascension and south polar distance in the system of J2000 at the epoch of the survey blue plate for each object, and lists an estimate of the blue and red magnitude. For POSS-I sources, the photometric system is the photographic system defined by the O and E emulsions and filters, while southern sources are measured in the photometric system defined by the IIIa-J and IIIa-F emulsions. It is believed that the typical astrometric error is about 0.25 arcseconds and that the typical photometric error is about 0.25 magnitudes. To avoid the necessity of consulting many catalogs, objects brighter than 11th magnitude that appear in the Guide Star Catalog (GSC) and that were not detected by the PMM were inserted. USNO-A covers the entire sky, and goes as deep as O=21, E=20, J=22, and F=21 for objects with appropriate colors. The limiting magnitude is brighter for objects with extreme colors, and follows from the requirement for a detection on both the blue and red survey plate. Although it covers the entire sky, there are holes in the catalog in the vicinity of bright stars, regions of nebulosity, crowded fields, etc.

The USNO-A is widely used by the astronomical community for many different applications, such as:

- statistical studies of sky regions. The extremely large size and scope of the data set opens up research possibilities that were previously impossible.
- searches for candidate optical counterparts of astronomical objects found at other wavelengths.
- mapping of object vicinities (in particular, finder charts for variable stars).

The catalog's distribution format (a set of ten CD-ROMs) requires minimum hardware and is well suited for all sorts of conditions, especially observations.

Unfortunately, the actual data in the catalog are not easily accessible. It is in the form of binary files, and the coordinates are given in one standard system (J2000.0). Thus, even generation of a simple finder chart is no trivial undertaking.

2. U-DARES

To help PC users solve the problem of data retrieval, we have created the USNO-A Catalog Data Retrieval Software, or U-DARES. This is a user-friendly program which lets one easily produce text samplings of the catalog and sky maps in Aitoff or celestial projections.

U-DARES requires a PC with DOS or Windows and a CD-ROM drive. It will work without a CD-ROM if the necessary USNO-A files are copied onto the hard disk.

The main function of U-DARES is to produce an ASCII table of object entries from a specified region, and a graphical sky map of the region. It can handle rectangular and circular regions in four different coordinate systems.

2.1. User Interface

To the user, U-DARES looks like an intuitive and self-explanatory menu that is used to enter the following information:

- The logical drive where the USNO files are located.
- The coordinate system in which the region is specified: equatorial, ecliptic, galactic or supergalactic. For equatorial and ecliptic coordinates, the equinox may also be set (the default is 2000).
- The shape of the region: rectangular or circular.
- Magnitude limits for the objects to be retrieved.

- The actual coordinates of the region (minimal and maximal or central and radius).

The software performs a lot of error control along the way. Illegal values of coordinates are impossible to enter, the command to start retrieval is unavailable until all the necessary data have been specified, inverted coordinates are implicitly swapped if necessary, etc.

2.2. Data Retrieval

During retrieval, the evolution of the process is shown on the screen. In addition, a sky map in the celestial or Aitoff projection is simultaneously plotted. Internally, U-DARES performs the following steps:

- Convert the region coordinates to the standard USNO-A system.
- Determine which USNO-A files overlap the user's region.
- Scan the USNO-A files for objects that fall within the user's region, convert the objects' coordinates back into the user's system.

If errors occur, or if the CD-ROM must be changed, U-DARES notifies the user via a message box with standard "Abort, Retry, Ignore" or "Continue, Cancel" options.

2.3. Output and Sky Maps

For every region, U-DARES creates an output file. This is an ASCII table with the following data for each object:

- position and magnitudes (blue and red).
- magnitude quality flag
- field ID
- indication on a correlated GSC entry

A U-DARES SkyMap can be produced both during retrieval and afterwards, by reading the sampling file. A coordinate grid with labels is displayed, and objects are plotted over it as circles. The radius of the circles is proportional to the magnitude. In the Aitoff projection, all objects are plotted as dots (this projection is intended for relatively large sky areas).

U-DARES automatically selects the highest resolution available and takes into account the screen aspect ratio, so that square regions look truly square at any resolution.

2.4. Distribution

To avoid the hassle of separate "Read Me" files, all the documentation has been built into the executable module. By selecting an option form the U-DARES menu, the user can produce a documentation file whenever he needs. Therefore, the whole package is distributed as a single file. U-DARES is a public domain product, available by FTP from

`ftp.inasan.rssi.ru/pub/UDARES`

Acknowledgments. We would like to thank the LOC of the ADASS'98 Conference for the financial support that made this presentation possible. We also acknowledge the Russian Academy of Sciences for the aid provided in the frame of the "Young scientists support program".

References

Malkov O. Y. & Smirnov O. M. 1994, in ASP Conf. Ser., Vol. 61, Astronomical Data Analysis Software and Systems III, ed. D. R. Crabtree, R. J. Hanisch, & J. Barnes (San Francisco: ASP), 183

Monet, D. 1996, USNO-A1.0: A Catalog of Astrometric Standards, (Washington: USNO)

Part 10. User Interfaces and Visualization

Astronomical Data Analysis Software and Systems VIII
ASP Conference Series, Vol. 172, 1999
D. M. Mehringer, R. L. Plante, and D. A. Roberts, eds.

Development of a Java-Based Image Browser System for the Subaru Telescope

Hajime Baba

Kyoto University, 606-8502, Japan

Naoki Yasuda, Shin-ichi Ichikawa, Masaru Watanabe

National Astronomical Observatory of Japan, 181-8588, Japan

Toshihiro Horaguchi

National Science Museum, Tokyo, 169-0073, Japan

Masaru Hamabe, Wako Aoki

University of Tokyo, 181-8588, Japan

Tomohiko Ozawa

Graduate University for Advanced Studies, 181-8588, Japan

Abstract. We have developed a Java-based image browser system for the Subaru telescope. The system will be used for a quick inspection of the archived images from remote institutes and of the observed images at a telescope operation room. The browser system supports functions to show spectral traces, statistics, etc., as well as an image itself to help users to easily evaluate the image quality. The browser system has been developed to reduce network traffic between a server and clients. The images, which are compressed for rapid transfer, are generated with care for each observational instrument in order to minimize the loss of the information.

1. Introduction

The Subaru Telescope[1] is equipped with three optical instruments which all have large-format (2k×4k) CCDs. They are operated under various observational modes: direct imaging with the *Subaru Prime Focus Camera* (*Suprime-Cam*, Miyazaki et al. 1998), echelle spectroscopy with the *High Dispersion Spectrograph* (*HDS*, Noguchi et al. 1998), and direct imaging, long-slit/multi-slit spectroscopy, polarimetry, and spectropolarimetry with the *Faint Object Camera and Spectrograph* (*FOCAS*, Iye et al. 1997).

[1] `http://www.subaru.nao.ac.jp`

These images will be archived in the *Subaru Telescope Archive System* (*STARS*, Takata et al. 1996). Because of the large size of individual frames, it is almost unrealistic for remote archive users to preview these archived raw images over the network, thus we need to prepare compressed images for remote browsing. Furthermore, we have to consider the fact that 2-D images do not always convey necessary and sufficient information for assessment of image quality; 1-D profiles or traces must be much more meaningful for spectral images, and even 0-D scalars, i.e., statistics of the pixel counts, may be sufficient for dark frames.

The usefulness of 1- and 0-D information applies similarly to image browsing in an observation room. If we consider that the number of frames obtained with these instruments will amount to more than several hundred per night, effective observations require a quick (automatic) display of 2-D images and their associated 1- and 0-D information so as to minimize the loss of actual observation time.

For these purposes we have developed an extended image browser system, which is hereafter referred to as the *OZEKI* system, for the Subaru Telescope. The system offers functions of processing 2-D images to draw and display 1- and 0-D information. For archiving it also includes a function that compresses raw 2-D image, draws 1- and 0-D information, and attaches them to the 2-D image to create a composite and compressed FITS file.

2. *OZEKI* System and Data Flow

OZEKI is to be used for the following three purposes with the Subaru Telescope (Fig. 1).

OZEKI/A	-	Quick look image (QLI) browser for archived images
OZEKI/B	-	Image browser for observed raw images
OZEKI/C	-	Performance evaluation tool for observational instruments (high speed version)

OZEKI/A consists of a data server and a QLI producer. QLIs are FITS files containing a 2-D image, a 1-D profile/trace, and a 0-D scalar, i.e., statistics of the image. The file size is moderately reduced and compressed to a few hundreds of kB from that of the original image for archiving. The reduction scheme depends on the instrument and/or observation mode with which the original image was taken, so that the essential information of the original image can be conserved as much as possible. Before the original image files are moved to a mass tape library and cleared from hard disks, QLI is created. The original image is separately processed by STARS so that its header information and ASCII Table Extension may be extracted to create files referred to as HDI and ATE files, respectively. When a request is submitted from a client to the *OZEKI/A* data server (via STARS), the server returns QLI, HDI, and ATE to the client, with the QLI further reduced to an appropriate size, if necessary, depending on the client machine performance and/or network capacity. The reduction rate can be explicitly specified from the client as well. QLI is displayed in the browser main window on a client machine.

OZEKI/B is essentially the same as the data server and the client of *OZEKI/A*, with the difference that the data which it receives is the original image that has just observed with the telescope.

OZEKI/C is a diversion of *OZEKI/B* for the special purpose of evaluating performance of observational instruments with visual inspection of the image data.

The browser windows for HDS are illustrated in Fig. 2. Besides the CCD images in the main window, spectral traces with rough wavelength calibration are automatically computed using HDI and ATE and are displayed in another window. This feature is useful particularly for inspecting multi-slit and echelle spectra taken with FOCAS and HDS, respectively. HDI and ATE can be also browsed upon request.

OZEKI is written in Java. We have used JDK 1.1.6, Swing 1.0.3, Horizon 1.3.1b, FITSWCS 2.4, Ptplot 1.3p1, and Horb 1.3b1p11a, with our original extension on Horizon so as to treat ASCII table extension and improvement on Ptplot to run speedily.

3. Present Status and Future Development

OZEKI is runnable in stand-alone mode as of October 1998. The following items are under development for the 'prototype' of *OZEKI*, as of December 1998;

- Server/client architecture
- Invocation foreign software (e.g., IRAF)
- Performance improvement

Next, we intend to plug *OZEKI* into STARS.

Acknowledgments. We thank Tadafumi Takata for valuable discussions and for his co-operation in developing the interface between *OZEKI* and STARS. H.B. is grateful to Prof. Shiro Nishimura for financial support to attend this meeting.

References

Iye, M., et al. 1997, in SPIE Proc., Vol. 2871, Optical Telescopes of Today and Tomorrow, ed. A. L. Ardeberg (Bellingham: SPIE), 1054

Miyazaki, S., et al. 1998, in SPIE Proc., Vol. 3355, Optical Astronomical Instrumentation, ed. S. D'Odorico (Bellingham: SPIE), 363

Noguchi, K., et al. 1998, in SPIE Proc., Vol. 3355, Optical Astronomical Instrumentation, ed. S. D'Odorico (Bellingham: SPIE), 354

Takata, T. et al. 1996, in ASP Conf. Ser., Vol. 101, Astronomical Data Analysis Software and Systems V, ed. G. H. Jacoby & J. Barnes (San Francisco: ASP), 251

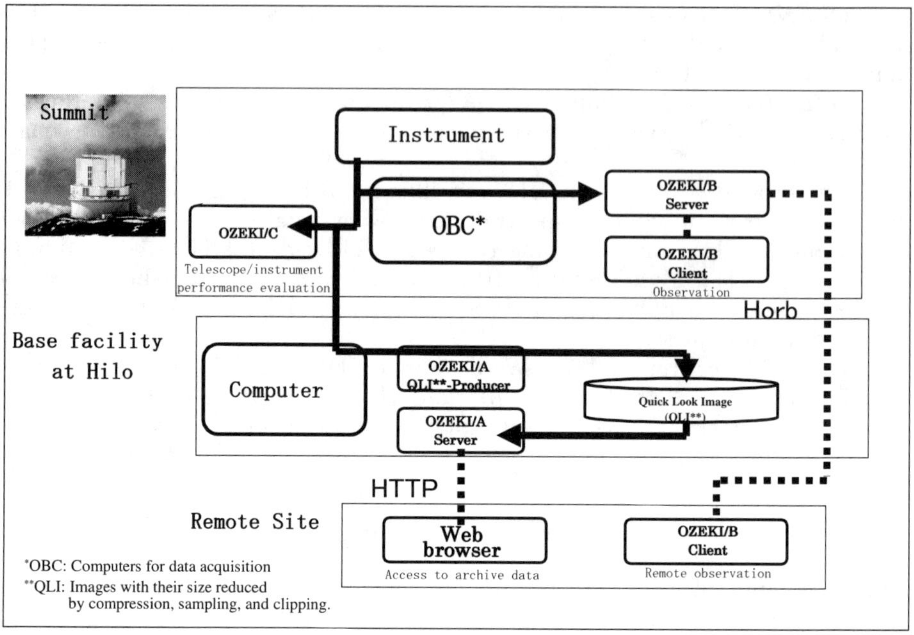

Figure 1. Data flow for OZEKI system

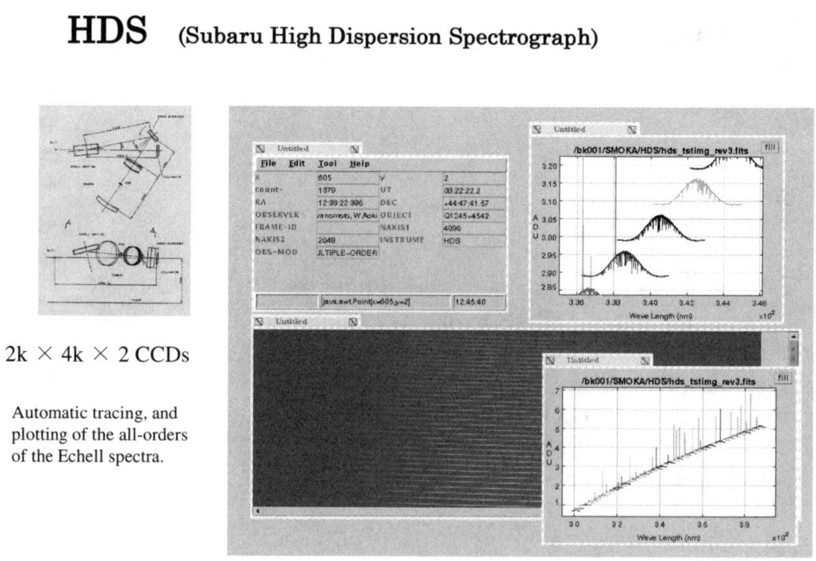

Figure 2. OZEKI windows for HDS

Astronomical Data Analysis Software and Systems VIII
ASP Conference Series, Vol. 172, 1999
D. M. Mehringer, R. L. Plante, and D. A. Roberts, eds.

A Java–IRAF Spectrum Visualization Tool

I. Busko

Space Telescope Science Institute, Baltimore, MD 21218, Email: busko@stsci.edu

Abstract. A Java application designed to display spectra from Hubble Space Telescope's STIS instrument relies on a Java–IRAF interface to access the data in a efficient way. This paper describes this interface implementation.

1. Introduction

A simple spectrum visualization tool is being developed by the Science Software Group at the Space Telescope Science Institute (STScI), basically to evaluate the capabilities (and weaknesses) of Java as a possible environment for development of data visualization and analysis tools tailored to Hubble Space Telescope (HST) data.

One of the problems with such a tool is to cope with the varied gamut of formats on which HST data exist. Since HST data in all their forms can be readily accessed by IRAF/TABLES I/O libraries, to save development time and to avoid current inefficiencies in Java I/O, we choose to link the Java code to the IRAF/TABLES I/O libraries using the Java Native Interface (JNI) facilities.

2. The Java–IRAF Interface Design

The interface must be implemented in two levels because currently the JNI only supports native code written in C/C++. The Java–IRAF interface functions must thus be coded in C. Since IRAF/TABLES libraries are written in SPP (a Fortran pre–processor), C bindings to allow the C native programs to access IRAF/TABLES libraries must be used as well. The availability of the C–VOS bindings developed by A. Farris at STScI conveniently solved this part of the problem. The layer structure of the resulting software is shown in Fig. 1.

The C–VOS is made of C–callable functions that map one–to–one to functionally equivalent IRAF/TABLE routines. C–VOS functions perform actions such as opening a table file, reading blocks of pixels from 2–D images, and so on. This can be accomplished with efficiency because both software layers, C and SPP, share the same memory address space (the physical machine memory where both run). Thus most of the actions involving data arrays can be performed by just passing memory pointers. An additional benefit of such "language binding" concept is that it becomes relatively easy to automate the process of generating the actual binding code.

Application and Java–IRAF interface (Java code)
JNI (Java–callable C code)
Java–IRAF interface (C code)
C–VOS bindings (C code)
IRAF/TABLES I/O libraries (SPP code)

Figure 1. The layer structure of the software.

Because Java code is run by the Java Virtual Machine (JVM), the JNI has to cope with the problem of a virtual address space (the Java–accessible data) that interfaces with a physical address space (the C–accessible data). The methods provided in the JNI to handle data arrays rely on memory "pinning" and data copying, thus they can lead to inefficiencies, at least on some JVM implementations.

Because of these potential inefficiencies, the adopted approach in designing the Java–IRAF interface avoids the concept of language binding. Instead, the required interface functions to support the spectrum visualization application are defined at a much higher level. No low–level functions (in terms of IRAF/TABLES libraries) such as "open table" or "write array in row/column table cell" were implemented. Instead, functions such as "get spectrum" or "get table header contents" are used to minimize traffic between the Java and C portions of the code. Each native method is responsible, using IRAF/TABLES I/O functions, to access all required data in the file system, and, using JNI functions, to build or access high–level Java objects (such as "Spectrum") defined in the Java side. Thus they don't act as simple bindings between equivalent low–level functions, but perform quite a lot of processing and data formatting on their own.

3. Example: Reading a STIS 1–D Spectrum

The Application Side

The application code that reads a STIS 1–D spectrum resorts to a static method defined in a "SpectrumFactory" Java class. The spectrum is returned packaged inside a Spectrum object:

```
...
...
try {
    Spectrum sp = SpectrumFactory.MakeSpectrum (...);
} catch (Exception e) {
...
```

```
...
```

The Interface: The Java Side

The SpectrumFactory class performs two functions: it declares all native methods, and it loads the native library. The IRAFInit method is responsible for initialization of the IRAF/TABLES libraries after they are loaded. IRAFInit must be run only once throughout the library's life, thus it is called from the class' static initialization block:

```
public class SpectrumFactory {

    private native static void IRAFInit();

    public static synchronized native Spectrum MakeSpectrum (...)
        throws IOException, ArrayStoreException;

    static {
        try {
            System.loadLibrary ("iraf");
            IRAFInit();
        } catch (Exception e) {
            ...
            ...
        }
    }
}
```

The Interface: The C Side

The MakeSpectrum native method reads the spectral data from the STIS 1–D file and builds the Spectrum object. The IRAFInit method just calls the C-VOS initialization function. Notice the include file created by the *javah* header generator program.

```
# include <jni.h>
# include "spectrum_SpectrumFactory.h"          /* created by javah */
# include <stdio.h>
# include <c_iraf.h>
# include <xtables.h>

JNIEXPORT void JNICALL
Java_spectrum_SpectrumFactory_IRAFInit (JNIEnv *env, jobject obj) {
        c_irafinit(0,0);          /* IRAF initialization */
}

JNIEXPORT jobject JNICALL
Java_spectrum_SpectrumFactory_MakeSpectrum (JNIEnv *env, jobject obj, ...) {
    jclass clspec;              /* JNI data type   */
    jobject spectrum;           /* JNI data type   */
    jmethodID mid;              /* JNI data type   */
    IRAFPointer tp;             /* C-VOS data type */
    ...
    ...
    /* Open and read table using C-VOS functions. */
    tp = c_tbtopn (name, IRAF_READ_ONLY, 0);
    if (c_iraferr()) {
        (*env)->ThrowNew (env, "Input table could not be opened.");
        return ((jobject)NULL);
    }
    ...
    ...
    /* Build and populate Spectrum object using JNI functions. */
    clspec = (*env)->FindClass (env,"spectrum/Spectrum");
    mid  = (*env)->GetMethodID (env, clspec, "<init>", "(I)V");
    spectrum = (*env)->NewObject (env, clspec, mid);
        ...
        ...
    return (spectrum);
}
```

The Interface: The Makefile

Below is the Solaris makefile that builds the interface code in this example, using file path names from the STScI environment. The order in which the linker searches the libraries is very important.

```
INCLUDE=/usr/local/jdk1.2/include
LIBS=/usr/local/jdk1.2/jre/lib/sparc/libjava.so \
        /usr/ra/stsdasx/lib/libhstio.a  /usr/ra/stsdasx/lib/libcvos.a \
        /usr/ra/tables/lib/libtbtables.a /usr/ra/irafx/lib/libex.a \
        /usr/ra/irafx/lib/libsys.a /usr/ra/irafx/lib/libvops.a \
        /usr/ra/irafx/unix/hlib/libos.a  /usr/ra/irafx/unix/hlib/libboot.a \
        /opt/SUNWspro/SC4.0/lib/libF77.a /opt/SUNWspro/SC4.0/lib/libM77.a \
        /opt/SUNWspro/SC4.0/lib/libsunmath.a /usr/lib/libdl.so.1 \
        /usr/lib/libelf.so.1 /usr/lib/libnsl.so.1 -lm -lsocket

build: SpectrumFactory.h MakeSpectrum.o lib

SpectrumFactory.h:
        javah -jni spectrum.SpectrumFactory

MakeSpectrum.o: MakeSpectrum.c spectrum_SpectrumFactory.h
        cc -c -I$(INCLUDE) -I$(INCLUDE)/solaris -I/usr/ra/stsdasx/lib \
        MakeSpectrum.c -o MakeSpectrum.o

lib: MakeSpectrum.o
        ld -G -t MakeSpectrum.o $(LIBS) -o libiraf.so
```

The first step builds the include file with the header generator program *javah*. The C code is then compiled with appropriate paths for both Java and IRAF/TABLES include files. Finally, the object code is linked with the Java and IRAF/TABLES libraries into a single shareable library named libiraf.so. This name is used by the Solaris OS to find at run time a library identified in the Java code by "iraf" (as in the System.loadLibrary call).

Astronomical Data Analysis Software and Systems VIII
ASP Conference Series, Vol. 172, 1999
D. M. Mehringer, R. L. Plante, and D. A. Roberts, eds.

A Program Complex Equipped With a GUI for Investigation of 3D Structure of Solar Magnetic Fields

G. V. Rudenko, V. V. Grechnev

Institute of Solar-Terrestrial Physics, Lermontov St. 126, Irkutsk, Russia 664033, Email: rud@iszf.irk.ru

Abstract. A program packet is developed for the detailed research into the 3D structure of the magnetic fields of solar active regions as well as the overall solar magnetic field. The application realized in the IDL program environment is supplied with the graphical user interface (GUI) which ensures convenient and efficient manipulation with data.

1. Introduction

Solar activity is the result of motion and interaction of the coronal magnetic fields, so their study is of primary interest of solar physicists. We have developed a program packet for the detailed research into the 3D structure of the magnetic fields of solar active regions as well as the overall solar magnetic field.

The work with data includes two stages. On the first step, coefficients of decomposition are calculated for a selected magnetogram and recorded into special "weight" files. Then the viewing and processing of data after those coefficients is made using an application supplied with graphical user interface (GUI). The application is developed in the IDL[1] program environment.

2. Calculation of the coefficients of decomposition

2.1. Source data

Initial data are taken from daily magnetograms of arbitrary regions on the visible side of the solar disk, and synoptic magnetograms. On their basis, coefficients of decomposition are calculated.

Input data for the GUI-supplied application are the special files containing weight coefficients:

- of flat harmonic decomposition for the reconstruction of the magnetic field for selected active regions, and
- of spherical harmonic decomposition for the reconstruction from synoptic data.

Those "weight files" are computed and recorded separately by another FORTRAN program.

[1] IDL is the trademark of Research Systems, Inc.

2.2. Initial mathematical statements of boundary-value problems to be solved

a) for the reconstruction of partial frames:

$\nabla \times \mathbf{B} = \alpha\mathbf{B}$	+	boundary conditions of the magnetic component
$\nabla\mathbf{B} = 0$		along the line of sight (B_d) at the photopsheric level (at each point);

b) for the reconstruction of the whole surface of the Sun:

$\mathbf{B} = -\nabla\phi$	+	boundary conditions of the magnetic component along the
$\Delta\phi = 0$		line of sight (B_l) at the photopsheric level at the instant of passage of a point across the central meridian.

In the former case, it is possible to select the constant α of the force-free approximation. In the latter case, only the potential approximation is used.

The calculations of the coefficients of the plane decomposition are based on a new method (Rudenko, Altyntsev, & Lubyshev 1997) which allows, unlike the traditional ones, to reconstruct the magnetic field above small regions according to photospheric magnetic fields which are observed in an arbitrary site on the solar disk. In fact, it is possible to use the magnetographic data on active regions which are located close to the limb. This latter circumstance extends opportunities to study magnetic fields of active regions and the evolution while they passing across the solar disk.

The coefficients of decomposition are calculated from magnetograms. Magnetograms of separate active regions can also be processed. Precise scaling and positioning to fit the working coordinate system is accomplished automatically and can also be made in the interactive mode, as well as selection of a region of interest. Data of other types can be used.

3. GUI-supplied application

The program packet was developed in the IDL program environment. The program has GUI (Fig. 1) and provides the following facilities:

- Loading files containing coefficients of a decomposition of the magnetic field.
- Computation of all components of the magnetic field at arbitrary heights and in plane radial sections. Visualization of any (optional) Cartesian and spherical components of the magnetic field, modulus, B_l, and B_d. Images are represented either as a picture plane for any specified date and time, or as a synoptic map.
- Issue of all coordinates and the specified components of the magnetic field corresponding to the position of the mouse pointer.
- Rotation of images by the specified number of days from the current date, and setting the image to the current date and time.
- Selection of a region of interest by means of a flat rectangular box and zooming in to the selected part of the image.

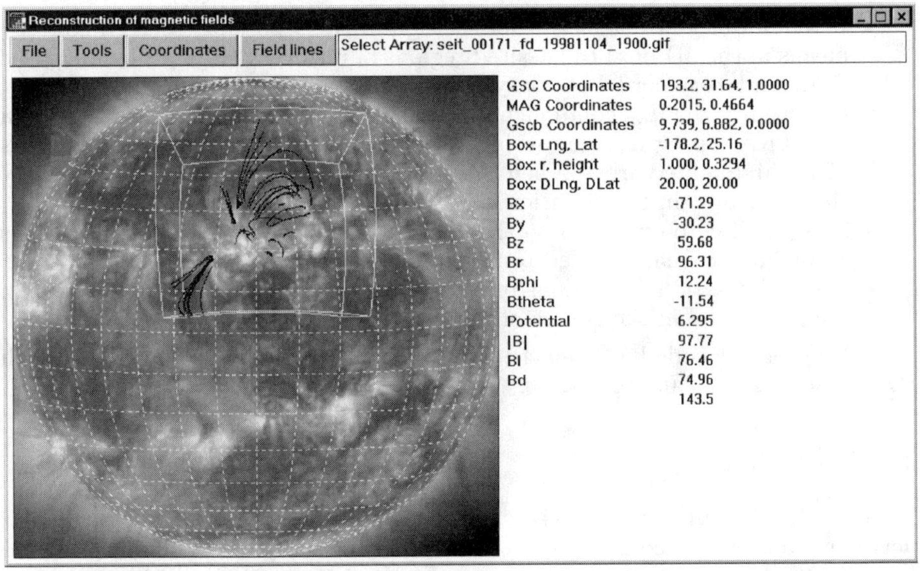

Figure 1. A screen dump of the GUI-supplied application.

- Computation and visualization of the field lines starting with the position on the photosphere specified by the mouse pointer.
- Highlighting an arbitrary field line, and issue of all coordinates and components of the magnetic field for any point along it. The point on the field line is specified by means of the mouse pointer, and its position is indicated by the marker.
- Erasing the highlighted field lines, retaining the 3D coordinates of all current starting points in a file, or loading them from a pre-existing file.
- Drawing and controlling the position and size of the 3D box for highlighting the bounded volume of computation and drawing field lines. All coordinates and magnetic components at the position of the mouse pointer are displayed which correspond to its either outer (depending on the mode) or inner surfaces. When computing the field lines, the starting points are taken from the boundary surfaces of the box.
- Provision is made for a joint processing with images loaded from standard format (FITS, GIF, BMP) files. For FITS files containing standard information about the centering of the image and the pixel size (Kitt Peak magnetograms, Nobeyama and SSRT radio maps, Yohkoh/SXT images), automatic attachment to the heliographical coordinates and to the current date and time is accomplished. For arbitrary images, the location of the disk center and the radius are determined in the interactive mode from three points on the limb. After that, rotational manipulations can be performed with this image, and it can be set to any date and time. It is also possible to overlay field lines computed from pre-loaded coefficients of harmonic decomposition of the magnetic field.

- If the additional image is a magnetogram, it is possible to highlight (by means of the 3D box) the desired regions of reconstruction of the magnetic field, and to record boundary conditions into a file. As the boundary conditions, the B_d component is used. This file is an input file for a FORTRAN program computing the harmonic decomposition coefficients of the highlighted area. When this file is generated, it is possible to specify the value of α in the equation $\nabla \times \mathbf{B} = \alpha\mathbf{B}$.
- It is possible to overlay onto the current image any other magnetic values as well as an image loaded from another file.
- Three-dimensional rotations of contours, all field lines, and of the 3D box are performed in the course of all transformations.

The program code has been checked under MS Windows 95 and UNIX with IDL versions since 3.0.1 to 5.1.

4. Conclusion

We consider a possibility to make calculations of magnetic fields for particular active regions after requests through Internet. In this case, files containing coefficients of decomposition would be calculated by our FORTRAN programs and made available.

We plan to develop a data bank and to supply it with files on the known active regions in the future.

Acknowledgments. We are indebted to Dr. D. Stern (Research systems, Inc.) for his assistance in familiarizing IDL 5.0. We are grateful to the ADASS'98 conference for the financial aid which has made possible the attendance of the authors at the conference. We are indebted to Mr. V. Mikhalkovsky (ISTP) for his assistance in preparation of English version of the text. This work was supported by grant of Russian Foundation RFFI.

Magnetograms which we use are produced by NSO/Kitt Peak in collaboration with NSF/NOAO, NASA/GSFC and NOAA/SEL.

References

Rudenko, G. V., Altyntsev, A. T., & Lubyshev, B. I. 1997, in Ann. Rep. of Joint Organization for Solar Observations, ed. A. Antalova & A. Kučera, 94

Astronomical Data Analysis Software and Systems VIII
ASP Conference Series, Vol. 172, 1999
D. M. Mehringer, R. L. Plante, and D. A. Roberts, eds.

Xguiphot: A New Aperture Photometry Tool for IRAF

Lindsey E. Davis

National Optical Astronomy Observatories, Tucson, AZ 85719

Abstract. Xguiphot is a new GUI driven interactive aperture photometry tool being developed to run under the IRAF environment.

1. Introduction

Xguiphot as a new aperture photometry tool being developed to run under the IRAF environment. In interactive mode Xguiphot is a GUI driven IRAF task; in non-interactive mode it is a parameter driven IRAF task. The Xguiphot GUI is based on the same widget server technology, object manager library, and messaging facilities employed by the IRAF XGterm graphics terminal emulator and XImtool display server (Tody 1995). Xguiphot is an extension of an earlier prototype written to aid in the development and testing of the IRAF GUI facilities (Davis 1994).

2. Operating Modes

Xguiphot can be run interactively with a GUI interface, interactively without the GUI, or non-interactively. On task startup Xguiphot requests a list of images, a list of objects files, a list of output files, initial values for the algorithm parameters, and the name of the GUI definition file. In interactive mode Xguiphot downloads the GUI definition file to the widget server to create the GUI, initializes the panels and widgets, displays the first image, then waits for instructions from the user. If the GUI definition file is unspecified, Xguiphot activates the usual IRAF XGterm window and graphics cursor, displays the first image, and waits for input from the user. In non-interactive mode, Xguiphot can be driven by a user supplied object list or by an object list generated internally at runtime.

3. Task Features

Xguiphot image and object list management features include: interactive image and object list browsing, interactive object list creation using the image display and image cursor, automated object list creation using a built-in object detection algorithm, graphics overlay of objects lists, and interactive editing of objects lists.

Xguiphot provides four object measurement modes: interactive object measurement using the image display and image cursor, interactive selection of objects from an existing object list, automated measurement of an existing object

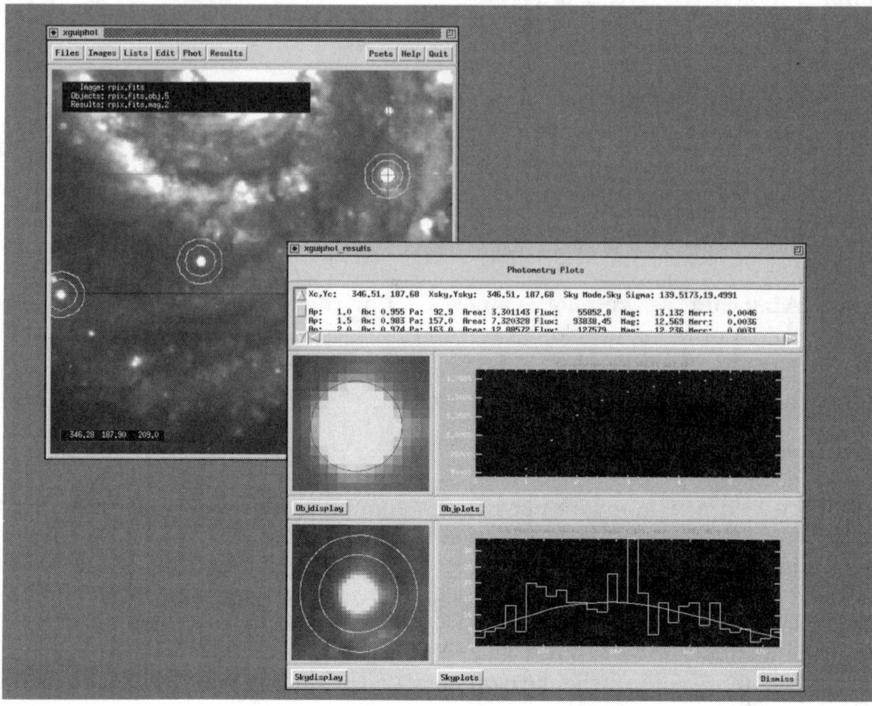

Figure 1. The Xguiphot command window showing the current image with the object and sky apertures for the measured objects superimposed on it. Detailed results for the last measured object, marked by the cursor in the image display window, are displayed in the graphical results panel.

list, and automated object detection and measurement. Quick-look output is written to the status line; full output is written to disk in a format compatible with other IRAF photometry software. Options are available for graphically displaying the individual object and sky regions and the object and sky photometry results as shown in Fig. 1.

The Xguiphot object and sky analysis algorithms support: multi-aperture fractional pixel photometry through circular, elliptical, rectangular, or polygonal object apertures, a concentric annular or offset circular, elliptical, rectangular, or polygonal sky aperture, and multi-aperture shape analysis using second order moment computations.

4. GUI Features

Xguiphot GUI support for browsing images includes an image selection panel, automatic image display, an image header display panel, cursor readout display, pan, zoom, and window functions, and built-in IRAF cursor mode functions. The image display window and associated panels are shown in Fig. 2.

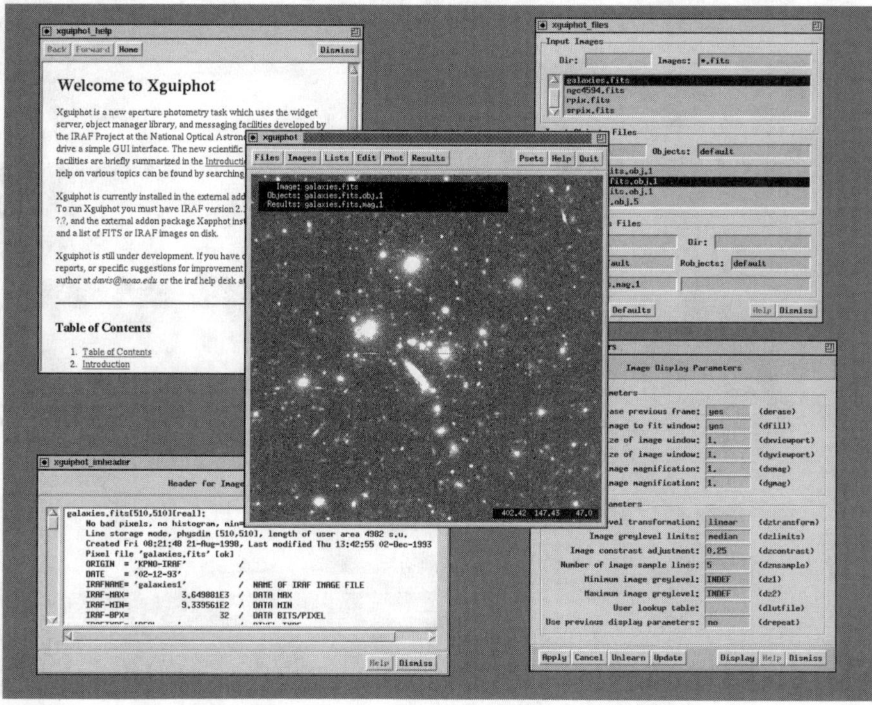

Figure 2. The Xguiphot command window containing the main menu buttons, the image display window, the image cursor, the file status box, and the coordinate readout box. Also shown are the on-line help, file selection, image header display, and display parameter set editing panels.

Xguiphot GUI features for object list creation and editing include an object file selection panel, an object list viewing and editing panel, and an object marker for tracking the current object. The object list viewing and editing panels are shown in Fig. 3.

Xguiphot GUI support for object measurement includes a photometry aperture marker for interactively defining object and sky regions, an image window status line display, and graphical and tabular results display panels. The image display window and graphics display panel are shown in Fig. 1. Keystroke accelerators are supplied for common measuring functions.

Parameter set editing panels are provided for all the Xguiphot parameter sets. Parameter sets can be viewed, edited, unlearned, or updated using a common interface. Individual parameters can also be edited using the usual IRAF colon command syntax. Figures 2 and 3 show sample parameter set editing panels. Xguiphot also has a built-in help facility which uses the standard HTML widget as shown in Fig. 2.

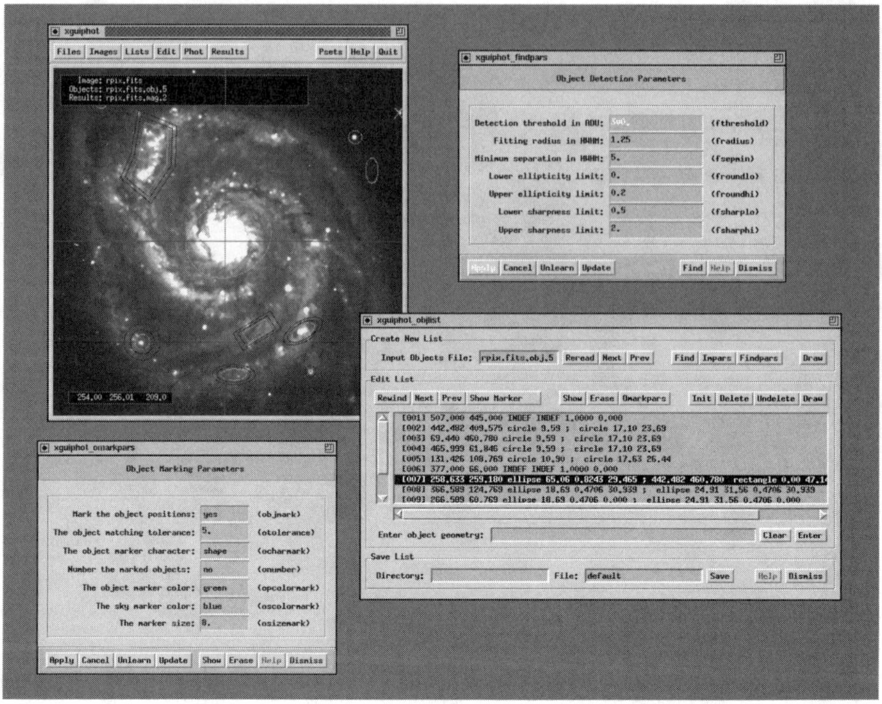

Figure 3. The Xguiphot command window showing the current image with the objects list superimposed on it. The image cursor marks the current object which is also highlighted in the object list viewing and editing panel. Also shown are the object detection and marking parameter set editing panels.

5. Current Status and Future Plans

Xguiphot adds new scientific analysis capabilities to IRAF and demonstrates that the current IRAF GUI facilities can be used for building complex science applications. Xguiphot should be ready for release in early 1999. Interested users should check the IRAF Web site at `http://iraf.noao.edu/` and the IRAF newsgroups for status updates.

References

Davis, L. E. 1994, in ASP Conf. Ser., Vol. 61, Astronomical Data Analysis Software and Systems III, ed. D. R. Crabtree, R. J. Hanisch, & J. Barnes (San Francisco: ASP), 75

Tody, D. 1995, in ASP Conf. Ser., Vol. 77, Astronomical Data Analysis Software and Systems IV, ed. R. A. Shaw, H. E. Payne, & J. J. E. Hayes (San Francisco: ASP), 89

Astronomical Data Analysis Software and Systems VIII
ASP Conference Series, Vol. 172, 1999
D. M. Mehringer, R. L. Plante, and D. A. Roberts, eds.

SAOTk: Tcl/Tk Widgets for Imaging and Data Visualization

William Joye, Eric Mandel

Smithsonian Astrophysical Observatory, Cambridge, MA 02138

Abstract. *SAOTk* is an integrated set of *Tcl/Tk* widgets for astronomical imaging and data visualization. The widget set is composed of an Image, Panner, Magnifier, and Colorbar widget. In addition to "classical" support for imaging *FITS* data, manipulating colormaps, region marking, and coordinate readout (including WCS), *SAOTk* widgets also support arbitrary image scaling and rotation, advanced printing and graphics, image mosaics, and shared memory.

The *SAOTk* widgets will be incorporated into the next major release of *SAOtng*, a release that will allow the astronomical community to begin moving away from its primary dependence on *Unix* and *X11*. They also can be incorporated into any standard *Tcl/Tk* environment to build custom data analysis and visualization applications.

1. Introduction

For more than fifteen years, software developers at Smithsonian Astrophysical Observatory have chased the elusive goal of providing a perfect image display program to the astronomical community. We started our quest in the early 1980's by developing *SAOimage*, one of the very first astronomy programs to utilize the *X Window System* within the *Unix* workstation environment. *SAOimage* broke new ground in astronomical image display by providing an intuitive and easy to use graphical menu/command button interface. It has since been used by thousands of astronomers around the world.

In the early 1990's, we introduced *SAOimage: The Next Generation (SAOtng)*. The aim of this new offering was to move beyond the concept of a static and stand-alone image display. We wanted to allow users to extend the program beyond the functionality that we had compiled and linked into it. We also wanted to give users the ability to control the program from external programs, scripts, etc. Toward this end, we developed the *X Public Access (XPA)* mechanism, which enabled *SAOtng* (and other programs) to define named "public access points" for communication of data and commands with other programs. *XPA* offered public access to traditionally internal *SAOtng* algorithms such as image file loading and scaling. It also supported user-configurable analysis of image data and bi-directional communication with other processes.

In this new phase of our work, we aim to move beyond the bounds of any particular image display program and even beyond the bounds of the *Unix* operating system that has been so central to astronomical software over the past decade and a half. We therefore have developed *SAOTk*, an integrated set of *Tcl/Tk* widgets for astronomical imaging and data visualization.

Table 1. SAOTk Widgets

Image	FITS viewer for a single image or a mosaic of images
Colorbar	control of imaging color environment
Panner	thumbnail image for position control
Magnifier	magnified view of a section of the image

The *SAOTk* widgets will be incorporated into the next major release of the *SAOimage* line of image display programs, a release we have code-named *DS9*. The availability of these widgets, along with *DS9*, will allow the astronomical community to begin moving away from its primary dependence on *Unix* and *X11*. The widgets themselves are designed to be incorporated easily in any standard *Tcl/Tk* environment to build custom data analysis and visualization applications.

2. The SAOTk Image Widget

The Image widget is used to display *FITS* data. It supports both *FITS* images and binary tables. Multiple extension *FITS* files are also supported. The widget supports arbitrary zoom, rotation, orientation, and panning. The *FITS* data are maintained in raw form. `BSCALE`, `BZERO`, and `BLANK` keywords are supported.

The Image widget is designed to support rendering of large *FITS* images and binary tables. For images, the rendering time is constant, and is based on the size of the screen, not the size of the data. Thus, the widget is capable of rendering very large images (8k $\times$ 8k or larger), or a large mosaic of images (such as 36 images, each 2k $\times$ 4k). Since all rendering is done directly from the raw *FITS* data, memory requirements are very small.

For *FITS* binary tables, the rendering time is based on the number of filtered rows or events. Since the image is binned *on the fly*, and then rendered to the screen, memory requirements again are very small. Of course, if the data are sorted by position, rendering time is much faster, since the widget can calculate which segments of a table contain the events that need to be processed.

The Image widget supports display of a mosaic of images, where each image segment is a separate *FITS* image. These segments may be stored in one multi-extension file, or may be located in a number of different files. The widget uses *FITS* mosaic keywords to place each image in its proper place in the overall image space.

One special feature of the Image widget is the ability to rotate, orient, and zoom an image to align its display to an arbitrary WCS. This feature is very useful for making comparisons between images from different instruments or telescopes that have different coordinate systems or WCS parameters.

As with all 8-bit, Pseudo-Color displays, each *FITS* pixel data value is scaled to a display value between 0 and 255. This scaling is done *in real time*, so that the widget may be used for real-time applications. The mapping from the raw data to the scaled value is based on the start/stop clip values and the scaling algorithm. Clipping values can be defined using minimum/maximum data values, *IRAF* zscale, or user-specified values. Scaling algorithms currently include `linear`, `ln` (from *SAOimage*), `log` (from *IRAF*), and `sqrt`. Users can easily add new scaling algorithms.

If the data are contained in a *FITS* binary table, the events are filtered and binned *in real time*, based on the current pan, zoom and rotation values. Events may be filtered using any column, and spatial region masking also is supported.

The Image widget can read *FITS* data from a file, `STDIN`, or a *TCL* Channel. It also supports a number of different memory management options: memory can be allocated as needed (and swapped to disk), it can be mapped directly to the *FITS* file *(mmap)*, or the *FITS* data can be accessed from a shared memory segment. The default is *mmap*.

The Image widget supports a variety of coordinate systems, as defined in the *FITS* header: `image, ccd, detector, amplifier,` and WCS `(fk4, fk5, galactic, ecliptic, linear)`.

The Image widget supports *PostScript* level 1 and level 2 printing. This is not a screen capture process, but a full-featured *PostScript* driver. A variety of document sizes are available, and grayscale and color imaging are supported. Level 2 output is compressed via `RLE` and encoded via `ASCII85`, so the size of the output *PostScript* data is usually very small.

The Image widget also supports a variety of markers or regions, including `circle, ellipse, rectangle, polygon, line,` and `point`. A marker's position is attached to the image and may be specified in any of the supported coordinate systems. Each marker has a number of properties, including color, text, and name. Each marker also allows specification of behavior such as whether the marker can be selected, edited, moved, rotated, deleted, etc.

3. The SAOTk Colorbar Widget

The Colorbar widget controls the color environment for the application. Currently, it requires an 8-bit Pseudo-Color visual, though this need not be the default visual. This requirement will be removed in a future release.

The Colorbar widget supports a variety of colormap formats, including colormaps from *SAOimage, SAOtng/ximtool,* and other display programs. There are a number of built-in maps, and new maps may be loaded, edited, and saved.

The widget will request a minimum number of colors in the default colormap. If these are not available, it will use a private colormap. The private colormap will be populated with colors from the default colormap to minimize color flashing.

4. The SAOTk Panner and Magnifier Widgets

The Panner and Magnifier widgets are designed to work with the Image widget. Each Image Widget can be assigned to work with one Panner widget and one Magnifier widget.

The Panner widget displays a thumbnail image of the data being displayed, in order to support of fine position control. It also overlays an image orientation compass and a WCS orientation compass. The WCS compass may be positioned at the center of the image or at the tangent point.

The Magnifier widget provides a magnified view into the data. Two magnification algorithms are available: `Fast` and `More Accurate`. The later is especially useful for non-integer zoom and rotation values. The magnification value may be user-specified, but usually is a power of two.

5. DS9: Putting It All Together

DS9 is our code-name for the next version of the *SAOimage* family of display programs. It is a *Tcl/Tk* application which utilizes the *SAOTk* widget set. It also incorporates a new version of *XPA* (an implementation which is independent of the *X Window System*) to allow external processes to access and control its data, GUI functions, and algorithms. *DS9* will support all capabilities in the current version of *SAOtng*, including direct display of *FITS* images and binary tables, multiple frame buffers, region/cursor manipulation, many scale algorithms and colormaps, and easy communication with external analysis tasks. It is highly configurable and extensible to meet the evolving needs of the astronomical community.

DS9 is a stand-alone application that requires no installation or support files. Preliminary versions of *DS9* exist for Sun Solaris 2.5/2.6 and Linux. Other Unix platforms will be supported in the near future. Ports to Apple Macintosh and Microsoft Windows also will be available in the next year. All versions and platforms will support a consistent set of capabilities.

DS9 supports advanced features such as multiple frame buffers, mosaic images, tiling, blinking, geometric markers, colormap manipulation, scaling, arbitrary zoom, rotation, pan, and a variety of coordinate systems (including Image, CCD, Detector, and WCS). *DS9* also supports *FTP* and *HTTP* access.

The GUI for *DS9* is user-configurable. GUI elements such as the coordinate display, panner, magnifier, horizontal and vertical graphs, and colorbar can be configured via menus or the command line.

An exciting new feature of *DS9* is the ability to display multiple images in tile mode, where each image is scaled, oriented, and rotated to match WCS parameters. By locking cross-hairs, features between images are easily compared.

6. Conclusion

More information about the *SAOTk* widgets, *DS9*, *XPA*, or *SAOtng* (including our latest software offerings) can be found on the World Wide Web at:

`http://hea-www.harvard.edu/RD/`

While our goal of developing a perfect image display program has been elusive thus far, we are confident of achieving it the next release.

Acknowledgments. This work was performed in large part under a grant from NASA's Applied Information System Research Program (NAG5-3996), with support from the AXAF Science Center (NAS8-39073).

Astronomical Data Analysis Software and Systems VIII
ASP Conference Series, Vol. 172, 1999
D. M. Mehringer, R. L. Plante, and D. A. Roberts, eds.

Converging Horizons: Collaborative 2/3D Visualization Tools for Astronomy

Raymond L. Plante, Paul J. Rajlich, Steve Pietrowitz, Wei Xie, Asif Qamar

National Center for Supercomputing Applications (NCSA), University of Illinois Urbana-Champaign, Urbana, IL 61801

Abstract. We present a variety of Java–based visualization tools for collaborative astronomical research which integrate a number of diverse technologies currently under development at NCSA. The NCSA Horizon Image Data Browser is a Java package for building Java applets that interact with 2D visualizations of multi-dimensional image data. The latest release of the package, Horizon 2.0, features a number of new capabilities including transparent support of collaboration and flexible support for multidimensional arrays. Astro3Vis is a pair of tools for creating and viewing 3D visualizations of astronomical images in a Web environment. The VRML Server is used to create 3D visualizations of FITS image cubes from the NCSA Astronomy Digital Image Library (ADIL) in VRML format. While any VRML 2.0 browser can be used to view the visualizations, interactivity is enhanced when the VRML Server is used in conjunction with a Astro3D, a Java VRML viewer with specialized features supporting editing and collaboration. Together, the two tools that make up Astro3Vis provide an environment for creating 3D figures that can be published in electronic journals.

1. 2D Visualization with NCSA Horizon

The NCSA Horizon Image Data Browser[1] is a Java package for creating applets and applications that browse scientific images. The background and scope of this toolkit was presented at ADASS VI (Plante et al. 1997), and some of key features important to astronomers—support for metadata and world coordinate systems—were presented at ADASS VII (Plante et al. 1998). This year, we are releasing Horizon 2.0 which provides several new features:

- Java 1.1 compatibility
- a more fully developed data access model
- transparent support for collaboration
- improved performance in speed and memory use

1.1. Supporting Multidimensional Arrays

Java's support for efficient access to multidimensional arrays is somewhat lacking. To make up for this shortcoming, Horizon provides a subpackage,

[1] `http://imagelib.ncsa.uiuc.edu/Horizon`

`ncsa.horizon.arrayND`, that gives fast, flexible access to multidimensional array data by:

- supporting a variety of storage schemes,
- providing fast, type-independent access to array elements,
- avoiding array copies whenever possible,
- when array copies are needed, maximizing the use of `System.arraycopy()`,
- minimizing redundant/costly access operations.

Access to multidimensional array data is usually made most efficient when it is done through via 1-D sub-arrays which we refer to as `ArrayChunks`. Special iterators called `Chunkerators` allow one to step through a multidimensional array in 1-D chunks. The `ncsa.horizon.arrayND` package is general purpose and can be used independently of the rest of the Horizon package.

1.2. Collaboration with Horizon and Habanero

Collaboration support allows a group of astronomers to explore the images together even when they are dispersed around world. Horizon supports collaboration through another package called NCSA Habanero[2]. When a user starts up the Habanero environment, she can contact her collaborators by email to join her session. Together, they have access to a variety of general purpose and application–specific collaborative applets, or *hablets*. When a hablet is run, all participants see a copy of it; when one person interacts with the hablet's GUI, everyone sees the result. General purpose hablets that help the participants work together include a chat window and white board. Horizon hablets, loaded into the Habanero environment, thus allow the collaborators to explore images together (Crutcher, Plante, & Rajlich 1998).

The central idea behind Habanero's collaborative model is event sharing: events generated in one version of a hablet is shared with the collaborating versions; thus, all collaborators see the same state. The high level GUI classes from Horizon 2.0 have been made Habanero–aware by defining the events that should be shared by default. This makes it easy create applets that can either be run collaboratively within the Habanero environment or stand–alone.

2. 3D Visualization Over the Web with Astro3Vis

Astro3Vis[3] is a pair of tools—a server and a client—which allow astronomers to create and interact with 3D visualizations of astronomical images over the Web. VRML is used to transport the visualization from the server to the client. Using VRML is advantageous for two reasons in particular: first, it brings high–end graphics to low–end workstations, and second, the visualizations can be easily edited and saved away for future display, say, over the Web. The two parts to Astro3Vis are the VRML Server and the Astro3d, a VRML browser.

---

[2] `http://www.ncsa.uiuc.edu/SDG/Software/Habanero/HabaneroHome.html`

[3] `http://monet.astro.uiuc.edu/Astro3Vis`

2.1. The VRML Server

The VRML Server[4] is used to create 3D visualizations from FITS image cubes over the Web. It is currently installed as a service of NCSA Astronomy Digital Image Library (ADIL)[5]. When this tool was first introduced (Crutcher, Plante, & Rajlich 1998), it featured an HTML forms interface. Today, visitors use a Java interface to request images from the ADIL, enter the visualization parameters, and download the results as a VRML file. When loaded into a VRML browser, users see the data visualized within a cube which can be turned and examined from any angle (the so-called "examine" navigation mode, see Fig. 1); alternatively, users can explore the data by "walking" through and around the data.

The Server supports a few visualization methods including isosurfaces, raster slices, and contour slices. Not only can a single visualization feature multiple methods, it can also feature multiple datasets. Thus, one can, for example, compare the emission from different chemical species in a single VRML file. The interface also allows one to request subregions, subsamplings, and relative scalings of the axes.

The VRML Server is based on the Visualization Toolkit (VTK)[6], a free, highly portable C++ library for rendering 3D graphics. The VRML files it creates can be viewed with any standard VRML 2.0 browser; thus, it is possible to take advantage of 3D graphics on a large variety of platforms including Windows, Macs, and Unix machines.

2.2. Astro3D

The client side of Astro3Vis is a pure–Java VRML browser called Astro3D. This application, shown in Fig. 1, allows one to explore any standard VRML 2.0 file using the traditional "walk" and "examine" navigation modes. Astro3D is built on top of the Java 3D API and NCSA Portfolio[7], an object library for building Java 3D applications.

Astro3D has several features that extend beyond traditional VRML browsers. These include a 3D cursor that can be moved around within the data cube and whose position is displayed in the browsers interface. The cursor can also be used to annotate the VRML files with lines, boxes, and text; the edited file can then be saved to local disk. 2D snapshots of the visualization can also be saved to disk as JPEG files. Thus, Astro3D allows astronomers to create 3D figures which can be posted on the Web or published in an electronic journal.

Other Astro3D features are enabled when it is used with the VRML Server (usually done by installing as a VRML helper application to the Web browser). One such feature is the ability to select subregions with the 3D cursor and request a new visualization of that region from the server. One can also track world coordinates with the cursor. This feature is possible because the VRML Server

---

[4] `http://imagelib.ncsa.uiuc.edu/VRMLServer`

[5] `http://imagelib.ncsa.uiuc.edu/imagelib`

[6] `http://www.kitware.com/vtk.html`

[7] `http://havefun.ncsa.uiuc.edu/Java3D/portfolio/`

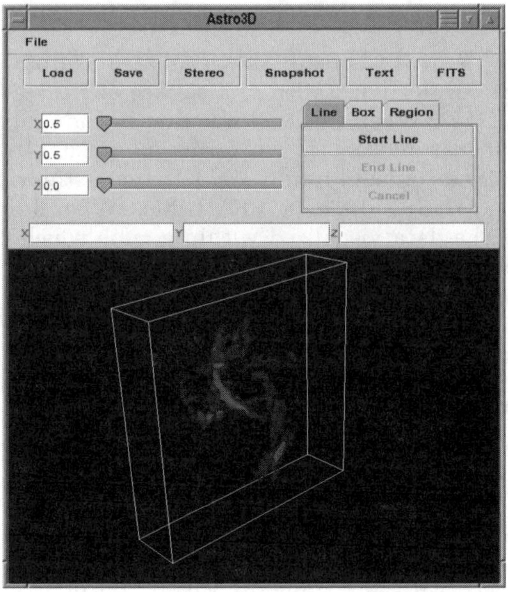

Figure 1. The Astro3D displaying a VRML visualization generated by the VRML Server of an image from the ADIL. This image shows hydrogen recombination line emission from the Galactic Center (Roberts & Goss 1993).

embeds the FITS header information within the VRML file; while the header is ignored by standard VRML browsers, Astro3D reads the header and interprets it with the help of the NCSA Horizon coordinate classes. Other features include support for collaboration and stereo (when stereo hardware is available).

Acknowledgments. NCSA Horizon is being developed with support from Project 30, a NASA AISRP grant (NAG 5-4228).

References

Crutcher, R. M., Plante, R. L., & Rajlich, P. 1998, in ASP Conf. Ser., Vol. 145, Astronomical Data Analysis Software and Systems VII, ed. R. Albrecht, R. N. Hook, & H. A. Bushouse (San Francisco: ASP), 3

Plante, R. L., Xie, W., Plutchak, J., McGrath, R. E., & Lu, X. 1998, in ASP Conf. Ser., Vol. 145, Astronomical Data Analysis Software and Systems VII, ed. R. Albrecht, R. N. Hook, & H. A. Bushouse (San Francisco: ASP), 99

Plante, R. L., Goscha, D., Crutcher, R. M., Plutchak, J., McGrath, R. E., Lu, X., Folk, M. 1997, in ASP Conf. Ser., Vol. 125, Astronomical Data Analysis Software and Systems VI, ed. G. Hunt & H. E. Payne (San Francisco: ASP), 341

Roberts & Goss 1993, ApJS, 86, 133

Astronomical Data Analysis Software and Systems VIII
ASP Conference Series, Vol. 172, 1999
D. M. Mehringer, R. L. Plante, and D. A. Roberts, eds.

DASHA-2: Improving Visualization and Processing of Photometric Data with IDL Objects

Oleg Smirnov

Institute of Astronomy of the Russian Academy of Sciences 48 Pyatnitskaya Str., Moscow 109017 Russia

Abstract. DASHA-2 is the next iteration of the DASHA package (Smirnov & Ipatov 1995) for processing data produced by the DAOPHOT II PSF photometry software (Stetson 1987, 1992). The original package was implemented under the PCIPS image processing system (Smirnov & Piskunov 1995), whereas DASHA-2 is completely written in IDL. The package is tailored for the reduction of large batches of CCD observations, where the same field is observed at different exposures and in different photometric bands. DASHA-2 is currently in heavy use at the Institute of Astronomy, where it is routinely employed to reduce observations of globular clusters, resulting in photometric data for upwards of 15,000 objects in a single field.

1. Introduction

DASHA-2 consists of four widget-based IDL applications, providing functionality for:

- Visualization of photometric data (e.g., photometric tables produced by DAOPHOT II – lists of stars detected in each CCD frame, along with their estimated PSF magnitudes.)
- Interactive filtering of photometric data using various goodness-of-fit statistics.
- Calibration of photometry using standards in the field.
- Cross-identification of objects between tables produced by DAOPHOT II for different CCD frames of the same field.
- Merging of data into a single master table, containing mean magnitudes for each detected object.
- Cross-identification of objects between master tables compiled for different photometric bands, and compilation of a single composite table containing magnitudes and color indexes.

The design of DASHA-2 made full use of the new object-oriented capabilities implemented in IDL[1] version 5. Development produced a powerful Foundation Class Layer (FCL), which allows the programmer to easily create new

[1]IDL is the trademark of Research Systems, Inc.

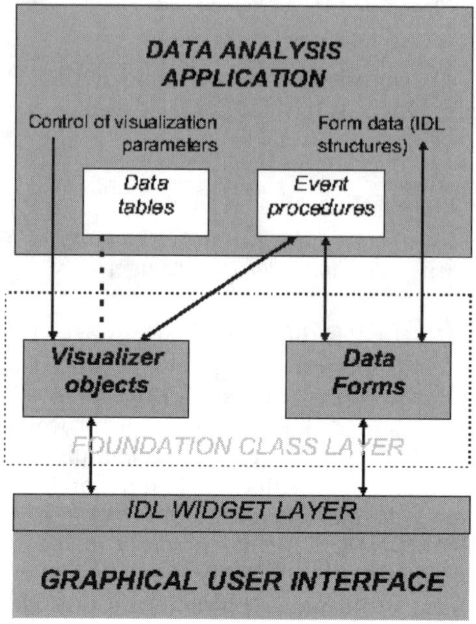

Figure 1. Structure of an application employing the FCL.

data analysis applications with sophisticated user interfaces and visualization capabilities.

The Foundation Class Layer hides the complexity of IDL widgets and graphics behind easy-to-use IDL classes. The two primary classes are the Visualizer, which provides a complex widget for interactive visualization of large data structures, and the DataForm, which simplifies the creation of complex user interfaces. This paper focuses on the benefits of the FCL for rapid development of scientific applications in IDL.

2. Structure of a DASHA-2 Application

The typical structure of an application employing the FCL is illustrated by Fig. 1. As one can see from the figure, the FCL completely isolates the GUI from the application. The rationale behind this design is as follows. With the ever-increasing amount of CPU power available to the average astronomer, interactivity is becoming the bottleneck for many data analysis applications. Whereas a decade ago the average user was mostly idle while the CPU was busy number crunching, today it is the average CPU that is mostly idle, waiting for the astronomer to prepare the data and evaluate the results. Thus, for many (though not all) data analysis applications, operational efficiency is determined more by a well-designed and functional GUI than by algorithmic efficiency. However, as any programmer can attest, GUI implementation is one of the more mundane and time-consuming tasks – and the more functional the GUI, the greater the pains required to make it all work. The FCL allows a developer to expend

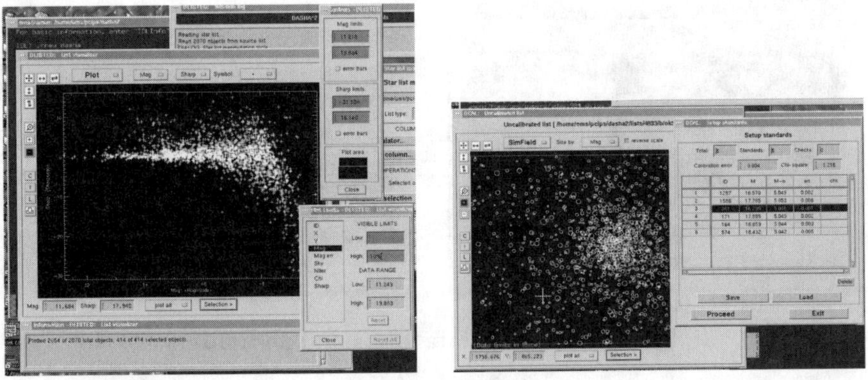

Figure 2. A Visualizer object (*left*). A sample DataForm object that can receive input from the Visualizer via an event procedure (*right*).

minimum effort on GUI implementation, and concentrate on the algorithmic aspect of an application. The FCL also ensures that all applications employing it have a similar look-and-feel (and allows improvements in the look-and-feel without touching the application code). While this concept is not new – it exists in many UI toolkits available today, the FCL is a very high-level toolkit specifically tailored to astronomical data analysis needs. For example, it provides powerful visualization capabilities, which can be very important for presenting results to the astronomer in an efficient way.

3. The Visualizer Object

Interactive visualization is available to any application, just by creating a Visualizer object and assigning it a data set. The visualizer object creates a complex widget-based GUI for visualizing the data set (Fig. 2, left), and then manages all user interaction transparently to the application. The Visualizer object encapsulates the following visualization functionality:

- Provides visualization (including hard copy) of any kind of data table (i.e., array of IDL structures).
- Allows the user to control visualizations on the fly, interactively generate plots, diagrams and histograms for any columns of the table, and control plot parameters.
- Zoom in and out of plot regions using the mouse pointer, track pointer position in data space.
- Set data limits on individual columns (as values or percentiles).
- Select data objects directly on the plot using click-and-drag, extend the selection across multiple plots, and filter displayed data based on the selection.
- Query objects by pointing to them on the plot.

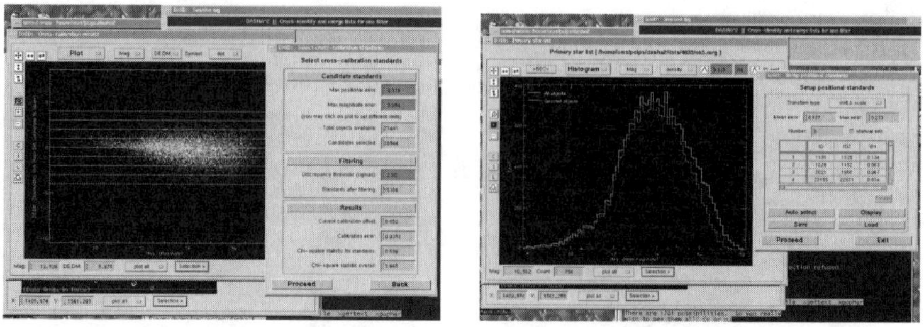

Figure 3. More examples of DataForms and Visualizers.

All functions of the Visualizer object may be controlled both by the user and the application. The application can interact with the Visualizer object to set up specific visualizations and receive user feedback.

4. DataForm Objects

DataForm objects allow an application to generate sophisticated user interfaces without the complexity of IDL widgets. Any application can create a DataForm object and fill it with text entry fields (which feature automatic input type checking), tables, and various standard controls (buttons, droplists, check-boxes, etc.), all in a few lines of code. Data exchange between the form and the application is simplified by mapping the form fields to an IDL structure; the form may be filled with initial values and later queried with a single call.

More sophistication can be added by employing event procedures. Event procedures allow forms to immediately respond to user input, and also handle events from Visualizer objects. The form in Fig. 2 (right), employed in the photometric calibration procedure of DASHA-2, features a table of photometric standards. The user can add standards to the table by simply selecting them on the Visualizer (here in "simulated field" mode) with the mouse. Other example DataForms and Visualizers are presented in Fig. 3.

5. Conclusion

While DASHA-2 is a complete package, the FCL is nowhere near that elusive state. For example, it still lacks a Visualizer object for 2D images and other data sets. Development of the FCL is continuing. It is planned to release an initial version to the public under a GPL-style license. For more information, please contact the author at `oms@inasan.rssi.ru`.

Acknowledgments. I would like to thank the LOC of the ADASS'98 Conference for the financial support that made this presentation possible. I would also like to acknowledge the Russian Academy of Sciences for the aid provided by the "Young Scientists Support Program".

References

Smirnov, O. & Ipatov, A. 1995, in ASP Conf. Ser., Vol. 77, Astronomical Data Analysis Software and Systems IV, ed. R. A. Shaw, H. E. Payne, & J. J. E. Hayes (San Francisco: ASP), 488

Smirnov, O. & Piskunov, N. 1995, in ASP Conf. Ser., Vol. 77, Astronomical Data Analysis Software and Systems IV, ed. R. A. Shaw, H. E. Payne, & J. J. E. Hayes (San Francisco: ASP), 133

Stetson, P. B. 1987, PASP, 99, 191

———, 1992, in ASP Conf. Ser., Vol. 25, Astronomical Data Analysis Software and Systems I, ed. D. M. Worrall, C. Biemesderfer, & J. Barnes (San Francisco: ASP), 297

Astronomical Data Analysis Software and Systems VIII
ASP Conference Series, Vol. 172, 1999
D. M. Mehringer, R. L. Plante, and D. A. Roberts, eds.

XSKYMAP 2: A Multi-Catalog Visualization Facility

Oleg Smirnov, Oleg Malkov

Institute of Astronomy of the Russian Academy of Sciences 48 Pyatnitskaya Str., Moscow 109017 Russia

Abstract. XSKYMAP is a widget-based IDL application for visualization of astronomical catalogs. Since then XSKYMAP has grown into a powerful sky catalog visualization tool.

1. Introduction

The version of XSKYMAP presented here supports the following catalogs:

ZGSC: a compressed version of the Guide Star Catalog version 1.1. ZGSC was created at the Institute of Astronomy (Smirnov & Malkov 1997). ZGSC employs a custom binary format and an adaptive compression algorithm to achieve 6:1 lossless compression of GSC 1.1 – down to about 200 MB (from 1.2 GB). ZGSC, like its uncompressed counterpart, contains roughly 20 million objects. The compressed size makes it entirely practical to keep the catalog on-line on a hard disk for a dramatic improvement in access times.

The PPM family of astrometric catalogs, namely the Catalog of Positions and Proper Motions, the Catalog of Positions and Proper Motions – South, the Bright Stars Supplement to the PPM and PPM South Catalog, Revised Edition, and the 90000 Stars Supplement to the PPM Star Catalog (Roeser & Bastian 1988; Roeser & Bastian 1993; Roeser, Bastian, & Kuzmin 1993) contain J2000 positions and proper motions of altogether 468861 stars. Their main purpose is to provide a convenient, dense, and accurate net of astrometric reference stars that represents the IAU coordinate system on the sky.

GSC's inherent depth of field is supplemented by extremely precise positions of relatively brighter stars from the PPM.

XSKYMAP provides a wide range of visualization tools for various applications. The current version has been integrated with the control software for the Galileo Italian National Telescope as an observation support tool (Pasian et al. 1998); the primary applications being generation of finder charts and preliminary telescope positioning.

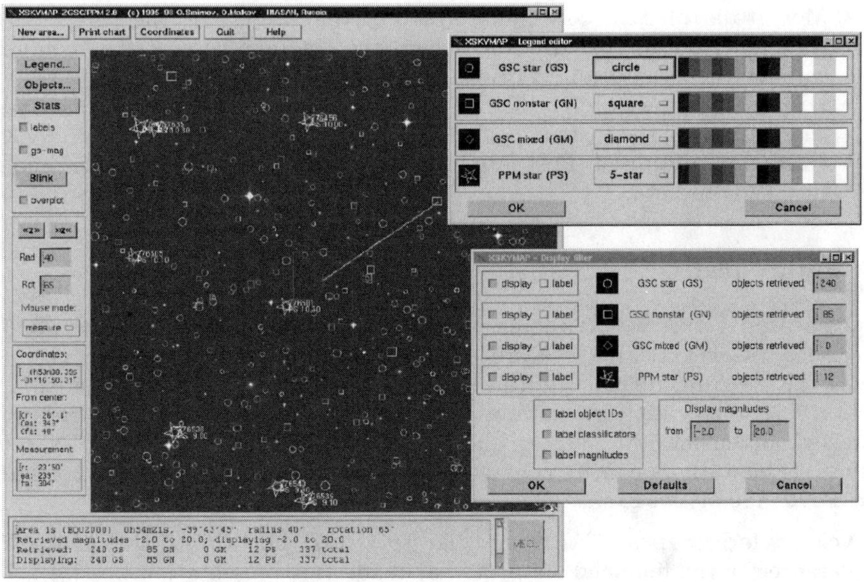

Figure 1. The XSKYMAP Interface

2. XSKYMAP Features

XSKYMAP's primary screen, along with some additional dialogs, is presented in Fig. 1. The software has the following important features:

- On-the-fly data retrieval for a user-supplied region. XSKYMAP can work in four coordinate systems: equatorial and ecliptic (for any equinox), galactic and supergalactic.
- Uses a custom map projection routine, allowing for accurate plotting of rectangular areas even in polar regions (something the map projection routines of IDL[1] currently lack), and supporting arbitrary rotation of the map relative to North on the sky.
- Mouse-based catalog feedback: click on an object to obtain its full catalog entry or entries.
- Tracking of pointer coordinates when over the map, with dynamic display of sky coordinates and separation/positional angle relative to center of area. Click and drag to measure separation and positional angle between two points on the map.
- Mouse operations for zoom in/zoom out and moving to new map center.
- User can interactively change the of map legend (i.e., symbol and color used for each type of object), and selectively display and label objects of a particular type.

---

[1]IDL is the trademark of Research Systems, Inc.

- Move to a different coordinate system while preserving the current area (i.e., new coordinates are established, and the map is rotated to compensate for the orientation of North in different coordinate systems).
- Display 2D images (e.g., directly from an instrument) in greyscale under the map. Plot the instrument's field-of-view box over the map.
- Hard copy output in PostScript format, both in map-only mode, and in image+map mode.

3. Future plans

In the nearest future we plan to extend XSKYMAP to support the new USNO-A catalog (Monet 1996). We are also redeveloping the catalog access module of the program to make full use of the object-oriented capabilities of IDL version 5, so that support for additional catalogs may be added to the software by implementing a catalog query object according to predefined specifications. The authors can be contacted by e-mail at `oms@inasan.rssi.ru` and `omalkov@inasan.rssi.ru`.

Acknowledgments. We would like to thank the LOC of the ADASS'98 Conference for the financial support that made this presentation possible. We would also like to acknowledge the Russian Academy of Sciences for the aid provided by the "Young Scientists Support Program".

References

Monet, D. 1996, USNO-A1.0: A Catalog of Astrometric Standards, (Washington: USNO)

Pasian, F. et al. 1998, in ASP Conf. Ser., Vol. 145, Astronomical Data Analysis Software and Systems VII, ed. R. Albrecht, R. N. Hook, & H. A. Bushouse (San Francisco: ASP), 433

Roeser, S. & Bastian, U. 1988, A&AS, 74, 449

———, 1993, Bull. Inform. CDS, 42, 11

Roeser, S., Bastian, U., & Kuzmin, A. 1993, A&AS, 105, 301

Smirnov, O. & Malkov, O. 1997, in ASP Conf. Ser., Vol. 125, Astronomical Data Analysis Software and Systems VI, ed. G. Hunt & H. E. Payne (San Francisco: ASP), 426

Astronomical Data Analysis Software and Systems VIII
ASP Conference Series, Vol. 172, 1999
D. M. Mehringer, R. L. Plante, and D. A. Roberts, eds.

Image Display Paradigm #3

Dyer Lytle

Lunar and Planetary Laboratory, University of Arizona, Tucson, AZ 85721

Elizabeth Stobie, Anthony Ferro, Irene Barg

Steward Observatory, University of Arizona, Tucson, AZ 85721

Abstract. Image Display Paradigm #3 (IDP3) is an IDL (Interactive Data Language) package developed by the NICMOS Software Group at the University of Arizona. Its original purpose was to provide a versatile graphics tool for PSF subtraction with NICMOS data. However, with a clever structured design IDP3 became a unique and very powerful image analysis tool. In this presentation we explore IDP3, its development and its features.

1. Introduction

IDP3 is a tool for manipulating data images that employs a powerful but easy-to-use graphical user interface with support for FITS and HDF input formats. It allows the user to work with a collection of images and display one or more in a graphics window simultaneously. Images may be individually moved, scaled, and rotated to bring features into registration. Each image may be subtracted from, added to, or combined with the composite display using one of several other functions (divide, multiply, XOR, etc). IDP3 provides a region-of-interest pop-up tool for intensive examination of sub-regions of the main display. There are cross-section plots, masks, surface plots, statistics, spreadsheet displays, radial profiles, and other tools at the user's disposal. Combination or individually registered images may be saved to FITS files.

2. Features of IDP3

- The main image display [work area] is resizable, scalable, zoomable, and shiftable displaying multiple images simultaneously.
- IDP3 employs Lytle's general principles of interactivity (see Appendix A).
- Data alignment [shift and rotation] by visual, statistical, and centroidal criteria.
- User definable and adjustable data masks and anti-masks.
- Region of interest display.
- Cross sections and radial profiles.
- Surface plots and surface views from multiple perspectives.

- Zooming and interpolation with conservation of flux.
- Arithmetic operations on selected images including addition, subtraction, division and XOR.
- Image blinking.
- Multiple mouse buttons perform separate functions in multiple windows.
- Documentation, all modifications to data recorded in header of saved image.
- Support for FITS and HDF formats on input, FITS format on output.
- User preferences allow customization.

3. Design and philosophy

The evolution of IDP3 was driven by a need for an interactive tool for subtracting point spread functions (PSFs) from data taken with the space telescope instrument, NICMOS. This evolutionary process had the following steps.

- A prototype tool (psfsubtract) was written and tested in IDL[1].
- A few users tried the tool and made suggestions for improvement.
- At a general meeting at which a large fraction of the potential users were present the prototype tool was presented and demonstrated. Attendees were encouraged to comment on and criticize the prototype and to suggest features they would like to see.
- The notes taken from this meeting and were used to outline a requirements document for the final program. The design for the new program was begun and the data structures that would be needed were specified.
- It was at this time that it was seen that there was a possibility that the software being written could be generalized to handle a broader spectrum of input data and to allow the user more general functionality than the original specification dictated. Thus, it was decided to make a program that would satisfy the PSF group (the needs of the few) but that would also be useful (hopefully) for the general user community (the needs of the many).
- The program was then written, tested, and released to users.
- In the months following the initial release, the program evolved based on user requests for added features. These requests were carefully evaluated and an attempt has been made, whenever possible, to interpret user requests in such a way that when the feature was added to the software, it was added in a sufficiently general way so as to be useful in many situations, not just for the particular problem that particular user was trying to solve at that particular time.

4. Internal structure

Structures vs. COMMON Statements In IDP3, an attempt was made to write the program without the use of common statements. This requirement was satisfied although it proved unnecessary. Common statements

---

[1] IDL is the trademark of Research Systems, Inc.

cause problems when multiple copies of the same program are run from inside a given instance of IDL. However, because named structures were used, multiple copies of IDP3 should not be run together. To make IDP3 truly self-independent, the named structures should be replaced with anonymous structures, that is, when structures are allocated, they should be inline in the program rather than being referenced from an external file, i.e., idp3_structs. This would make the code look a bit sloppier (in our opinion) but would allow the user to run multiple copies of IDP3 simultaneously from a single IDL session without unwanted interactions between copies.

Structure and Pointer usage The first parts of the program written were the data structures. Pointers were used to allow easier memory management and more elegant structures and information passing.

Root structure, Updates, Coordinating windows With a number of interactive windows on the screen, the event handlers for these windows must coordinate in such a way as to keep the internal data up to date and valid at all times.

Active windows, Automatic updates Some of the windows on the screen display current internal data values. When any of this internal data changes, the displayed values and/or displayed 1D and 2D plots must be automatically updated to reflect these changes.

Flags and Signals The various data structures in the program contain a variety of variables used to pass current values of flags around inside the program. Some of these flags can be thought of as signals to other parts of the program.

Event handling Various event handlers handle events from various windows. Sometimes, events from a particular widget must be turned off to allow another subroutine to turn on and interpret events for that widget. This is a way of overloading events for a particular widget.

5. Conclusions and Future Plans

The time invested in the software engineering of IDP3 was time very well spent. Advancements in IDL Version 5, particularly the support for data pointers, made it possible to develop a very solid but flexible framework for IDP3 that easily facilitates future enhancement.

IDP3 has been a huge success with members of the NICMOS Project and continues to be enhanced according to user requests. It may also be evolved as part of a suite of analysis routines for other NASA missions including the Imaging Science Subsystem of the Cassini Mission to Saturn.

IDP3 is available from the NICMOS website[2] at the University of Arizona.

---

[2] `http://nicmos.as.arizona.edu/software`

6. Appendix A

LYTLE'S GENERAL PRINCIPLES OF INTERACTIVITY

I User Help

- Online help will be available.
- Make active button and field help (cursor tracking events) available and selectable if possible.

II General computer display layout considerations

- Multiple windows for flexibility
- Resizable graphics where appropriate
- Limit number of windows so they don't overfill the display

III Active window updates

- If the window layout must change with the update, kill and redraw the window
- If the window layout remains the same with the update, just update the fields in the window and/or replot the 1-D or 2-D data.

IV 'Modes' considered harmful

- When possible, use separate mouse buttons to perform multiple input functions in a window.
- Avoid using 'modes' where mouse buttons perform different functions depending on which 'mode' button has been pressed.

V Pull-down menu buttons vs display screen real estate

- If an application involves image display and lots of function buttons, it is often best to put the function buttons in pull-down menus, including walking menus.

VI Grey out buttons when their use would be inappropriate.

VII Use 'bulletin board' bases for differentiating grouped functions.

VIII Implement 'user preference' files when appropriate.

IX Do not build a GUI that requires changes to the user's .Xdefaults file.

X Catch errors and report them rather than allowing the program to bomb.

Part 11. Software Systems and Tools

Astronomical Data Analysis Software and Systems VIII
ASP Conference Series, Vol. 172, 1999
D. M. Mehringer, R. L. Plante, and D. A. Roberts, eds.

Starlink's Software Distribution System

Martin J. Bly

Starlink Project[1,2]*, Rutherford Appleton Laboratory, Chilton, DIDCOT, Oxfordshire OX11 0QX, United Kingdom. Email: bly@star.rl.ac.uk*

Abstract. The Starlink Software is a collection of astronomical packages and utilities comprising over 120 separate components. When built, the system needs between 500 and 650 MB disk space (without source) making it ideal for distribution on CD-ROM. Each component has been individually packaged into an easy-to-install bundle for users to obtain over the Internet via a "point and click" WWW interface from the Starlink Software Store.

1. Introduction

The Starlink Software is a large collection of applications, subroutine libraries and utilities for processing and reducing astronomical data. Traditionally the Starlink Project distributed the collection via magnetic tape and frequent network updates. However, with three different hardware and operating system types to support and the increasing size of the collection and frequency of updates, the Project decided to offer two releases per year and move over to using CD-ROM as the primary distribution medium, backed up by a repository of the packages which is accessible from the WWW.

CD-ROM offers the convenience of speed of access, random access to software on disk and a more reliable transport medium. Its capacity of 650 MB is ideally suited to offering a complete software system that can be run directly from the CD-ROM. The WWW offers ease of use and user-selection of packages for an installation tailored to a user's requirements.

2. Starlink CD-ROM Distribution

The Starlink Software is supported on three different hardware/operating systems: Alpha/Digital Unix, PC/RedHat Linux and SPARC/Solaris. A set of CD-ROMs is issued twice a year, in Spring and Autumn. Each issue consists of one CD for each system containing the directly runnable software, and an extra Updates CD-ROM which may be used to tailor installations to specific data

[1] Managed by the Council for the Central Laboratory of the Research Councils on behalf of the Particle Physics and Astronomy Research Council

[2] http://www.starlink.rl.ac.uk/

processing and reduction needs. The software on the directly runnable disks may also be copied to hard disk for faster access.

Each CD has brief instructions on the inlay card, and has a more detailed installation note in an `INSTALL` file on the CD itself. In addition, there is a comprehensive guide to the CD-ROMs: *Starlink Software CD-ROMS User's Guide*. This document[3] contains detailed instructions to enable users to use the tools provided on the CD-ROMs to install complete copies of the Starlink Software or create tailored installations.

Starlink CD-ROMs are distributed by post from the Rutherford Appleton Laboratory (RAL). Users should send an email request to the distribution request address at `cdrom@star.rl.ac.uk` and should include their name, post and email addresses, and details of which CD-ROMs are required. Each CD set dispatched includes a copy of the User's Guide.

3. Starlink Software Store

To enhance ease of access to individual components of the Starlink Software Collection the Starlink Project has produced the Starlink Software Store[4]. The Software Store is a WWW-based "point-and-click" series of menus linked to some CGI scripts and a repository of package distribution sets, all residing on the Starlink WWW server at RAL.

To use the Software Store, users start at the top page and follow the hyperlinks to the index pages. These list the inventory of packages available in three sets: Subroutine Libraries, Applications and User Interfaces, and Utilities. There is also a Latest News page which gives details of the most recent CD issues.

When a user selects a package to download, he or she is presented with a form which asks him to chose which of the available hardware/software combinations he wants. Information stored within the Software Store system at RAL also allows the Store scripts to tailor the form to offer a menu of addition packages that might be needed to run or build the software chosen. The user may opt to obtain the additional packages then, or can come back and get them individually later. In the case of applications packages which require parts of the Starlink infrastructure software in order to run, a second form is presented which gives the user the opportunity to download the required components.

Once the selection process is complete, a request is sent to a CGI script on the server at RAL which makes a bundle of the components for download. The bundle is a self-unpacking shell script with binary components `uuencoded` within it. The bundle is offered to the browser for download and the user saves it to disk with an appropriate name.

To unpack the bundle (script), the user runs it. The bundle unpacks itself to produce the compressed tar file distribution sets and a script which is used to complete the installation. At each stage the user is asked questions about the intended installation configuration to allow the unpacking of the compressed tar

---

[3] `http://www.starlink.rl.ac.uk/star/docs/sun212.htx/sun212.html`

[4] `http://www.starlink.rl.ac.uk/cgi-store/storetop`

files and the installation process to proceed. The installation script may also be used to uninstall the software.

The applications and utilities are available in ready built form without source, for the three supported systems. The subroutine libraries are available ready built with and without source, and in source only form. The distribution sets for the Starlink applications can be quite large, up to some tens of megabytes in some cases, so the download bundles are also quite large. It may not be possible for sites with poor or slow network connectivity to obtain such large packages via download in one try, so the Store also provides "point-and-click" access to the components of a bundle via standard FTP.

The repository of package distribution sets is updated twice yearly to coincide with the CD-ROM distributions. Individual distribution sets are updated to provide major bug fixes between CD-ROM releases.

4. Conclusion

Together, the Starlink Software Store and the CD-ROM distributions provide an efficient, easy to use and reliable method of distribution for multi-component software collections such as the Starlink Software.

Astronomical Data Analysis Software and Systems VIII
ASP Conference Series, Vol. 172, 1999
D. M. Mehringer, R. L. Plante, and D. A. Roberts, eds.

Design Study on Integration of StarView II and JIPA Using Advanced Java Mechanisms

Markus Dolensky

Space Telescope – European Coordinating Facility, Karl-Schwarzschild-Str. 2, D-85748 Garching, Germany

Bruce Mayhew, Bridget Kennedy

Space Telescope Science Institute, 3700 San Martin Drive, Baltimore, MD 21218

Abstract. This case study examines the use of the latest software techniques for merging independent Java applets and applications. STScI's new Java interface to the archive, StarView II, will make use of the ST-ECF stand-alone FITS image viewer JIPA. We discuss how state-of-the-art software techniques like Remote Method Invocation (RMI), multi-threading, and the Reflection API influence the design of a common interface between StarView II and JIPA. Furthermore, it is demonstrated how the JDBC interface will provide the two applications with concurrent access to the archive.

1. Merging Java Components – StarView II & JIPA

STScI's prototype of the new Java interface to the archive, Starview II (SV II) makes use of ST-ECF's[1] stand-alone FITS image viewer JIPA[2]. The interface between these independent Java components is based on two design patterns:

- Observer Pattern (Fig. 1)
- Design Inheritance (Fig. 2)

2. Observer Pattern

The intent of the observer pattern is a way to define a one-to-many dependency between objects so that when one object changes state, all its dependents are notified and update automatically (see, for example, `http://www.cs.wustl.edu/~schmidt/cs242/patterns.html`).

[1] `http://ecf.hq.eso.org/`

[2] `http://archive.eso.org/java/jipa/`

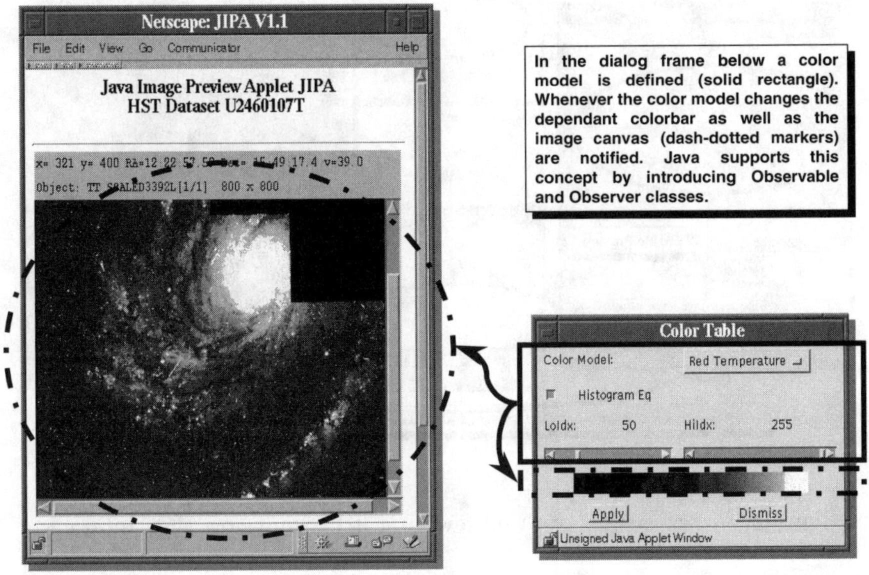

Figure 1. Snapshot of JIPA using the Observer Pattern

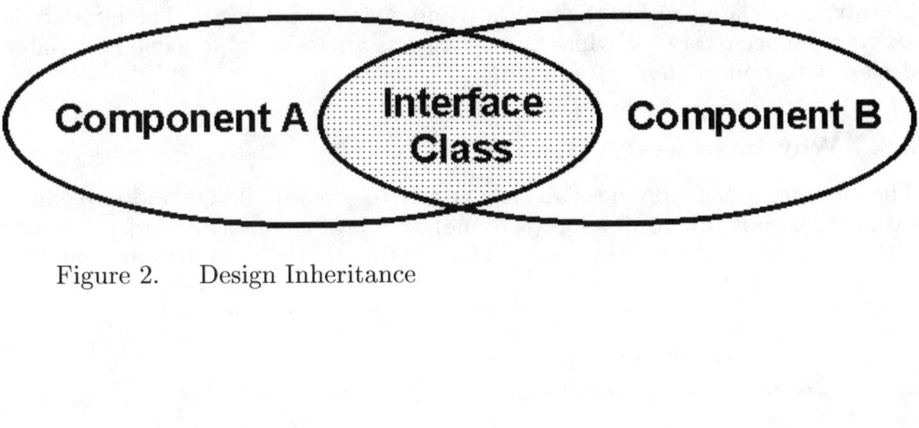

Figure 2. Design Inheritance

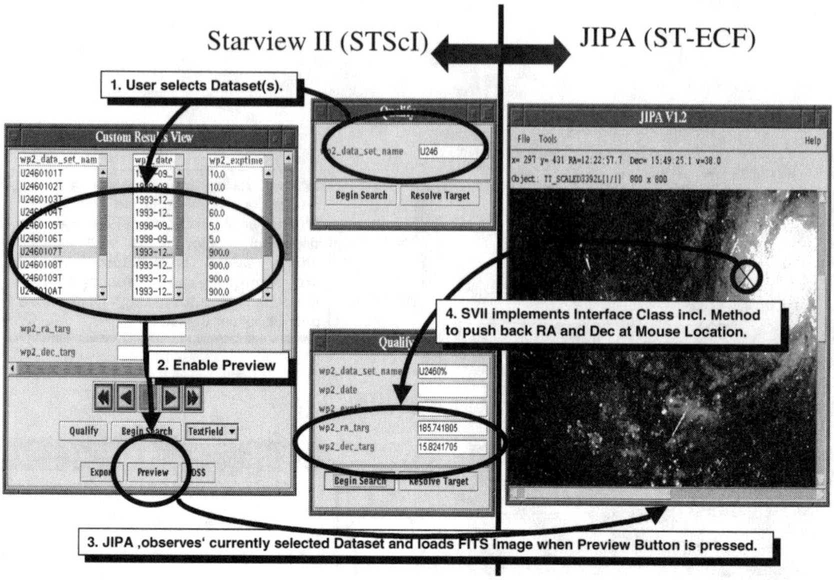

Figure 3. Screenshot of StarView II & JIPA and how they interact

3. Design Inheritance

3.1. What Are Java Interfaces?

An interface is a collection of method definitions (like C function prototypes) and constant values (Campione & Walrath 1998)[3].

3.2. Implementing and Using Interfaces

A class implementing an interface provides a method implementation for all the prototypes declared within the interface and its superinterfaces. The interface is a new reference data type which can be used as any other type name in variable declarations and method parameters.

3.3. Why Interfaces?

The interface scheme provides an elegant way to separate design and implementation. It is possible to develop (compile) two Java components independently – they just share the interface source file containing the constants and method declarations (Fig. 2).

[3] `http://java.sun.com/docs/books/tutorial/java/more/interfaces.html`

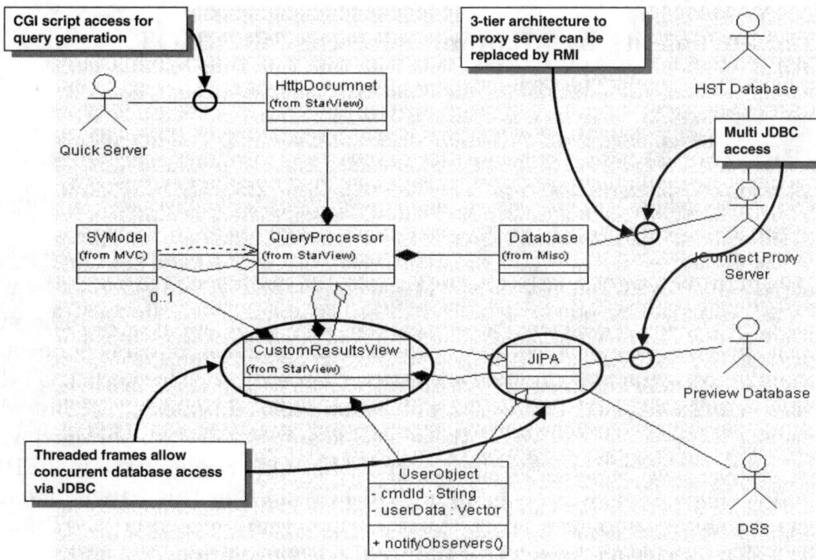

Figure 4. Server Access Architecture of StarView II

4. Server Access Architecture

The archive browser StarView II has various interfaces (Fig. 4). Apart from the user interface, there is the Quick Server which translates queries into SQL statements. Furthermore, there is the JDBC interface to the actual database, and finally there are two more interfaces which communicate with other Java components as described above. Again, these two are the observer/observable pattern (implemented as class UserObject) and design inheritance. As illustrated, JIPA has its own means to access images from the preview DB and DSS catalogue.

Currently, a mixture of CGI (Common Gateway Interface) and JDBC (Java Database Connectivity) is used for network connections. In the future, the three-tier JDBC architecture to a proxy server, which is imposed by Java security restrictions, might be replaced by RMI (Remote Method Invocation).

References

Campione, M. & Walrath, K. 1998, The Java Tutorial (2nd ed.), (Reading: Addison-Wesley), 145

Astronomical Data Analysis Software and Systems VIII
ASP Conference Series, Vol. 172, 1999
D. M. Mehringer, R. L. Plante, and D. A. Roberts, eds.

Enhancements by Mapping with the ISOPHOT Interactive Analysis (PIA)

Carlos Gabriel

ISO Data Centre, ESA Astrophysics Division, Villafranca del Castillo, Spain

Abstract. The Infrared Space Observatory (ISO), launched by ESA in November 1995, ended its operations successfully in April 1998. The subsequent post-operational phase, which will last for 3.5 years, is devoted mainly to re-analysis and re-calibration of the data. The ISOPHOT Interactive Analysis (PIA) is not only the main calibration tool but also the software tool of choice for the scientific analysis of data obtained with the imaging spectrophotopolarimeter ISOPHOT. Some of the main PIA enhancements related to mapping are presented in this paper, namely, the development of new algorithms for the creation of ISOPHOT maps as well as a sub-package for beam profile deconvolution.

1. Introduction

The end of the ISO mission (Kessler et al. 1996) in April 1998 due to Helium boil-off does not mean the end of the scientific work related to the data gathered by the scientific instruments on board. This is true for all four ISO instruments. ISOPHOT (Lemke et al. 1996), the imaging spectrophotopolarimeter of the mission, covered the full ISO wavelength range (2.5-240 μm) and was composed of several sub-instruments. The main software tool for data analysis used both for calibrating ISOPHOT as well as for producing scientific results is the ISO**P**HOT **I**nteractive **A**nalysis (PIA), referred to in several earlier ADASS conferences (Gabriel et al. 1997a, 1997b, 1997c; Gabriel, Acosta-Pulido, & Heinrichsen 1998), in which different aspects of this package and its development were presented. PIA is entirely based on the Interactive Data Language (IDL). The ISO Data Centre at Villafranca, Spain, the central institution for maximising exploitation of the ISO data of a network of data centres in Europe and the USA, is also responsible for the further development and maintenance of PIA.

The PIA capabilities were increased continuously during and after the operational phase of the ISO mission. One of the main reasons for the continuing development is due to the feedback received from astronomers in the more than 150 scientific institutes around the world. The other reason is due to the deeper understanding of the very complex ISOPHOT instrument and all its different operation modes by the ISOPHOT instrument experts, who are steadily increasing the reliability and accuracy of the calibration (Schulz et al. 1999; Klaas et al. 1997). The continuous improvement is reflected in new releases of the package which are freely distributed to the users via the PIA homepage[1]. The homepage also contains documentation, help libraries, publications on PIA, FAQs, a mailbox for receiving bug reports, comments, etc.

Here we will concentrate only on the latest improvements related to mapping with PIA. A very recent article (Gabriel & Acosta-Pulido 1999) summarizes the most important enhancements of PIA in the last year.

2. Alternative Mapping Algorithms in PIA

A detailed description of the different ISOPHOT mapping modes can be found in the ISOPHOT Observer's Manual (Klaas et al. 1994) while the basic mapping capabilities of PIA are described in Gabriel et al. 1997c.

ISOPHOT performs mapping for all the different detector subsystems, namely, the single pixel photometers (P1, P2, P3) in the range 2.5-100 μm with a variable aperture diameter(5″-180″), the far infrared detector arrays C100 (a 3×3 array with a pixel size of 43.5″×43.5″) and C200 (a 2×2 array with a pixel size of 89.4″×89.4″) in the range 50-240 μm, and the spectrometer grating array with 128 pixels in the range 2.5-11.7 μm with an entrance aperture of 24″×24″.

Mapping was done basically using the raster capability of the ISO spacecraft. Distances between raster legs and raster points were freely specified (under certain restrictions) leading to different oversampling factors. For the C100 and C200 long wavelength detector arrays, the observer could also take advantage of the chopper for increasing the spatial resolution in a special observing mode. The spacecraft's jitter during a raster observation adds a further complication. The measurements of single sky positions are therefore not regularly gridded, and the spatial resolution is a complex function of all the elements mentioned above as well as the wavelength of the filter which was used.

PIA offers different methods for obtaining a map from an observation which are included in the general mapping GUI which are summarized below.

1. The "full coverage" mapping algorithm. The original mapping algorithm used in PIA ("full coverage method") is based on a simple gridding function for binning the signals obtained with the individual detector pixel beams[2] at different raster/chopper positions into map pixels. Every sky position is observed several times by an individual pixel in the normal case of oversampling and by different detector pixels when mapping with one of the detector arrays. Due to oversampling, the achieved spatial resolution can be higher than the simple detector beam. Therefore, the map pixel size (freely choosable by PIA) can be smaller than the beam.
   The contribution of the signal to a map pixel is calculated using the overlap area of beam and map pixel. Co-adding the information from the different beams into an image element results in a smearing trapezoidal function which depends on the distance between beam center and center of the computed map pixel. This has the benefit of smoothing the noise but the

[1] `http://www.iso.vilspa.esa.es/manuals/PHT/pia/pia.html`

[2] The term "detector beam" refers to the detector pixel size in the case of an array detector, or to the aperture size in the case of a single pixel detector (P1, P2 or P3).

disadvantage of reducing the probability of observing point sources with low signal to noise ratios.

2. The "distance weighting" algorithm. This method is just a variation of the "full coverage" mapping algorithm, but instead of taking the area of overlap between beam center and map pixel center for calculating the contribution, a different method is used. For a given map pixel, every contributing signal is weighted using the distance between beam center and center of the map pixel. The individual weighting factor is given by $(1-D)^3$ with D being the distance in beam size units (the exponent 3 was empirically found to produce good results).
3. The "trigrid interpolation" In this method, the central positions of all the signals from the different pointings are used as a planar set of points which are then Delaunay triangulated[3]. The main property of Delaunay triangulation is that the circumcircle of any triangle contains no other vertices in its interior. After having obtained the triangulation, a regular grid of interpolated values is obtained in which only the nearby points are used[4]. Linear or smooth quintic polynomial interpolation can be selected. Since every sky position is generally observed several times within the raster map, re-sampling is necessary prior to applying the triangulation. This is done by co-adding all signals which have beam centers within a certain radius (e.g., 1/4 of the chosen map pixel size) and assigning to the co-added value a sky position, which is calculated from the average position of all contributing signals.

Both methods 2) and 3) achieve higher spatial resolution but increase the noise. Although this statement is qualitatively clear, it is not easily quantified. Only detailed simulations, which take into account the very different possibilities for mapping with the diverse ISOPHOT sub-instruments and observing modes, can give an answer. These studies have already started and will be conducted in the near future together with the analysis of real data for statistical comparisons of the results achieved by the different algorithms.

3. Map Deconvolution

Footprint matrices have been established for the ISOPHOT filters, by folding the corresponding point spread function with the optical aperture. Their validity has been confirmed by comparison with data from very detailed raster scans. The establishment of the footprint matrices make possible a reliable deconvolution of the maps by using deconvolution techniques such as the Maximum Likelihood Method or the Maximum Entropy Method.

Since the application of deconvolution techniques requires a high level of interaction for judging the obtained results, PIA has included a GUI for deconvolution. Using this GUI, the different deconvolution methods can be chosen, the footprint matrix used can be displayed, initial parameters for the deconvolution can be changed, and the results can be inspected.

---

[3]as performed by the IDL routine TRIANGULATE, IDL Ref.Guide V5, p922

[4]as performed by the IDL routine TRIGRID, IDL Ref. Guide V5, p.925

4. Summary

New algorithms for the creation of ISOPHOT maps and a sub-package for their deconvolution have been developed within the framework of PIA. Testing using simulated data as well as using a large sample of real data is necessary for a final quantitative assessment. This is facilitated through the flexibility of the user-friendly GUIs.

References

Gabriel, C., & Acosta-Pulido, J. 1999, Proc. of The Universe as Seen by ISO, (ESA SP-427), ed. P. Cox & M. Kessler, in press

Gabriel, C., Acosta-Pulido, J., & Heinrichsen, I. 1998, in ASP Conf. Ser., Vol. 145, Astronomical Data Analysis Software and Systems VII, ed. R. Albrecht, R. N. Hook, & H. A. Bushouse (San Francisco: ASP), 165

Gabriel, C., Acosta-Pulido, J., Heinrichsen, I., Morris, H., & Tai, W.-M. 1997a, in ASP Conf. Ser., Vol. 125, Astronomical Data Analysis Software and Systems VI, ed. G. Hunt & H. E. Payne (San Francisco: ASP), 108

Gabriel, C., Acosta-Pulido, J., Kinkel, U., Klaas, U., & Schulz, B. 1997b, in ASP Conf. Ser., Vol. 125, Astronomical Data Analysis Software and Systems VI, ed. G. Hunt & H. E. Payne (San Francisco: ASP), 112

Gabriel, C., Heinrichsen, I., Skaley, D., & Tai, W.-M. 1997c, in ASP Conf. Ser., Vol. 125, Astronomical Data Analysis Software and Systems VI, ed. G. Hunt & H. E. Payne (San Francisco: ASP), 116

Kessler, M. F., et al. 1996, A&A, 315, L27

Klaas, U., Krüger, H., Heinrichsen, I., Heske, A., Laureijs, R. (ed.), 1994, ISOPHOT Observer's Manual, ISO Data Centre

Klaas, U., et al. 1997, Proc. of the First ISO Workshop on Analytical Spectroscopy, (ESA SP-419), ed. A. Heras, K. Leech, N. R. Trams, & M. Perry, 113

Lemke, D., et al. 1996, A&A, 315, L64

Schulz, B., et al. 1999, Proc. of The Universe as seen by ISO, (ESA SP-427), ed. P. Cox, & M. Kessler, in press

Astronomical Data Analysis Software and Systems VIII
ASP Conference Series, Vol. 172, 1999
D. M. Mehringer, R. L. Plante, and D. A. Roberts, eds.

DISH: The Single Dish Environment in AIPS++

Robert W. Garwood, Joseph P. McMullin

NRAO[1], *520 Edgemont Road, Charlottesville VA 22903*

Abstract. We present a short description of the single dish analysis environment in AIPS++, or DISH. DISH is available in the recent third AIPS++ beta release. DISH and AIPS++ will be the primary analysis tools for data from the Green Bank Telescope. Snapshots of DISH in use are presented. The initial focus of DISH has been on traditional single dish spectral line data analysis. Plans for additions to DISH both in tools for 1-D data analysis as well as single dish imaging are described.

1. Introduction

AIPS++[2] is a system for astronomical data processing which is now in its third beta release. The first full release is expected in 1999. The heart of AIPS++ is a powerful command line interpreter called glish (Paxson & Saltmarsh 1993; Schiebel & Paxson 1998).

DISH is a collection of glish scripts and clients which provides an environment within AIPS++ intended to be used for single dish radioastronomy data analysis. Its initial aim is to be a worthy replacement for traditional single dish analysis programs such as UniPOPS. Eventually it will provide access to more advanced data calibration and imaging utilities which will share the same basic design as the synthesis calibration and imaging utilities in AIPS++. The primary design goal has been to provide a graphical user interface that is intuitive, unsurprising, and responsive. Users should feel that results go to obvious places and, whenever possible, are displayed as they occur. Some uses of the single dish aspects of AIPS++ are reported by McMullin, Garwood, & Minter (1999).

2. The DISH Graphical User Interface

Figure 1 shows the major components of the DISH GUI. The Results Manager is the core of DISH. All glish variables which DISH creates in response to user actions are stored in the Results Manager and listed in the `Variable` window. These variables are all available at the glish command line so the user can interact with them. DISH immediately plots the currently selected variable if

---

[1] The National Radio Astronomy Observatory is a facility of the National Science Foundation operated under cooperative agreement by Associated Universities, Inc.

[2] `http://aips2.nrao.edu`

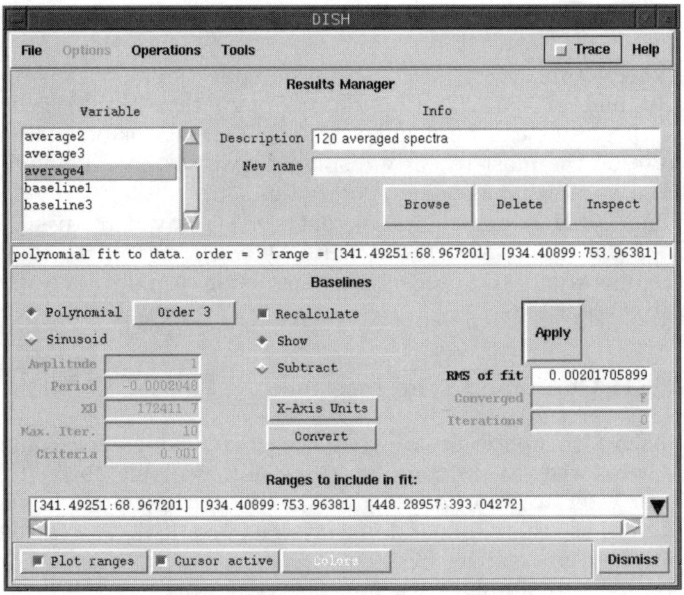

Figure 1. DISH: Showing the Results Manager and one Operation.

it can be plotted. A description is associated with each variable. The user can change the description or the name of a variable in the Results Manager. Certain variables can also be browsed.

The Results Manager recognizes three different types of variables. Individual scans are two-dimensional data arrays plus any associated information (including analysis history information). The first dimension of the data array can be anything but is typically a frequency or velocity axis while the second dimension is a Stokes axis. This type of data is called an SDRECORD. These can be displayed as line graphs and the associated information can be browsed. The second type of variable is a collection of SDRECORDS. These can also be browsed, but not displayed. The selection operation in DISH applies user defined selection criteria to such a collection, producing a new collection. The third type of variable is everything else. The Results Manager does little more than keep track of this data for the user. Users can store any variable in the Results Manager from the glish prompt. Other types of special purpose data types may be added in the future.

The Menu bar at the top of the DISH GUI provides access to the File, Operation, Help and other features of DISH. New data files are created and existing data files are opened through the File menu. The GUI panels for each available operation are enabled and dismissed through the Operations menu. Multiple operation panels may be present at any time although screen size limitations are generally such that only a few will typically be displayed at one time. When enabled, each operation panel appears at the bottom of the DISH window. The operation shown here is for fitting baselines to the currently displayed spectra. The most recent status message is shown in the text immediately below the Results Manager. Messages displayed here are also echoed to the AIPS++ logger.

Operations within DISH may operate on collections of scans or they may operate on individual scans. All operations place their results in the Results Manager. If the result can be displayed, it is. Operations on individual scans always operate on the most recently displayed scan so that, hopefully, the user is not confused as to what is being operated on.

Each time DISH is exited, its current state is saved to disk. This state includes all of the contents of the Results Manager as well as the parameter values for the operations and user interface preferences. DISH is restored to this state when it is restarted.

3. The DISH Command Line Interface

Underlying each GUI operation are glish commands which are available to the user. Users who wish to do complex tasks not available through the DISH Graphical User Interface or who want to capture existing operations into a glish script will need to use this command line interface to DISH. The trace tool, when active, shows these underlying glish commands as each GUI operation occurs. These can be used to simply learn how the GUI works or the text from that window can be copied to an editor and used as the basis for a glish script.

All of the values in the Results Manager can be manipulated at the glish command line. These results can then be returned to the Results Manager. A simple example which illustrates this is given below.

```
rawdata.setlocation(1)                      # 1
on := rawdata.get()                         # 2
rawdata.setlocation(3)                      # 3
off := rawdata.get()                        # 4
result := on                                # 5
result.data.arr :=                          # 6
   (on.data.arr - off.data.arr)/
   off.data.arr
dish.rm().add('result'                      # 7
              'Difference of rows 1 and 3',
              result,
              'SDRECORD')
```

The goal in this example is to reduce an on-off total power scan pair by extracting the "on" and the "off" source scans from an already opened data set, constructing the difference scan, and inserting the result in the Results Manager. In this example, `rawdata` is a previously opened collection of SDRECORDs available in the Results Manager. In `rawdata`, an on-source scan is located at the first record and an off-source scan is located at the third record. Line 1 makes sure that `rawdata` points at the first record. The data are then retrieved from that record and assigned to `on`. This is repeated for the off-source data which are assigned to `off`. The result is initialized by simply making `result` a copy of `on`. This sets the data structure and header words appropriate for these data. The difference of the two data arrays divided by the `off` data array is calculated in line 6. Finally, in line 7, `result` is added to the Results Manager. The final argument identifies this result as an SDRECORD. It is then available to be plotted automatically when the user selects it.

4. Future Plans

The first public release of DISH in AIPS++ occurred in the most recent AIPS++ beta release. We are now responding to user feedback. For the first full release of AIPS++, operations such as multi-component fitting and FFTs will be added to DISH. The current underlying data format used by DISH is a simple AIPS++ table created by direct conversion of single dish FITS data. By the first release of AIPS++, DISH will be able to interact with an AIPS++ MeasurementSet, the AIPS++ table structure used to hold astronomical data. The command line interface needs to be cleaned up so that users can more easily insert their own specialized operations into DISH.

The long range plan for single dish calibration and imaging in AIPS++ is to follow the same model used for synthesis data within AIPS++ as discussed by A. Kemball at this conference. An initial test of single dish calibration is expected in early 1999. An on-the-fly single dish imaging tool is planned for later in 1999.

References

McMullin, J. P., Garwood, R. W., & Minter, A. H. 1999, this volume, 466

Paxson, V. & Saltmarsh, C. 1993, Proc. of the 1993 Winter USENIX Technical Conference (Berkeley: USENIX Assoc.), 141

Schiebel, D. & Paxson, V. 1998, "The Glish 2.7 User Manual", National Astronomy Radio Observatory, `http://aips2.nrao.edu/aips++/docs/reference/Glish/Glish.html`

Astronomical Data Analysis Software and Systems VIII
ASP Conference Series, Vol. 172, 1999
D. M. Mehringer, R. L. Plante, and D. A. Roberts, eds.

Some Astrophysical and Instrumental Applications of AIPS++

Joseph P. McMullin, Robert W. Garwood

NRAO, 520 Edgemont Road, Charlottesville, VA 22903

Anthony H. Minter

NRAO, P.O. Box 2, Green Bank, WV 24944

Abstract. We present several applications of the AIPS++ environment as applied to the development of the Green Bank Telescope as well as current astrophysical problems. In particular, we demonstrate the use of AIPS++ in building tools for analyzing data at various points along the data stream, from diagnosing receivers and calibration, to data reduction and scientific modeling.

1. Introduction

AIPS++ (Astronomical Information Processing System, written in the C++ language) is an ambitious project to standardize the processing of (primarily radio) astronomical data, both interferometric and filled aperture. Currently under development by an international consortium of observatories, AIPS++ is now under beta release, with several other 'experimental' releases available to consortium organizations.

In particular, the Green Bank Telescope (GBT) project requires AIPS++ as an analysis environment. To accommodate and facilitate the GBT development, several sets of tools have been fashioned within AIPS++ at several points along the data path, from general calibration routines and data quality assessment to analysis and modeling. The comprehensive suite of mathematical applications available under the beta release offers a facile means of interacting and manipulating data for scientific purposes. In the following section, we detail some of the applications of AIPS++ for the GBT project and point to the future.

2. Current Usage

The following examples illustrate some of the many uses of AIPS++ at NRAO Green Bank. The majority of these examples were written in the Glish language (e.g., Paxson & Saltmarsh 1993; Schiebel & Paxson 1998) which is the main CLI to AIPS++ and allows access to the various libraries and clients within.

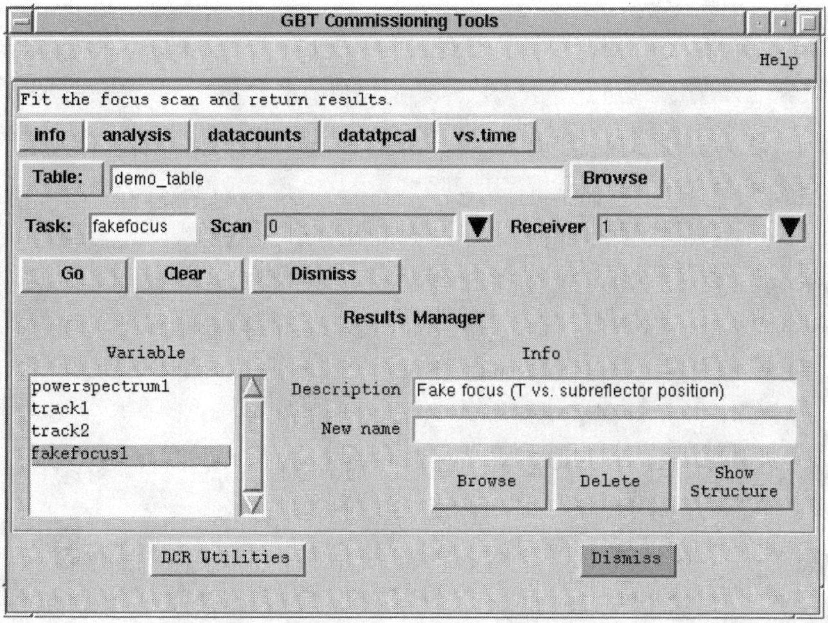

Figure 1. Example Commissioning Tool GUI.

2.1. Data Calibration

A series of tools have been created to aid in understanding and calibrating the equipment involved in taking observations.

gbtlogview A diagnostic tool for engineers to assess and monitor the receivers, and weather stations. It is composed of a GUI interface which allows selection of a series of devices over a selected time range which are subsequently 'filled' to an AIPS++ table. The plotting interface allows multiple columns to be intercompared.

bandpass analysis Several of the GBT backends are first being tested on the NRAO 140-foot telescope. Total power integrations of different continuum calibrators at many frequencies are being obtained for comparison with GBT performance and as diagnostic tests of the hardware. The goal is to be able to quantitatively describe the improvements of an unblocked aperture on the non-random frequency structure of instrumental baselines.

continuum commissioning An in-house commissioning package has been written in Glish using the AIPS++ modules to provide a set of standard tasks for characterization and commissioning of the telescope. Tasks include: T_{sys} calculation, T_{Ant} calculation, τ_{zenith} calculation, focus measurements, pointing measurements, and power spectra calculations. In addition, plotting of all quantities (e.g., individual phase data) is possible. A pointing model module is currently under development.

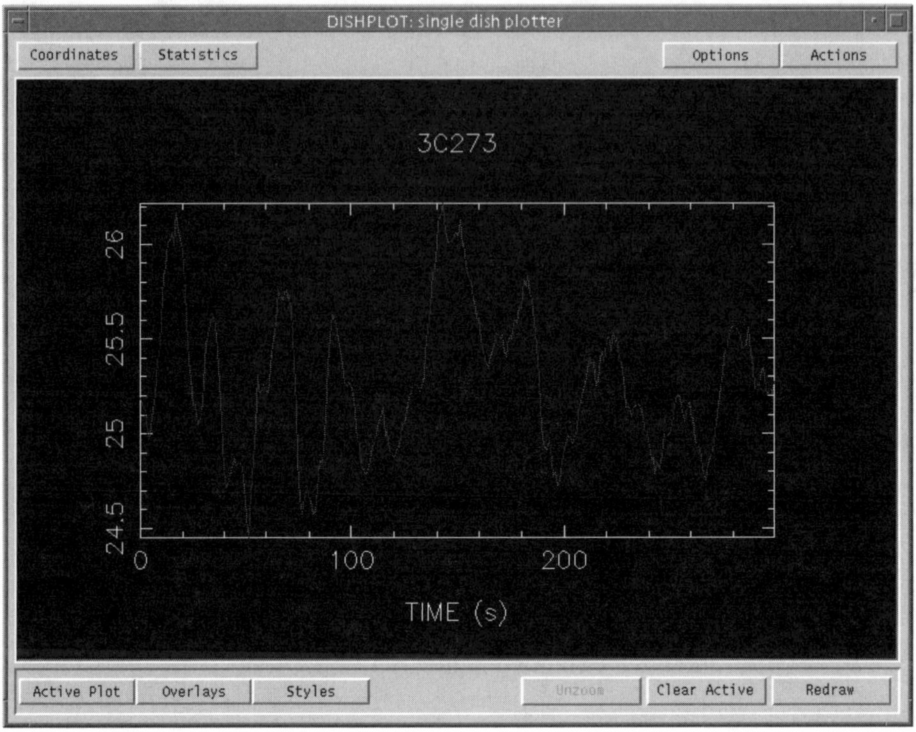

Figure 2. Example Commissioning Tool Plot. Display shows plot of a tracking scan on 3C273.

2.2. Data Quality

Several tasks have been developed for general assessment of data quality during observing.

monitoring radiometer observations An 86 GHz tipping radiometer is installed in Green Bank for monitoring atmospheric conditions. This data is analyzed every hour through execution of an AIPS++ program. The program writes out information in HTML format for updating a web page of information. Current and archived data are available along with an atmospheric transmission curve over a relevant range of frequencies.

monitoring and archiving radio frequency interference Low frequency observations are particularly plagued by radio frequency interference. In response, an archive tool for submitting interference events and allowing table searches to aid identifying these events has been created. The availability of Tk widgets within the Glish programming environment along with sophisticated table querying within AIPS++ make this possible.

2.3. Science

Although AIPS++ is only in Beta testing, the large number of analysis tools allows a fairly high level of analysis. Some examples follow.

DISH The DISH software, a single-dish subset of AIPS++, is also under Beta release. Data from several telescopes have been loaded into the system and reduced (see Garwood & McMullin 1999).

Pulsar Tools Several general tools for pulsar analysis are also under development. These are intended to be generic tools for displaying and calculating basic parameters (e.g., pulse time-of-arrival, dynamic spectra, de-dispersal).

Modeling Gas Structure in the Interstellar Medium The suite of mathematical functions allows simulating a data cube with a Kolmogorov spectrum of turbulence. This data cube can then be used to simulate observations; in particular, a model was used to compare observed properties in Molecular Clouds to discern statistical biases in different types of analysis probing molecular cloud structure. These results are currently in preparation (Minter & McMullin 1997).

3. Future Directions

With the GBT expected to be delivered in late 1999, there will be increased development in the coming year. In particular, we are in the midsts of developing the on-line reduction system for the GBT as well as aiding in the holographic measurements.

References

Garwood, R. W. & McMullin, J. P. 1999, this volume, 462

Minter, A. H. & McMullin, J. P. 1997, BAAS, 191, 2005

Paxson, V. & Saltmarsh, C. 1993, Proc. of the 1993 Winter USENIX Technical Conference (Berkeley: USENIX Assoc.), 141

Schiebel, D. & Paxson, V. 1998, "The Glish 2.7 User Manual", National Astronomy Radio Observatory, `http://aips2.nrao.edu/aips++/docs/reference/Glish/Glish.html`

Astronomical Data Analysis Software and Systems VIII
ASP Conference Series, Vol. 172, 1999
D. M. Mehringer, R. L. Plante, and D. A. Roberts, eds.

Automated Programming for Physics and Astrophysics

Michele Maris

Astronomical Observatory of Trieste (OAT), Via G.B. Tiepolo 11, I-34131, Trieste, Italy

Maria Staniszkis

Mathematics and Informatics Department, Udine University, Via delle Scienze 206, I-33100, Udine, Italy

Abstract. The present communication describes the effort to develop a model of Automated Program Generator (APG) proposed for the production of the simulation software for the ESA Planck/LFI mission.

1. Introduction

Considering that the core of a simulation project is the production of the simulation software, the problems related to its creation, development and documentation represents a major part of the whole work. In order to facilitate these tasks, in the past years many tools have been proposed (Cook, Painter, & Brown 1991; Cook & Painter 1992; Brown 1992; Dubois 1994; Dubois & Yang 1996; Beazley & Lomdahl 1997), but from the user's point of view the ideal solution would be one in which the computer is given an abstract description of the required simulation and is left to the task of producing the code. This is possible if most operations required to support and execute a simulation could be standardized, allowing the related code to be specified by a high level language; this leads us to *Automatic Programming* and *Program Generation* concepts. A sort of "software robot" called the *Automated Program Generator* (*APG*) is presented, which generates simulation software starting from a specification given by the user. Moreover, if we suppose that the simulations are partitioned into *simulation classes*, it is possible to build a library of standard reusable simulations.

The approach to automatic programming presented here was first proposed for the Planck/LFI collaboration (Mandolesi et al. 1998) where it was applied to explore solutions to the on-board data compression problem (Maris et al. 1999), but in its simplest form it was also applied to solar neutrino (astro)physics giving excellent results (Maris 1999).

2. A Simulation Model

The design of an APG system has to be founded on a *model of scientific simulations* whose role is to define both the basic components of a simulation project and their purposes and relations in the management of the simulation cycle.

Non-interactive scientific simulations are considered in the present communication: the *simulator* receives as input a sequence of simulations to be executed and, without any interaction with the user, produces as output numerical results in forms of tables, data files, or graphs. The real world *experiments* may use the same or slightly modified *experimental apparatus* and varying *experimental conditions* often in dependence from obtained *results.* Considering a simulator run a *computer world experiment,* we can view the *input parameters* as experimental conditions and an implementation of *simulation functions* and *output forms* as an experimental apparatus. Furthermore, we can consider a *computer experiment* fully documented when all its parts are described in the relative documentation.

The components of a simulation model are: the *simulation parameters space* – the set of parameters which guide the simulator behavior; *the simulation results space* – the values of the simulation outputs; the *parameters hierarchies* which order parameters and results; the *sampling mesh for parameters* defined as the set of discrete sampling functions; *the constraints over parameters* which define the computation regions in a mesh and at last the *parameters interdependencies.* A simulation, then, may be seen as a process applied to parameters producing requested results and controlled by the exploration strategy defined by the sampling mesh.

3. The APG

The APG is supposed to perform the *automatic program generation* from high level specifications supplied by the *user* and/or by a *database.* Compiling them (with the aid of external compilers and linkers when required), it assembles an executable code including specialized functions and building blocks. The produced code should satisfy the following requirements: is specialized for a given scope and independent from hardware and software platforms, merges many software packages and libraries in an hybrid architecture through *cascades of script files*, and is self–documented.

In order to produce a useful and working tool, a risk assessment analysis was conducted considering three risk areas: Users, Technology, Design. The resulting constraints led to the definition of a user requirement document and of the implementation strategy. Between the basic elements of the strategy the most remarkable ones are: the design of a specialized specification language for simulations and the use of the *multi-language strategy* which means that tasks may be described in different programming languages.

In summary an APG for simulation purposes should be a software which, given a high level simulation specification, hierarchies and constraints on parameters, is able to construct the parameter space and to generate in a consistent order the requested computations. In particular the process of the mesh generation and its exploration can be seen as a kind of *intelligent scanning process* of the parameters space. The proposed APG system is mainly composed by the following elements:

Specification Language which allows the user to describe a specific simulation through a specification document;

APG Engine which combines the information from the specification document with that from the database to obtain the low level code.

Data Base which stores information about: procedures, functions, methods and modules, software documentation, previous simulations;

Simulations Data Base which stores information about: starting data, intermediate and final results, log files and simulations documentation;

Execution Manager which supervises the execution of the generated simulation;

Report Generator which generates simulation reports;

Error and Messages Manager reacts to errors, exceptions, traps and messages between modules.

4. Simulation Specification

Assuming that the specification language should be easily understood and used by a scientist whose main concern is the study of a physical process reproduced by the simulation and its numerical convergence, the developers of the proposed APG wish to design an expressive specification language and at the same time to take care of the programming.

The specifications of simulations in the APG under construction are divided into three parts:

1. definitions of parameters and functions. In the initial version the parameters could be specified by a single value, or by a list of values, a list of expressions containing loops and application of functions which could be fetched from the data bases, a reference to a file (typically for parameters obtained as output of a previous simulation), another simulation. All the functions and transformations could be defined in any of the foreseen languages or through references to data bases, where the APG compiler is going to fetch them;
2. optional definition of an order in which the parameters contribute to the mesh construction;
3. the final part, where it is possible to indicate the functions for the simulation, presents the possibility of the incremental design of simulations.

At this point the compiler produces the actual code of the simulator, by ordering the code for the "intelligent" scanning of the mesh and then by calling the simulation functions according to parameters dependencies and hierarchies.

5. Conclusions

The principles of an Automated Program Generator (APG) system useful to manage simulations at a high level of abstraction were discussed in their various aspects. The realization of such a system requires the development of a specific language, whose design has to be based on a model of the tasks commonly accomplished by simulations in both physics and astrophysics. The language and its compiler is presently being developed and a beta release is expected in a short time.

The first application planned for such system will be the management of the simulations needed by the microwave Low Frequency Instrument consortium for the European Space Agency (ESA) Planck mission.

Acknowledgments. Many people supported the effort of the APG development and deserve acknowledgments. In particular Dr. M. Pucillo, Dr. F. Pasian, Dr. R. Smareglia and Dr. M. Callegari of OAT. M. M. is indebted scientifically with J.F. Painter of Lawrence Livermore Laboratories at UCLA and with Dr. G. Maris, Dr. A. Bignoli, Dr. G. Bollani and Mr L. Terzaroli of the Research and Development Division of Etnoteam, Milan.

References

Beazley, D. M. & Lomdahl, P. S. 1997, Comp. in Phys., 11, 230

Brown, S. A. 1992, Comp. in Phys., 6, 262

Cook, G. O., Jr. & Painter, J. F. 1992, in Expert Systems for Scientific Computing, ed. E. N. Houstis, J. R. Rice, & R. Vichnevestsky (North-Holland: Elsevier), 401

Cook, G. O., Jr., Painter, J. F., & Brown, S. A. 1991, J. Sci. Comp., 6, 193

Dubois, P. F. 1994, Comp. in Phys., 8, 70

Dubois, P. F. & Yang, T.-Y. 1996, Comp. in Phys., 10, 359

Mandolesi N., et al. 1998, Low Frequency Instrument for Planck, Proposal submitted to ESA

Maris, M. 1999, this volume, 50

Maris, M., Pasian, F., Smareglia, R., Maino, D., Burigana, C., Staniszkis, M., & Barriga, J. 1999, this volume, 145

Astronomical Data Analysis Software and Systems VIII
ASP Conference Series, Vol. 172, 1999
D. M. Mehringer, R. L. Plante, and D. A. Roberts, eds.

The Faint Object Spectrograph Post-COSTAR Spectropolarimetry Correction

M. De La Peña, H. Bushouse, A. Storrs, A. Koratkar, C. Keyes

Space Telescope Science Institute, Baltimore, MD 21218

R. Allen

Steward Observatory, University of Arizona, Tucson, AZ 85721

Abstract. While the COSTAR mirror assembly installed on the Hubble Space Telescope (HST) during the second servicing mission effectively restored HST optical capabilities, off-axis reflections from the COSTAR mirrors also introduced a measurable amount of instrument polarization. We developed a methodology to quantify and remove the induced polarization as detected by the FOS spectropolarimeter (Allen 1995), and modified the FOS calibration pipeline to accommodate post-COSTAR FOS Blue spectropolarimetry observations.

1. Introduction

Since the COSTAR mirrors have different reflectivities for light vibrating parallel and perpendicular to the line of centers between the mirrors, the polarization state of the incoming light to the FOS spectropolarimeter was altered. In addition, the parallel vibrations were shifted in phase relative to the perpendicular vibrations, so some portion of incoming linear polarization was converted to circular polarization, and vice versa.

The effects of the COSTAR mirrors were determined from observations of known unpolarized and polarized standard stars. While FOS Blue observations of an unpolarized standard, BD+28D4211, provided the relative measures of the parallel and perpendicular reflectivities, corresponding observations of a polarized standard, BD+64D106, were used to determine the phase shift. Once the induced polarization effects were determined, the proper polarization for an unknown source could then be recovered by numerically inverting the reflection process during pipeline processing.

2. Determination of the Correction

Quantifying the induced polarization was accomplished by processing observations for both the unpolarized and polarized standards through a special version of the FOS calibration pipeline (Storrs et al. 1998). In this special processing, the data were calibrated with the typical pre-COSTAR calibrations, but the final output data were forced to remain in the instrument frame, and were not rotated onto the sky frame which would be done in typical processing. While the standard version of the FOS calibration pipeline, `calfos`, is implemented as an *IRAF/STSDAS* task, the special version was used strictly to derive the post-COSTAR corrections. Consequently, the post-COSTAR corrections are bootstrapped from the pre-COSTAR calibrations. Once post-COSTAR corrections were generated by the special pipeline, it was necessary to smooth and interpolate the data to derive the final corrections. The post-COSTAR corrections comprise a new FOS calibration file, PCPHFILE. Table 1 lists the spectropolarimetric calibrated modes. There are no flux and polarimetry calibrations defined for the FOS Red detector.

Table 1. Calibrated Spectropolarimetry Modes.

Detector	Grating	Aperture (Pre-COSTAR)	Aperture (Post-COSTAR)
FOS/BLUE	G130H	4.3	—
	G190H	4.3	1.0
	G270H	4.3	1.0
	G400H	—	1.0
FOS/RED	G190H	4.3	—
	G270H	4.3	—

3. Implementation

The post-COSTAR polarization correction has been implemented for FOS Blue spectropolarimetry data in Version 3.0 and higher of both the calibration pipeline, `calfos`, as well as in the stand-alone polarimetry processing task, `calpolar`. The advantage of `calpolar` is that it provides the user with the additional flexibility to manipulate the flux calibrated data prior to polarization processing.

In order to obtain *any* calibrated spectropolarimetric data with `calfos`, the keyword MOD_CORR must be set to PERFORM in the raw data header file (D0H); this is true for both pre- and post-COSTAR epoch processing. If the spectropolarimetry data were acquired during the post-COSTAR epoch as indicated by the header keyword "KYDEPLOY = T", which stands for COSTAR deployed for FOS (T or F), *and* a post-COSTAR calibration is available as noted in Table 1, the post-COSTAR polarimetry correction obtained from the PCPHFILE reference file will be applied. If the D0H file does not contain the new PCPHFILE keyword, this keyword must be added for proper post-COSTAR spectropolarimetric processing. The D0H file can be updated to add this new keyword using the `chcalpar` task.

If no post-COSTAR polarimetry calibration file were provided for the PCPHFILE calibration file keyword (or an incorrect file were supplied), `calfos` will process all spectropolarimetric data with only the pre-COSTAR corrections. In this instance if the data were from the post-COSTAR epoch, warning messages will be issued regarding the application of only pre-COSTAR corrections.

A proper post-COSTAR calibration depends upon the data being processed using the standard pre-COSTAR corrections, with the exception that the data

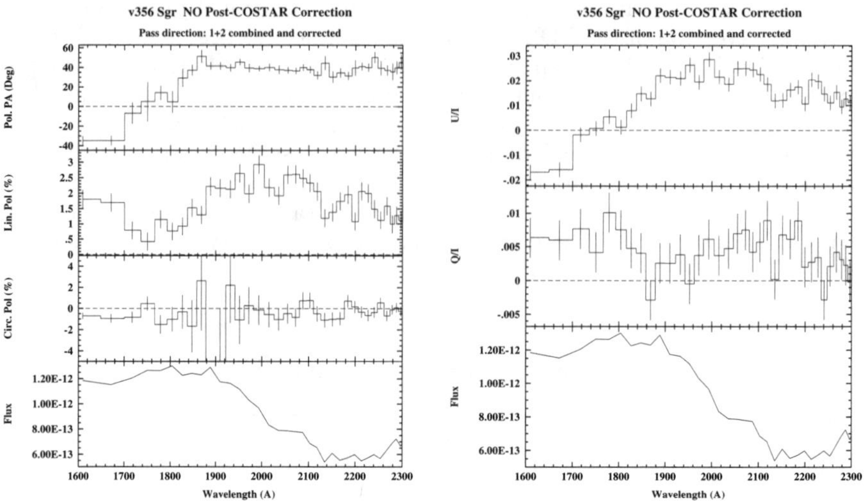

Figure 1. v356 Sgr with only the Pre-COSTAR Correction.

are left in the instrument frame rather than being rotated onto the sky frame. The Stokes parameters are then corrected to remove the additional polarization imposed by the COSTAR mirrors; the linear and circular polarization, as well as the phase angle are re-derived, and the post-COSTAR corrected final results are rotated onto the sky frame.

4. Results

The post-COSTAR corrections essentially remove the wavelength-dependent COSTAR-induced instrumental polarization in the FOS Blue observations. Figure 1 shows the polarization and Stokes vectors for the situation when no post-COSTAR correction has been applied to observations of v356 Sgr. The wavelength regime from ~1900 – 2300 Å exhibits weak linear polarization on the order of 2%; this polarization is COSTAR-induced. Figure 2 shows the polarization and Stokes vectors for the situation when the post-COSTAR correction has been applied.

In self-consistency tests it was found that even after the observations which were used to determine the post-COSTAR corrections were themselves corrected for post-COSTAR induced polarization, there remained a residual of ~0.08% in Stokes Q.

5. Conclusions

In the post-COSTAR epoch, the combined effects of residual geomagnetically-induced motion (GIM) and spacecraft jitter do not produce polarization greater than 1%. Additional sources of error in polarization are due to the photon

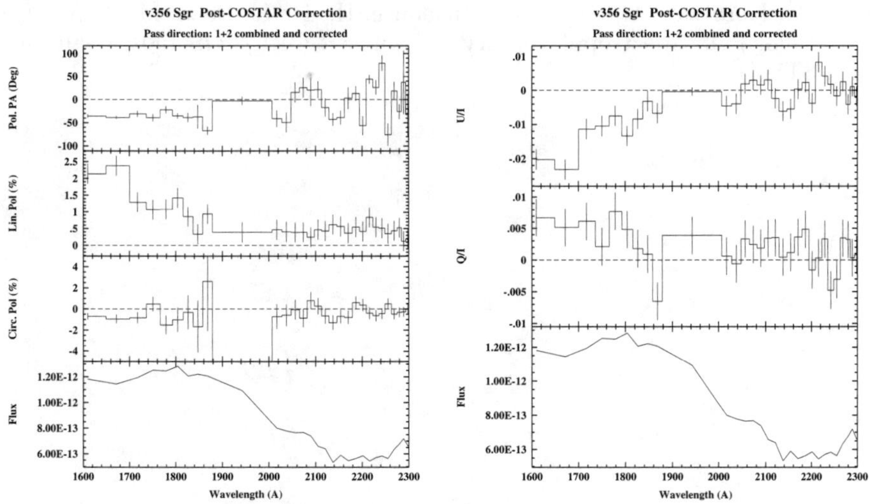

Figure 2. v356 Sgr with the Post-COSTAR Correction applied.

statistics of the observation and the error in the retardation calibration (Allen & Smith 1992). Since post-COSTAR observations made with only four polarizer positions cannot be corrected for COSTAR-induced circular polarization, these observations contain an additional 0.4% uncertainty in Q. Only observations which have been made with eight or sixteen polarizer positions can be adequately corrected for the effects of the COSTAR mirrors.

Individual pre-COSTAR spectropolarimetric data should always be checked for systematic errors due to motion perpendicular to the dispersion and the large point spread function. Light loss from one polarizer position to the next can introduce systematic effects that mimic the effect of a polarized signal. In general, since the linear polarization is biased towards positive values, observers of weakly polarized objects should correct for this bias; this can be accomplished using the `polbias` task. The `calfos` processing messages should always be monitored. If the polarimetry processing is requested, but an incorrect post-COSTAR calibration file is specified or none exists, the processing will issue warning messages and continue. In this case, only the pre-COSTAR calibrations will be applied.

References

Allen, R. G. 1995, in Calibrating Hubble Space Telescope Post-Servicing Mission, ed. A. Koratkar and C. Leitherer, (Baltimore: STScI), 60

Allen, R. G. & Smith, P. S. 1992, FOS Polarimetry Calibrations (STScI FOS IDT Report CAL/FOS-078), (Baltimore: STScI)

Storrs, A., Koratkar, A., Keyes, C., Bushouse, H., De La Peña, M., & Allen, R. 1998, FOS Spectropolarimetry (STScI ISR CAL/FOS-150), (Baltimore: STScI)

Astronomical Data Analysis Software and Systems VIII
ASP Conference Series, Vol. 172, 1999
D. M. Mehringer, R. L. Plante, and D. A. Roberts, eds.

Python in Astronomy

Norbert Pirzkal, Richard N. Hook

ST-ECF, Karl-Schwarschild Str.2, Garching bei Munchen D-85748, Germany

Abstract. We report on the use of Python to perform basic astronomical tasks. Python is a powerful, object-oriented scripting language that is easy to extend, free, and available for most computer platforms. Python, together with its Numerical module, and with the pFitsio module that is being developed by one of the authors, provides an attractive alternative to other languages such as Perl, IDL, and the IRAF CL, which are already widely used by the community. Simple examples of how Python can be used to explore and manipulate real astronomical data are given.

1. Why Python?

Python is an interpreted, interactive, object-oriented programming language. Invented in 1990 by Guido van Rossum, it shares many of its features with other modern languages such as Tcl, Perl and Java. Python has a simple, easy to learn syntax which emphasizes readability, and it draws many of its features from other popular languages such as C, C++, Modula-3, ABD, and others. We have found that Python satisfies the need for a high level programming language which fills the gap between regular shell scripting languages and compiled languages such as C. For tasks that are too complicated and too large to implement with regular shell scripts, but for which the extra overhead of coding everything in C is not warranted, Python offers a quick and elegant option. Some of the attractive features of Python include the use of transparent byte-code compilation for speed, automatic memory management and garbage collection, and a very powerful object oriented and modular design. Python ships with a large number of modules (See the Python standard documentation), and can be easily extended in C. With modules such as pFitsio or PyFits and the Numerical modules, astronomical images and tables can be easily accessed and manipulated as numerical arrays. We present small Python examples as illustrations of the language's power and encourage the reader to learn more about Python from the book by Lutz (1996).

2. Examples

2.1. Reading a List from a File and building Dictionaries

In this example (Fig. 1), we assume that we have a file called *assoc.clear*. This file contains a list of datasets from the HST STIS camera which have been previously determined to be combinable into longer integration time images that we wish to refer to as "Associations". The file is as follow:

```
ID        sub dataset    xoff  yoff   ra      dec   r d  exp
046P3X010 1 046P3X010  0.00  0.00 271.37 -19.90 0 1 300.00 ...
046P3X010 1 046P3Y010 -0.07 -0.02 271.37 -19.90 0 1 300.00 ...
...
```

We would like to be able to easily query all the members of a given association ("ID" column), and to be able to access and potentially modify the information of the individual datasets that make up an association. We therefore build a Dictionary of datasets where each dataset's name serves as a key in this dictionary. Each entry of the dataset dictionary is in fact itself a dictionary which uses the original ASCII file column name as keys to hold the ASCII file original data. Hence, dataset['O46P47010']['xoff'] would return the value of 0.0. Another dictionary is then created and uses the names of the associations (Column 1 in the ASCII file) as a key. Each association key is then assigned a list of datasets from the dataset dictionary. Once the dictionaries are set up, which only takes a few lines (grey overlay region), one has the ability to manipulate and "query" those dictionaries in a very high level, natural, way.

2.2. Reading FITS data into Arrays and manipulating those Arrays

Starting from the set of dictionaries we have built in our first example, we now proceed to read in each individual FITS file (Fig. 2). Notice that we now load both the pFitsio and the Numerical modules and how, once an image is stored in a two dimensional array, we can compute statistics on any part of this array. Our second example also shows the use of the powerful Numerical module's where() command which the IDL[1] user will recognize.

3. Availability and References

Python runs on virtually all computer platforms, including the JAVA virtual machine (JPython). One can download the compiled Python binaries or the Python source code from the Python Homepage[2]. Exhaustive Python documentation, tutorials, JPython, and third party modules can also be found on the Python homepage. Information concerning the pFitsio and PyFits modules should be directed to the authors of this paper, and Paul Barrett (see Barrett & Bridgman 1999), respectively.

References

Barrett, P. E. & Bridgman, W. T. 1999, this volume, 483

Lutz, M. 1996, Programming Python, (Sebastopol: O'Reilly)

---

[1]IDL is the trademark of Research Systems, Inc.

[2]http://www.python.org

```
import futils
```
Import the author's file utility module

```
d = futils.parseasciifile("assoc.clear"
```
Parse the ASCII file and store the result in an array d

```
d[0]
```
The first row of the file contained the column descriptions

```
['assoc_id', 'sub_assoc_id', 'dataset', 'xoff', 'yoff', 'ra_targ',
'dec_targ', 'ra_sd', 'dec_sd', 'actual_duration', 'ass_release_date_dmf']
```

```
d[2]
```
The association ID is in the first column, the dataset member's name in the third column

```
['O46P3X010', 1.0, 'O46P3X010', 0.0, 0.0, 271.378077, -19.90682, 0.0, 1.0,
300.0, 558543382.0]
```

```
len(d)
14
```
Get the length of the array d.

```
dataset={}
dic={}
```
Initialize an empty Dictionary to contain all the datasets (each line in the original ASCII file) and another one to contain the dataset associations

```
for i in d[2:]:
      tmp={}
```
Loop over the rows of d

Initialize an empty Dictionary to hold the associations

```
      for j in range(3,len(i)):
            tmp[d[0][j]]=i[j]
```
Assign each element of a row to a Dictionary using the column description of the original ASCII file as a key for this element

```
      dataset[i[2]] = tmp
```
In the dataset Dictionary, create a new element using the dataset's name as the key, and assign to it the Dictionary that was just created previously for this dataset

```
      if dic.has_key(i[0]):
            dic[i[0]][i[2]]=dataset[i[2]]
      else:
           dic[i[0]]={}
           dic[i[0]][i[2]]=dataset[i[2]]
```
If an entry for the current association of which the current row/dataset is a member exists then add this dataset to what is already in the association dictionary. Otherwise, create a new one.

The association Dictionary should now contain a table of known associations, which we display here by querying the valid keys of this Dictionary

```
dic.keys()
['O46P3X010', 'O46P46010', 'O46P49010', 'O46P40010', 'O46P47010',
'O46P4D010', 'O46P4A010']
```

We can now query the valid keys of a particular association, hence returning the list of datasets in this association

```
dic['O46P49010'].keys()
['O46P4C010', 'O46P4B010', 'O46P49010']
```

```
dic['O46P49010']['O46P4B010']['xoff']
0.04
```
Find out the x-offset of the dataset O46P4B010 member of the association O46P49010

```
dataset['O46P4B010']['xoff']=-1.0
dic['O46P49010']['O46P4B010']['xoff']
-1.0
dataset['O46P4B010']['xoff']
-1.0
```
Changing a value in one of the entries of the dataset dictionary properly affects the content of the association Dictionary

```
for item in dic['O46P49010'].keys():
    print item,dic['O46P49010'][item]['xoff'],\
              dic['O46P49010'][item]['yoff']

O46P4C010 0.06 10.76
O46P4B010 -1.0 10.78
O46P49010 0.0 0.0
```
Once the association dictionary is populated, we can easily query information about association, loop over its member datasets etc...

Figure 1. Dictionary and List creation and manipulation, (see text for more description).

```
import Numeric
import string
import pfitsio
import Statistics
```

Load the Numerical module to handle array algebra, the string module for string manipulation, the pfitsio module to access FITS files, and the Statistics module to provide some basic statistical functions.

```
for item in dic['O46P49010'].keys()
```

Following the previous example, iterate over the member datasets of the association O46P49010

Read the data from the second extension of the file "item"_crj.fits, and store the result in the array data

```
   data = pfitsio.read_data(string.lower(item)+"_crj.fits",2)
```

Get the x and y direction offsets from our dataset Dictionary, and convert those to integer values

```
   xoff = -int(dic['O46P49010'][item]['yoff'])
   yoff = -int(dic['O46P49010'][item]['xoff'])

   print "Cleaning the image",item
```

Compute the standard deviation and the median of the data, ignoring the borders of the image which are known to be noisy

```
   stdv =Statistics.standardDeviation(Numeric.ravel(data[50:950,50:950]))
   median = Statistics.median(Numeric.ravel(data[50:950,50:950]))
```

Replace all elements of data which have values higher than one sigma above the median of the entire image

```
   data = Numeric.where(Numeric.greater(data,median+1*stdv),median,data)
```

Get the shape of the data array and store the x and y sizes in xm and ym respectively

```
   xm,ym = Numeric.shape(data)


   print "Adding ",item," xoff=",xoff," yoff=",yoff
```

Compute the x and y coordinates of the overlap regions

```
   x1,x2 = max(0,xoff),min(xm-1,xm-1+xoff)
   y1,y2 = max(0,yoff),min(ym-1,ym-1+yoff)
   v1,v2 = max(0,-xoff),min(xm-1,xm-1-xoff)
   w1,w2 = max(0,-yoff),min(xm-1,xm-1-yoff)
```

Try to add the computed overlap region of the image that was just read to the current image sum. If this fail (if the image sum does not already exist, then create a new combined image of the appropriate size and store the current image in it

```
   try:
      comb[v1:v2,w1:w2] = comb[v1:v2,w1:w2] + data[x1:x2,y1:y2]
   except:
      comb = Numeric.zeros((xm,ym),Numeric.Float)
      comb[v1:v2,w1:w2] = data[x1:x2,y1:y2]
```

Once all the images have been summed up into the array comb, write this array to a file called test.fits

```
Cleaning the image O46P49010
Adding O46P49010 0 0
Cleaning the image O46P4B010
Adding O46P4B010 0 10
Cleaning the image O46P4C010
Adding O46P4C010 0 0


pfitsio.write_data(comb,"test.fits")
```

Figure 2. Array and FITS file access using Python, (see text for more description).

Astronomical Data Analysis Software and Systems VIII
ASP Conference Series, Vol. 172, 1999
D. M. Mehringer, R. L. Plante, and D. A. Roberts, eds.

PyFITS, a FITS Module for Python

P. E. Barrett[1,2], W. T. Bridgman[1]

NASA/Goddard Space Flight Center, Greenbelt, MD 20771

Abstract. PyFITS is a module for reading, writing, and manipulating FITS files using the interactive, object-oriented language, Python. The module is composed of two files: a generic low-level C library for manipulating multidimensional arrays of C-type structures and a high-level Python module. FITS files can be manipulated at several different levels, beginning with the header-data unit at the highest level to rows and columns of binary tables at the lowest level. In addition, header-data units and columns of binary tables are accessible by index or name. PyFITS also interfaces to NumPy, the Python numerical array module.

1. What is Python?

python, (*Gr. Myth.* An enormous serpent that lurked in the cave of Mount Parnassus and was slain by Apollo). 1. any of a genus of large, non-poisonous snakes of Asia, Africa and Australia that crush their prey to death. 2. popularly, any large snake that crushes its prey. 3. totally awesome, bitchin' language that will someday crush the $s out of certain other so-called VHLL's ;-)

Python is an interpreted, interactive, object-oriented programming language and is often compared to Tcl, Perl, Scheme, or Java. It can be used either interactively, by typing `python` at the shell prompt, or as a script, by typing `python script.py`. It combines remarkable power with very clear syntax and has modules, classes, exceptions, very high level dynamic data types, and dynamic typing. There are interfaces to many system calls and libraries, as well as to various windowing systems (X11, Motif, Tk, Mac, MFC). New built-in modules are easily written in C or C++. Python is also usable as an extension language for applications that need a programmable interface.

The Python implementation is portable: it runs on many brands of UNIX, on Windows, DOS, OS/2, Mac, Amiga, etc. If your favorite system isn't listed here, it may still be supported, if there's a C compiler for it. Ask around on the `comp.lang.python` newsgroup – or just try compiling Python yourself. Python is copyrighted but freely usable and distributable, even for commercial use. Access to the source code can be found through the Python Home Page[3].

---

[1]Universities Space Research Association, Seabrook, MD

[2]Current address: Space Telescope Science Institute, Baltimore, MD

[3]`http://www.python.org/`

For more information on extending Python, see Pirzkal & Hook (1999) in these proceedings.

2. What is PyFITS?

PyFITS is python module for reading, writing, and manipulating FITS files. A FITS file is treated as a 'list' of *header-data units* or HDUs. Access to the HDU is by an integer index or using the extension name as a dictionary key. The primary HDU can be accessed using the keyword 'PRIMARY'. This keyword or dictionary feature of PyFITS enables access to a HDU without having to know the absolute index of the extension in the FITS file. Hence, software which uses this feature of PyFITS is likely to be more robust and portable than software that does not.

Each HDU contains two parts, a *header* and *data* unit, though the data unit may contain no data. The header part of the HDU is a list of records containing the keyword, the value, and a comment. Access to the header unit is via list behavior, so records can be accessed, assigned, appended, deleted, inserted, etc. by using either its integer index or its keyword. The benefits of keyword access to the header are as noted above.

The data part of the HDU is treated as an array or a list of records depending on the extension type. For example, the primary array obviously has array behavior, while the binary table extension is treated as a 1-dimensional array of C-structures or records. It is therefore possible to access an entire array or extension or just a small slice. For example, when accessing a binary table extension, a single record or list of records can be accessed using array notation. It is also possible to access just a single column of a binary table by using an integer index or the column name. All of these features allow quick and efficient access to all aspects of the data in a FITS file.

The listing below is a sample session demonstrating the use of Python with the PyFITS module.

```
>>> from fitsio import *
>>> fits = Ffile('a_fits_file.fits', 'r')
>>> len(fits)                       # number of HDUs in file
4
>>> fits[0].hdr                     # print primary header
SIMPLE  =                    T  /  FITS STANDARD
BITPIX  =                   16  /  Binary Data
NAXIS   =                    2  /  Number of image axes
NAXIS1  =                  512  /  Dimension of axis 1
NAXIS2  =                  512  /  Dimension of axis 2
BSCALE  =           1.000000E0  /  Real = tape*BSCALE + BZERO
BZERO   =           0.000000E0  /  Real = tape*BSCALE + BZERO
DATE    = '10/05/93'  /  FITS creation date
IRAFNAME= './rp30019301_im1.imh'  /  Output IRAF image name
TELESCOP= 'ROSAT   '  /  telescope (mission) name
INSTRUME= 'PSPC    '  /  instrument (detector) name
RADECSYS= 'FK5     '  /  WCS for this file (e.g., Fk4)
EQUINOX =           2.000000E3  /  equinox (epoch) for WCS
CTYPE1  = 'RA---TAN'  /  axis type for dim. 1 (e.g., RA---TAN)
CTYPE2  = 'DEC--TAN'  /  axis type for dim. 2 (e.g., DEC--TAN)
...
END
>>> fits[0].hdr['BITPIX']           # print value of BITPIX
16
>>> a = fits[0].data                # assign primary array to
>>>                                 #   variable a
>>> x = fits[1].data[:][0]          # assign column 1 of extension 1
```

```
>>>                             #   (a binary table) to array x
>>> event = fits['EVENT']       # assign EVENT extension to
>>>                             #   variable event
>>> y = event[:]['Y']           # assign column Y of binary
>>>                             #   table EVENTS to array y
>>>
```

This sample demonstrates opening a FITS file and displaying the header keywords, along with the basic syntax of access the binary data tables in the file.

3. Implementing PyFITS

PyFITS is mostly coded in Python using Python classes, since this allows for the fastest development and most portability. PyFITS contains 12 classes: FBool, Fcard, Fhead, Fdata, Ffield, FHDU, FPrimary, FRandomGroupHDU, FASCIITableHDU, FBinTableHDU, FITS, and Ffile. FBool, Fcard, Fhead, Fdata, and Ffield are low level classes and provide access to header card records, data arrays, etc.. The FHDU class is a base header-data unit class. The other HDU classes inherit the basic behavior of the FHDU class. The highest level class is the FITS class itself. An associated class that allows file access is the Ffile class. Note that a FITS object does not necessarily have to be associated with a FITS file. The file is only used for permanent storage.

The one part of PyFITS that has not been coded in Python is the record array module. This module is imported by PyFITS and is used to read and write data in binary tables. During the development of PyFITS, Python did not have an efficient mechanism to read and write binary tables and a Python implementation, though usable, was not very fast or efficient. Hence, a C-extension module was written to read and write an array of records from native machine format to big- or little-endian byte formats. This is a general purpose module capable of reading multi-dimensional arrays of C-structures or records. The module can be used independently of PyFITS to read and write binary data and has links to import and export data to the NumPy module for matrix and array math.

Since this module is fully integrated into Python, it allows fast and efficient access to a binary data file using array notation. It is possible to access the data as a single record, a list of records, a single item, a list of items, or any combination of these. The module also allows items in a record to be named and then accessed by name, instead of by column or field index.

4. Conclusion

PyFITS is a Python extension module which enables astronomers to easily and efficiently manipulate FITS files either interactively using the Python command-line prompt, or as part of a large executable program. Most of the program is coded in Python, making it easy to enhance and maintain, while the data access layer has been coded in C, making it fast and efficient. A possible next step in the development of this project would be to add a GUI for file browsing.

The development of PyFITS provides a good example of modern programming principles and design. The implementation uses 1) an object-oriented programming language which enables a modular design, reuse of code via inheritance, and operator overloading; 2) a very-high-level language (VHLL), like Python, for rapid program development (development times are typically a factor of ten faster than using compiled languages), while using a low-level language, like C, for speed and efficiency (only for those parts of the code that really need it), and 3) an interpreted (scripting) language for fast development and ease of use.

References

Pirzkal, N. & Hook, R. N. 1999, this volume, 479

Astronomical Data Analysis Software and Systems VIII
ASP Conference Series, Vol. 172, 1999
D. M. Mehringer, R. L. Plante, and D. A. Roberts, eds.

CFITSIO, v2.0: A New Full-Featured Data Interface

William Pence

HEASARC, NASA/Goddard Space Flight Center, Greenbelt, MD 20771

Abstract. Version 2.0 of the CFITSIO subroutine interface for reading and writing FITS format data files is now available from the HEASARC web site. Some of the major new features in this version are the ability to read FITS files over the network, support for reading IRAF format images, and options for filtering the input file at run time.

1. New CFITSIO Features

CFITSIO is a software library that contains both C and Fortran-callable interface routines for reading and writing data files in FITS format. It is fast and efficient and typically produces data throughput rates of 5 – 20 MB/s on modern workstations and PCs. Version 2.0 of the CFITSIO library is available at `http://heasarc.gsfc.nasa.gov/fitsio` and contains many new features:

- Y2000 compliant: supports the new `YYYY-MM-DDThh:mm:ss` FITS DATE keyword format.
- Any FITS file on the internet can be read by supplying the FTP or HTTP URL as the name of the file.
- IRAF format images (*.imh files) can be read. They are converted into FITS format on the fly before being opened by CFITSIO.
- FITS files can be read or written in memory or shared memory for better I/O performance than on magnetic disk.
- FITS files can be piped between tasks on the stdin and stdout streams, thus reducing the need to write temporary intermediate files on disk.
- Compressed FITS files (gzip or Unix compress formats) can be read.
- FITS Tables can be filtered or modified at run time by adding qualifiers to the file name using a general C-like syntax that supports arbitrarily complex expressions.
- Tables can be filtered at run time using Good-Time-Interval files (GTIs) and SAOimage region files. The application only sees the table rows that satisfy the filtering expression.
- Any single table column can be binned into a 1-D FITS image, and any pair of table columns can be binned into a 2-dimensional FITS image at run time. The application program then opens the FITS image, not the original table.
- A single FITS file can be opened multiple times and treated as independent files, e.g., the application program can read or write to different extensions in the FITS file simultaneously.

- New routines for dealing with large sets of hierarchically organized data files have been added.
- New FITS tables or images can be created using ASCII template files which describe the file structure.

2. Extended Filename Syntax

Support for different types of files and for filtering of the input file is provided through an extended filename syntax, as illustrated in the following examples:

- `myfile.fits`: opens the FITS file on disk in the current directory.
- `myfile.imh`: opens the IRAF format image file by converting it into a temporary FITS format image in memory (by default) which is then opened and passed to the application program.
- `myfile.fits.gz[events, 2]`: uncompresses the gzipped file myfile.fits into memory (by default) and moves to the extension which has the keywords `EXTNAME = 'EVENTS'` and `EXTVER = 2`.
- `-`: a dash (minus sign) signifies that the input file is to be read from the stdin file stream, or that the output file is to be written to stdout.
- `ftp://legacy.gsfc.nasa.gov/test/vela.fits`: FITS files in any FTP archive site on the internet can be opened with read-only access. Files with HTTP URLs can be opened in the same way.
- `shmem://h2[events]`: opens the FITS file in a shared memory segment.
- `mem://`: creates a scratch output file in computer memory. The resulting 'file' is deleted when the program exits, so this is mainly useful for testing purposes when a permanent copy of the output file is not needed.
- `myfile.fits[EVENTS][col Rad = sqrt(X**2 + Y**2)]`: creates a temporary file (in memory or on disk) that is identical to myfile.fits except it contains a new column in the EVENTS extension called `Rad` whose value is computed using the indicated expression which is a function of the values in the X and Y columns.
- `myfile.fits[EVENTS][PHA > 5]`: creates and opens a temporary FITS files that is identical to `myfile.fits` except that the EVENTS table will only contain the rows that have values of the PHA column greater than 5. In general, any arbitrary boolean expression using a C or Fortran-like syntax can be used to select rows from a table. GTI and regions files can also be use to select rows.
- `myfile.fits[EVENTS][bin (X,Y)=1,2048,4]`: creates a temporary FITS primary array image by binning (i.e., computing the 2-dimensional histogram) of the values in the X and Y columns of the EVENTS extension. In this case the X and Y coordinates range from 1 to 2048 and the image pixel size is 4 units in both dimensions, so the resulting image is 512 $\times$ 512 pixels in size.
- `myfile.fits[EVENTS][PHA  > 5][bin (X,Y)=1,2048,4]`: The column modifiers, row selection, and column histogramming filters can be combined into more complex operations. In this case the EVENTS extension is filtered for rows that have PHA greater than 5, then the X and Y columns are binned into an image.

Acknowledgments. Many people have contributed to the development of version 2.0 of CFITSIO. In particular, members of the Integral Science Data Center (Jurek Borkowski, Bruce O'Neel, and Don Jennings) designed the concept of plug-in I/O drivers now used by CFITSIO, and provided the drivers for accessing files in shared memory and via the FTP, HTTP, and ROOT protocols. Also, Don Jennings wrote the hierarchical grouping routines, and Jurek Borkowski wrote the ASCII template file parser.

Uwe Lammers, from the XMM project at ESTEC, provided the very fast expression parsing algorithm that is used by CFITSIO when calculating new column values or filtering table rows at run time. Peter Wilson, (NASA/GSFC, RSTX) combined this algorithm with CFITSIO's iterator function for even fast I/O throughput. Peter Wilson also wrote the Fortran-callable wrappers for CFITSIO.

And, last but not least, Jonathan McDowell and collaborators at the AXAF Science Center developed a file filtering and binning syntax which formed the basis for the CFITSIO extended filename syntax, and Doug Mink (Smithsonian Astrophysical Observatory) provided the routines for converting IRAF format images into FITS format.

Astronomical Data Analysis Software and Systems VIII
ASP Conference Series, Vol. 172, 1999
D. M. Mehringer, R. L. Plante, and D. A. Roberts, eds.

The STSDAS Table Selector Library

Bernard Simon, Philip Hodge

Space Telescope Science Institute, Baltimore, MD 21218

Abstract. The 2.0.1 release of Tables included the ability to access table subsets. To the user and the programmer such a subset appears to have fewer rows and/or columns than the entire table. The user defines a subset by appending expressions to the table name. This is transparent to the programmer, as the program can just pass the table name to the usual table open routine. Internally, the procedure that evaluates the expressions calls two interpreters: one to evaluate the row selector expression, and another to evaluate the column selector. The syntax for the column selector follows usual Unix pattern matching, and the syntax of the row selector follows the syntax first used in qpoe filters.

1. What Is It?

The library implements a set of procedures that allows the user to select a subset of rows or columns from Stsdas tables. There are two selectors: one for selecting a subset or rows and one for selecting a subset of columns. The information on which rows or columns to select is passed as a string, which is compiled and evaluated by the selector library. The selector library has been embedded in the Stsdas tables library, so that existing tables tasks will allow you to transparently operate on a subset of a table instead of a full table.

Selectors may be applied only to existing tables. They may be used with tables opened read-write, but with some restrictions. When you add rows or columns to a read-write table, for example with tedit, all the new rows and columns are automatically selected. You can add or delete rows at the end of a table that has a row selector in effect, but you cannot add or delete rows before the last row. When you add a row beyond the last selected row, the new row is actually written after the last row in the underlying table. If a row selector is in effect, inserting rows before the end is not allowed because it can be ambiguous as to where they should be inserted. The option of row and column selectors with read-write tables can be very useful with such tasks as tedit or tcalc, but it should be used with caution when reordering the table, such as with tsort.

2. How Can I Use It?

Any program linked with the current version of the tables library can use the column and row selector. Just append the selector expression enclosed in square brackets, with a prefixed "r:" or "c:" to indicate if the row or column selector is being used.

Programmers can also call the selector library directly. The row selector has two interfaces: one which evaluates the expression for a single row and one which evaluates the expression for the entire table. Here is a typical example of the use of the row at a time selector:

```
tp = tbtopn (table, READ_ONLY, NULL)
numrow = tbpsta (tp, TBL_NROWS)
pcode = trsopen (tp, filter)
do irow = 1, numrow {
    if (trseval (tp, irow, pcode)) {
        # Do something neat here
    }
}
call trsclose (pcode)
call tbtclo (tp)
```

The interface which evaluates the expression for an entire table returns a set containing the row numbers for which the expression is true. The set can then be accessed and manipulated by other procedures in the row selector interface. One example of how to use this interface is:

```
set = trsrows (tp, filter)
nset = rst_nelem (set)
do iset = 1, nset {
    irow = rst_rownum (set, iset)
    # do something with the row here
}
call rst_free (set)
```

3. What Is the Syntax?

The column selector filter is a comma separated list of column patterns. A column pattern is either a column name, a file name containing a list of column names, or a pattern using the usual IRAF pattern matching syntax. For example, the string `[c:b,time*,@column.lis]` would be expanded as the column names b, any column name beginning with "time", and all the column names in the file column.lis. If the column list is entirely white space, the array of column descriptors will include all the columns in the table, as this seems the most reasonable default. If the first non-white character is the negation character (!), the array of column descriptors will include all columns not matched by the list. The negation character only has this meaning at the beginning of the list.

Column names may also contain array sections having the same format as image sections. The sections are surrounded by parentheses. For example `[c:spec(1:200:2),image(*,30),spec(20:*)]`. Tasks which use the selector library through the tables library do not support array sections. Array sections are only supported by tasks like tstat, which call the selector library directly.

The row selector uses the same syntax as qpoe filters. For sake of an example, suppose we have a star catalog with the columns Name, Ra, Dec, V, B-V, and U-B. The simplest row selector is an equality test. The name of the column appears on the left of an equals sign and the column value appears on the right. For example, `[r:name=eta_uma]`. Column numbers can be used in place of the column name. This is especially useful for ASCII tables. Values can be either numbers or strings. It is usually not necessary to place strings in quotes. However, if any string (including a column name) contains embedded blanks or characters significant to the qpoe filter, such as equal signs, commas, or colons, it should be placed in quotes.

Ranges of values can be specified by giving the endpoints of the ranges separated by a colon. For example, `[r:v=10:15]` selects all rows with visual magnitude between 10 and 15. Ranges include their endpoints. Ranges can also be used with strings as well as numbers. Ranges can also be one sided. The filter `[r:dec=80:]` selects all rows with declination greater than or equal to eighty degrees and the filter `[r:dec=:-40]` selects all declinations less than or equal to forty degrees south. A filter can contain a list of single values and ranges. The values in the list should be enclosed in parentheses. For example, `[r:name=(eta_uma,alpha_lyr)]` or `[r:b-v=(-1:0,0.5:1)]`.

Individual values or ranges can be negated by placing a ! in front of them. For example, `[r:name=!eta_uma]` selects every row except the star named eta_uma and `[r:ra=!0:6]` selects all rows except those with right ascension between zero and six hours. An entire list can be negated by placing a ! in front of the column name or the parentheses enclosing the list. The filters `[r:!name=(eta_uma,alpha_lyr)]` and `[r:name=!(eta_uma,alpha_lyr)]` are both equivalent.

Filters can test more than one column in a table. The individual tests are separated by commas or semicolons. All tests in the filter must succeed for the filter to be accepted. For example, `[r:ra=1.3:1.4,dec=40:42]` selects a rectangular region in the catalog. A range of row numbers can also be selected by placing the word row on the left side of the equals sign. For example, `[r:row=10:20]` selects rows from ten to twenty inclusive and `[r:row=50:]` selects all rows from fifty on. Row selection can be combined with any other test in a filter. A filter, can also be placed in an include file, for example `[r:@filter.lis]`. Include files can be a part of a larger expression and include files can contain other files, up to seven levels deep.

4. How Does It Work?

The column selector expands a list of column names or patterns into an array of column descriptors. First it checks for a leading negation character, stripping it if found. Then it opens the list of column patterns as a file. It then processes each pattern in the list in a loop. If the pattern is a file name, it pushes the current file descriptor on a stack and opens the new file in its place. If it is a column pattern, it uses the IRAF pattern matching procedures to match the pattern against all column names in the table. All matches are checked against the array of columns already matched and appended if they are not found. If it is a column name, it is checked to see if it is in the table and then added to the array of columns matched if not already in the array. At the end of the file, the previous file descriptor is popped from the stack until no more descriptors are left in the stack. If the pattern was negated, the match array is inverted by comparing it to the columns in the table and including columns not in the match array in the output array. If the pattern is not negated, the match array is the output array.

The tables row selector takes a string which describes the rows to be selected and compiles it into pseudo-code. It then evaluates the pseudo-code to build a data structure containing an array of row numbers. When the task accesses a row, the row number is mapped to the corresponding number in the data structure. Thus to the task the table looks like it only contains the rows selected by the expression.

The row selector expression is generated by a parser generated by yacc. The parser converts the expression into a binary tree. It then checks to see if the expression can be optimized by pre-evaluating row number range subexpressions. If the expression can be optimized, the parser walks the binary tree, evaluating the row number subexpressions and snipping them out of the tree. The modified tree is then converted into postorder pseudo-code. The result of the row number expression is stored as a set of row ranges. If there was no row number subexpression in the original expression, the row number range includes all the rows in the table.

The expression is evaluated for a single row by first checking to see if the row number is in the set of row ranges. If not, the expression is false for that row. If it is, the pseudo-code is evaluated. The evaluator is implemented as a simple select statement in a loop. Values in the table columns are compared to the constants stored in the pseudo-code. The result of the comparison is pushed on a stack. At the end of the evaluation of the pseudo-code, only a single value remains on the stack, which is returned as the result of the evaluation.

Each pseudo-code instruction contains six fields: the operation field, the column field, the true and false jump fields, and the low and high value fields. Not every field is used in each instruction. If a field is not used it is set to zero. The operation field contains an integer code used to choose the branch of the select statement in the evaluator to execute. The column field contains the table column descriptor of the column operated on by the instruction. If set to zero, the row number is used in the operation instead. The jump fields contain the index of the next instruction to execute after the current instruction, depending on the result of the current operation. If the corresponding field is zero, the index is incremented by one instruction instead. The use of jump fields enables the expression evaluator to only compare values to a list until the first success or only check column values until the first failure, speeding up the evaluation of the expression. The value fields contain pointers into a string buffer containing the constant values used by the operation. Although all constants are stored as strings, the evaluator performs numeric as well as string comparisons. The type of comparison is encoded in the operation field.

Astronomical Data Analysis Software and Systems VIII
ASP Conference Series, Vol. 172, 1999
D. M. Mehringer, R. L. Plante, and D. A. Roberts, eds.

Perl at the Joint Astronomy Centre

Tim Jenness, Frossie Economou, Remo P. J. Tilanus[1], Casey Best[2], Richard M. Prestage, Pam Shimek[2]

Joint Astronomy Centre, 660 N. A'ohōkū Place, Hilo, HI 96720

Karl Glazebrook, Tony J. Farrell

Anglo-Australian Observatory, P.O. Box 296, NSW 2122, Australia

Abstract. Perl is used extensively at the JAC (UKIRT and JCMT) and because of the language's flexibility (enabling us to interface perl to any library) we are finding that it is possible to write all of our utilities in it. This simplifies support and aids code reuse (via the module system and object oriented interface) as well as shortening development time.

Currently we have developed interfaces to messaging systems (ADAM and DRAMA), I/O libraries (NDF, GSD), astronomical libraries (SLALIB) and the Starlink noticeboard system (NBS). We have also developed tools to aid in data taking (the JCMT observation desk) and data processing (SURF and ORAC-DR)

This paper will briefly review the facilities available, with an emphasis on those which might be of interest to other observatories.

1. Introduction

At JAC, we make extensive use of the perl language, beyond its popular uses in system administration and web authoring. Described as the "Swiss Army Chainsaw" of computing, the language's power and versatility provides appropriate solutions to a wide range of problems.

Here we describe some of the utilities and libraries we have developed and will be making available to the astronomical community. A web-site is currently being developed that will act as a repository for astronomical software written in perl (`http://astro.perl.org`).

2. Data Reduction

Perl has been used extensively to help with data reduction at the JAC. The ORAC-DR data reduction pipeline has been developed as a modular extensible

[1]Netherlands Foundation for Research in Astronomy (NFRA)

[2]University of Victoria

pipeline system that can be used to reduce data from any instrument or telescope with only minor changes to the infrastructure. This has been achieved by using perl object-oriented programming and the ability to generate data reduction code on-the-fly but executed by the current perl interpreter in a specified namespace. More information on the ORAC-DR data reduction pipeline can be found in Jenness & Economou (1999).

The SCUBA data reduction package (SURF) (Jenness & Lightfoot 1998) has also made use of perl in order to simplify some tasks for log generation, file header manipulation and automated data reduction.

3. Observation Planning

Perl/Tk is now the standard GUI language at the JAC. We are using it primarily in developing applications to aid in the observing process (the JCMT 'Observation Desk'). Examples are:

sourceplot: graphical representation of source positions.

scusky: used to display the current configuration of SCUBA overlaid on an image of the region obtained from the Digital Sky Survey.

odfgen: observations preparation tool for SCUBA

obslog: allows observers to create observation logs and comment on the usefulness of the data

Of these, `sourceplot` is the most general and is not specifically designed for use at the JCMT (Fig. 1).

4. Modules Developed at the JAC

In many cases, development of these applications has required that extension modules be written to provide access to C/FORTRAN libraries from perl. It has been necessary to develop the following perl modules:

4.1. Messaging systems

Currently, both JCMT and UKIRT make use of the ADAM messaging system for telescope control and for data reduction systems. As part of the ORAC-DR system it was necessary to add the ability to send and receive ADAM messages from perl.

The JCMT is now in the process of upgrading its telescope system to DRAMA (the AAO's Distributed Real-Time Application programming system) (Farrell, Shortridge, & Bailey 1993). Given our previous experience with perl it has been decided to use perl for the high level interface to the telescope system. The first step in achieving this is complete and it is now possible to write DRAMA tasks in perl as well as sending and receiving DRAMA messages from perl user interfaces. This module also includes an interface to the SDS (self-defining data structure) format.

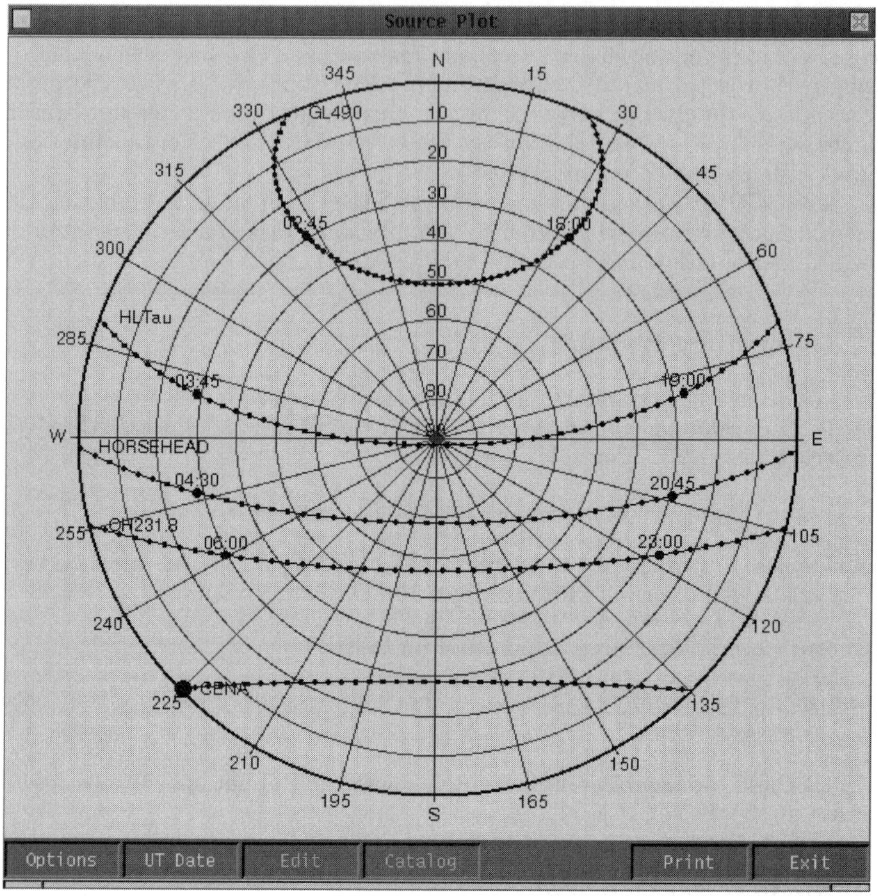

Figure 1. An azimuth-elevation image created from `sourceplot`. Zero elevation is at the edge of the image and 90 degrees elevation at the centre.

4.2. File I/O

Perl interfaces to access data files have been developed for the Starlink N-dimensional data format (NDF) and the Global Section Data (GSD) format in use at JCMT and UKIRT. Both these modules provide access to the complete set of library routines plus simplified wrappers using the PerlDL extension.

4.3. Astrometry

For astrometry calculations we have built an interface to Pat Wallace's SLALIB library.

4.4. Miscellaneous modules

Other perl modules developed at the JAC include:

- NBS – this is the Starlink NoticeBoard system and provides a machine independent interface to shared memory systems.
- XMGR – a display module for the popular XMGR 1-D graph plotting program (now called GRACE).

5. Other Modules

Full use is made of perl modules written outside of the JAC. Most notable of these are:

5.1. CFITSIO

This module, written by Pete Ratzlaff, provides full access to the standard CFITSIO library allowing FITS files to be read into and written from perl programs.

5.2. PGPLOT

`pgperl` (Glazebrook 1996) is the perl interface to Tim Pearson's PGPLOT library. This package provides the full set of PGPLOT routines to the perl programmer. An example of using the perl PGPLOT interface and PDL can be found in the WORF system (Jenness, Economou, & Tilanus 1997) used at the JAC for remote eavesdropping of observing sessions.

5.3. PDL

From the perldl home page (`pdl.perl.org`, hosted by the JAC):

> PDL ("Perl Data Language") gives standard perl the ability to *compactly* store and *speedily* manipulate the large N-dimensional data arrays which are the bread and butter of scientific computing.

Of note for astronomy is that PDL includes a basic FITS reader as well as simple commands for reading NDF, GSD and other image formats. PDL also includes display interfaces to KARMA, SAOtng, PGPLOT and OpenGL.

References

Farrell, T. J., Shortridge K., & Bailey, J. A. 1993, BAAS, 25, 954

Glazebrook, K. 1996, in ASP Conf. Ser., Vol. 101, Astronomical Data Analysis Software and Systems V, ed. G. H. Jacoby & J. Barnes (San Francisco: ASP), 307

Jenness, T., & Economou, F. 1999, this volume, 171

Jenness, T., Economou, F., & Tilanus, R. P. J. 1997, in ASP Conf. Ser., Vol. 125, Astronomical Data Analysis Software and Systems VI, ed. G. Hunt & H. E. Payne (San Francisco: ASP), 401

Jenness, T., & Lightfoot, J. F. 1998, in ASP Conf. Ser., Vol. 145, Astronomical Data Analysis Software and Systems VII, ed. R. Albrecht, R. N. Hook, & H. A. Bushouse (San Francisco: ASP), 216

Astronomical Data Analysis Software and Systems VIII
ASP Conference Series, Vol. 172, 1999
D. M. Mehringer, R. L. Plante, and D. A. Roberts, eds.

WCSTools: An Image Astrometry Toolkit

Douglas J. Mink

Harvard-Smithsonian Center for Astrophysics

Abstract. Very few people want to deal with the details of positional astronomy, but many astronomers want to know precise positions of objects in the sky. New instruments, such as fiber-fed spectrographs, require more precise positions than are available for many objects. WCSTools provides programs and subroutines which allow both scientists and programmers to access astrometric information using the world coordinate system (WCS) of an image. WCSTools programs translate between image and sky coordinates, list catalog stars in the area covered by an image, find stars in an image and list them with their sky positions, and fit a WCS by matching image objects to catalogs. SAOimage, SkyCat, and SAOtng use WCSTools subroutines to display sky coordinates while browsing an image. The software reads and writes FITS and IRAF .imh images and deals easily with Space Telescope and Digitized Sky Survey FITS images. The proposed FITS WCS standard is being tracked, and interim formats are being supported.

1. Introduction

An astronomical image world coordinate system (WCS) links each pixel in an astronomical image to a specific direction in the sky. Thus one can either find objects with known sky positions in an image or find the sky positions of objects discovered in an image. Spacecraft images, such as those from the Infrared Astronomy Satellite (IRAS) or the Hubble Space Telescope come with WCS information in their header. The images of the Digitized Sky Survey also come with WCS information. Using very large reference catalogs such as the HST Guide Star Catalog or the USNO-A Catalogs (Monet 1996), one can get enough stars to fit a fairly accurate WCS to almost any ground-based image using the **imwcs** program (Mink 1997).

What can you do with an image with a world coordinate system attached to it? You can work on specific applications, such as tracking each pixel's sky position as a cursor moves across an image browser (Mink 1996). Or you can make WCS information easy to use and find out what happens. WCSTools provides both modular subroutines and simple programs which provide straightforward access to world coordinate information to programmers and astronomers as transparently as possible. Subroutines need only a FITS or IRAF image header and coordinates to transform. Programs need only a FITS or IRAF image and coordinates.

2. Community-Driven Software

Given that a software package does something useful, there are two principles which need to be followed for it to be used by the astronomical community. It must be easy to use or adapt, that is it must have what Mandel & Murray (1998) called a "minimal buy-in", and it must work with real data, requiring adaptations by the package developer which could be called in a reciprocal term, "sell-out".

To make WCSTools subroutines easy to use, there are simple interfaces which require minimal information input–image headers and coordinates–and provide straightforward output: both formatted strings and binary numbers. Derived information is contained in a single data structure which can easily be passed between subroutines. In the most simple case, a subroutine user must read a FITS header, using their own subroutine or one included in this package, call *wcsinit()* to fill the WCS data structure, and call *pix2wcs()* or *pix2wcst()* for string output, to compute sky coordinates from image pixel coordinates or *wcs2pix()* to do the reverse.

As the WCSTools subroutines have been adopted as part of the SAOtng (Mandel 1997), its successor (as discussed by Joye & Mandel at this conference), and SkyCat (Albrecht et al. 1997) image browsers, many new features have been added at the request of the developers of those programs, and many bugs have been found by their users. Resolving these requests has taken a significant amount of the WCSTools developer's time, the aforementioned "sell-out". Spending that time, paid for by the developer's institution must be justified. Local use of the programs being developed elsewhere is one good reason to share code. More users find more bugs, the repair of which makes the programs more robust and thus more useful to everyone, including users at the developers institution.

The most difficult part of software development is to make it work with real data. WCSTools can read and write not only FITS image files, but IRAF .imh files, accessing most images made by astronomers. The image-reading subroutines can read the version 2 IRAF .imh files introduced in 1997 as well as version 1, and, unlike IRAF, can read images of either byte order on computers with either native byte order. The **i2f** program uses this capability to convert IRAF image files, no matter what their system of origin, to FITS files which can be more widely used.

To get WCS information out of as many kinds of images as possible, the WCSTools package supports 30 map projections, including the 8 AIPS projections (-SIN, -TAN, -ARC, -NCP, -GLS, -MER, -AIT, and -STG), 19 additional projections proposed as FITS WCS standards by Griesen & Calabretta (1996), the plate projection used by the Digitized Sky Survey, and IRAF and SAO polynomial projections. If no other WCS is present in the header, an approximate tangent plane projection WCS is created with the tangent point at the center of the image using the RA and DEC specified in the header, and a plate scale in arcseconds per pixel given in the image header or on the command line. As soon as a FITS standard WCS polynomial is proposed, it will be included as an option.

3. Use as Script Components

WCSTools programs, including **SAOimage**, are meant to be run with all of their arguments on the command line, making it easy to use them with any scripting language. Like the subroutines, they require minimal input–image file names and coordinates–and provide output in ways which are easily transmitted to other programs.

The use of WCSTools programs in scripts is illustrated by two recent projects undertaken at the Center for Astrophysics to improve galaxy positions. The first was an attempt to make sure that our database of spectra had positions good to an arcsecond or so. A Unix shell script was written which read lists totaling several thousand file names, extracted right ascensions and declinations from their headers using WCSTools **gethead**, extracted a Digitized Sky Survey field centered on that position into a local FITS file, and displayed that image using **SAOimage**, where the new -f1, -f2, and -f3 commands were used to set actions to be performed when the F1, F2, and F3 keys were used. These actions either sent the cursor position to the WCSTools **sethead** program which reset the position in the archive spectrum file or added the file name to a list of problem files or files which had really bad positions.

A second project was to get accurate positions of galaxies for the Updated Zwicky Catalog (Falco et al. 1999). Many of these positions could be obtained by matching the original Zwicky Catalog against catalogs with positions which were known to be good enough, but a list of ambiguous Zwicky Catalog entries remained. Region files containing labeled shapes at image coordinates computed from the sky positions of cataloged objects in Digitized Sky Survey images centered on the supposed position of a galaxy were written by the WCSTools **imcat** program and displayed by **SAOimage**. Using the function key commands described in the previous example, the coordinates of the correct position were added to a list. Figure 1 shows part of an **SAOimage** display of one of these images.

4. Internet Access

WCSTools documentation, with many examples, is available on the World Wide Web at `http://tdc-www.harvard.edu/software/wcstools`.

A future release of WCSTools will also include Internet access to catalogs and images. Code to read the tab tables format used by SkyCat catalog servers is already incorporated into WCSTools.

Acknowledgments. Bill Joye, Allan Brighton, and Eric Mandel, who use my subroutines in their image browsers, have passed on their problem data to make my interpretations of real image headers more robust. Andreas Jaunsen has been patient while I fixed the bugs he found. Susan Tokarz, Emilio Falco, Norman Grogin, and Jan Kleyna have tested the catalog search and display programs and used them as tools to do science.

Figure 1. A portion of an **SAOimage** display showing three galaxy positions from two catalogs

References

Albrecht, M. A., Brighton, A., Herlin, T., Biereichel, P., & Durand, D. 1997, in ASP Conf. Ser., Vol. 125, Astronomical Data Analysis Software and Systems VI, ed. G. Hunt & H. E. Payne (San Francisco: ASP), 333

Falco, E. E., et al. 1999, PASP, in press

Griesen, E. & Calabretta, M. 1996, A&A, submitted

Mandel, E. 1997, in ASP Conf. Ser., Vol. 125, Astronomical Data Analysis Software and Systems VI, ed. G. Hunt & H. E. Payne (San Francisco: ASP), 253

Mandel, E. & Murray, S. S. 1998, in ASP Conf. Ser., Vol. 145, Astronomical Data Analysis Software and Systems VII, ed. R. Albrecht, R. N. Hook, & H. A. Bushouse (San Francisco: ASP), 142

Mink, D. J. 1996, in ASP Conf. Ser., Vol. 101, Astronomical Data Analysis Software and Systems V, ed. G. H. Jacoby & J. Barnes (San Francisco: ASP), 249

———, 1997, in ASP Conf. Ser., Vol. 125, Astronomical Data Analysis Software and Systems VI, ed. G. Hunt & H. E. Payne (San Francisco: ASP), 249

Monet, D. 1996, USNO-A1.0: A Catalog of Astrometric Standards, (Washington: USNO)

Astronomical Data Analysis Software and Systems VIII
ASP Conference Series, Vol. 172, 1999
D. M. Mehringer, R. L. Plante, and D. A. Roberts, eds.

CIRSI Data Reduction System – CIRDR

S. J. Chan[1], S. E. Persson[2], R. G. McMahon[1], C. D. Mackay[1], R. Ellis[1], M. G. Beckett[2], M. Hoenig[1]

Abstract. We report on our on-going software project *CIRSI Data Reduction System*, which has been developed under the IRAF environment. This is a compact infrared processing and analysis software system for a wide-field infrared camera, the Cambridge Infrared Survey Instrument (CIRSI). It contains four subpackages for quick-look analysis during real-time observing, preprocessing, basic reduction and useful tools. This system can handle a large amount of raw data which can be pre-processed and reduced automatically during real-time observations.

1. CIRSI - Cambridge Infrared Survey Instrument

CIRSI is based on 4 Rockwell HAWAII 1024×1024 detectors. It contains no cold optics other than a cold filter wheel inside a dewar. It operates from 0.8 to 1.8 μm (Z to H band) on non-infrared optimized telescopes, with a planned upgrade for K band operation in 1999. The image scales are dependent on the telescopes to be used (e.g., $0''.45$/pixel on the 2.5 m Isaac Newton Telescope (INT) f3.3 Prime and $0''.32$/pixel on the 4.2 m William Herschel Telescope f2.8 Prime). A single channel CCD controller is used to multiplex the 16 outputs (4 outputs per chip) into a single pre-amplifier and signal processing data chain. The camera control software, *PIXCEL*, runs under Windows 95 (Beckett et al. 1998). There are two read-out modes from the camera which are Reset-Read-Read Sampling (RRR) and Non-Destructive Read or Sloping Sampling (NDR, see Fig. 1, Fowler & Gatley 1991).

2. CIRDR - CISRI Data Reduction System

2.1. General Description

CIRDR is a data reduction and analysis system for reducing and analysing data obtained from CIRSI. It is written under the IRAF environment. The programming languages which are being used in this system are IRAF-CL, IRAF-SPP (IRAF subset preprocessor language), Fortran and C-shell scripts. In the current version (version 0.0.3), there are four packages (see Fig. 2) which are CQLOOK (CIRSI Quick Look Data Reduction Package) for producing rough

[1]Institute of Astronomy, University of Cambridge, Cambridge, U.K.

[2]Carnegie Observatories, Pasadena, California, U.S.A.

sky subtracted images from sets of dithered observations during an observation run, CPLINE (CIRSI Pipeline Data Reduction Package) for producing second pass/final-version of sky subtracted images, CIRUTIL (CIRSI General Toolkit Package) and CIRCONTRIB (CIRSI General Contribution Package).

On-line help is available in all active subpackages and sub-subpackages. Two on-line introductory user guides, *CIRSI User Guide* and *CQLOOK User Guide*, are also available.

CQLOOK is the most developed package in this system. It is used for producing rough sky subtracted images from sets of dithered observations during an observation run. Because the observations are in the near-infrared, the strong and variable sky background requires dithering of many short exposures and the recombination of the images with careful sky subtraction to produce deep images.

There are two active subpackages which are for basic preprocessing in order to produce *real astronomical* images (CPREPROC) and for quick (first pass) data reduction to produce rough sky subtracted, dithering-mosaic images (CREDUCT), as well as tasks for the first INT run (CINT1).

2.2. Special Functionality

File Handling System. The file handling system uses a *system-level approach* which is an *object-oriented like infrastructure*. The advantages of this design are that it is simple and portable, as well as that it has a compact structure on a solid foundation. The main functionality is in the first level. Handling files from quadrants and/or chips as well as read-out modes are in the second and third levels. The automatic processing is in the 4th level.

Quick Processing. Quick processing tasks can be used during real time observation, normal data reduction at a telescope, or at a home institute. There are two master programs (CNDRPROC, CRRRPROC) to process data automatically for quick look purposes. They can process data from 1 to 4 chips, 1 to 16 quadrants, one selected chip, or one selected quadrant. There are several display tasks to examine fields and two tasks to check saturation levels. Tasks for pre-processing can get raw data from a *data reservoir* in which raw data are stored. There are two tasks for making a mosaic image (dithering or offset) (see Fig. 3). Both are interactive and semi-automatic.

Calibration. Tasks for making flats are based on all currently available methods which are domeflat, twilight-sky flat, and moving-sky flat. There are at least two tasks for making bad pixel masks. For extra flexibility, users may enter their own parameters for making calibration images and sky images. Furthermore, tasks are also available to automatically *suppress* objects in images and to make second pass sky images.

2.3. Topic of interest

Uniform stripe noise appears in images (Fig. 4a). The current correction methods in the system are based on a linear median filtering method. There are two typical problems occurring in the current algorithms which are ghost stripes

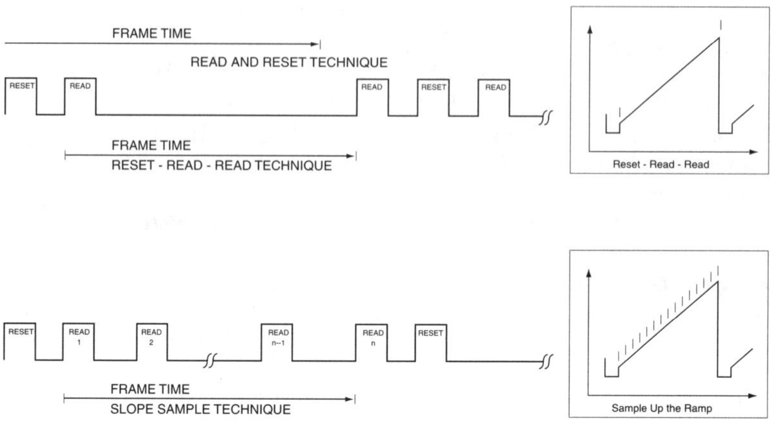

Figure 1. The read-out mode in CIRSI.

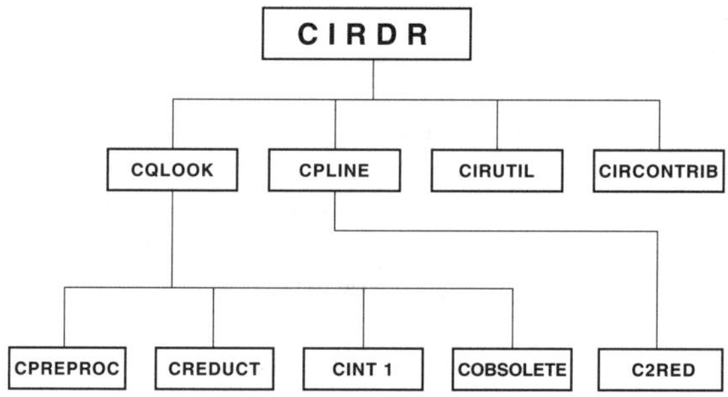

Figure 2. CIRSI Data Reduction System flowchart.

caused by bright objects within images (Fig. 4b) and the size of bright objects are decreased and faint objects are smoothed out (Fig. 4c). An improved method, which can eliminate the above problems, is based on a second pass linear median filtering method (Fig. 4d).

Acknowledgments. S. J. Chan thanks the Program Organizing Committee of the Eighth Annual Conference on Astronomical Data Analysis Software and Systems for offering her full financial support to attend the conference.

References

Beckett, M. G., Mackay, C. D., McMahon, R. G., Parry, I. R., Ellis, R. S., Chan, S. J., & Hoenig, M. 1998, in SPIE Proc., Vol. 3354, ed. A. M. Fowler (Bellingham: SPIE), 431

Fowler, A. M., & Gatley, I. 1991, in SPIE Proc., Vol. 1541, Infrared Sensors: Detectors, Electronics, and Signal Processing, ed. T. S. Jayadev, (Bellingham: SPIE), 127

Figure 3. The center field of M51 in the H band. It was obtained during June 1998 CIRSI commissioning run at the INT with 4 offset pointings (447″) and 3 ditherings (23.″5). Exposure time is 12 minutes.

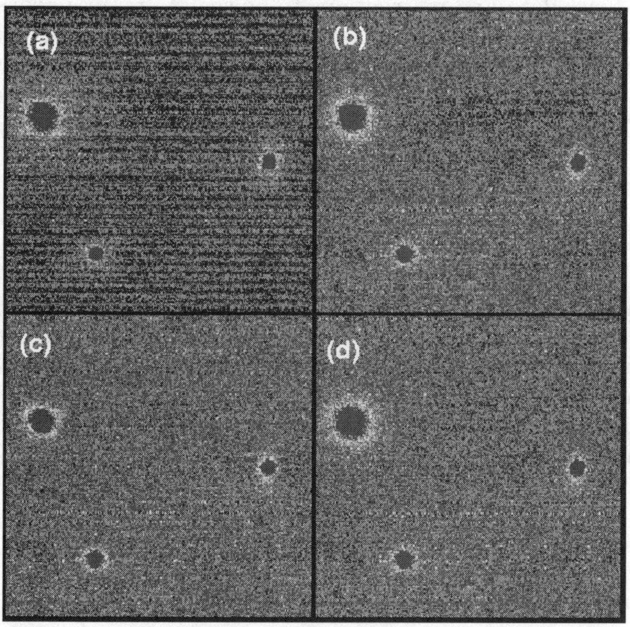

Figure 4. Noise correction: (*a*). no correction, (*b*) and (*c*). current noise correction methods in the system, (*d*). second pass linear median filtering method.

Astronomical Data Analysis Software and Systems VIII
ASP Conference Series, Vol. 172, 1999
D. M. Mehringer, R. L. Plante, and D. A. Roberts, eds.

Demonstration of Parallel AIPS++ Deconvolution Application

Wes Young

National Radio Astronomy Observatory, P.O. Box O, Socorro, NM, 87801

Douglas A. Roberts

National Center for Supercomputing Applications, University of Illinois Urbana-Champaign, Urbana, IL 61801

Abstract. A parallel algorithm applicator has recently been added to the AIPS++ system. This applicator class can be used for algorithms that can carry out significant work across multiple processors with little or no communication between the processors. This demonstration will show the first use of this class to deconvolve a large spectral line cube within AIPS++. The demonstration will be running remotely on one of the NCSA Origin2000 systems. The implementation of parallel processing is with the Message Passing Interface (MPI). MPI is a portable system that allows data and instructions to be sent to remote processors (either on the same machine or on different machines). Thus, the potential speed-up demonstrated on the highest-end systems available at large centers, such as NCSA, can be realized in part on a modest multiple-processor computer systems as well.

1. Introduction

Data reduction in radio astronomy has several processing steps which are amenable to parallelization. Spectral-line deconvolution is an obvious candidate since processing can be carried out on each plane independently.

We demonstrate here a "parallelized" Clark CLEAN (Clark 1980) algorithm implemented in the AIPS++ environment.

The demonstration by necessity uses small image cubes (512×512 pixels by 100 channels) for timeliness. We make significant time-savings gains when processing large image cubes with many planes and pixels.

2. Implementing Parallelization in AIPS++ Goals

1. Use the standard AIPS++ User Interface (for familiarity).
2. Overlay a parallelization infrastructure on existing AIPS++ algorithm implementations.
3. Minimize changes to existing code.

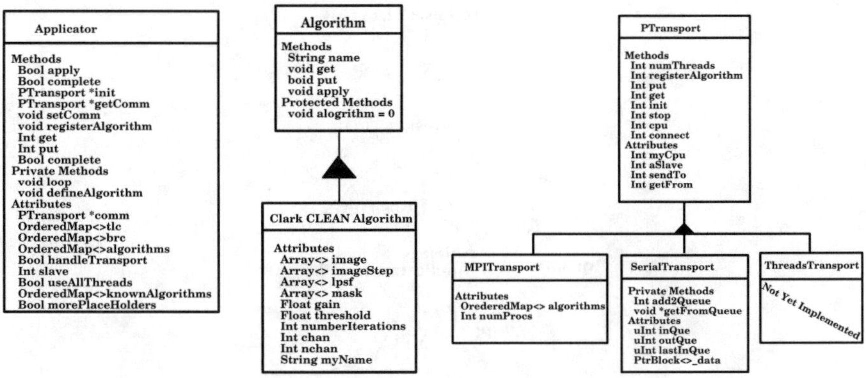

Figure 1. Object diagramas of the Applicator and Algorithm classes (*left*) and PTransport classes (*right*).

3. Design

We chose an Algorithm-Applicator scheme for the embarrassingly parallel problems. To communicate between the Algorithm and Applicators we use a PTransport object. This scheme is well suited to the AIPS++ programming environment.

The Algorithm runs n-copies of itself, each are independent of its siblings (in our case, the deconvolution of one plane in a dirty image). The Applicator is the controller, setting up the problem, sends data to the Algorithm processes, and receives the results.

PTransport is the conduit for transferring data between the Algorithms and the Applicator. Using this design provides a way to take advantage of parallel processing if it's available. Object diagrams of the Applicator, Algorithm, and PTransport classes are shown in Fig. 1.

4. Implementation

We use a simple master/slave scheme for doing the parallelization. The Applicator (pimager) does the bookkeeping and controls the IO to and from the AIPS++ system. The Algorithm (Clark CLEAN) processes receive a plane from the dirty image and deconvolution parameters from the Applicator, deconvolves the image, and send the results back to the Applicator.

The Applicator determines what type of PTtransport to use at run time, parallel, serial or in the future threads. Currently we have implemented serial and MPI PTransports[1]. A thread based PTransport will be implemented (once the AIPS++ libraries are thread-safe).

---

[1]While it may appear that there are n pimagers running, only one of the pimagers is a pimager object (applicator), the rest are ClarkCleanAlgorithm objects (algorithms).

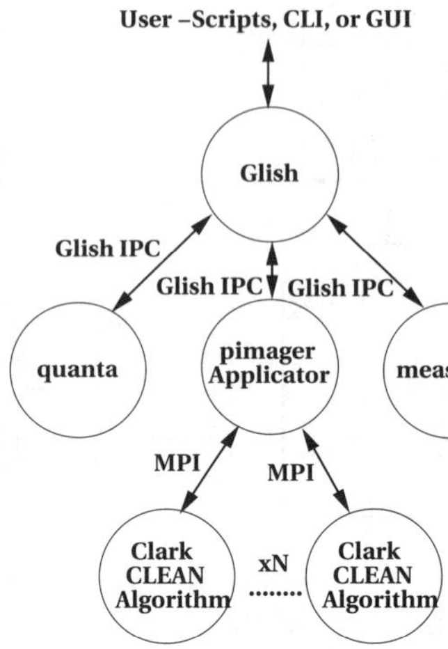

Figure 2. Applicator/Algorithm process schematic.

5. Operating Environment

Our system is evolving. To process on the "large" multiprocessor SGI systems at NCSA we create a command script and pass it into a batch queue. Notification of when the processing is complete is done via e-mail.

6. Results

Preliminary testing of parallelization in AIPS++ show promising speed ups for multi-channel deconvolution. Figure 6. shows the speed up falling away from linear at 15 processors. We are investigating why this is happening.

Dave Westpfahl (New Mexico Tech), supplied us with four pointings of VLA HI data of M33 (B, C, and D arrays) to create the images. Data in the demo are from one of these pointings.

We intend to combine the four pointings into a mosaic of 6000×6000 pixels (about two degrees on a side) by 140 channels.

Acknowledgments. Thanks to Brian Glendenning for AIPS++ and many hours of discussions. Dave Westpfahl provided the M33 datasets. Athol Kemball who provide many useful comments. We dedicate this work to the memory of our friend and colleague Dan Briggs. Blue skies Dan.

References

Clark, B. G. 1980, A&A89, 377

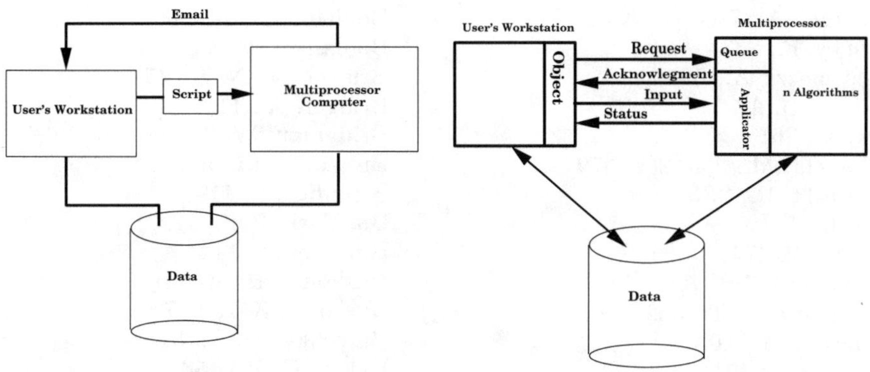

Figure 3. Caption (*left*). A more transparent scheme for interacting with batch queues (*right*).

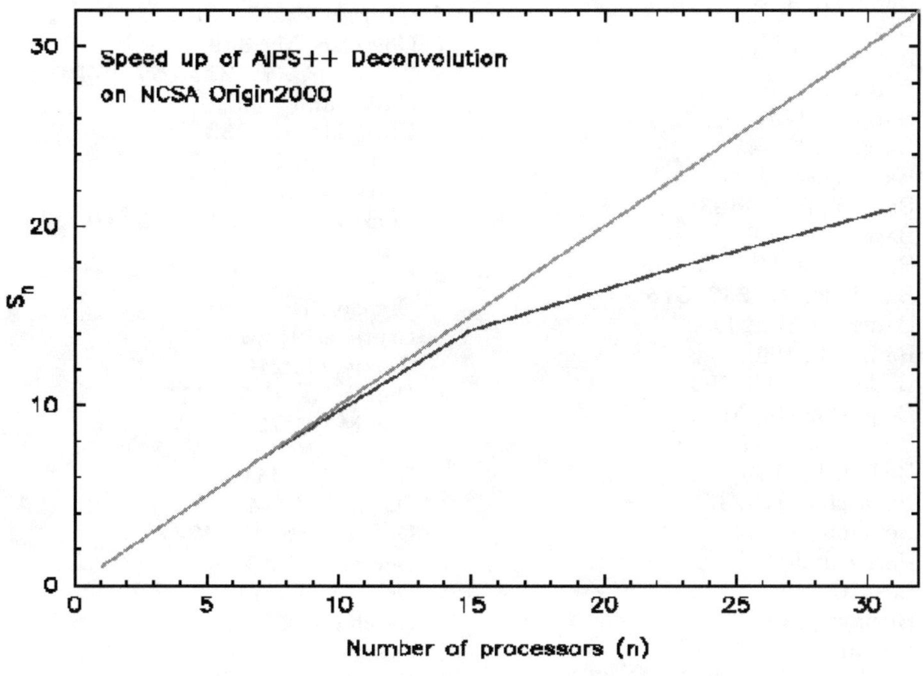

Figure 4. Clark CLEAN speed up of M33 HI 512×512 pixels by 100 channels using AIPS++ Applicator/Algorithm classes.

Author Index

Subject Index

Colophon

The Proceedings were created using tools from previous ADASS conferences. The style file provided by ASP was based largely on the work of Jeannette Barnes and George Jacoby for ADASS V. The style file was augmented with HTML macros from the html style file provided with the latex2html[1] package so that users would not have to additionally include the html style file. We also rewrote some of the macros to better handle underscores and tildes within URLs. Additional astronomical symbols and macros were added from the AAS-TeX style, created by Chris Beinesderfer. Finally, we added several new macros to aid indexing and cross-referencing. We would like to acknowledge the efforts of the editors of previous volumes, especially Harry Payne, Gareth Hunt, and Jeannette Barnes who happily discussed the challenges of editing an ADASS volume. This year, all submissions were received electronically via ftp.

The book was created from a master latex file that included the individual contributions on-the-fly during the document processing. Several tools written in Perl were used to filter and search submissions. GNU make was used to manage the creation of the book, which included a combination of filtering, latex, and dvips to make the final output file. An HTML version of the Proceedings is being made available on-line by the use of latex2html (version 98.2 beta5).

The work of editing the manuscripts was divided among the three editors with each responsible for editing about forty papers. Access control was provided by Revision Control System (RCS), which allowed the editors to simultaneously edit the individual manuscripts. The book itself is the first ADASS proceedings to be submitted electronically as a PostScript file to ASP for printing.

In 1999, a permanent repository (adass.org) is being set up for all ADASS materials including information of ADASS conferences, pointers to ADASS related news groups and web pages and past proceedings. The editing tools we used, including the Perl scripts, styles, and our Makefile as well as editorial notes will be made available at the new ADASS web site.

[1] `http://www-dsed.llnl.gov/files/programs/unix/latex2html/manual/`